Mathematical Analysis
and Applications
Part A
ADVANCES IN MATHEMATICS
SUPPLEMENTARY STUDIES, VOLUME 7A

ADVANCES IN
Mathematics
SUPPLEMENTARY STUDIES

EDITED BY Gian-Carlo Rota

EDITORIAL BOARD:

Mathematical Analysis
and Applications

ESSAYS DEDICATED TO LAURENT SCHWARTZ
ON THE OCCASION OF HIS 65TH BIRTHDAY

Part A

ADVANCES IN MATHEMATICS
SUPPLEMENTARY STUDIES, VOLUME 7A

EDITED BY

Leopoldo Nachbin

Departamento de Matemática Pura
Universidade Federal do Rio de Janeiro
Rio de Janeiro
Brazil

and

Department of Mathematics
The University of Rochester
Rochester, New York

1981

ACADEMIC PRESS

A Subsidiary of Harcourt Brace Jovanovich, Publishers

New York   London   Toronto   Sydney   San Francisco

ACADEMIC PRESS, INC.
111 Fifth Avenue, New York, New York 10003

United Kingdom Edition published by
ACADEMIC PRESS, INC. (LONDON) LTD.
24/28 Oval Road, London NW1 7DX

Library of Congress Cataloging in Publication Data
Main entry under title:

Mathematical analysis and applications.

(Advances in mathematics. Supplementary
studies; v. 7A)
English or French.
Includes bibliographical references.
1. Mathematical analysis--Addresses, essays,
lectures. 2. Schwartz, Laurent. I. Nachbin,
Leopoldo. II. Series.
QA300.M294   515   80-1780
ISBN 0-12-512801-0 (pt. A)

PRINTED IN THE UNITED STATES OF AMERICA

81 82 83 84   9 8 7 6 5 4 3 2 1

Laurent Schwartz

Contents

Notice sur les travaux scientifiques de Laurent Schwartz

Laurent Schwartz

PART A

On the Symbol of a Distribution

W. Ambrose

On Analytic and C[∞] Poincaré Lemma

Aldo Andreotti and Mauro Nacinovich

Weighted Inequalities in *L²* and Lifting Properties

Rodrigo Arocena, Mischa Cotlar, and Cora Sadosky

Green's Functions for Self-Dual Four-Manifolds

M. F. Atiyah

Contingent Derivatives of Set-Valued Maps and Existence of Solutions to Nonlinear Inclusions and Differential Inclusions

Jean Pierre Aubin

Sur des équations intégrales d'évolution

M. S. Baouendi and C. Goulaouic

A Local Constancy Principle for the Solutions of Certain Overdetermined Systems of First-Order Linear Partial Differential Equations

M. S. Baouendi and F. Treves

A Note on Isolated Singularities for Linear Elliptic Equations

Haim Brézis and Pierre-Louis Lions

Une étude des covariances mesurables

Pierre Cartier

Topological Properties Inherited by Certain Subspaces of Holomorphic Functions

Seán Dineen

Sur le quotient d'une variété algébrique par un groupe algébrique

J. Dixmier and M. Raynaud

Forced Oscillations of Nonlinear Hamiltonian Systems, II

Ivar Ekeland

Von Neumann's Uniqueness Theorem Revisited

Gérard G. Emch

Symmetry of Positive Solutions of Nonlinear Elliptic Equations in R^n

B. Gidas, Wei-Ming Ni, and L. Nirenberg

A Remark on the Representation Theorems of Frederick Riesz and Laurent Schwartz

J. Horváth

List of Contributors

Numbers in parentheses indicate the pages on which the authors' contributions begin.

W. AMBROSE (29), Department of Mathematics, Massachusetts Institute of Technology, Cambridge, Massachusetts 02139

ALDO ANDREOTTI† (41), Scuola Normale Superiore, Pisa, Italia 56100

RODRIGO AROCENA (95), Facultad de Ciencias, Universidad Central de Venezuela, Caracas, Venezuela

M. F. ATIYAH (129), Mathematical Institute, University of Oxford, Oxford, England

JEAN PIERRE AUBIN (159), Centre de Recherche de Mathématiques de la Decision, Université de Paris–Dauphine, 75775 Paris, France

M. S. BAOUENDI (231, 245), Department of Mathematics, Purdue University, West Lafayette, Indiana 47007

HAIM BRÉZIS (263), Département de Mathématiques, Université de Paris VI, 75230 Paris, France

PIERRE CARTIER (267), I.H.E.S., Bures sur Yvette, France

MISCHA COTLAR (95), Facultad de Ciencias, Universidad Central de Venezuela, Caracas, Venezuela

SEÁN DINEEN (317), Department of Mathematics, University College Dublin, Dublin, Ireland

J. DIXMIER (327), Département de Mathématiques, Université de Paris VI, 75230 Paris, France

IVAR EKELAND (345), Centre de Recherche de Mathématiques de la Decision, Université de Paris–Dauphine, 75775 Paris, France

GÉRARD G. EMCH (361), Departments of Mathematics and Physics, The University of Rochester, Rochester, New York 14627

B. GIDAS (369), Institute for Advanced Study, Princeton, New Jersey 08540

C. GOULAOUIC (231), Centre de Mathématiques, École Polytechnique, Palaiseau, France

J. HORVÁTH (403), Department of Mathematics, University of Maryland, College Park, Maryland 20742

PIERRE-LOUIS LIONS (263), Département de Mathématiques, Université de Paris VI, 75230 Paris, France

MAURO NACINOVICH (41), Università di Pisa, Istituto di Matematica "L. Tonelli," 56100 Pisa, Italia

†Deceased.

xiii

WEI-MING NI† (369), Institute for Advanced Study, Princeton, New Jersey 08540

L. NIRENBERG (369), Courant Institute of Mathematical Sciences, New York University, New York, New York 10012

M. RAYNAUD (327), Département de Mathématiques, Université de Paris XI, Orsay, France

CORA SADOSKY‡ (95), Facultad de Ciencias, Universidad Central de Venezuela, Caracas, Venezuela

LAURENT SCHWARTZ (1), Centre de Mathématiques, École Polytechnique, Paris, France

F. TREVES (245), Department of Mathematics, Rutgers University, New Brunswick, New Jersey 08903

†Present address: Department of Mathematics, University of Pennsylvania, Philadelphia, Pennsylvania 19104.

‡Present address: Department of Mathematics, Howard University, Washington, D.C. 20059.

Preface

Laurent Schwartz became 65 years old on March 5, 1980. A number of his admirers are contributing to the present collection of mathematical writings in his honor. It is his unusual stature, from both the scientific and the human viewpoints, that motivates our paying this homage to him.

There is no need to introduce Laurent Schwartz to the general mathematical public. He has won a justified fame for his research and expository contributions to mathematical analysis and its applications, particularly his epoch-making and long-lasting development of distributions. His synthesis of previous work and the creation of distribution theory are turning points in differential and integral calculus. He has also taken a deep and continued interest in the progress of mathematics teaching and research in many of the developing countries. He has stood as a real leader in what concerns human rights in the scientific community.

From the historical viewpoint, it is interesting to point out on this occasion that Laurent Schwartz is a relative of two famous mathematicians, Paul Lévy and Jacques Hadamard.

His wife, Marie-Hélène Schwartz, who is a research mathematician devoted to geometry, is the daughter of Paul Lévy. As Laurent Schwartz told me in 1980, it was under the direct, personal influence of Paul Lévy that he became so deeply interested in functional analysis and probability. Paul Lévy was a great master in both fields, as one of their founders, in spite of the fact, or maybe precisely due to the circumstance, that he was a self-taught man (in the opinion of Laurent Schwartz).

Jacques Hadamard was his great-uncle; Hadamard's wife was the sister of the mother of Schwartz's mother. Laurent Schwartz knew Jacques Hadamard well and called him simply "Oncle Jacques." As we know, Hadamard was a great research mathematician also interested in the teaching of, and psychological aspects associated with, mathematics. When Laurent Schwartz was only in his early teens, Jacques Hadamard pointed out to him the existence of the Riemann zeta function and its relevance to mathematics. Teenager Laurent knew nothing of that "simple" mathematics, but surprised Uncle Jacques in view of his "excellent" acquaintance with equations of second degree. When Laurent Schwartz told me that story in 1980, he carefully pointed out that it was a recollection of his youth. Well, it may not be accurate from the viewpoint of Uncle Jacques, but it certainly must be true as far as teenager Laurent is concerned, and it is delicious at any rate.

While Laurent Schwartz described to me his family relationships with Paul

Lévy and Jacques Hadamard, he mentioned their wonderful influence on him, but at the same time modestly added that he had received a heavy heritage of responsibility from his ancestors. However, if we look now to Laurent Schwartz's present and past performance in mathematics, we conclude that he has turned out to be one of the greatest mathematicians of his generation, certainly up to the high level of such a tradition and culture.

LEOPOLDO NACHBIN

Contents of Part B

Notice sur les travaux scientifiques de Laurent Schwartz[†]

Centre de Mathématiques
École Polytechnique
Paris, France

Contents

A. Curriculum vitae
B. Exposé général des travaux. leurs motivations
C. Exposé detaillé des travaux: I. Polynômes, sommes d'exponentielles, fonctions moyenne-périodiques, analyse et synthèse harmoniques; II. Théorie des distributions; III. Analyse fonctionnelle, distributions à valeurs vectorielles et théorème des noyaux; IV. Physique théorique; V. Théorie de l'intégration, probabilités, probabilités cylindriques et applications radonifiantes.
D. Publications

A. CURRICULUM VITAE

Laurent SCHWARTZ
Né le 5 Mars 1915 à Paris
Domicilié 37, rue Pierre Nicole, 75005 Paris

1934–1937	Ecole Normale Supérieure
1937	Agrégé de Mathématiques
1937–1940	Service militaire et guerre
1940–1942	Attaché de recherche au C.N.R.S.
1943	Docteur-ès-Sciences (Université de Strasbourg réfugiée à Clermont-Ferrand)
1943–1944	Boursier de l'Aide à la Recherche Scientifique
1944–1945	Chargé d'enseignement à Grenoble
1945–1952	Maître de Conférences, puis Professeur à la Faculté des Sciences de Nancy
1946	Chargé du Cours Peccot
1953–1959	Maître de Conférences, puis Professeur à la Faculté des Sciences de Paris
1959–1969	Professeur à la Faculté des Sciences, puis à l'Université Paris VII. Professeur à l'Ecole Polytechnique

[†] Notice imprimée à l'occasion de la candidature à l'Académie des Sciences de Paris et mise à jour en Octobre 1980.

1969–1980   Professeur à l'Ecole Polytechnique (détaché de l'Enseignement Supérieur)
1980–   Professeur à l'Université Paris VII, Directeur du Centre Mathématique de l'Ecole
            Polytechnique

Prix

1950   Médaille FIELDS, Congrès International des Mathématiciens, Cambridge, Massachusetts, U.S.A.
1955   Prix Carrière de l'Académie des Sciences, Paris, France.
1964   Grand Prix de Mathématiques et de Physique, Académie des Sciences, Paris, France.
1972   Prix Cognac-Jay, Académie des Sciences, Paris, France, (avec J. L. Lions et B. Malgrange).

Distinctions diverses

1956   Professeur Honoraire de l'Université d'Amérique, Bogota, Colombie et Professeur Honoraire de l'Université de Buenos-Aires, Argentine.
1957   Membre correspondant de la Société Royale des Sciences de Liège, Belgique.
1958   Membre Honoraire de l'Union Mathématique Argentine. Académicien Honoraire, Académie des Sciences Exactes, Physiques et Naturelles, Buenos-Aires, Argentine.
1960   Docteur Honoris Causa de l'Université de Humboldt, Berlin, R.D.A.
1962   Docteur Honoris Causa de l'Université Libre de Bruxelles, Belgique.
1964   Membre Correspondant de l'Académie des Sciences du Brésil.
1965   Professeur Honoraire de l'Université Nationale des Ingénieurs du Pérou, Lima, Pérou.
1971   Membre Honoraire du Tata Institute of Fundamental Research, Bombay, Inde.
1972   Membre Correspondant, Académie des Sciences, Paris, France.
1975   Membre de l'Académie des Sciences.
1977   Membre Etranger de l'Académie Indienne des Sciences.

B.   EXPOSE GENERAL DES TRAVAUX

Pendant mes années de lycée, je me suis intéressé d'abord au latin et au grec, puis à la géométrie et à l'analyse, mais aussi à la physique, la chimie, la biologie. Je cherchais déjà à me faire de tout ce que je connaissais en mathématiques des théories cohérentes ou des exposés cohérents de théories existantes, à la fois pour des raisons d'esthétique mathématique et en vue de créer des outils maniables dans les applications.

A l'Ecole Normale, j'ai étudié l'intégrale de Lebesgue, les probabilités, les fonctions d'une variable complexe, les équations aux dérivées partielles elliptiques, la géométrie infinitésimale, mais aussi l'analyse fonctionnelle (cours de J. LERAY au Collège de France). P. LEVY a eu dès lors une grande influence sur moi, tant pour les probabilités que pour l'analyse classique. J'ai réfléchi dès cette époque sur la fameuse fonction de Dirac, mais sans aboutissement. On retrouve dans tous mes travaux ultérieurs les marques de cette formation initiale.

Mes premières années de chercheur (1940–1942) se sont passées à Clermont-Ferrand où était réfugiée la Faculté des Sciences de Strasbourg. Là, il y avait grande concentration de mathématiciens: J. DIEUDONNÉ, H. CARTAN, J. DE POSSEL, CH. EHRESMANN, J. DELSARTE, A. WEIL, S. MANDELBROJT, A. LICHNEROWICZ. La rencontre de N. BOURBAKI m'a initié à des idées toute nouvelles après ma formation d'analyste classique et m'a orienté vers l'algèbre et la topologie, pas tellement pour elles-mêmes que pour leurs applications à l'analyse. Le cours d'analyse fonctionnelle de J. DIEUDONNÉ a été à l'origine de ma thèse; celle-ci, consacrée á l'étude des sommes d'exponentielles, utilise des méthodes d'analyse fonctionnelle d'une manière nouvelle pour résoudre des problèmes d'approximation de type classique.

A la fin de la guerre, travaillant tout seul, je me suis fait une théorie complète de la dualité dans les espaces vectoriels topologiques généraux, théorie qui m'a paru alors sans application et que j'ai gardée pour moi; elle devait être la clef de la théorie des distributions. C'est cette formation antérieure qui fait que la "découverte" des distributions fut en fait presque instantanée au début de 1945. Trouver une *théorie* qui rendait toutes les fonctions indéfiniment dérivables et permettait la dérivation terme à terme des séries convergentes, c'était exactement le genre de recherche qui me convenait: théorie cohérente, mais restant près des réalités et des applications. J'ignorais alors les travaux de S. BOCHNER, S. L. SOBOLEV, de tous ceux qui avaient travaillé sur des objets analogues; par contre, je savais déjà que les solutions d'une équation aux dérivées partielles elliptique pouvaient se définir sans mettre de dérivées dans la definition et qu'elles étaient automatiquement indéfiniment dérivables, et qu'au contraire, pour les équations hyperboliques, il devait y avoir une définition n'utilisant pas de dérivées et donnant des solutions effectivement non dérivables.

Mes travaux sur les distributions ont été ensuite constamment accompagnés de travaux d'analyse fonctionnelle, chacun motivant l'autre. Dans les dernières années, l'analyse fonctionnelle, l'intégration et mes souvenirs de probabilités, en même temps que la difficulté de concilier les points de vue hostiles des divers groupes de mathématiciens sur la théorie de la mesure, m'ont naturellement conduit á une synthèse, qui est donc elle aussi une théorie, celle des mesures de Radon sur les espaces topologiques arbitraires. Il en est découlé très naturellement des recherches sur les probabilités cylindriques, les applications radonifiantes, la désintégration des mesures et ses applications aux processus stochastiques, qui sont le sujet de mes

recherches actuelles, aujourd'hui complètement orientées vers les probabilités.

C. Expose detaille des travaux

I. *Polynomes, sommes d'exponentielles, fonctions moyenne-periodiques, analyse et synthese harmoniques*

(1) *Polynomes et sommes d'exponentielles reelles.* Ch. H. Muntz avait géne-
ralisé le théorème d'approximation de Weierstrass comme suit. Soit Λ:
$\lambda_0 = 0, \lambda_1, \lambda_2, \ldots \lambda_n, \ldots$ une suite donnée de nombres réels ≥ 0. Alors toute
fonction continue sur $[0, 1]$ est limite uniforme de polynômes généralisés
(sommes finies $\sum a_n x^{\lambda_n}$, $\lambda_n \in \Lambda$) si et seulement si la série $\sum_{n=1}^{+\infty} 1/\lambda_n$ diverge.
Si alors la série $\sum_{n=1}^{+\infty} 1/\lambda_n$ converge, quelles sont les fonctions continues qui
peuvent être approchées par les polynômes précédents? La réponse à cette
question fut l'objet de ma thèse. Les fonctions approchables sont exactement
les fonctions continues sur $[0, 1]$, qui sont analytiques et développables en
série de Taylor (lacunaires) généralisées $\sum_{n=0}^{+\infty} a_n x^{\lambda_n}$, convergentes par groupe-
ment de termes pour $|x| < 1$.

Le changement de variables $x = \exp(-X)$ ramène à un problème plus
maniable sur des sommes d'exponentielles réelles $\sum a_n \exp(-\lambda_n X)$ sur la
demi-droite $\mathbb{R}_+$. On trouve alors des théorèmes sur les séries de Dirichlet
lacunaires.

(2) *Sommes d'exponentielles imaginaires.* Ce travail [5] fait suite aux précédents et la substitution des sommes d'exponentielles aux polynômes lui donne
tout son intérêt, car il va concerner des séries de Fourier généralisées
$\sum a_n \exp(i\lambda_n X)$, où les λ_n ne sont pas tous des multiples de l'un d'entre eux.
On se donne encore une suite Λ de nombres réels (de signe quelconque) et
un intervalle $[a, b]$ de la droite réelle. Ou bien les polynômes trigono-
métriques généralisés, c'est-à-dire les sommes finies $\sum a_n \exp(i\lambda_n X)$, permettent d'approcher toute fonction continue, ou bien on peut caractériser
les fonctions continues qui peuvent être approchées; ce sont exactement
celles qui ont un développement en série de Fourier généralisée, convergeant par groupement de termes et facteurs exponentiels d'Abel: $F(X) =$
$\lim_{t \to 0} \sum a_n \exp(i\lambda_n X - |\lambda_n|t)$. Ainsi un procédé de sommation classique des
séries de Fourier, celui d'Abel, est valable, moyennant en plus des groupe-
ments de termes, pour toutes les séries d'exponentielles imaginaires. Les
groupements de termes sont ici inévitables, dans le cas où il y a des groupements de λ_n très voisins les uns des autres [par exemple si l'on prend la suite
de couples n, $n + 1/n!$, on sera obligé de faire des groupements de deux
termes pour qu'il y ait convergence].

Ces articles ont donné naissance à plusieurs travaux ultérieurs, d'autres
mathématiciens et de moi-même, sur les fonctions moyenne-périodiques et

l'analyse ou synthèse harmonique. Ils donnent une convergence de développements suivant des harmoniques dont les fréquences sont distribuées un peu n'importe comment, ce qui est le cas de tous les systèmes vibratoires non périodiques; donc ils pourraient être susceptibles d'applications physiques.

(3) *Theorie des functions moyenne-periodiques* [9]. Ces fonctions ont été introduites par J. DELSARTE (qui a d'ailleurs eu une très grande influence sur moi, ainsi que J. DIEUDONNÉ pendant mes 7 années à Nancy (1945–1952)). Il avait posé le problème en termes d'équations intégrales, dont il avait trouvé les solutions par des sommes d'exponentielles. Il avait eu des difficultés de convergence, d'où la nécessité d'hypothèses simplificatrices; je me suis aperçu qu'en fait la convergence avait lieu dans le cas le plus général, pourvu qu'on utilise à la fois les groupements de termes et les facteurs exponentiels d'Abel du travail précédent. Mais en même temps, j'ai trouvé une autre définition des fonctions moyenne-périodiques, inspirée des idées nouvelles sur la théorie des groupes (théorèmes taubériens de WIENER, livre d'A. WEIL sur les groupes topologiques, paru en 1940). Il faut faire jouer un rôle fondamental au groupe des translations sur la droite. Une fonction continue F sera dite moyenne-périodique si l'ensemble des combinaisons linéaires finies de ses translatées ne permet pas l'approximation, uniformément sur tout compact, de toute fonction continue. On définit alors un spectre de F: c'est l'ensemble des λ (ici nombres complexes quelconques) tels que $\exp(\lambda X)$ soit approchable par des combinaisons linéaires de translatées de F. On est obligé d'introduire des éléments multiples dans le spectre, et pour cela de considérer non seulement des exponentielles $\exp(\lambda X)$, mais des exponentielles-polynômes $\exp(\lambda X)P(X)$, P polynôme. La recherche du spectre sera l'analyse spectrale ou analyse harmonique; on montrera qu'il n'est jamais vide si $F \neq 0$. Ensuite on cherchera la synthèse spectrale ou synthèse harmonique: F est-elle approchable par les combinaisons linéaires finies d'exponentielles-polynômes de son spectre, est-elle une série $\sum a_n \exp(\lambda_n X)$ $P_n(X)$ formée des exponentielles polynômes de son spectre, avec un mode de convergence convenable? Il se trouve que c'est vrai, et la convergence cherchée est toujours par groupements de termes avec facteurs exponentiels d'Abel.

J'ai aussi donné une autre démonstration, indépendante de ce mode de convergence, pour l'analyse et la synthèse spectrales. J. P. KAHANE en a donné ultérieurement une plus simple. En termes physiques, on peut donner de l'analyse et de la synthèse spectrales l'énoncé simplifié suivant: le spectre d'une onde est l'ensemble des fréquences des ondes monochromatiques qui peuvent être obtenues par combinaison de l'onde et de ses déphasées; l'onde est une série des ondes monochromatiques correspondant à son spectre de fréquences. Mais cette image cache évidement les principales difficultés mathématiques du problème.

Il n'est pas nécessaire de partir *d'une* fonction et on peut, plus généralement, faire l'analyse et la synthèse harmonique d'un sous-espace fermé, invariant par les translations, de l'espace des fonctions continues sur la droite.

B. MALGRANGE, dans sa thèse (1955), a donné une généralisation partielle du théorème précédent au cas de fonctions de plusieurs variables. Malgré de nombreuses recherches depuis, la possibilité de généraliser à plusieurs variables est longtemps restée un problème ouvert; Gurewitch a montré récemment que le résultat était faux, il y a pour les dimensions ≥ 2 des sous espaces fermés invariants par translation, qui ne contiennent aucune exponentielle.

(4)   *Analyse et synthese harmoniques dans divers espaces.*   La théorie précédente est relative à l'espace des fonctions continues, pour la convergence uniforme sur tout compact. En faisant varier l'espace fonctionnel et toujours intervenir les translations, on obtient des problèmes très variés. L'un d'eux a longtemps retenu l'attention: il est relatif aux espaces L^1 et L^∞; l'analyse harmonique, consistant simplement à dire que le spectre n'est pas vide, est le célèbre théorème taubérien de N. WIENER ou son équivalent dual, le théorème de BEURLING. On a longtemps cherché si la synthèse harmonique était possible aussi dans L^1 et L^∞. J'ai donné le premier exemple montrant qu'elle n'était pas possible dans $\mathbb{R}_n$, pour $n \geq 3$, dans [13]. Ce résultat était inattendu. Le cas de $\mathbb{R}^1$ et de $\mathbb{R}^2$, beaucoup plus difficile, est resté longtemps non résolu. C'est P. MALLIAVIN qui l'a résolu et généralisé.

J'ai étudié ensuite l'analyse et la synthèse harmoniques dans l'espace $\mathscr{S}'$ des distributions tempérées sur $\mathbb{R}^n$. La transformation de Fourier des distributions tempérées, dont je parlerai plus loin, me permettait de ramener ce problème à un problème d'algèbre différentielle: dans l'algèbre des fonctions indéfiniment différentiables sur $\mathbb{R}^n$, un idéal fermé est-il déterminé par ses idéaux ponctuels? Je n'ai pas pu répondre à cette question et me suis adressé à H. WHITNEY qui avait beaucoup étudié les fonctions différentiables. Il a pu résoudre le problème, avec une réponse affirmative, ce qui m'a permis de résoudre les problèmes harmoniques dans l'espace $\mathscr{S}'$. Le théorème de WITHNEY, ainsi né d'un problème d'analyse harmonique, a donné liue à de nombreux travaux ultérieurs sur les fonctions différentiables, notamment de B. MALGRANGE et J. C. TOUGERON.

II.   *Theorie des distributions*

J'ai trouvé les principaux résultats sur les distributions vers 1945–1946, mais mes deux livres [19] ne sont parus qu'en 1950 et 1951. C'est cette théorie qui m'a valu la Médaille Fields du Congrès International des Mathématiciens en 1950. Beaucoup de mathématiciens ont pressenti les distributions; le calcul symbolique d'Heaviside, dès la fin du siècle dernier, était, sous forme particulièrement incorrecte et qui fut très incomprise, un calcul de distributions. N. WIENER, dès 1926, a "régularisé" par des fonctions indéfiniment dérivables à support compact, pour définir des images de Fourier qui n'existaient pas au sens usuel. J. LERAY, K. FRIEDRICHS, tous les mathématiciens de l'école de R. COURANT ont utilisé des dérivées faibles, qui sont en fait des dérivées au sens des distributions, lorsqu'elles sont elles-mêmes des fonctions. Je crois qu'on peut particulièrement citer les travaux de S. BOCHNER, T. CARLEMAN et S. L. SOBOLEV (dont en fait je n'avais pas connaissance en 1945).

S. BOCHNER a introduit, dans son livre sur l'intégrale de Fourier en 1932, sur la droite réelle, des "dérivées formelles de produits de polynômes par des fonctions de L^2"; ce sont mes futures distributions tempérées. Il en fait la transformation de Fourier. C'est un appendice de quelques pages à son livre; il ne semble pas lui-même y attacher beaucoup d'importance. Le support n'y est pas, ni aucune topologie, ni les conditions de transformation du produit multiplicatif en produit de convolution, et finalement la distribution de Dirac δ n'y est pas nommée. L'effort de T. CARLEMAN va dans le même sens que celui de S. BOCHNER: trouver un formalisme pour la transformation de Fourier. Il prend des "différences formelles de fonctions holomorphes à croissance lente dans le demi-plan complexe supérieur et dans le demi-plan inférieur"; ce sont encore mes futures distributions tempérées et T. CARLEMAN en fait la transformation de Fourier. En fait, ce travail prépare plutôt les hyperfonctions de M. SATO, sous la forme exposée par A. MARTINEAU, que les distributions. S. L. SOBOLEV a introduit ses fonctionnelles de façon très conséquente, dans un long article en 1936, en vue d'étudier les équations aux dérivées partielles. Les fonctions plusieurs fois différentiables à support compact sont les fonctions tests nécessaires à la définition: il y a une dérivée généralisée, donc on peut appliquer à ces fonctionnelles des opérateurs différentiels; et les problèmes aux limites sont modifiés, en ramenant les conditions aux limites dans le second membre, par l'intervention de couches portées par la surface limite. Il a introduit aussi les espaces fonctionnels H^s qui portent son nom, encore pour étudier les équations aux dérivées partielles. Mais δ n'y figure pas, ni la convolution ni la transformation de Fourier, ni aucune des questions vectorielles topologiques soulevées par les distributions. Le calcul des distributions le plus audacieux a été incontestablement fait par les physiciens théoriciens. La "fonction de Dirac" date de 1926; mais les physiciens ont été considérablement plus loin, bien avant que les mathématiciens n'aient commencé à entrevoir une approche du problème. Par exemple, ni S. BOCHNER, ni

T. Carleman, ni S. L. Sobolev n'ont abordé les rapports avec la physique. En 1950, quand mon livre sur les distributions est paru, les physiciens avaient déjà introduit toutes les fameuses fonctions singulières de la physique théorique relativiste; or ce sont-là des distributions malaisées á définir mathématiquement de facon convenable et qui n'ont été approfondies complètement qu'à partir de la thèse de P. D. Methee en 1954.

Il a donc fallu construire une théorie complète, cohérente, avec tous les outils d'analyse fonctionnelle nécessaires: définir correctement les distributions, leurs dérivées, le produit tensoriel, la convolution, la transformation de Fourier, et surtout en fait donner *le bon point de vue* susceptible d'être utilisé dans les domaines d'applications les plus divers, et qui a été universellement adopté immédiatement.

Je ne résumerai que très rapidement la théorie des distributions.

- Le chapitre I définit les distributions comme formes linéaires continues sur l'espace $\mathscr{D}$ des fonctions indéfiniment dérivables à support compact. Il apparaît ainsi aussitôt que, d'une part les distributions généralisent les fonctions un peu comme les nombres réels généralisent les nombres rationnels ou comme les nombres complexes généralisent les nombres réels, mais aussi qu'elles donnent une définition mathématiquement correcte des "distributions" de charges électriques ou de masses de la physique ou de la mécanique, d'où leur nom.

- Le chapitre II définit les dérivées des distributions: toute distribution est indéfiniment dérivable, et la dérivation est une opération continue, donc on pourra désormais dériver terme à terme des séries convergentes, au sens des distributions naturellement. Ce chapitre donne en outre de nombreux exemples et on retrouve aussi les "parties finies" de J. Hadamard. On trouve aussi l'exemple qui servira de fondement aux formules de l'électrostatique: $\Delta(1/r) = -4\pi\delta$, dans $\mathbb{R}^3$ (formule de Poisson pour les potentiels).

- Le chapitre III étudie les structures topologiques de l'espace $\mathscr{D}$ et de l'espace $\mathscr{D}'$ des distributions. C'est là qu'il y avait à résoudre les plus grandes difficultés, la théorie de la dualité dans les espaces vectoriels topologiques localement convexes n'ayant pas encore à ce momentlà tous les outils nécessaires.

- Le chapitre IV étudie la multiplication; elle n'est pas toujours possible et c'est là que se trouvent certaines des plus grandes difficultés des calculs de la théorie quantique des champs.

- Le chapitre V étudie le produit tensoriel.

- Le chapitre VI traite de la convolution, où de remarquables résultats peuvent être obtenus. Les principales applications aux équations aux dérivées partielles proviennent de la convolution; c'était visible pour les équations à coefficients constants, et j'ai donné au chapitre VI un grand nombre d'applications (par exemple problème de Cauchy pour des équations hyperboliques). Mais la découverte des opérateurs pseudo-différentiels en 1964 (A. UNTERBERGER et J. BOKOBZA–UNTERBERGER, L. NIRENBERG, L. HÖRMANDER) a même montré qu'en fait la convolution intervenait aussi fondamentalement dans les équations à coefficients variables.

- Le chapitre VII traite de la transformation de Fourier des distributions tempérées, avec l'introduction des espaces $\mathscr{S}$ et $\mathscr{S}'$, leurs propriétés topologiques, la transformation de la multiplication en convolution et vice-versa, avec les espaces $\mathscr{O}_M$ et $\mathscr{O}'_C$, les théorèmes de BOCHNER et de PALEY–WIENER généralisés.

Dans l'édition de 1966 de mon livre, ont été ajoutés deux chapitres:

- le chapitre VIII sur la transformation de Laplace
- le chapitre IX sur les courants, ou formes différentielles-distributions sur les variétés, étendant ainsi les distributions aux champs de tenseurs, aux courants électriques, de ligne, surface ou volume tels qu'ils interviennent dans les équations de Maxwell, etc . . . L'idée des courants avait été pressentie par de RHAM qui, aussitôt paru mon article de 1945 sur les distributions, en fit indépendamment de moi la théorie qu'il a ensuite remarquablement appliquée à la théorie des formes harmoniques sur les variétés riemanniennes et à la topologie algébrique sur les variétés.

J'ai publié quelques autres articles sur les distributions:

- [17], qui les applique aux équations d'évolution avec problème de Cauchy (chaleur, ondes, etc. . .);
- [22], sur les multiplicateurs de $\mathscr{F}L^p$ (étude qui a été poursuivie par de nombreux auteurs);
- [24], qui résout les problèmes de Cousin sur une variété analytique complexe par des méthodes explicites que ne fournit pas la théorie des faisceaux et va dans le sens des premiers travaux de K. KODAIRA;
- [34], qui résout, dans un cas très particulier, le problème de la division des distributions, problème que j'avais posé dans mon livre en vue des équations de convolution tempérées et qui n'a été résolu que bien plus tard par des méthodes très difficiles par L. HÖRMANDER et surtout par S. ŁOJASIEWICZ. (Ce dernier a d'ailleurs poursuivi très loin les méthodes qu'il avait introduites pour résoudre le problème de la division);
- [37], qui introduit les distributions régulières et intégralement régulières dans les changements de coordonnées;
- enfin [48] et [59] qui traitent de problèmes particuliers.

Les distributions, et les méthodes introduites dans la théorie, ont eu des applications dans les domaines les plus variés des mathématiques. En analyse surtout: analyse de Fourier, groupes de Lie (avec F. BRUHAT et HARISH–CHANDRA), fonctions analytiques mises en dualité avec les fonctionnelles analytiques et hyperfonctions, théorie du potentiel (M. BRELOT et J. DENY), théorie des semi-groupes distributions (J. L. LIONS et J. CHAZARAIN), mais avant tout dans la théorie des équations aux dérivées partielles (et équations de convolution ou pseudo-différentielles). Cette théorie, qui fait la liaison des mathématiques avec la physique et la mécanique, eut depuis toujours la faveur des plus grands mathématiciens. En France, H. POINCARÉ et J. HADAMARD y consacrèrent une grande part de leur vie. Il serait trop long, et d'ailleurs déplacé, d'en faire ici l'historique. Beaucoup d'articles sur les équations aux dérivées partielles ont directement influencé ma rédaction des distributions entre 1945 et 1950 (I. FREDHOLM, J. HADAMARD, S. L. SOBOLEV, Marcel RIESZ, N. ZEILON et L. GARDING, J. LERAY, R. COURANT, K. FRIEDRICHS, B. LEVI, etc...). Et je crois que ce sont en bonne partie les distributions qui ont permis, ou pour le moins largement favorisé, des travaux comme ceux de J. L. LIONS et B. MALGRANGE, L. NIRENBERG, L. HÖRMANDER, G. STAMPACCHIA, E. MAGENES, etc..., et le très grand développement mondial actuel des équations aux dérivées partielles qui occupent une place essentielle dans les séminaires, colloques et publications d'analyse.

Les distributions ont continué à être utilisées partout en physique théorique; continué, puisque les physiciens, fort justement, n'avaient pas attendu que les mathématiciens créent l'outil rigoureux dont ils avaient besoin et s'en étaient heureusement déjà servi avant! De très difficiles problèmes sur les distributions sont posés par la théorie quantique des champs et beaucoup sont sans doute encore loin d'être résolus. Mais on utilise aussi les distributions en électricité, ondes hertziennes (filtrage), en diffraction, etc... Le fait que les distributions soient enseignées dans les cours de 2ème cycle des universités pour les mathématiciens, et souvent pour les physiciens, mérite d'être mentionné. Mon livre de "Méthodes Mathématiques pour les Sciences Physiques", [46], est destiné à un cours de licence pour mathématiciens et physiciens et a été traduit en anglais, espagnol, russe et japonais.

Enfin, il faut signaler que bien d'autres êtres mathématiques nouveaux sont nés à la suite des distributions: fonctionnelles analytiques et hyperfonctions (M. SATO–A. MARTINEAU), duals d'espaces de GEVREY et d'espaces d'autres classes de MANDELBROJT (J. LERAY–Y. OHYA, CH. ROUMIEU, E. MAGENES, J. L. LIONS, CH. GOULAOUIC, M. S. BAOUENDI), fonctions généralisées (I. M. GELFAND et G. E. CHILOV), distributions généralisées de N. ARONSZAJN et M. S. BAOUENDI.

III. *Analyse fonctionelle, distributions a valeurs
     vectorielles et theoreme des noyaux*

(1) *Espaces vectoriels topologiques localement convexes.* Les outils d'analyse fonctionelle nécessaires à la théorie des distributions étaient insuffisants et il a ete nécessaire de les créer. Les remarquables travaux de S. Banach, à part le théorème de Hahn–Banach, ne sortaient guère des espaces de Banach et de Frechet. Or le dual d'un espace de Frechet n'est pas un espace de Frechet. J. Dieudonne avait étendu bien des résultats aux espaces localement convexes arbitraires (et j'avais été fortement influencé par lui); G. W. Mackey avait fait de remarquables travaux, mais restés assez ignorés. Or il était indispensable de développer correctement les parties topologiques de la théorie des distributions. J'ai toujours été intéressé par les applications de l'analyse fonctionnelle à l'analyse. Ma thèse [4] utilisait déjà largement des méthodes d'analyse fonctionnelle. Que j'aie fait de l'analyse fonctionnelle "pour elle-même" et qu'ensuite j'en utilise les résultats, il n'y a rien là que de très naturel.

Il fallait étendre les espaces de Frechet et définir leurs limites inductives. La notion de limite inductive, en algèbre ou en théorie des catégories, est aujourd'hui universellement connue et considérée comme élémentaire, mais elle était ignorée en 1945 (alors que les limites projectives avaient été introduites dans le livre d' A. Weil sur les groupes topologiques). C'est J. Dieudonne qui en a eu l'idée, à la suite de ma description topologique de l'espace $\mathscr{D}$ des fonctions indéfiniment dérivables à support compact, et nous avons dégagé dans [14] less principales propriétés des espaces de Frechet et de leurs limites inductives, que j'avais utilisées et démontrées directement dans le cas particulier des espaces de distributions. C'est à partir de notre travail que N. Bourbaki a introduit les notions d'espace bornologique, espace tonnelé, etc. . . A partir de ces années 50, et pendant une dizaine ou une quinzaine d'années, les espaces vectoriels localement convexes généraux ont remplacé partout les espaces de Banach, et ont donné lieu à de nombreux travaux, dont les plus importants furent ceux d'A. Grothendieck sur les produits tensoriels topologiques ε et π et les espaces nucléaires, qui ont d'ailleurs été l'objet de mon séminaire 1953–1954. On doit même signaler que beaucoup de maîtrises de mathématiques de nos Universités contiennent un certificat sur les espaces vectoriels topologiques. A l'heure actuelle, on abandonne un peu, à juste titre je pense, les espaces vectoriels topologiques pour revenir aux espaces de Banach dont on étudie les propriétés géométriques fines.

(2) *Le theoreme des noyaux et les distributions a valeurs vectorielles.* La   physicien P. A. M. Dirac avait senti intuitivement le théorème des noyaux, disant

que toute application linéaire continue de $\mathscr{D}_y$ dans $\mathscr{D}'_x$ peut s'exprimer, d'une manière unique, par un noyau-distribution, $K_{x,y}$, sous la forme: $T_x = \int K_{x,y} \varphi(y) dy$. Il ne pouvait l'exprimer qu'en termes extrêmement vagues, c'est néanmoins lui qui m'en a donné l'idée et ma démonstration dans [18], assez compliquée du point de vue de l'analyse fonctionnelle telle qu'elle existait alors, fit l'object de mon exposé au Congrès International des Mathématiciens de 1950. J'ai ensuite très longuement développé la théorie des fonctions différentiables et des distributions à valeurs vectorielles dans [30, 39, 42], généralisant ainsi considérablement le théoréme des noyaux et utilisant massivement les travaux précités de A. GROTHENDIECK.

(3) *Perturbations d'homomorphismes par des operateurs compacts dans les espaces de Frechet.* Il s'agit là d'une Note aux Comptes-Rendus de l'Académie des Sciences [23], qui étend aux espaces de FRECHET les théorèmes classiques de Frédéric RIESZ sur les opérateurs compacts dans les espaces de BANACH. Le but de cette note était de permettre à H. CARTAN et J. P. SERRE de montrer que la cohomologie d'une variété analytique compacte, par rapport à un faisceau analytique cohérent, est de dimension finie. L'extension aux espaces analytiques et plusieurs autres théorèmes utilisent toujours [23].

(4) *Sous-espaces hilbertiens d'espaces vectoriels topologiques et noyaux reproduisants.* [51] étend des résultats de S. BERGMAN et N. ARONSZAJN. Les meilleures applications qu'il peut avoir sont relatives aux probabilités gaussiennes dans les espaces vectoriels topologiques. Je n'ai vu ces applications que plus tard et ne les ai jamais rédigées, mais elles se voient sans peine et sont implicitement connues, et utilisées par tous les probabilistes. D'autre part, [52] développe à partir de là une théorie des "points distributions" d'une représentation unitaire d'un groupe de Lie et montre, par exemple, que toute distribution de type positif d'un groupe de Lie semi-simple est intégrale de distributions de type positif élémentaires (théorème de BOCHNER généralisé).

(5) *Le Theoreme du graphe ferme.* BANACH, avait énoncé un théorème du graphe fermé pour les espaces de FRECHET. A. GROTHENDIECK l'avait considérablement étendu, mais il restait inapplicable à certains espaces comme l'espace $\mathscr{D}'$ des distributions. Cependant A. GROTHENDIECK avait émis comme conjecture la validité du théorème du graphe fermé pour cet espace. La conjecture est restée ouverte plus de 10 ans. D. RAIKOV, mathématicien soviétique, a prouvé cette conjecture en 1966 et j'en ai trouvé une démonstration beaucoup plus simple relative même à un graphe borélien, utilisant la théorie de la mesure, quelques semains après lui. J'utilisais le fait que $\mathscr{D}'$ est souslinien. Ma démonstration a encore été simplifiée par A. MARTINEAU

qui l'a rendus indépendante de la théorie de l'intégration, et c'est M. de WILDE qui a, semble-t-il, trouvé la forme définitive du théorème en introduisant les espaces à réseau absorbant, qu'on appelle d'ailleurs espaces de WILDE.

IV. *Physique theorique*

Je m'étais intéressé depuis longtemps à la physique théorique. A la suite d'un séminaire commun entre mathématiciens et physiciens, tenu par Maurice LEVY et moi-même en 1956–57, j'ai essayé d'utiliser les distributions et les noyaux reproduisants [51] pour étudier les particules élémentaires en mécanique quantique relativiste. Je me suis borné aux particules sans interaction. C'est là un problème physique d'un intérêt évidemment très limité, mais je pense que la compréhension complète de ce cas est indispensable à toute étude du cas général avec interaction. On peut dire, si on veut, que des travaux comme ceux de A. WIGHTMAN sur la théorie quantique des champs doivent d'abord partir du cas sans interaction. Cela revient encore à se demander quelles sont exactement les particules élémentaires et on sait que ce problème devient de plus en plus compliqué! Il est lié à l'étude des représentations unitaires irréductibles de groupes de Lie, ayant comme quotient le groupe de Lorentz inhomogène ou groupe de Poincaré. Il se trouve que les noyaux reproduisants des sous-espaces hilbertiens [51] permettent une étude très complète et trouvent les "fonctions singulières" de la mécanique quantique (qui sont des distributions et non des fonctions) précisément comme des noyaux reproduisants; on les définit en fait par leur transformée de Fourier. On trouve aussi, de cette manière, une "densité de probabilité de présence", déjà trouvée par E. WIGNER, et une "probabilité de spin". Le livre que j'ai écrit à ce sujet est un "Lecture Notes" de séminaires tenus, l'un à Buenos-Aires [43], l'autre à Berkeley [44]. Il a d'abord été publié en U.R.S.S., avec une préface de N. N. BOGOLIUBOV, puis en français et en anglais [70].

J'ai aussi publié un petit article [49], à la suite d'une conférence à un séminaire de Physique à Rio de Janeiro. Les physiciens ont besoin, dans le cas d'une dimension, d'utiliser les propriétés d'analyticité dans le demi-plan supérieur de l'image de Fourier d'une distribution à support dans la demi-droite positive, pour déduire des relations entre sa partie réelle et sa partie imaginaire, dites relations de causalité. Ces relations utilisent la convolution avec $vp.$ $(1/x)$ et exigent des conditions très strictes pour pouvoir être appliquées; c'est ce que fournit cet article. Les cas importants pour la physique sont pluridimensionnels, la demi-droite positive est remplacée par le cône de Lorentz d'avenir, ce qui donne idée des difficultés de l'application correcte des formules cherchées.

V. *Theorie de l'intégration, probabilités, probabilités cylindriques et applications radonifiantes*

(1) *Mesures de radon sur des espaces topologiques arbitraires.* J'ai été formé, pendant mes années d'Ecole Normale Supérieure, à l'étude de l'intégration et du calcul des probabilités par P. LEVY à qui je dois énormément (voir mes publications [1, 2, 3]). J'ai travaillé ensuite dans d'autres directions, mais je suis revenu à l'intégration et aux probabilités à partir de 1964. En réalité, les distributions, l'analyse fonctionnelles, l'intégration, devaient m'y conduire normalement.

La théorie de l'intégration a toujours été l'objet d'un conflit assez aigu entre deux tendances. D'une part, on peut faire une théorie abstraite de la mesure, partant d'un ensemble muni d'une tribu sur lequel une mesure est une fonction dénombrablement additive. D'autre part, notamment à la suite des travaux de H. CARTAN et A. WEIL, N. BOURBAKI a défini une mesure de Radon comme une forme linéaire continue sur l'espace $C(X)$ des fonctions continues à support compact sur X, et l'espace X doit alors être un espace topologique localement compact. J'ai introduit en 1964, [55], une notion de mesure de Radon sur des espaces topologiques séparés arbitraires, qui peut servir d'unification entre les deux conceptions. Si l'espace est localement compact, on retrouve les mesures de Radon de BOURBAKI. Les mesures abstraites restent plus générales, mais toutes les tendances actuelles des probabilistes vont vers la recherche d'un espace probabilisé des épreuves (Ω, μ) *particulier*, qui est en fait équivalent à un espace topologique souslinien muni d'une mesure de Radon au sens que j'ai défini. Cette théorie des mesures de Radon sur les espaces arbitraires est parue au "Tata Institute of Fundamental Research" de Bombay [80], rédigée en collaboration avec J. CHOKSI, K. N. GOWRISANKARAN et J. HORVATH. Entre temps, les mêmes idées étaient aussi adoptées par N. BOURBAKI et publiées en 1969. J'ai bénéficié pour cette recherche d'idées de P. A. MEYER et de P. CARTIER.

(2) *Probabilités cylindriques et applications radonifiantes.* Ces nouvelles mesures de Radon que j'ai définie permettent d'avoir une bonne théorie des probabilités cylindriques, introduites par I. SEGAL, L. GROSS et l'école soviétique, J. V. PROHOROV, V. V. SAZONOV, I. M. GELFAND, R. A. MINLOS. On peut ainsi démontrer dans toute sa généralité le théorème de MINLOS [80], mais surtout faire une théorie complètement nouvelle, celle des applications p-radonifiantes.

Une probabilité cylindrique sur un espace vectoriel topologique localement covexe E, est un système projectif de probabilités de Radon relatif aux projections de E sur ses quotients de dimension finie. Ainsi, sur un espace hilbertien E, il existe une probabibilité cylindrique fondamentale, dite de Gauss, qui se projette orthogonalement sur tout sous-espace de dimension finie F de E, suivant la probabilité canonique de Gauss de F. Une probabilité cylindrique sur E définit aussi, non pas une variable aléatoire á valeurs dans E, mais un système projectif de variables aléatoires ou encore une application linéaire de E' dans un espace $L°(\Omega, \mathcal{O}, \pi)$ de variables aléatoires réelles. Ainsi beaucoup de lois de probabilités classiques s'expriment en termes de probabilités cylindriques; le système projectif de variables aléatoires, associé á la probabilité cylindrique de Gauss de l'espace hilbertien L^2 sur $\mathbb{R}$, s'appelle aussi le bruit blanc, qui est en quelque sorte un bruit aléatoire aux fréquences également distribuées sur toute la droite.

Si λ est une probabilité de Radon sur un espace de BANACH E, son p-ordre $(0 < p < +\infty$; en fait, on doit aussi considérer des valeurs de p nulles et même négatives) est $(\int \|x\|^p \, d\lambda(x))^{1/p} = \|\lambda\|_p$. Si maintenant λ est une probabilité cylindrique sur E, son p-type est

$$\operatorname*{Sup}_{\xi \in E', \, \|\xi\| \leq 1} \left(\int t^p \, d(\xi\lambda)(t) \right)^{1/p} = \|\lambda\|_p^*.$$

On dit alors qu'une application linéaire continue u d'un BANACH E dans un BANACH F est p-radonifiante si, pour toute probabilite cylindrique λ de type p sur E, $u(\lambda)$ est de Radon d'ordre p sur F.

On voit que cette notion est une généralisation de celle des applications p-sommantes, introduite par S. KWAPIEN et étudiée par A. PIETSCH, A. PELCZINSKY, A. PERSSON, à partir de 1963. Il est évident qu'une application p-radonifiante est p-sommante, mais rien n'était connu sur la réciproque. J'ai montré que u est p-sommante si et seulement si elle est approximativement p-radonifiante de E dans le bidual de F. Pour $1 < p < +\infty$, u est p-sommante si et seulement si elle est p-radonifiante. Plusieurs de ces résultats ont été obtenus avec la collaboration de S. KWAPIEN.

Après avoir trouvé toutes les applications 0-randonifiantes dans les espaces de suite [75], j'ai systématiquement étudié les applications p-radonifiantes dans mon séminaire de l'Ecole Polytechnique 1969–1970 [79], puis dans un long article [31]. Un des principaux théorèmes de cet article est le théorème de dualité, permettant de montrer que certaines applications sont radonifiantes, et donnant en fait la plupart des exemples connus dans la pratique. Il fait jouer un rôle fondamental aux probabilités de P. LEVY, généralisant celle de Gauss, et dites lois stables. On peut, par le théorème de dualité, trouver plusieurs des théorèmes sur les séries de Fourier aléatoires connues antérieurement (énumérées par J. P. KAHANE dans son livre sur ce sujet), mais aussi le théorème de MENCHOV sur les séries orthogonales et en donner des généralisations (par exemple par A. NAHOUM et B. MAUREY).

Le fait que la fonction aléatoire du mouvement brownien, primitive du bruit blanc, soit presque sûrement continue et même holdérienne de tout ordre $< \frac{1}{2}$, peut s'exprimer comme suit: l'injection canonique de l'espace H^1 de SOBOLEV dans l'espace C^β des fonctions holdériennes d'ordre β est, localement sur $\mathbb{R}$, p-radonifiantes pour $\beta < \frac{1}{2} - 1/p$. La théorie des applications radonifiantes a été appliquée par T. BENSOUSSAN au filtrage de systèmes gouvernés par des équations aux dérivées partielles, et par Paul KREE et Bernard LASCAR à l'étude des équations aux dérivées partielles et opérateurs pseudo-différentiels en dimension infinie. Mais ce sont surtout les applications p-sommantes qui ont en fait connu un grand développement relativement indépendant de celui des applications radonifiantes, dans les dernières années, par les notions de type et de cotype des espaces de Banach, introduites par Bernard MAUREY, et les travaux remarquables de nombreux chercheurs français et étrangers, parmi lesquels je me bornerai à citer, au Centre de Mathématiques de l'Ecole Polytechnique, MAUREY, puis Gilles PISIER et Bernard BEAUZAMY.

(3) *Desintegrations regulieres.*   Il y a longtemps que les probabilistes, voyant que les probabilités conditionnelles n'existent pas toujours, ont introduit les espérances conditionnelles, avec une utilisation systématique des temps d'arrêt. Cependant M. JIRINA a montré que les probabilités conditionnelles (désintégration d'une mesure) existent dans des conditions très générales. J'ai étendu le théorème de JIRINA, en introduisant l'espérance conditionnelle d'une fonction à valeurs mesures; la désintégration d'une probabilité λ sur une sous-tribu $\mathcal{T}$ est alors l'espérance conditionnelle relative à $\mathcal{T}$ et λ de la fonction à valeurs mesures $\omega \mapsto \delta(\omega)$. Mais on peut aller beaucoup plus loin et démontrer, pour les surmartingales à valeurs mesures, des propriétés analogues à celles des surmartingales scalaires; ce qui permet de définir la désintégration régulière d'une mesure par rapport à une famille de tribus dépendant du paramètre temps, croissante et continue à droite comme on en rencontre dans l'étude des processus stochastiques. Ces désintégrations régulières ont des propriétés très remarquables, que j'ai dégagées dans [84], avec un complément court mais intéressant dans [86]. Voici par exemple un des théorèmes les plus difficiles de [84]: si $(\lambda_\omega^t)_{(t,\omega)}$ est une désintégration régulière de λ pour la famille de tribus $(\mathcal{T}^t)_{t \in \mathbb{R}}$ et si X est un processus optionnel à valeurs dans un espace souslinien, alors, pour λ-presque tout ω, pour tout t, λ_ω^t est porté par l'ensemble des ω' dont la trajectoire coincide avec celle de ω jusqu'à l'instant t. Cet article introduit aussi la notion nouvelle de tribus fortes. Si les $\mathcal{T}^t$ sont des tribus fortes, ce qui est toujours le cas dans les processus, alors, pour λ-presque tout ω, pour tout t, λ_ω^t admet comme désintégration régulière, pour les temps $\geq t$, la même que λ elle-même. Cela permet de caractériser toutes les mesures qui ont une désintégration ou une désintégration régulière donnée, comme intégrales des mesures extremales

ayant cette propriété, quoiqu'on ne soit absolument pas dans les conditions d'applications du théorème de CHOQUET sur les points extrémaux.

Les désintégrations régulières me paraissent susceptibles d'avoir de nombreuses et fécondes applications. On peut les appliquer aux martingales et surmartingales régulières à valeurs scalaires ou mesures. Par exemple, si les tribus sont fortes, on peut changer la probabilité de base d'une surmartingale régulière en la remplaçant par une intégrale arbitraire de ses désintégrantes, sans cesser d'avoir une surmartingale régulière; la même propriété est valable pour un processus fortement markovien à la place d'une surmartingale régulière. Les projections optionnelle, accessible, prévisible d'un processus réel mesurable s'écrivent remarquablement bien à partir d'une désintégration régulière (par exemple la projection optionnelle de X est $(t, \omega) \mapsto \lambda_\omega^t(X^t)$). La désintégration de λ par rapport à la tribu du passé d'un temps d'arrêt T s'exprime aussi très bien: c'est $\omega \mapsto \lambda_\omega^{T(\omega)}$.

(4) *Applications des destintegrations regulieres aux processus de Markov.* Il est bien connu que, si l'espace des états E est localement compact polonais, et si $(P_t)_{t \in \mathbb{R}_+}$ est un semi-groupe fortement continu d'opérateurs sur l'espace $C_0(E)$ des fonctions continues sur E tendant vers 0 à l'infini, positif ($P_t f \geq 0$ pour $f \geq 0$), de norme ≤ 1, il définit un processus de Markov, homogène dans le temps, d'espace d'états E; l'espace Ω est l'espace canonique des trajectoires, et $\mathbb{P}^x$ est la loi sur Ω pour laquelle le processus part à l'instant 0 de $x \in E$. J'ai montré dans [100] qu'on pouvait, sans plus de difficulté, supposer E souslinien complètement régulier quelconque, remplacer le semi groupe des P_t par un pseudo-groupe de $P_{s,t}$ (avec $P_{r,s} \circ P_{s,t} = P_{r,t}$), et remplacer la démonstration habituelle par une autre utilisant les désintégrations régulières, $(t, \omega) \mapsto \lambda_\omega^t$; par ailleurs $\mathbb{P}^x$ est remplacé par $\mathbb{P}^{\sigma,x}$, loi sur Ω lorsque le processus part de $x \in E$ à l'instant $\sigma \in \mathbb{R}$. On exprime explicitement $\mathbb{P}^{\sigma,x}$ et λ_ω^t à partir des $P_{s,t}$; pour les temps $\geq s$, λ_ω^s coincide avec $\mathbb{P}^{s,X_s(\omega)}$, ou encore $\theta^s \lambda_\omega^s = \mathbb{P}^{s,X_s(\omega)}$; c'est cette formule (vraie pour tout s et tout ω) qui remplace la propriété de Markov forte.

(5) *Semi-martingales a valeurs dans une variete differentielle et martingales conformes a valeurs dans une variete analytique complexe* [105]. La formule d'Itô sur l'intégrale stochastique montre qu'une fonction C^2 d'une semi-martingale est une semi-martingale; il est donc raisonnable de penser qu'on peut étudier les semi-martingales à valeurs dans les variétés différentielles de classe C^2. On dira simplement que X est une semi-martingale à valeurs dans V si, pour toute fonction réelle φ de classe C^2 sur V, $\varphi(X)$ est une semi-martingale réelle. On a les propriétés qu'on peut en attendre: par exemple la stabilité par les applications C^2; et, si $\tilde{V}$ est un revêtement de V, X continue et qu'on

relève X en $\tilde{X}$ de manière que X_0 soit $\mathscr{T}_0$-mesurable, $\tilde{X}$ est une semi-martingale à valeurs dans $\tilde{V}$. On peut alors définir des intégrales stochastiques remarquables: si X est continue, si J est un processus optionnel cotangent ($J(s, \omega)$ cotangent à V au point $X(s, \omega)$), on peut essayer de définir $\int_{]0,t]}(J_s|dX_s)$ (J_s est cotangent, dX_s "à peu près" tangent). Cette intégrale n'est pas définie, parce que dX_s n'est pas vraiment tangent, mais sa partie martingale locale continue est bien définie; on la note $J.X^c$. Par exemple, si $\bar{\omega}$ est une forme différentielle borélienne sur V, de degré 1, on pourra définir son intégrale $\omega(X).X^c$ sur les trajectoires. On en donne diverses propriétés intéressantes, par exemple: l'ensemble des $J.X^c$, où J est "intégrable", est exactement l'espace stable $\mathscr{M}(X)$ de martingales locales continues engendré par les $(\varphi(X))^c$, φ réelle C^2 sur V; donc celui-ci admet un système de générateurs formé de $N = \dim V$ martingales réelles orthogonales, ce qui est un peu inattendu car V n'est plongeable que dans $\mathbb{R}^{2N}$, non dans $\mathbb{R}^N$. Je traite l'exemple particulièrement intéressant où X est une diffusion définie par un opérateur différentiel

$$L = \frac{1}{2} \sum_{i,j} a^{i,j} \frac{\partial^2}{\partial x^i \partial x^j} + \sum_i b^i \frac{\partial}{\partial x^i}$$

d'ordre 2, elliptique sur V. On retrouve diverses formules liées á l'équation différentielle stochastique définissant X; on trouve en particulier qu'il y a correspondance bijective entre les systèmes de N mouvements browniens indépendants $(B_k)_{k=1,\ldots,N}$ appartenant à l'espace canonique des trajectoires, et les champs optionnels de N vecteurs cotangents $(J_k)_{k=1,\ldots,N}$ orthonormés (pour la forme quadratique des $a^{i,j}$), précisément par $B_k = J_k.X^c$.(C'est ce qui explique que la méthode des équations différentielles stochastiques (qui exige des coefficients lipschitziens) soit impuissante à traiter globalement le problème de la diffusion, parce qu'une variété riemannienne n'admet pas de champ lipschitzien de vecteurs orthonormés!)

Les martingales conformes (complexes) sont celles dont le carré est encore une martingale; elles sont stables par les applications holomorphes. Il est donc naturel de penser qu'on peut définir des martingales conformes sur des variétés analytiques complexes. Mais il y a ici une difficulté considérable qui ne se présentait pas dans le cas réel: si V n'est pas de Stein, elle a trop peu de fonctions holomorphes. Il faut localiser le problème; en gros, X est une martingale conforme à valeurs dans V analytique si, pour toute fonction complexe C^2 sur V (C^2 au sens réel), $\varphi(X)$ est une martingale conforme "pendant sa vie" dans tout ouvert de V où φ est holomorphe. On étend alors à ce cas complexe les résultats précédents, notamment les intégrales stochastiques $J.X^c$. Si L est le laplacien Δ d'une structure kählérienne sur V, le processus markovien associé est une martingale conforme.

Je termine en ce moment deux articles. Le premier [112] est relatif aux semi-martingales formelles. On sait que, si X est une semi-martingale réelle, elle définit une mesure sur la tribu prévisible, à valeurs dans l'espace des classes de fonctions mesurables, par l'intégrale stochastique,

$$H \mapsto \int_{[0, +\infty]} H_s\, dX_s;$$

DELLACHERIE a donné une caractérisation des semi-martingales par les propriétés de ces mesures. Ces mesures sont l'équivalent d'une mesure ≥ 0 finie, comme la mesure de Lebesgue sur le segment $[0, 1]$. Mais on peut introduire des semi-martingales formelles, qui sont l'analogue des mesures ≥ 0 non finies, comme la mesure de Lebesgue sur $\mathbb{R}$. Si alors X est une semi-martingale formelle, H un processus prévisible non borné arbitraire, on peut toujours définir son intégrale stochastique $H.X$, sans condition d'intégrabilité, et c'est encore une semi-martingale formelle. On développe ainsi un calcul intégral sur les semi-martingales formelles, qui est un peu au calcul intégral sur les semi-martingales vraies ce qu'est le calcul des distributions par rapport au calcul des fonctions: tout existe toujours, pas comme semi-martingale vraie, mais comme semi-martingale formelle. Il est alors possible [113] de combiner les résultats indiqués à (5) et l'intégration stochastique formelle pour faire une belle théorie des semi-martingales sur les variétés et de l'intégrale stochastique des processus cotangents d'ordre 2. La formule d'Ito montre en effet que les éléments du second ordre interviennent inévitablement. Dans (5) a été défini $J.X^c$, mais pas $J.X$, pour J processus optionnel cotangent; mais $J.X$ est bien définie si J est un processus optionnel cotangent d'ordre 2. On déduit de là des représentations tangentielles d'ordre 2 des semi-martingales, la notion de sous-espace vectoriel tangent d'ordre 2 en chaque point à une semi-martingale sur la variété, et finalement une expression intrinsèque (sans cartes) des équations différentielles stochastiques et des propriétés de leurs solutions, basées sur la géométrie différentielle du second ordre sur une variété différentielle. Ce travail rejoint des travaux de Paul–André MEYER, Jean–Michel BISMUT et Paul MALLIAVIN. On démontre ainsi qu'on peut relever toute semi-martingale continue, par une connexion d'un espace fibré.

D. PUBLICATIONS

1936
1. Sur une question de calcul des probabilités, *Bull. Sci. Math.* (2) **60** (1936), 1.

1941
2. Sur les fonctions à variation bornée et les courbes rectifiables, *C. R. Acad. Sci. Paris* **212** (3 Mars 1941), 331–333.
3. Sur le module de la fonction caractéristique du calcul des probabilités, *C. R. Acad. Sci. Paris* **212** (17 Mars 1941), 418–421.

1942

4. Etude des sommes d'exponentielles réelles; thèse de doctorat, Actualités Scientifiques et Industrielles, Hermann, Paris, 1943; deuxième édition, Hermann, Paris, 1959.

5. Approximation d'une fonction quelconque par des sommes d'exponentielles imaginaires, *Ann. Fac. Sci. Univ. Toulouse* **VI** (1942, paru en 1943), 111–176.

1944

6. Sur certaines familles non fondamentales de fonctions continues, *Bull. Soc. Math. France* **72** (1944), 141–145.

1945

7. Généralisation de la notion de fonction, de dérivation, de transformation de Fourier, et applications mathématiques et physiques, *Ann. Univ. Grenoble, Sect. Sci. Math. Phys. (N. S.)*, **21** (1945, paru en 1946), 57–74.

1946

8. Sur les fonctions moyenne-périodiques, *C. R. Acad. Sci. Paris* **223** (8 Juillet 1946), 68–70.

1947

9. Théorie générale des fonctions moyenne-périodiques, *Ann. of Math.* **48** (1947), 857–929.

10. Théorie des distributions et transformation de Fourier, *Ann. Univ. Grenoble, Sect. Sci. Math. Phys. (N. S.)* **23** (1947–1948), 7–24.

11. Théorie des distributions et transformation de Fourier, "Analyse Harmonique," *Colloque C.N.R.S.*, *N° 15, Nancy, 15–22 Juin 1947*, pp. 1–8, Gauthier-Villars, Paris, 1949.

1948

12. Généralisation de la notion de fonction et dérivation: théorie des distributions, *Ann. Télécommunications* **3** (1948), 135–140.

13. Sur une propriété de synthèse spectrale dans les groupes non compacts, *C. R. Acad. Sci. Paris* **227** (1er Juillet 1948), 424–426.

1949

14. En collaboration avec. J. Dieudonné, La dualité dans les espaces ($\mathscr{F}$) et ($\mathscr{LF}$), *Ann. Inst. Fourier (Grenoble)* **I** (1949, paru en 1950), 61–101.

15. Les mathématiques en France pendant et après la guerre, *Proc. Canadian Mathematical Congress*, 2nd, *Vancouver, 16 Août–10 Sept. 1949*, p. 49–67, Univ. of Toronto Press, Toronto, 1951.

1950

16. "Théorie des distributions," t. 1, Actualités Scientifiques et Industrielles, Hermann, Paris, 1950.

17. Un lemme sur la dérivation des fonctions vectorielles d'une variable réelle. Les équations d'évolution liées au produit de composition, *Ann. Inst. Fourier (Grenoble)* **II** (1950, paru en 1951), 18–49.

18. Théorie des noyaux, *Proc. Internat. Congr. Mathematicians, 1950, Cambridge (U.S.A.)* Vol. I, p. 220–230.

1951

19. "Théorie des distributions," t. 2, Actualités Scientifiques et Industrielles, Hermann, Paris, 1951; troisième édition, 1966 (t. 1 et 2 en un seul volume).

20. Analyse et synthèse harmoniques dans les espaces de distributions, *Canad. J. Math.* Vol. **III** (4) (1951), 503–512.

1952

21. "Transformation de Laplace des distributions," Séminaire Mathématique de l'Université de Lund, tome dédié à Marcel Riesz, pp. 196–206, Lund, 1952.

22. Sur les multiplicateurs de $\mathscr{F}L^p$, Kungl. Fysiografiska Sällskapets I Lund Forhändlingar, **22** (21) (1952), pp. 1–5.

1953

23. Homomorphismes et applications complètement continues, *C. R. Acad. Sci. Paris* **236** (29 Juin 1953), 2472–2473.

24. Courant associé à une forme différentielle méromorphe sur une variété analytique complexe, "Géométrie Différentielle," *Colloque C.N.R.S.*, N° 52, Strasbourg, 26 Mai–1 er Juin 1953, p. 185–195, C.N.R.S., Paris, 1953.

1954

25. Problèmes aux limites dans les équations aux dérivées partielles elliptiques, *Colloque sur les équations aux dérivées partielles*, (*CBRM*), 2nd, Bruxelles, 24–26 Mai 1954, p. 13–24, Masson, Paris, 1955.

26. L'oeuvre de Poincaré : équations différentielles de la physique, Journée Internationale Henri Poincaré à La Haye, 11 Sept. 1954, Le Livre du Centenaire de la Naissance de Henri Poincaré, 1854–1954, p. 219–225, Gauthier-Villars, Paris, 1955.

27. Sur l'impossibilité de la multiplication des distributions, *C. R. Acad. Sci. Paris* **239** (11 Octobre 1954), 847–848.

28. Produits tensoriels topologiques d'espaces vectoriels topologiques. Espaces vectoriels topologiques nucléaires. Applications. Séminaire Schwartz, Institut Henri Poincaré, Paris, 1953–1954.

1955

29. En collaboration avec J. L. Lions, Problèmes aux limites sur les espaces fibrés, *Acta Math.* **94** (1955), 155–159.

30. Espaces de fonctions différentiables à valeurs vectorielles, *J. Analyse Math.* **4** (1954–1955), 88–148.

31. Equations aux dérivées partielles, Séminaire Schwartz, Institut Henri Poincaré, Paris, 1954–1955.

32. L'énumération transfinie et l'oeuvre de M. Denjoy, *Bull. Sci. Math.* (2), **t. 79** (1955), 1–18.

33. "Lectures on complex analytic manifold," Tata Institute for Fundamental Research, Bombay, 1955.

34. Division par une fonction holomorphe sur une variété analytique complexe, *Summa Brasiliensis Math.* **3** (9) (1955), 181–209.

1956

35. Problèmes mixtes pour l'équation des ondes, Séminaire Schwartz, Institut Henri Poincaré, Paris, 1955–1956.

36. Variedades analiticas complejas; Ecuaciones diferenciales parciales elipticas. Université Nationale de Colombie, Bogota, 1956.

1957

37. Distributions semi-régulières et changements de coordonnées, *J. Math. Pures Appl.* **36** (1957), 109–127.

38. "Mixed problems on partial differential equations. Representations of semi-groups," Tata Institute for Fundamental Research, Bombay, 1958.

39. Théorie des distributions à valeurs vectorielles, chap. I, *Ann. Inst. Fourier*, (*Grenoble*) **VII** (1957), 1–141.

1958

40. Su alcuni problemi della teoria delle equazione differenziali lineari di tipo ellittico, *Rend. Sem. Mat. Fis. Milano* **27** (1958), 1–41.
41. Généralisation des espaces L^p, *in* "Tome jubilaire dédié à M. Paul Lévy," pp. 241–250.
42. Théorie des distributions à valeurs vectorielles, chap. II, *Ann. Inst. Fourier, (Grenoble)* **VIII** (1958, paru en 1959), 1–209.

1959

43. "Matematica y Fisica Cuantica," Fac. des Sc. de Buenos-Aires, p. 1–266, 1959.

1960

44. "Applications of Distributions to the Study of Elementary Particles in Relativistic Quantum Mechanics," pp. 1–207, Univ. de California Press Berkeley, 1960.
45. Density of Probability of Presence of Elementary Particles, *Proc. Berkeley Symp. on Mathematics, Statistics, and Probability, 4th, June 20–July 30, 1960,* Vol. III, p. 307–314, Univ. of California Press.

1961

46. "Méthodes mathématiques pour les sciences physiques," deuxième édition, Hermann, Paris, 1965.
47. Transformata di Fourier delle Distribuzioni; spazi di Hilbert e nuclei associati, pp. 1–80, Institut Mathématique de l'Université de Rome—Centro Internazionale Matematico Estivo (C.I.M.E.), 1961.

1962

48. Convergence de distributions dont les dérivées convergent, En hommage à G. Polya. Studies in Mathematical Analysis and Related Topics, pp. 364–372, Stanford Univ. Press, Stanford, California 1962.
49. Causalité et Analyticité, *An. Acad. Brasil. Ciênc.*, Rio de Janeiro, Vol. **34**, (1) 1–21, (1962).

1963

50. Some applications of the theory of distributions, Lectures on Modern Mathematics, No. 1, pp. 23–58, Wiley, New York, 1963.

1964

51. Sous-espaces hilbertiens d'espaces vectoriels topologiques et noyaux associés (noyaux reproduisants), *J. Analyse Math.* **XIII** (1964), 115–256.
52. Sous-espaces hilbertiens et noyaux associés; applications aux représentations des groupes de Lie, *Deuxième colloque d'Analyse Fonctionnelle, (CBRM), Liège 4–6 Mai 1964,* p. 153–163, Gauthier-Villars, Paris, 1964.
53. "Functional Analysis," pp. 1–212, Courant Institute of Mathematical Sciences, New York University, New York, 1964.
54. Théorème d'Atiyah-Singer sur l'indice d'un opérateur différentiel elliptique, Séminaire Cartan-Schwartz, Fasc. 1 et 2, Institut Henri Poincaré, Paris, 1963–1964.
55. Mesures de Radon sur des espaces non localement compacts, *Proc. Internat. Summer Institute on Theory of Distributions, Lisbon, Sept. 1964,* p. 1–21, Centro de Calculo Cientifico, 1964.

1965

56. En collaboration avec S. Mandelbrojt, Jacques Hadamard (1865–1963), *Bull. Amer. Math. Soc.* **71** (1965), 107–129.
57. Mesures de Radon sur des espaces topologiques arbitraires, Cours de 3ème cycle, Inst. Henri Poincaré, Paris, 1965–1966.

1966

58. Sur le théorème du graphe fermé, *C. R. Acad. Sci. Paris* **263** (2 Novembre 1966), 602–605.
59. Un nouveau théorème sur les distributions, *C. R. Acad. Sci. Paris* **263** (19 Décembre 1966), 899–901.

1967

60. Radon Measures on Souslin Spaces, *Symp. Analysis*, Kingston Ontario, June 1967.
61. Le modèle d'une théorie des ensembles, *Conférence Institut Henri Poincaré, Paris, 26 Octobre 1967*.
62. Extension du théorème de Sazonov-Minlos à des cas non hilbertiens, *C. R. Acad. Sc. Paris* **265** (18 Décembre 1967), 832–834.
63. "Cours d'Analyse," tomes 1 et 2, Hermann, Paris, 1967.

1968

64. Réciproque du théorème de Sazonov-Minlos dans des cas non hilbertiens, *C. R. Acad. Sci. Paris* **266** (3 Janvier 1968), 7–9.
65. Démonstration de deux lemmes sur les probabilités cylindriques, *C. R. Acad. Sci. Paris* **266** (8 Janvier 1968), 50–52.
66. Désintégration régulière d'une mesure par rapport à une famille de tribus, *C. R. Acad. Sci. Paris* **266** (12 Février 1968), 424–425.
67. Applications des désintégrations régulières, *C. R. Acad. Sci. Paris* **266** (19 Février 1968), 467–469.
68. Un théorème sur les suites de variables aléatoires, Symposia Mathematica, Vol. II, p. 204–209, Istituto Nazionale di Alta Matematica, 1968.
69. "Application of Distributions to the Theory of Elementary Particles in Quantum Mechanics," Gordon and Breach, New York, 1968.

1969

70. "Application des distributions à l'étude des particules élémentaires en mécanique quantique relativiste," Gordon and Breach, Paris, 1969.
71. Probabilités cylindriques et applications radonifiantes, *C. R. Acad. Sci. Paris* **268** (24 Mars 1969), 646–648.
72. Un théorème de convergence dans les L^p, $0 \leq p < \infty$, *C. R. Acad. Sci. Paris* **268** (31 Mars 1959), 704–706.
73. Un théorème de dualité pour les applications radonifiantes, *C. R. Acad. Sci. Paris* **268** (9 Juin 1969), 1410–1413.
74. Applications du théorème de dualité sur les applications p-radonifiantes, *C. R. Acad. Sci. Paris* **268** (30 Juin 1969), 1612–1615.
75. Mesures cylindriques et applications radonifiantes dans les espaces de suites, *Proc. Internat. Conf. on Functional Analysis and Related Topics, Tokyo, April 1969*, pp. 1–59.
76. Un théorème de dualité pour les applications p-radonifiantes. Applications de ce théorème, Colloque C.N.R.S. N° 186: "Les probabilités sur les structures algébriques", Clermont-Ferrand, 30 Juin–5 Juillet 1969, pp. 319–326, C.N.R.S., Paris, 1970.
77. Applications p-radonifiantes et théorème de dualité, Colloquium on Nuclear Spaces and Ideals in Operator Algebras, *Studia Math.* **38** (1970), 203–213. (Cf. article N° 82).

1970

78. "Analyse. Deuxième partie: topologie générale et analyse fonctionnelle," Hermann, Paris, 1970.
79. Applications radonifiantes, Séminaire d'Analyse de l'Ecole Polytechnique, Paris, 1969–1970.

80. En collaboration avec J. Choksi et J. Horvath, "Radon Measures on Arbitrary Topological Spaces and Cylindrical Measures," Tata Institute of Fundamental Research, Bombay.
81. Probabilités cylindriques et applications radonifiantes, *J. Fac. Sci. Univ. Tokyo Sect. IA Math.* **18** (2) (1970), 139–286.
82. Applications *p*-radonifiantes et théorème de dualité, *Troisième colloque d'Analyse Fonctionnelle. (CBRM), Liège, 14–16 Septembre 1970,* pp. 153–163, Vander, Louvain, 1971.
83. Produits tensoriels g_p et d_p, applications *p*-sommantes, applications *p*-radonifiantes, Séminaire Bourbaki, Novembre 1970, Paris, 1970–1971.

1971

84. Surmartingales régulières à valeurs mesures et désintégrations régulières d'une mesure, *J. Analyse Math.* **XXVI** (1973), 1–168.
*85. Séminaire Goulaouic–Schwartz 1970–1971, Equations aux dérivées partielles linéaires et analyse fonctionnelle, Ecole Polytechnique, Paris, 1971.
86. En collaboration avec B. Maurey et A. Nahoum, Etude du transformé d'un processus par une surmartingale régulière de mesures, *C. R. Acad. Sc. Paris* **274** (3 Mai 1972), 1365–1368.

1972

87. La fonction δ et les noyaux, *in* "Aspects of Quantum Theory," Cambridge Univ. Press, London and New York, 1972.
*88. Séminaire Goulaouic–Schwartz, Ecole Polytechnique, Paris.

1973

*89. Séminaire Goulaouic–Schwartz, Ecole Polytechnique, Paris
*90. Séminaire Maurey–Schwartz, Ecole Polytechnique, Paris.
91. Seminar Schwartz, L. Schwartz and M. H. Schwartz, Australian National University, Canberra, Australie.

1974

*92. Séminaire Goulaouic–Schwartz, Ecole Polytechnique, Paris.
*93. Séminaire Maurey–Schwartz, Ecole Polytechnique, Paris.
94. Désintégrations régulières, tribus fortes et probabilités extrémales, *Rev. Columbiana Mat.* **VIII** (1974), 39–45.

1975

95. "Tenseurs," Hermann, Paris, 1975.
*96. Séminaire Goulaouic–Schwartz Ecole Polytechnique, Paris.
*97. Séminaire Maurey–Schwartz, Ecole Polytechnique, Paris.

1976

*98. Séminaire Maurey–Schwartz, Ecole Polytechnique, Paris.
*99. Séminaire Goulaouic–Schwartz, Ecole Polytechnique, Paris.
100. Processus de Markov et désintégrations régulières, *Ann. Inst. Fourier, (Grenoble)* **27** (3) (1977), 211–277.
101. Disintegrations of measures, Tata Institute of Fundamental Research, Bombay, 1976.

1977

*102. Séminaire Goulaouic–Schwartz, "Equations aux dérivées partielles," Ecole Polytechnique, Paris.

1978

*103. Séminaire Goulaouic–Schwartz, "Equations aux dérivées partielles," Ecole Polytechnique, Paris.

*104. Séminaire sur la géométrie des espaces de Banach (ex Maurey-Schwartz), Ecole Polytechnique, Paris.

105. Semi-martingales sur des variétés et martingales conformes sur des variétés analytiques complexes, Lecture Notes in Mathematics, No. 780, Springer-Verlag, Berlin and New York.

1979

106. Geometry and probabilities in Banach spaces, *Conférence franco-asiatique, Singapour, Mai 1979.*

107. Survey of the theory of martingales, *Bull. Malaysian Math. Soc.* **2** (2) (1979), 61–74.

*108. Séminaire Goulaouic–Schwartz, "Equations aux dérivées partielles," Ecole Polytechnique, Paris.

109. Séminaire d'analyse fonctionnelle (ex séminaire sur la géométrie des espaces de Banach), Ecole Polytechnique, Paris.

1980

*110. Séminaire Goulaouic–Schwartz "Equations aux dérivées partielles," Ecole Polytechnique, Paris.

111. Séminaire d'analyse fonctionnelle, Ecole Polytechnique, Paris.

112. Semi-martingales formelles, *in* "Séminaire de Probabilités," XV, 1979–1980, Lecture Notes in Mathematics, Springer-Verlag, Berlin and New York, to appear in 1981.

1980–1981

113. Géométrie différentielle et semi-martingales, in "Séminaire de Probabilités," XVI, 1980–1981, Lecture Notes in Mathematics, Springer-Verlag, Berlin and New York, to appear in 1982.

(Les séminaires précédés de * sont des séminaires tenus au Centre de Mathématiques, organisés collectivement et qui ne doivent pas véritablement être attribués à L. Schwartz).

Part A

On the Symbol of a Distribution

W. Ambrose

Department of Mathematics
Massachusetts Institute of Technology
Cambridge, Massachusetts

Dedicated to Laurent Schwartz

We define here, for any Riemannian manifold X, a linear subspace $\mathscr{D}'_\sigma(X)$ of the space $\mathscr{D}'(X)$ of distributions on X, which is intrinsically defined (i.e., without reference to such things as phase functions and amplitudes) and contains the Fourier integral distributions, and then we define a full symbol for distributions in $\mathscr{D}'_\sigma(X)$. Our symbol will be defined on a bundle closely related to the bundle of second-order tangent vectors to X. In doing this we make no contribution to analysis for we use, as defining property of $\mathscr{D}'_\sigma(X)$, a property proved in [4] and [3] for Fourier integral distributions. Our contribution is to establish an isomorphism between the collection of Lagrange subspaces of $T(X)_{(x,z)}$ [where $T(X)_{(x,z)}$ is the tangent space to the tangent bundle $T(X)$ at the point $(x, z) \in T(X)$] with a space obtained from the second-order tangent vectors at x, and to use this isomorphism to obtain a full symbol for distributions in $\mathscr{D}'_\sigma(X)$.

We use the Riemannian metric of X in two ways: (1) it gives a volume element so that functions can be integrated (and half-densities eliminated), and (2) it gives the Riemannian connection, which provides at each $(x, z) \in T(X)$ a "horizontal" complement to the "vertical" subspace of $T(X)_{(x,z)}$.

The Riemannian connection is essential in the following ways: (1) it makes possible a dissection of second-order tangent vectors to X into pure first and pure second-order components; (2) through it, the pure second-order tangent vectors at $x \in X$ are canonically isomorphic to the symmetric linear transformations of X_x into itself, and (3) via (1) and (2) it gives a canonical $1:1$ correspondence between the Lagrange subspaces of $T(X)_{(x,z)}$ and the set of all pairs (A, E) where A is a symmetric linear transformation of X_x into X_x, and E is a linear subspace of the null space N_A of A. Thus we are led to consider the bundle of all (x, z, A, E) where (x, z) is any point of $T(X)$, A is any symmetric linear transformation of X_x into X_x, and E is any linear subspace of N_A. We denote this bundle by $\Sigma(X)$. For each T in $\mathscr{D}'_\sigma(X)$ σ_T will be defined on a subset $\Sigma_T(X)$ of $\Sigma(X)$ whose complement is "negligible" in a sense explained below.

29

The values of our symbols will lie in the space $S^\infty(\mathbf{R}^+)/S^{-\infty}(\mathbf{R}^+)$, where $S^m(\mathbf{R}^+)$ denotes the Hörmander space $S^m(X \times Z)$ in the case where X is a point and $Z = R^+$. [The values of the symbol p of a pseudo-differential operator can be considered to lie in $S^\infty(\mathbf{R}^+)/S^{-\infty}(\mathbf{R}^+)$ if one considers $p(x, \xi)$ to be the map: $(x, \xi) \to p(x, \cdot \xi)$ mod $S^{-\infty}(\mathbf{R}^+)$.] Because the symbol is defined on second-order elements we do not need the identifications used in [4] and hence do not need the line bundle L which occurs there. For this reason we do not encounter Maslow indices and related matters.

As previously remarked, the analysis used here comes from [4]. In particular, our symbol for a distribution T will be obtained, following [4], by considering certain functions g on the tangent spaces X_x, lifting T to a distribution T_x on X_x, forming the function g^+ defined on R^+ by: $g^+(s) = e^{isg(x)}T_x(e^{-isg(\cdot)})$, and then defining the symbol to be g^+ mod $S^{-\infty}(\mathbf{R}^+)$. We denote the symbol of T by σ_T. To obtain $\sigma_T(x, z, A, E)$ we define a certain g, depending on (x, z, A, E), and use this procedure to define $\sigma_T(x, z, A, E)$. Hörmander obtains only a principal symbol because the method of stationary phase, which he uses, brings in second-order elements and because of lack of uniqueness of appropriate g's. Because we accept dependence on second-order elements [by using the bundle $\Sigma(X)$] and because we can choose a unique g for each (x, z, A, E) (using the Riemannian structure) we can take g^+ mod $S^{-\infty}(\mathbf{R}^+)$ to be the symbol, instead of using only equivalence classes formed from the first-order part in its expansion. Our geometric correspondence mentioned above ensures that this symbol will be "correct" in the case of Fourier integral distributions. In particular, it gives the Hörmander symbol, if one makes his identifications, because it is constructed in the same way.

This paper has two sections. The first discusses the necessary geometry. The second defines the class $\mathscr{D}'_\sigma(X)$ of distributions for which we define a symbol, and gives the definition of the symbol.

1. On Lagrange Subspaces and Second-Order Tangent Vectors

Throughout this paper X will be an n-dimensional Riemannian manifold, and if $x \in X$ then X_x will be the (real) tangent space to X at x. The scalar product on $X_x \times X_x$ (given by the Riemannian metric) will be denoted by $\langle \cdot, \cdot \rangle$. We shall also consider the vector space (over $\mathbf{R}$) of second-order tangent vectors at x, which we denote by X''_x. We now discuss X''_x.

Let $I_x = [\phi \in C^\infty(X) | \phi(x) = 0]$. By definition, X_x is the set of linear maps t of $I_x \to \mathbf{C}$ such that (1) $t\phi \in \mathbf{R}$ if ϕ is real valued and (2) $t = 0$ on I_x^2. Similarly, X''_x is the linear space (over $\mathbf{R}$) of all linear maps t of $I_x \to \mathbf{C}$ such that (1) $t\phi$

is real if ϕ is real valued and (2) $t = 0$ on I_x^3. If $x_1, \ldots, x_n$ is any coordinate system of X around x then each $t \in X_x$ can be expressed as

$$t = \sum a_i \frac{\partial}{\partial x_i}(x), \qquad a_i \in \mathbf{R},$$

and each $t \in X_x''$ can be expressed as

$$t = \sum a_i \frac{\partial}{\partial x_i}(x) + \sum a_{ij} \frac{\partial^2}{\partial x_i \partial x_j}(x), \qquad a_i \in \mathbf{R}, \quad a_{ij} \in \mathbf{R}, \quad a_{ij} = a_{ji}. \quad (1.1)$$

Because X is Riemannian we can define the notion of a *pure* second-order tangent vector and, letting $X_x^p = $ the subspace of X_x'' of all pure second-order vectors, any $t \in X_x''$ can be uniquely expressed as

$$t = t_1 + t_p, \qquad t_1 \in X_x, \qquad t_p \in X_x^p. \quad (1.2)$$

This is explained in [2] for any affine connection but we discuss it here only for the Riemannian case.

To obtain (1.2) we consider *normal* coordinate systems at x. By definition, a *normal coordinate system* at x is any coordinate system carried over, via $\exp_x$, from an orthonormal coordinate system on X_x. That is, consider any orthonormal base $e_1, \ldots, e_n$ of X_x, define the functions u_i on X_x by

$$u_i(y) = \langle y, e_i \rangle, \qquad y \in X_x.$$

Consider any open subset Q of X_x such that $\exp_x$ is a diffeomorphism of Q onto $\exp_x Q$. Then the coordinate system $\{x_i\}$ defined on $\exp Q$ by $x_i = u_i \circ \exp_x^{-1}$ is normal. To make the domain unique we always suppose that $Q = B(0, R)$ where R is the biggest positive number (or infinity) such that $\exp_x$ is a diffeomorphism on $B(0, R)$. Thus any two normal coordinate systems at x, $\{x_i\}$ and $\{x_i'\}$, differ by an orthogonal transformation, i.e., there exists a constant orthogonal $n \times n$ matrix (c_{ij}) such that

$$x_j' = \sum_i c_{ji} x_i. \quad (1.3)$$

Also $x_i(x) = 0$ for all i.

Now suppose $\{x_i\}$ and $\{x_i'\}$ are normal coordinate systems at x. Then if $t \in X_y$, for any y in their domain,

$$t = \sum a_i \frac{\partial}{\partial x_i} + \sum a_{ij} \frac{\partial^2}{\partial x_i \partial x_j} = \sum a_i' \frac{\partial}{\partial x_i'} + \sum a_{ij}' \frac{\partial^2}{\partial x_i' \partial x_j'}, \quad (1.4)$$

where all these derivatives are evaluated at y, $a_{ij} = a_{ji}$, $a_{ij}' = a_{ji}'$. Applying t to the x_i, x_i', $x_i x_j$, $x_i' x_j'$ we find

$$a_i = t x_i, \quad a_{ij} = t(x_i x_j)/2, \quad a_i' = t x_i', \quad a_{ij}' = t(x_i' x_j')/2, \quad (1.5)$$

and hence, using (1.3),

$$a'_j = tx'_j = t\left(\sum_i c_{ji}x_i\right) = \sum_i c_{ji}a_i, \tag{1.6}$$

$$a'_{ij} = t(x'_i x'_j)/2 = t\left(\left(\sum_k c_{ik}x_k\right)\left(\sum_l c_{jl}x_l\right)\right)\bigg/2 = \sum_{kl} c_{ik}c_{jl}a_{kl},$$

which shows, if the $\{x_i\}$ and $\{x'_i\}$ are related by (1.3), then, at all points in their domain,

$$\sum a_j \frac{\partial}{\partial x_j} + \sum a_{ij} \frac{\partial^2}{\partial x_i \partial x_j} = \sum_j \left(\sum_i c_{ji}a_i\right)\frac{\partial}{\partial x'_j} + \sum_{ij}\left(\sum_{kl} c_{ik}c_{jl}a_{kl}\right)\frac{\partial^2}{\partial x'_i \partial x'_j}. \tag{1.7}$$

This shows in particular that if $t \in X''_x$ then all $a_i = 0$ if and only if all $a'_i = 0$, so we define, using any normal coordinate system $\{x_i\}$ at x,

$$X^p_x = \left[t \in X''_x \Big| t = \sum_{ij} a_{ij} \frac{\partial^2}{\partial x_i \partial x_j}(x)\right], \tag{1.8}$$

and the preceding shows this definition independent of the particular normal coordinate system at x which is used. It also shows the existence of a unique intrinsic decomposition, depending only on the Riemannian metric, of the form (1.2).

From the matrix (a_{ij}) we also get an intrinsic (depending only on the Riemannian metric) linear isomorphism onto

$$\alpha : X^p_x \to S(X_x), \tag{1.9}$$

where

$$S(X_x) = \text{all symmetric linear transformations of } X_x \to X_x. \tag{1.10}$$

Here α is defined as follows. Choose any normal coordinate system $\{x_i\}$ at x and define $A = \alpha(t) \in S(X_x)$ by the following. If $t = \sum a_{ij}\partial^2/\partial x_i \partial x_j$ with $a_{ij} = a_{ji}$ then

$$A \frac{\partial}{\partial x_j}(x) = \sum a_{ij} \frac{\partial}{\partial x_i}(x), \tag{1.11}$$

i.e., $A = \alpha(t)$ is the element of $S(X_x)$ whose matrix with respect to the $\partial/\partial x_i$ is (a_{ij}). Again (1.7) shows [since (c_{ij}) is orthogonal] that A and α are independent of the normal coordinate system at x which is used.

Let $T(X)$ be the tangent bundle to X, minus its 0-section and $T'(X)$ the cotangent bundle to X, minus its 0-section; let π denote the natural projection of either of these into X. One usually considers the natural 1-form σ' on $T'(X)$, defined by the following. If $\tilde{t} \in T'(X)_{(x,\theta)}$ then $\sigma'(t) = \theta(\pi_* t)$. However the Riemannian metric which we assume on X gives a map M:

$$M: T(X) \to T'(X): (x,t) \to (x, \langle \cdot, t \rangle),$$

and we use this map to transfer to $T(X)$ all considerations of Lagrange subspaces and Lagrange submanifolds of $T'(X)$. Thus we define $\sigma = M^*\sigma'$, $\tau = d\sigma$, and shall deal with σ, τ instead of σ', τ'.

If $\{x_i\}$ is any coordinate system of X we have an associated coordinate system $\{y_i, z_i\}$ of $T(X)$ defined by

$$\begin{aligned} y_i(x, z) &= x_i(x), \\ z_i(x, z) &= dx_i(z) = zx_i. \end{aligned} \tag{1.12}$$

We use this notation consistently below.

At each $(x, t) \in T(X)$ we define

$$V_{(x,t)} = [\tilde{t} \in T(X)_{(x,t)} | \pi_* \tilde{t} = 0] \tag{1.13}$$

and, choosing any normal coordinate system $\{x_i\}$ at x, define

$$H_{(x,t)} = \left[\tilde{t} \in T(X)_{(x,t)} \Big| \tilde{t} = \sum b_i \frac{\partial}{\partial y_i}(x, t) \right], \tag{1.14}$$

where $\{y_i, z_i\}$ is the associated coordinate system of the $\{x_i\}$. It is clear from (1.20) that this definition is independent of the choice of the normal coordinate system at x. Clearly

$$V_{(x,t)} = \left[\tilde{t} \in T(X)_{(x,t)} \Big| \tilde{t} = \sum a_i \frac{\partial}{\partial z_i}(x, t) \right],$$
$$T(X)_{(x,t)} = V_{(x,t)} \oplus H_{(x,t)}. \tag{1.15}$$

We now define, using the same coordinate systems at x:

$$H: X_x \to H_{(x,t)}: \sum a_i \frac{\partial}{\partial x_i}(x) \to \sum a_i \frac{\partial}{\partial y_i}(x, t),$$

$$V: X_x \to V_{(x,t)}: \sum a_i \frac{\partial}{\partial x_i}(x) \to \sum a_i \frac{\partial}{\partial z_i}(x, t),$$

$$J: T(X)_{(x,t)} \to T(X)_{(x,t)}: \begin{cases} \sum a_i \dfrac{\partial}{\partial y_i}(x, t) \to \sum a_i \dfrac{\partial}{\partial z_i}(x, t), \\[2ex] \sum a_i \dfrac{\partial}{\partial z_i}(x, t) \to \sum a_i \dfrac{\partial}{\partial y_i}(x, t), \end{cases} \tag{1.16}$$

$$P_V: T(X)_{(x,t)} \to V_{(x,t)}: \text{defined by decomposition in (1.15)},$$

$$P_H: T(X)_{(x,t)} \to H_{(x,t)}: \text{defined by decomposition in (1.15)}.$$

It is easily seen that these are independent of the choice of the normal coordinate system at x.

We put on $T(X)$ the Riemannian metric defined by the following. $V_{(x,t)}$ and $H_{(x,t)}$ shall be orthogonal and H and V shall be isometries. It is then clear that for $v \in V_{(x,t)}$, $h \in H_{(x,t)}$,

$$\tau(v, h) = \langle Jv, h \rangle = \langle v, J^{-1}h \rangle = \langle \pi_* Jv, \pi_* h \rangle \tag{1.17}$$

for v, v' in $V_{(x,t)}$, h and h' in $H_{(x,t)}$ and that

$$\tau(v, h) = \langle Jv, h \rangle = \langle \pi_* Jv, \pi_* h \rangle,$$
$$\tau(v, v') = \tau(h, h') = 0. \tag{1.18}$$

DEFINITION:

$$\tilde{S}(X_x) = [(A, E) \mid A \in S(X_x) \text{ and } E \text{ is a linear subspace of the null space } N_A \text{ of } A],$$

and

$$\mathcal{L}(x, t) = \text{all Lagrange subspaces of } T(X)_{(x,t)}.$$

We now establish, for each $(x, t) \in T(X)$, a 1:1 map $\beta = \beta_{(x,t)}$ of $\tilde{S}(X_x)$ onto $\mathcal{L}(x, t)$. β is defined by

$$\beta(A, E) = HE \oplus V(N_A \cap E^\perp) \oplus [Vs + HAs \mid s \in N_A^\perp]. \tag{1.19}$$

We now prove that

$$\beta(A, E) \in \mathcal{L}(x, t), \tag{1.20}$$

$$\beta \text{ is 1:1}, \tag{1.21}$$

and

$$\beta \text{ is onto.} \tag{1.22}$$

Proof (of (1.20). We first show $\dim \beta(A, E) = n$. Let $n_A = \dim N_A$. Clearly $\dim JHE + \dim(N_A \cap E^\perp) = n_A$ and $\dim[Vs + HAs \mid s \in N_A^\perp] = n - n_A$, so $\dim \beta(A, E) = n$. Now we show $\beta(A, E)$ is Lagrangian. Let $L = \beta(A, E)$. Using a normal coordinate system $\{x_i\}$ at x with the associated $\{y_i, z_i\}$, we have

$$\tau = \sum dz_i \, dy_i \quad \text{on } \bigcup_t T(X)_{(x,t)}.$$

From this it is trivial that

$$\tau(h, h') = 0 \quad \text{if} \quad h, h' \in H_{(x,t)},$$
$$\tau(v, v') = 0 \quad \text{if} \quad v, v' \in V_{(x,t)}.$$

If $h \in HE$ and $v \in V(N_A \cap E^\perp)$ then by (1.18), $\tau(v, h) = \langle \pi_* Jv, \pi_* h \rangle = 0$. It remains to show, for $h \in H_{(x,t)}, v \in V_{(x,t)}$, and $s \in N_A^\perp$, that $\tau(h, VS + HAs) = 0$ and $\tau(v, Vs + HAs) = 0$. The first of these follows from $\tau(h, Vs + HAs) = \tau(h, Vs) = \langle \pi_* h, \pi_* Jv \rangle = 0$, and the second is essentially the same. ∎

Proof of (1.21). Trivial because if $\beta(A, E) = \beta(A', E')$ then A and A' have the same null space and $A = A'$ on the orthogonal complement of that null space. ∎

Proof of (1.22). Let $L \in \mathscr{L}(x, t)$. Define

$$L_H = L \cap H_{(x,t)}, \qquad L_V = L \cap V_{(x,t)}, \qquad L' = L \cap (L_H + H_V)^\perp,$$
$$E = \pi_* L_H, \qquad F = \pi_* J L_V. \tag{1.23}$$

Clearly

$$JL_V \perp L_H, \quad JL_H \perp L_V, \quad JL_V \perp L', \quad JL_H \perp L', \quad L_V \perp JL', \quad L_H \perp JL'.$$

Hence any $l \in L$ is uniquely expressible as

$$l = l_V + l_H + l', \quad l_V \in L_V, \quad l_H \in L_H, \quad l' \in L'. \tag{1.24}$$

We shall need

$$V_{(x,t)} = L_V \oplus JL_H \oplus P_V L',$$
$$P_V L' \perp L_V, \quad P_V L' \perp JL_H, \tag{1.25}$$
$$P_V \text{ is } 1{:}1 \text{ on } L'.$$

Proof of (1.25). We have $P_V L' \perp L_V$ because if $l_V \in L_V$ and $l' + h \in P_V L'$ with $l' \in L'$, $h \in H_{(x,t)}$ then $\langle l', l_V \rangle = 0$ (by definition of L') and $\langle h, l_V \rangle = 0$ (because $H_{(x,t)} \perp V_{(x,t)}$), thus $\langle l' + h, l_V \rangle = 0$. Similarly we have $P_V L' \perp JL_H$. Let $n_H = \dim L_H$ and $n_V = \dim L_V$. We see that P_V is $1{:}1$ on L' because if $P_V l' = 0$ ($l' \in L'$) then $l' \in H_{(x,t)}$, so $l' \in L' \cap L_H = (0)$. Hence $\dim PV_L' = \dim L' = n - n_H - n_V$. And $\dim(L_V \oplus JL_H) = n - n_H - n_V$, proving (1.25). ∎

By (1.25), each $v \in V_{(x,t)}$ is uniquely expressible as

$$v = l_1(v) + l_2(v) + l'(v) + h(v)$$

where

$$l_1(v) \in L_V, \quad l_2(v) \in JL_H, \quad l'(v) \in L', \quad h(v) \in H_{(x,t)}, \quad l'(v) + h(v) \in V_{(x,t)}, \tag{1.26}$$

and we use this decomposition to define linear transformations, A and A' by

$$A' : V_{(x,t)} \to H_{(x,t)} : v \to h(v), \qquad A : X_x \to X_x : A = \pi_* A' V. \tag{1.27}$$

We now prove A symmetric. Let $z, w \in X_x$, $u = Vz$, $v = Vw$. Hence we have

$$
\begin{aligned}
\langle Az, w \rangle &= \langle \pi_* A'Vz, w \rangle = \langle \pi_* A'u, \pi_* Hw \rangle \\
&= \langle \pi_* A'u, JVw \rangle = \langle \pi_* A'u, \pi_* Jv \rangle \\
&= \langle \pi_* h(u), Jv \rangle = \tau(v, h(u)).
\end{aligned} \tag{1.28}
$$

Thus to prove A symmetric it will be sufficient to prove

$$
\tau(v, h(u)) = \tau(u, h(v)) \qquad \text{for all} \quad u, v \in V_{(x,t)}. \tag{1.29}
$$

We prove (1.29) by

$$
\begin{aligned}
\tau(v, h(u)) &= \tau(v, u - l_1(u) - l_2(u) - l'(u)) = \tau(v, -l_2(u) - l'(u)) \\
&= \tau(l_V(v) + l_2(v) + l'(v) + h(v), -l_2(u) - l'(u)) = \tau(h(v), -l_2(u) - l'(u)) \\
&= \tau(h(v), l_1(u) + h(u) - u) = -\tau(h(v), u) + \tau(h(v), l_1(u)) \\
&= \tau(u, h(v)) + \tau(h(v) + l'(v), l_1(u)) = \tau(u, h(v)).
\end{aligned}
$$

We now prove

$$
N_A = E \oplus F. \tag{1.30}
$$

Clearly $N_A \supseteq E + F$. Now suppose $x \in N_A$. Let $v = Vx$, and it will be sufficient to prove that $A'v = 0$ implies $v \in L_V + JL_H$. Write $v = l_1(v) + l_2(v) + l'(v) + h(v)$ as above and we need to show $h(v) = 0$ implies $l'(v) = 0$. But $h(v) = 0$ implies $l'(v) \in V_{(x,t)}$, hence $l'(v) \in L_V \cap L'$, thus $l'(v) = 0$.

Let R_A be the range of A. It is an elementary fact that $R_A = N_A^\perp$ (since A is symmetric). This gives the corresponding fact for A': $R_{A'} = (JL_V + L_H)^\perp$.

We now define

$$
\gamma: \mathscr{L}(x, t) \to \tilde{S}(X_x): L \to (A, E),
$$

where A is defined by (1.27) and E in (1.23). It is clear that β and γ are inverses, which proves (1.22). ∎

If $L \in \mathscr{L}(x, t)$ we write $(A(L), E(L))$ for the corresponding element of $\tilde{S}(X_x)$ and if $(A, E) \in \tilde{S}(X_x)$ we write $L(A, E)$ for the corresponding element of $\mathscr{L}(x, t)$.

2. DEFINITION OF THE SYMBOL

As before, X will always be an n-dimensional Riemannian manifold. We first define the bundle $\Sigma(X)$ over X on which our symbols will be defined, then define the class $\mathscr{D}'_\sigma(X)$ of distributions for which we shall define a symbol and then define the symbol.

The bundle $\Sigma(X)$ will be a disjoint union of bundles $\Sigma_p(X)$, for p an integer with $0 \leq p \leq n$. For the definition of $\Sigma_p(X)$ we define, for such p,

$$\tilde{S}_p(X_x) = [(A, E) \in \tilde{S}(X_x) | \dim E = p], \tag{2.1}$$

then define

$$\Sigma_p(X) = [(x, t, A, E) | t \in X_x, (A, E) \in \tilde{S}_p(X)],$$
$$\Sigma(X) = \bigcup_p \Sigma_p(X). \tag{2.2}$$

The C^∞ bundle structures of $\Sigma(X)$ and $\Sigma_p(X)$ are conventional and we do not discuss them. We remark, however, that each $\Sigma_p(X)$ is open in $\Sigma(X)$.

By *Euclidean space* we shall mean a linear space U over $\mathbf{R}$ for which a positive definite scalar product is given on $U \times U$. U will usually be X_x, although later U may be $\mathbf{R}^n$. We define

$\tilde{S}_p(U) = [(A, E) | A$ is a symmetric linear transformation of U into U and E is a p-dimensional linear subspace of $N_A]$,

where N_A is again the null space of A. If $\{e_i\}$ is any orthonormal base of U we say $\{e_i\}$ is *adapted* to E if $E = \text{span}\{e_1, \ldots, e_p\}$, and we say $\{e_i\}$ is *adapted* to (A, E) if it is adapted to E and $N_A = \text{span}\{e_1, \ldots, e_p, e_{p+1}, \ldots, e_{p+q}\}$. If $\{e_i\}$ is any orthonormal base of U we let $\{u_i\}$ by the linear coordinate system of U defined by

$$u_i(u) = \langle u, e_i \rangle, \qquad u \in U, \tag{2.3}$$

and $\{v_i, w_i\}$ be the associated coordinate system of $T(U)$, defined by

$$v_i(u, w) = u_i(u), \qquad w_i(u, w) = du_i(w), \qquad (u, w) \in T(U). \tag{2.4}$$

We know from (1.19) that if $\{e_i\}$ is adapted to E and (a_{jk}) is the matrix of A with respect to $\{e_i\}$ then

$$L(A, E) = \text{span}\left[\frac{\partial}{\partial v_i} \middle| 1 \leq i \leq p\right] \oplus \text{span}\left[\frac{\partial}{\partial w_i} + \sum_j a_{ij} \frac{\partial}{\partial v_j} \middle| p + 1 \leq i \leq n\right], \tag{2.5}$$

and if $\{e_i\}$ is adapted to (A, E) then we have both (2.5) and

$$a_{ij} = 0 \qquad \text{for} \quad 1 \leq i, j \leq p + q \tag{2.6}$$

where $\dim N_A = p + q$.

For each $(A, E) \in \tilde{S}_p(U)$ we define a symmetric linear transformation A_E of U into U by:

$$A_E u = \begin{cases} u & \text{if} \quad u \in E \\ 0 & \text{if} \quad u \in N_A \cap E^\perp \\ Au & \text{if} \quad u \in N_A^\perp. \end{cases}$$

We shall use below the following variants of the classes of functions $S^m(X \times Z)$ (see [4]), but only for the case where X is a point, so we shall write $S^m(Z)$, and in particular we write $S^m(\mathbf{R}^+)$ when $Z = \mathbf{R}^+$. The Z in the definition we now give will always be understood to be an open cone, not containing 0, in some Euclidean space and we let $Z_1 = [z \in Z \mid \|z\| = 1]$. With these conventions we now define, for $m \in \mathbf{R}$,

$$S^m(\mathbf{R}^+) = [f \in C^\infty(\mathbf{R}^+) \mid \text{for every integer } n \geq 0 \text{ there exists } C_n \in \mathbf{R}^+$$
$$\text{such that } |(\partial^n f)(s)| \leq C_n s^{m-n} \text{ for all } s \geq 1]$$

$$S^\infty(\mathbf{R}^+) = \bigcup_{m \in R} S^m(\mathbf{R}^+)$$

$$S^{-\infty}(\mathbf{R}^+) = \bigcap_{m \in R} S^m(\mathbf{R}^+)$$

$$S^m(Z) = [f \in C^\infty(Z) \mid f(\cdot z) \in S^m(\mathbf{R}^+) \text{ for all } z \in Z_1],$$

where this "$\cdot$" denotes a variable in $\mathbf{R}^+$. The last line will be understood for $m = \pm\infty$ as well as for $m \in \mathbf{R}$.

For $x \in X$ and $(A, E) \in \tilde{S}_p(X_x)$ we define a function $g_x^{A,E}$ on $X_x \times X_x$ by:

$$g_x^{A,E}(y, z) = \langle y, z \rangle - \|z\| \langle A_E y, y \rangle$$

$$= \sum_{i=1}^n u_i(y) u_i(z) - \|z\| \sum_{i=1}^p u_i(y)^2 - \|z\| \sum_{jk} a_{jk} u_j(y) u_k(y) \quad (2.7)$$

where the expression on the second line of (2.7) holds under the assumption that the u_i are obtained (via (2.3)) for an $\{e_i\}$ adapted to E.

For $x \in X$ let R_x be the largest R in $\mathbf{R}^+ \cup (\infty)$ such that $\exp_x$ is a diffeo of $B(O_x, R)$ (O_x = the origin in X_x) onto $B(x, R)$. We write $\exp_x^{-1}$ for the inverse of the restriction of $\exp_x$ to $B(x, R_x)$.

If $T \in D'(X)$ we define its (A, E)-*transform* to be the function f_T, with values in $\mathbf{C}$, defined on $[(x, z, A, E) \in \Sigma(X) \mid \text{supp } T \subset B(x, R_x)]$ by

$$f_T(x, z, A, E) = ((\exp_x^{-1})_* T)(e^{-ig_x^{A,E}(\cdot, z)}) = T(e^{-ig_x^{A,E}(\exp^{-1}\cdot, z)}). \quad (2.8)$$

Note that if $A = 0$ and $E = (0)$ then

$$f_T(x, z, 0, (0)) = ((\exp_x^{-1})_* T)\,\widehat{}\,(z) = T(e^{-i\langle \exp_x^{-1} \cdot, z \rangle}). \quad (2.9)$$

The domain of f_T depends on T and may be empty so it is not very useful in this form (for the case of general Riemannian manifolds). And even in the case where X is $\mathbf{R}^n$ it is not what is wanted for a symbol because its value at (x, z, A, E) depends on the behavior of T away from x. However we get the symbol of T from f_T, for appropriate T, in the following way. The symbol σ_T will be a function on a certain subset $\Sigma_T(X)$ of $\Sigma(X)$ with values in

$S^\infty(\mathbf{R}^+)/S^{-\infty}(\mathbf{R}^+)$. To define it we choose $\phi \in \mathscr{D}(X)$ with supp ϕ near x and $\phi = 1$ on some neighborhood of x, and define

$$\sigma_T(x, z, A, E) = f_{\phi T}(x, \cdot z, A, E) \bmod S^{-\infty}(\mathbf{R}^+)$$

(this "·" representing a variable in $\mathbf{R}^+$). For T in the class $\mathscr{D}'_\sigma(X)$ defined below and (x, z, A, E) in $\Sigma_T(X)$ $f_{\phi T}(x, \cdot z, A, E)$ will be in $S^\infty(\mathbf{R}^+)$ and will be independent, modulo $S^{-\infty}(\mathbf{R}^+)$, of the choice of such ϕ.

DEFINITION 2.1. Let $T \in \mathscr{D}'(X)$ and $(x, z) \in T'(X)$. We define $Q(x, z, T)$ to be the set of all $(A, E) \in \tilde{S}(X_x)$ for which there exists an $R \in \mathbf{R}^+$ such that both:

(1) for each $\phi \in \mathscr{D}(B(x, R))$ the function $f_{\phi T}(x, \cdot z, A, E)$ is in $S^\infty(\mathbf{R}^+)$, and
(2) if $\phi \in \mathscr{D}(B(x, R))$ and $\phi = 0$ in some neighborhood of x then $f_{\phi T}(x, \cdot z, A, E) \in S^{-\infty}(\mathbf{R}^+)$.

DEFINITION 2.2. A subset S of $\Sigma(X)$ is *essential* if and only if for each (x, z) in $T'(X)$ the set of all $(A, E) \in \tilde{S}(X_x)$ such that $(x, z, A, E) \in S$ is open and dense in $\tilde{S}(X_x)$. A subset S of $\Sigma(X)$ is *negligible* if and only if its complement is essential.

DEFINITION 2.3. If $T \in \mathscr{D}'(X)$ we define

$$\Sigma_T(X) = [(x, z, A, E) \in \Sigma(X) | (A, E) \in Q(x, z, T)]$$

DEFINITION 2.4. $\mathscr{D}'_\sigma(X) = [T \in \mathscr{D}'(X) | \Sigma_T(X) \text{ is essential}]$.

DEFINITION 2.5. If $T \in \mathscr{D}'_\sigma(X)$ then σ_T, the *symbol* of T, is the function from $\Sigma_T(X)$ to $S^\infty(\mathbf{R}^+)/S^{-\infty}(\mathbf{R}^+)$ defined by: if $(x, z, A, E) \in \Sigma_T(X)$ then

$$\sigma_T(x, z, A, E) = f_{\phi T}(x, \cdot z, A, E) \bmod S^{-\infty}(\mathbf{R}^+)$$

where $\phi \in \mathscr{D}(B(x, R))$ and $\phi = 1$ in some neighborhood of x (and R is as in definition 2.1).

$\mathscr{D}'_\sigma(X)$ is a linear subspace of $\mathscr{D}'(X)$ because the intersection of two essential sets is essential. σ_T is determined by its value on any essential subset of $\Sigma(X)$ contained in $\Sigma_T(X)$. And $\sigma_{aS+bT} = a\sigma_S + b\sigma_T$, if S and T are in $\mathscr{D}'_\sigma(X)$, in the sense that the two sides are equal except on a negligible subset of $\Sigma(X)$.

There are many possible variants of these definitions and it is too early to know which variant is most appropriate. For example, we might have defined $\mathscr{D}'_\sigma(X)$ to be the larger class obtained by dropping condition (1) in Definition 2.1, and/or replacing $S^{-\infty}(\mathbf{R}^+)$ by those functions which tend rapidly to 0 at ∞ (that is, dropping the conditions on the derivatives). And one might impose various regularity conditions on the $f_{\phi T}(x, \cdot z, A, E)$ when the x, z, A, E vary. G. Uhlmann has pointed out to us that in some cases it will probably be desirable to consider symbols with values in $D'(\mathbf{R}^+)/S^{-\infty}(\mathbf{R}^+)$, which can perhaps be obtained by taking limits in $D'(\mathbf{R}^+)$ of the symbols considered here. (Presumably some more general distributions will have symbols which will be functions on higher (than second) order tangent bundles.) Obviously σ_T is a full symbol in the sense that $\sigma_T = 0$ if and only if $T \in C^\infty$. It is not hard to show, using the technique of [4], that all Fourier Integral distributions in $\mathscr{D}'(\mathbf{R}^n)$ are in $\mathscr{D}'_\sigma(\mathbf{R}^n)$.

If $T \in \mathscr{D}'_\sigma(X)$ and $(x, z, A, E) \in \Sigma(X)$ we denote by $m_T(x, z, A, E)$ the inf of all $m \in R$ such that $f_{\phi T}(x', \cdot, A, E) \in S^m(\Gamma)$ for some open cone Γ in $(X_x)_0$ and all x' in some neighborhood of x.

It is not hard to prove that if Λ is a Lagrange submanifold of $T(X)$ whose tangent space $\Lambda_{(x,z)}$ corresponds to $(A, E) \in S(X_x)$ under the correspondence of Section 1 then graph $dg_x^{A,E}(\cdot, z)$ is transversal to $\Lambda_{(x,z)}$ at (x, z). Hence we can define the spaces of distributions, $I^m(\Lambda)$ of Hörmander by:

$$I^m(\Lambda) = \big[T \in \mathscr{D}'_\sigma(X) \,\big|\, m_T(x, z, A, E) \leq m \text{ for all}$$
$$(x, z, A, E) \text{ such that } (x, z) \in \Lambda \text{ and}$$
$$\Lambda_{(x,z)} \text{ corresponds to } (A, E); \text{ and}$$
$$m_T(x, z, A, E) = -\infty \text{ if } (x, z) \notin \Lambda \big].$$

[The $m_T(x, z, A', E')$ for other (A', E'), if $(x, z) \in \Lambda$ and $\Lambda_{(x,z)}$ corresponds to (A, E), are easily seen to be determined by the $m_T(x, z', cA, E)$, for $z' \in (X_x)_0$, $c \in \mathbf{R}^+$, and the $A_E - A_{E'}$.]

REFERENCES

1. W. AMBROSE, Products of distributions with values in distributions, *J. Reine Angew. Math.* **315**, 73–91 (1980).
2. W. AMBROSE, R. S. PALAIS, AND I. M. SINGER, Sprays, *An. Acad. Brasil. Ciênc.* **32** (1960), 163–178.
3. J. J. DUISTERMAAT, "Fourier Integral Operators," Courant Inst., New York, 1973.
4. L. HÖRMANDER, Fourier integral operators, I, *Acta Math.* **127** (1971), 79–83.
5. A. WEINSTEIN, The order and symbol of a distribution, *Trans. Amer. Math. Soc.* **241** (1978), 1–54.

MATHEMATICAL ANALYSIS AND APPLICATIONS, PART A
ADVANCES IN MATHEMATICS SUPPLEMENTARY STUDIES, VOL. 7A

On Analytic and C^∞ Poincaré Lemma

ALDO ANDREOTTI[†]

*Scuola Normale Superiore
Pisa, Italia*

AND

MAURO NACINOVICH

*Università di Pisa
Istituto di Matematica "L. Tonelli"
Pisa, Italia*

DEDICATED TO LAURENT SCHWARTZ

Given a complex of differential operators with C^∞ or analytic coefficients on some open set Ω of $\mathbf{R}^n$ one can consider various forms of Poincaré lemma. In a previous paper [4] we have introduced for every point $x_0 \in \Omega$ a complex of differential operators with constant coefficients in $\mathbf{R}^n$, the symbolic complex. We have proved then that if the symbolic complex at x_0 is exact, then the given complex admits the Poincaré lemma at x_0 for formal power series (formal Poincaré lemma). In this paper we refine with a multigrading the notion of principal symbol of a differential operator at a point $x_0 \in \Omega$, this in accordance with notions introduced by Douglis and Nirenberg [7]. Then we extend the definition of the symbolic complex at a point of Ω.

Let us assume, from now on, that the operators of the given complex have real analytic (complex-valued) coefficients. One can then consider the problem to decide if the Poincaré lemma is valid for germs of real analytic (complex-valued) functions (analytic Poincaré lemma). We prove here that if at x_0 the symbolic complex is exact then the analytic Poincaré lemma is valid for the given complex at x_0. As a corollary we deduce then that if moreover the given complex is elliptic at x_0 (see definition in Section 1), then under the above specified assumptions, also the C^∞ Poincaré lemma is valid at x_0.

It is worth noticing that the ellipticity assumption cannot be dropped. Indeed if

$$L(x, D) = \frac{\partial}{\partial x_1} + i \frac{\partial}{\partial x_2} - 2i(x_1 + ix_2) \frac{\partial}{\partial x_3}$$

[†] Deceased.

41

is the operator of H. Lewy [11] we can consider the complex in $\mathbf{R}^3$:

$$\mathscr{E}(\Omega) \xrightarrow{L} \mathscr{E}(\Omega) \to 0$$

for every Ω open in $\mathbf{R}^n$. This complex admits the formal and the analytic Poincaré lemma at any point $x_0 \in \mathbf{R}^n$ as one easily verifies and its symbolic complex is exact, but it does not admit the C^∞ Poincaré lemma; this complex is not elliptic.

For the proof of the analytic Poincaré lemma mentioned above we use a decomposition of the given complex already employed in [2].

In the last two sections we introduce the notion of *correct complex* which is the ideal situation to which one can hope to apply the criteria established before. We prove that for a correct complex (with analytic operators) the symbolic complex fails to be exact only on proper analytic subsets (Proposition 5).

Finally we introduce the useful notion of involutive operators and we establish that for any involutive operator, in a neighborhood of a semi-analytic compact set, we can find correct involutive resolutions of finite length (Proposition 6).

1. Multigraded Symbols of Differential Operators

(a)   Let Ω be an open set in $\mathbf{R}^n$, let $\mathscr{E}(\Omega)$ denote the space of C^∞ functions on Ω, and let

$$A(x, D): \mathscr{E}^p(\Omega) \to \mathscr{E}^q(\Omega)$$

be a differential operator with C^∞ coefficients defined on Ω. Here $x = (x_1, \ldots, x_n)$ denote coordinates in $\mathbf{R}^n$ and, as usual, D stands for the symbol of differentiation: $D = (\partial/\partial x_1, \ldots, \partial/\partial x_n)$. The operator $A(x, D)$ is represented by a matrix

$$A(x, D) = (a_{ij}(x, D))_{1 \le i \le q, \, 1 \le j \le p}$$

of differential operators

$$a_{ij}(x, D) = \sum_\sigma a_{ij,\sigma}(x) D^\sigma,$$

where $a_{ij,\sigma}$ are C^∞ functions on Ω and where for the multi-index $\sigma = (\sigma_1, \ldots, \sigma_n) \in \mathbf{N}^n$ we have set, as it is customary,

$$D^\sigma = \partial^{\sigma_1 + \cdots + \sigma_n}/\partial x_1^{\sigma_1} \cdots \partial x_n^{\sigma_n},$$

the sum on the right-hand side being a finite sum.

We select integers a_j for $1 \leq j \leq p$ and b_i for $1 \leq i \leq q$ such that we can write

$$a_{ij}(x, D) = \sum_{|\sigma| \leq a_j - b_i} a_{ij,\sigma}(x)D^\sigma, \qquad |\sigma| = \sigma_1 + \cdots + \sigma_n.$$

This is always possible in many ways; for instance, one can take the b_i's all zero and the a_j's equal to the largest order of derivation k appearing in the operators $a_{ij}(x, D)$.

Then we define, for $\xi = (\xi_1, \ldots, \xi_n)$ a set of indeterminates,

$$\hat{a}_{ij}(x, \xi) = \sum_{|\sigma| = a_j - b_i} a_{ij,\sigma}(x)\xi^\sigma,$$

and set

$$\hat{A}(x, \xi) = (\hat{a}_{ij}(x, \xi))_{1 \leq i \leq q,\, 1 \leq j \leq p}.$$

We will call the matrix of polynomials $\hat{A}(x, \xi)$ *the symbol of type* (a_j, b_i) of the given operator $A(x, D)$. In particular, for the special choice of (a_j, b_i) mentioned above we get for $\hat{A}(x, \xi)$ a matrix of polynomials homogeneous of the same order k and we obtain the usual notion of symbol of a differential operator (see [4]). We note that for every choice of the type (a_j, b_i) the symbol $\hat{A}(x, \xi)$ is for every fixed x a homogeneous matrix of polynomials in ξ (see [5, Section 9]).

Given a second differential operator with C^∞ coefficients on Ω

$$B(x, D): \mathscr{E}^q(\Omega) \to \mathscr{E}^r(\Omega)$$

we can select integers c_h, $1 \leq h \leq r$, so that

$$B(x, D) = \left(\sum_{|\sigma| \leq b_i - c_h} b_{hi,\sigma}(x)D^\sigma \right)_{1 \leq h \leq r,\, 1 \leq i \leq q},$$

and we can consider the symbol of type (b_i, c_h) of $B(x, D)$:

$$\hat{B}(x, \xi) = \left(\sum_{|\sigma| = b_i - c_h} b_{hi,\sigma}(x)\xi^\sigma \right).$$

We can then consider the operator

$$B(x, D) \circ A(x, D): \mathscr{E}^p(\Omega) \to \mathscr{E}^r(\Omega)$$

and its symbol of type (a_j, c_h); we do have

$$\widehat{B \circ A}(x, \xi) = \hat{B}(x, \xi)\hat{A}(x, \xi).$$

(b)   In particular, if

$$\mathscr{E}^p(\Omega) \xrightarrow{A(x,D)} \mathscr{E}^q(\Omega) \xrightarrow{B(x,D)} \mathscr{E}^r(\Omega) \tag{1}$$

is a *complex*, i.e., if $B(x, D)A(x, D) = 0$, then for any $x_0 \in \Omega$ we do have

$$\hat{B}(x_0, \xi)\hat{A}(x_0, \xi) = 0$$

and therefore we do have, for any choice of $x_0 \in \Omega$, a new complex of differential operators with constant coefficients

$$\mathscr{E}^p(\mathbf{R}^n) \xrightarrow{\hat{A}(x_0, D)} \mathscr{E}^q(\mathbf{R}^n) \xrightarrow{\hat{B}(x_0, D)} \mathscr{E}^r(\mathbf{R}^n), \tag{2}$$

which shall be called *the symbolic complex of* (1) *at the point* x_0.

Let $\mathscr{P} = \mathbf{C}[\xi_1, \ldots, \xi_n]$ denote the ring of polynomials in the indeterminates ξ. To the complex (2) we associate the complex of $\mathscr{P}$-homomorphisms:

$$\mathscr{P}^p \xleftarrow{{}^t\hat{A}(x_0, \xi)} \mathscr{P}^q \xleftarrow{{}^t\hat{B}(x_0, \xi)} \mathscr{P}^r. \tag{3}$$

If $\mathscr{H} = C_0[\xi_1, \ldots, \xi_n]$ is the graded ring of homogeneous polynomials in the indeterminates ξ, we do have also a complex of "multigraded" $\mathscr{H}$-homomorphisms:

$$\mathscr{H}^p \xleftarrow{{}^t\hat{A}(x_0, \xi)} \mathscr{H}^q \xleftarrow{{}^t\hat{B}(x_0, \xi)} \mathscr{H}^r, \tag{4}$$

where the matrices $\hat{A}(x_0, \xi)$ and $\hat{B}(x_0, \xi)$ are homogeneous of types (a_j, b_i) and (b_i, c_h), respectively.

(c) We now make the assumption that in the complex (1) the operators A and B *have real analytic* (complex-valued) *coefficients* defined on the open set Ω.

We denote by $\mathscr{E}_{x_0}$, $x_0 \in \Omega$, the set of germs of C^∞ functions at x_0, and by A_{x_0} the set of germs of real analytic functions at x_0. We say that the complex (1) *admits the C^∞ Poincaré lemma at* x_0 if

$$\mathscr{E}^p_{x_0} \xrightarrow{A(x, D)} \mathscr{E}^q_{x_0} \xrightarrow{B(x, D)} \mathscr{E}^r_{x_0}$$

is an exact sequence. Similarly we will say that the complex (1) *admits the analytic Poincaré lemma at* $x_0 \in \Omega$ if

$$\mathscr{A}^p_{x_0} \xrightarrow{A(x, D)} \mathscr{A}^q_{x_0} \xrightarrow{B(x, D)} \mathscr{A}^r_{x_0}$$

is an exact sequence. Note that this last statement is meaningful since A and B have real analytic coefficients.

We say that *the complex* (1) *is elliptic at the point* $x_0 \in \Omega$ if, for any choice of $\xi^0 \in \mathbf{R}^n - \{0\}$, the sequence

$$\mathbf{C}^p \xrightarrow{\hat{A}(x_0, \xi^0)} \mathbf{C}^q \xrightarrow{\hat{B}(x_0, \xi^0)} \mathbf{C}^r$$

is an exact sequence. We want to establish the following theorems:

THEOREM 1. *Let* (1) *be a complex of differential operators with real analytic* (*complex-valued*) *coefficients in* Ω. *Assume that at a given point*

$x_0 \in \Omega$ *the symbolic complex* (2) *is exact. Then the complex* (1) *admits the analytic Poincaré lemma at* x_0.

We note explicitly that the necessary and sufficient condition for the exactness of the symbolic complex at x_0 is the exactness, at x_0, of the sequence (3) or (4).

THEOREM 2. *Let* (1) *be a complex of differential operators in* Ω, *with real analytic (complex-valued) coefficients. We assume that* (i) *at a given point* $x_0 \in \Omega$ *the symbolic complex* (2) *is exact;* (ii) *the given complex* (1) *is elliptic at* x_0. *Then the complex* (1) *admits the* C^∞ *Poincaré lemma at* x_0.

The proof of these two theorems will be given in the subsequent sections.

(d) Let Φ_{x_0} denote the ring of formal power series in $x - x_0$, i.e., centered at x_0. Given a complex (1) of differential operators in Ω with C^∞ coefficients and a point $x_0 \in \Omega$, we will say that the complex (1) *admits the formal Poincaré lemma at* x_0 if

$$\Phi_{x_0}^p \xrightarrow{A(x,D)} \Phi_x^q \xrightarrow{B(x,D)} \Phi_{x_0}^r$$

is an exact sequence. We have the following:

THEOREM 0. *Let* (1) *be a complex in* Ω *of differential operators with* C^∞ *coefficients, assume that for a given point* $x_0 \in \Omega$ *the symbolic complex* (2) *admits the formal Poincaré lemma at* x_0. *Then the given complex* (1) *also admits the formal Poincaré lemma at* x_0.

We note that if at a point x_0 the symbolic complex is exact then it also admits the formal Poincaré lemma. This is a consequence of [4, Section 2, last Corollary].

The proof of Theorem 0 is also contained in the same paper [4, Proposition 3], for the particular case in which the integers a_j, b_i, c_h of the multigrading are so chosen that

$$a_j - b_i = \text{const} = \sigma \quad \forall i, j \qquad \text{and} \qquad b_i - c_h = \text{const} = \tau \quad \forall j, h.$$

The extension to the more general case considered here needs only small adjustments and a brief indication will suffice. Let

$$u = \begin{pmatrix} u_1 \\ \vdots \\ u_p \end{pmatrix} \in \Phi_{x_0}^p, \qquad f = \begin{pmatrix} f_1 \\ \vdots \\ f_q \end{pmatrix} \in \Phi_{x_0}^q, \qquad g = \begin{pmatrix} g_1 \\ \vdots \\ g_r \end{pmatrix} \in \Phi_{x_0}^r$$

and let us assume (as it is permitted without loss of generality) that $x_0 = 0$ is at the origin of the coordinates in $\mathbf{R}^n$. We will write

$$u = \sum_{h_0}^{\infty} u^{(h)} \qquad \text{where} \quad u^{(h)} = \begin{pmatrix} u_1^{(h)} \\ \vdots \\ u_p^{(h)} \end{pmatrix}$$

with $u_j^{(h)}$ a homogeneous polynomial of degree $a_j + h$ $(1 \le j \le p)$. In the sum h_0 is a convenient integer $\gtrless 0$. Similarly we write

$$f = \sum_{h_0}^{\infty} f^{(h)} \qquad \text{with} \quad f^{(h)} = \begin{pmatrix} f_1^{(h)} \\ \vdots \\ f_q^{(h)} \end{pmatrix}$$

and $f_i^{(h)}$ a homogeneous polynomial of degree $b_i + h$ $(1 \le i \le q)$, and

$$g = \sum_{h_0}^{\infty} g^{(h)} \qquad \text{with} \quad g^{(h)} = \begin{pmatrix} g_1^{(h)} \\ \vdots \\ g_r^{(h)} \end{pmatrix}$$

and $g_s^{(h)}$ a homogeneous polynomial of degree $c_s + h$ $(1 \le s \le r)$.

If $\hat{A}(0, \xi)$ and $\hat{B}(0, \xi)$ are the multigraded symbols of the operators A and B at the origin, from the equations

$$\hat{A}(0, D)u = f, \qquad \hat{B}(0, D)f = g$$

we derive the relations

$$\hat{A}(0, D)u^{(h)} = f^{(h)} \qquad \text{and} \qquad \hat{B}(0, D)f^{(h)} = g^{(h)} \quad \forall h.$$

The proof of Theorem 0 consists in showing that, if $f \in \Phi_0^q$ is such that $B(x, D)f = 0$, then there exists a $u \in \Phi_0^p$ such that $A(x, D)u = f$. Let $f = \sum_{h_0}^{\infty} f^{(h)}$ with the above notations. From $B(x, D)f = 0$ we derive that $\hat{B}(0, D)f^{(h_0)} = 0$.

By the hypothesis of the theorem and the above remark we derive that we can find a vector $u^{(h_0)}$ of convenient homogeneous polynomials such that $\hat{A}(0, D)u^{(h_0)} = f^{(h_0)}$.

We have $B(x, D)(f - A(x, D)u^{(h_0)}) = 0$ and moreover $f_1 = f - A(x, D)u^{(h_0)}$ starts with a homogeneous part of type $h_0 + 1$. Then we have $\hat{B}(0, D)f_1^{(h_0 + 1)} = 0$ and therefore we can find $u^{(h_0 + 1)}$ such that $\hat{A}(0, D)u^{(h_0 + 1)} = f_1^{(h_0 + 1)}$. Then we have $B(x, D)(f_1 - A(x, D)u^{(h_0 + 1)}) = 0$ and $f_2 = f_1 - A(x, D)u^{(h_0 + 1)}$ starts with a homogeneous part of type $h_0 + 2$. Proceeding in this way we obtain a sequence of homogeneous vectors $u^{(h_0)}, u^{(h_0 + 1)}, \ldots, u^{(h_0 + k)}, \ldots$ and a sequence of equations:

$$f_1 = f - A(x,D)u^{(h_0)}$$
$$f_2 = f_1 - A(x,D)u^{(h_0+1)}$$
$$\vdots$$
$$f_{k+1} = f_k - A(x,D)u^{(h_0+k)}$$

where $f_{k+1}^{(h)} = 0$ if $h \le h_0 + k$. Setting $u = \sum_{h_0}^{\infty} u^{(h)}$, summing up the relations above, we obtain $A(x,D)u = f$, as we wanted.

2. ON THE CHARACTERISTIC VARIETY OF A HILBERT RESOLUTION

(a)   We consider a Hilbert resolution of a $\mathscr{P}$-module N:

$$0 \leftarrow N \leftarrow \mathscr{P}^{p_0} \xleftarrow{M_0(\xi)} \mathscr{P}^{p_1} \xleftarrow{M_1(\xi)} \cdots \leftarrow \mathscr{P}^{p_d} \leftarrow 0. \qquad (\alpha)$$

Let $\mathscr{R} = \mathbf{C}(\xi_1, \ldots, \xi_n)$ be the field of rational functions in the n variables ξ, and let $\rho = \operatorname{rank}_{\mathscr{R}} M_0(\xi)$.

We define the *characteristic variety* of the Hilbert resolution (α) as the *proper* algebraic subvariety of $\mathbf{C}^n$

$$V = \{\xi \in \mathbf{C}^n \,|\, \operatorname{rank}_{\mathbf{C}} M_0(\xi) < \rho\}.$$

LEMMA 1.   *Let $M(\xi)$ be a matrix with entries in $\mathscr{P}$. For every $\xi^0 \in \mathbf{C}^n$ we have* $\operatorname{rank}_{\mathbf{C}} M(\xi^0) \le \operatorname{rank}_{\mathscr{R}} M(\xi)$.

Proof.   Indeed, if a minor determinant of rank s of the matrix $M(\xi)$ is not equal to zero for $\xi = \xi^0$, it is in particular a polynomial not equal to zero.   ■

PROPOSITION 1.   *Given a Hilbert resolution (α), a point $\xi^0 \in \mathbf{C}^n$, the sequence*

$$\mathbf{C}^{p_0} \xleftarrow{M_0(\xi^0)} \mathbf{C}^{p_1} \xleftarrow{M_1(\xi^0)} \mathbf{C}^{p_2} \leftarrow \cdots \leftarrow \mathbf{C}^{p_d} \leftarrow 0 \qquad (\beta)$$

is exact if and only if $\xi^0 \notin V$ [i.e., $\operatorname{rank}_{\mathbf{C}} M_0(\xi^0) = \operatorname{rank}_{\mathscr{R}} M_0(\xi)$].

Proof.   From the exact sequence (α), since $\mathscr{R}$ is flat over $\mathscr{P}$, we deduce an exact sequence

$$\mathscr{R}^{p_0} \xleftarrow{M_0(\xi)} \mathscr{R}^{p_1} \xleftarrow{M_1(\xi)} \cdots \leftarrow \mathscr{R}^{p_d} \leftarrow 0.$$

Therefore, if we set $\sigma_i = \operatorname{rank}_{\mathscr{R}} M_i(\xi)$ $(\sigma_0 = \rho)$ we deduce that

$$p_i = \sigma_i + \sigma_{i-1} \qquad \text{for} \quad i = 1, 2, \ldots, d.$$

Assume that (β) is an exact sequence, in particular that (β) is exact at $\mathbf{C}^{p_1}$. Then we have $p_1 = \operatorname{rank}_{\mathbf{C}} M_1(\xi^0) + \operatorname{rank}_{\mathbf{C}} M_0(\xi^0)$. Because of the previous lemma it follows that we must have $\operatorname{rank}_{\mathbf{C}} M_0(\xi^0) = \sigma_0 = \rho$ (and $\operatorname{rank}_{\mathbf{C}} M_1(\xi^0) = \sigma_1$), hence $\xi^0 \notin V$.

Conversely, let us assume that $\xi^0 \notin V$, i.e., that

$$\operatorname{rank}_{\mathbf{C}} M_0(\xi^0) = \rho = \operatorname{rank}_{\mathscr{R}} M_0(\xi).$$

By renumbering the components of $\mathscr{P}^{p_0}$ and $\mathscr{P}^{p_1}$ we may assume that in the matrix $M_0(\xi)$ the determinant $D(\xi)$ of the first ρ rows and columns is different from zero for $\xi = \xi^0 : D(\xi^0) \neq 0$. Let $M_0(\xi) = (a_{ij}(\xi))_{1 \le i \le p_0, \, 1 \le j \le p_1}$. For $i = 1, \ldots, \rho$ all minor determinants of order $\rho + 1$ of the matrix

$$\begin{pmatrix} a_{11} & \cdots & a_{1p_1} \\ \vdots & & \vdots \\ a_{p_1} & \cdots & a_{pp_1} \\ a_{i1} & \cdots & a_{ip_1} \end{pmatrix}$$

are zero because two rows of that matrix are equal. In particular all minor determinants obtained by bordering the first rows and columns by the last row and another column are zero. This will give $\sigma_1 = p_1 - \rho$ vectors in $\ker M_0(\xi)$ of the form

$$Y_1(\xi) = \begin{pmatrix} c_{11}(\xi) \\ \vdots \\ c_{\rho 1}(\xi) \\ D(\xi) \\ 0 \\ \vdots \\ 0 \end{pmatrix}, \quad Y_2(\xi) = \begin{pmatrix} c_{12}(\xi) \\ \vdots \\ c_2(\xi) \\ 0 \\ D(\xi) \\ \vdots \\ 0 \end{pmatrix}, \quad \ldots, \quad Y_{\sigma_1}(\xi) = \begin{pmatrix} c_{1\sigma_1}(\xi) \\ \vdots \\ c_{\rho\sigma_1} \\ 0 \\ 0 \\ \vdots \\ D(\xi) \end{pmatrix}.$$

Since (α) is an exact sequence, $Y_1(\xi), Y_2(\xi), \ldots, Y_{\sigma_1}(\xi)$ belong to $\operatorname{Im} M_1(\xi)$ and therefore, in particular, $Y_1(\xi^0), \ldots, Y_{\sigma_1}(\xi^0)$ are in the image of $M_1(\xi^0)$.

Now, since $D(\xi^0) \neq 0$, these last σ_1 vectors are linearly independent over $\mathbf{C}$ and therefore, $\operatorname{rank}_{\mathbf{C}} M_1(\xi^0) \ge \sigma_1$. By the preceding lemma we deduce then that $\operatorname{rank}_{\mathbf{C}} M_1(\xi^0) = \operatorname{rank}_{\mathscr{R}} M_1(\xi) = \sigma_1$. This proves the exactness of (β) at $\mathbf{C}^{p_1}$. Now we have an exact sequence

$$\mathscr{P}^{p_1} \xleftarrow{\;M_1(\xi)\;} \mathscr{P}^{p_2} \xleftarrow{\;M_2(\xi)\;} \mathscr{P}^{p_3} \leftarrow \cdots \leftarrow \mathscr{P}^{p_d} \leftarrow 0 \qquad (\alpha_1)$$

with $\operatorname{rank}_{\mathbf{C}} M_1(\xi^0) = \operatorname{rank}_{\mathscr{R}} M_1(\xi)$, and arguing as before we deduce the exactness of the sequence

$$\mathbf{C}^{p_1} \xleftarrow{\;M_1(\xi^0)\;} \mathbf{C}^{p_2} \xleftarrow{\;M_2(\xi^0)\;} \mathbf{C}^{p_3} \leftarrow \cdots \leftarrow \mathbf{C}^{p_d} \leftarrow 0 \qquad (\beta_1)$$

at $\mathbf{C}^{p_2}$ and that $\operatorname{rank}_{\mathbf{C}} M_2(\xi^0) = \operatorname{rank}_{\mathscr{R}} M_2(\xi)$. We can repeat again the same argument and deduce the exactness of (β) at $\mathbf{C}^{p_3}$ and that $\operatorname{rank}_{\mathbf{C}} M_3(\xi^0) = \operatorname{rank}_{\mathscr{R}} M_3(\xi)$. In this way, iterating the argument, we obtain the desired conclusion. ∎

(b)   We end up this section with some straightforward remarks that will be of use in the sequel.

Let $M(x, \xi)$ be a matrix of polynomials in ξ with coefficients C^∞ functions on an open set Ω of $\mathbf{R}^n$. For every $x \in \Omega$ we set

$$\rho(x) = \operatorname{rank}_{\mathscr{R}} M(x, \xi).$$

We claim that $\rho(x)$ is a lower semicontinuous function on Ω.

For this, one has to show that, given $x_0 \in \Omega$, we can find a neighborhood U of x_0 in Ω such that $\rho(x) \geq \rho(x_0) \; \forall x \in U$. Indeed we can find $\xi^0 \in \mathbf{C}^n$ such that $\operatorname{rank}_{\mathbf{C}} M(x_0, \xi^0) = \operatorname{rank}_{\mathscr{R}} M(x_0, \xi)$. Then in a sufficiently small neighborhood U of x_0 we will have

$$\operatorname{rank}_{\mathbf{C}} M(x, \xi^0) \geq \operatorname{rank}_{\mathbf{C}} M(x_0, \xi^0) = \operatorname{rank}_{\mathscr{R}} M(x_0, \xi) = \rho(x_0).$$

By Lemma 1 at the beginning of this section we have for $x \in U$

$$\rho(x) = \operatorname{rank}_{\mathscr{R}} M(x, \xi) \geq \operatorname{rank}_{\mathbf{C}} M(x, \xi^0) \geq \rho(x_0).$$

Let now

$$\mathscr{E}^p(\Omega) \xrightarrow{\;A(x,D)\;} \mathscr{E}^q(\Omega) \xrightarrow{\;B(x,D)\;} \mathscr{E}^r(\Omega) \tag{1}$$

be a complex of differential operators with C^∞ coefficients in Ω. We choose a multigrading a_j ($1 \leq j \leq p$) for $\mathscr{E}^p$, b_i ($1 \leq i \leq q$) for $\mathscr{E}^q$, c_h ($1 \leq h \leq r$) for $\mathscr{E}^r$ and we consider for a given $x_0 \in \Omega$ the corresponding symbolic complex

$$\mathscr{E}^p(\mathbf{R}^n) \xrightarrow{\;\hat{A}(x_0,D)\;} \mathscr{E}^q(\mathbf{R}^n) \xrightarrow{\;\hat{B}(x_0,D)\;} \mathscr{E}^r(\mathbf{R}^n). \tag{2}$$

Assume that (2) is exact, i.e., assume that the algebraic complex of $\mathscr{P}$-homomorphisms

$$\mathscr{P}^p \xleftarrow{\;{}^t\hat{A}(x_0,\xi)\;} \mathscr{P}^q \xleftarrow{\;{}^t\hat{B}(x_0,\xi)\;} \mathscr{P}^r \tag{3}$$

is exact.

We claim that we can find a neighborhood U of x_0 in Ω such that, for every $x \in U$ we have

$$\operatorname{rank}_{\mathscr{R}} \hat{A}(x, \xi) = \operatorname{rank}_{\mathscr{R}} \hat{A}(x_0, \xi), \qquad \operatorname{rank}_{\mathscr{R}} \hat{B}(x, \xi) = \operatorname{rank}_{\mathscr{R}} \hat{B}(x_0, \xi),$$

$$\operatorname{rank}_{\mathscr{R}} \hat{A}(x, \xi) + \operatorname{rank}_{\mathscr{R}} \hat{B}(x, \xi) = q.$$

Indeed from the exactness of (3) we derive the exact sequence

$$\mathscr{R}^p \xleftarrow{\;{}^t\hat{A}(x_0,\xi)\;} \mathscr{R}^q \xleftarrow{\;{}^t\hat{B}(x_0,\xi)\;} \mathscr{R}^r$$

so that

$$\text{rank}_{\mathscr{R}}\,\hat{A}(x_0, \xi) + \text{rank}_{\mathscr{R}}\,\hat{B}(x_0, \xi) = q.$$

By lower semicontinuity in a sufficiently small neighborhood U of x_0 we have $\text{rank}_{\mathscr{R}}\,\hat{A}(x, \xi) \geq \text{rank}_{\mathscr{R}}\,\hat{A}(x_0, \xi)$ and $\text{rank}_{\mathscr{R}}\,\hat{B}(x, \xi) \geq \text{rank}_{\mathscr{R}}\,\hat{B}(x_0, \xi)$. But for every $x \in \Omega$ we must have

$$\text{rank}_{\mathscr{R}}\,\hat{A}(x, \xi) + \text{rank}_{\mathscr{R}}\,\hat{B}(x, \xi) \leq q$$

because for every $x \in \Omega$

$$\mathscr{R}^p \xleftarrow{\ {}^t\hat{A}(x,\xi)\ } \mathscr{R}^q \xleftarrow{\ {}^t\hat{B}(x,\xi)\ } \mathscr{R}^r$$

is a complex. This entails our statement.

3. FIBER TRANSFORMATIONS

(a)   Let Ω be open in $\mathbf{R}^n$ and let $\mathscr{E}(\Omega)$ denote the ring of C^∞ functions on Ω; let $\mathscr{A}(\Omega)$ denote the ring of real analytic (complex-valued) functions in Ω. Set

$$\mathfrak{A}_{\mathscr{E}}(\Omega) = \mathscr{E}(\Omega)[D_1, \ldots, D_n], \qquad \mathfrak{A}_{\mathscr{A}}(\Omega) = \mathscr{A}(\Omega)[D_1, \ldots, D_n].$$

These are the noncommutative rings of differential operators with C^∞ or, respectively, real analytic, coefficients on Ω.

Let $\mathscr{M}_{p \times p}(\mathfrak{A})$ be the set of matrices of type $p \times p$ with elements in $\mathfrak{A} = \mathfrak{A}_{\mathscr{E}}(\Omega)$ or $\mathfrak{A} = \mathfrak{A}_{\mathscr{A}}(\Omega)$. Every element $M \in \mathscr{M}_{p \times p}(\mathfrak{A})$ defines an endomorphism

$$M : \mathscr{E}^p(\Omega) \to \mathscr{E}^p(\Omega)$$

or, respectively,

$$M : \mathscr{A}^p(\Omega) \to \mathscr{A}^p(\Omega)$$

By composition of two elements M, N in $\mathscr{M}_{p \times p}(\mathfrak{A})$ we get a third element $N \circ M$, corresponding to the composition of the endomorphisms associated to M and to N, respectively. We have that $N \circ M \in \mathscr{M}_{p \times p}(\mathfrak{A})$, so that $\mathscr{M}_{p \times p}(\mathfrak{A})$ is an associative algebra.

Let a_j, $1 \leq j \leq p$, α_i, $1 \leq i \leq p$, be integers such that for the element $M \in \mathscr{M}_{p \times p}(\mathfrak{A})$ we can write

$$M = (m_{ij}(x, D))_{1 \leq i,\, j \leq p} \qquad \text{with} \quad m_{ij}(x, D) = \sum_{|\sigma| \leq a_j - \alpha_i} m_{ij,\sigma}(x) D^\sigma,$$

where $m_{ij,\sigma}$ are in $\mathscr{E}(\Omega)$ or $\mathscr{A}(\Omega)$, respectively. We define then the symbol of type (a_j, α_i) of M as the matrix

$$\hat{M}(x, \xi) = \left(\sum_{|\sigma| = a_j - \alpha_i} m_{ij,\sigma}(x) \xi^\sigma \right)$$

with elements in the commutative ring $\mathscr{E}(\Omega)[\xi_1, \ldots, \xi_n]$ or respectively, $\mathscr{A}(\Omega)[\xi_1, \ldots, \xi_n]$. This is in accordance with the definition of the principal symbol of the differential operator $M = M(x, D)$. As usual we do have

$$\widehat{N \circ M}(x, \xi) = \hat{N}(x, \xi)\hat{M}(x, \xi)$$

for $M, N \in \mathscr{M}_{p \times p}(\mathfrak{A})$, when we take for M the symbol of type (a_j, α_i) and for N the symbol of type (α_i, β_h), while for the composition $N \circ M$ we take the symbol of type (a_j, β_h).

Let $M \in \mathscr{M}_{p \times p}(\mathfrak{A})$ be associated to the multigrading (a_j, α_i). We will call the integer

$$h = \sum_{j=1}^{p} a_j - \sum_{i=1}^{p} \alpha_i$$

the *total degree* of M; it is the degree as a polynomial in ξ of the determinant of the symbol of $M : h = \deg_\xi(\det \hat{M}(x, \xi))$.

An element $M \in \mathscr{M}_{p \times p}(\mathfrak{A})$ of total degree 0 is called *invertible* if we can find $N \in \mathscr{M}_{p \times p}(\mathfrak{A})$ with multigrading of type (α_j, a_h) (and thus also of total degree 0) such that $N \circ M = $ identity.

Taking the symbols of type (a_j, a_h) in both sides we derive then $\hat{N}(x, \xi) \hat{M}(x, \xi) = $ identity, and therefore in particular

$$\det \hat{N}(x, \xi) \det \hat{M}(x, \xi) = 1.$$

This shows that *if a matrix $M \in \mathscr{M}_{p \times p}(\mathfrak{A})$ of total degree 0 is invertible, then the determinant of the symbol $\det \hat{M}(x, \xi)$ is independent from ξ and different from zero.*

We obtain the same conclusion if we define invertibility by means of a right inverse instead of a left inverse.

PROPOSITION 2.   *Let $M \in \mathscr{M}_{p \times p}(\mathfrak{A})$ correspond to the multigrading (a_j, α_i) of total degree zero; $\sum a_j = \sum \alpha_i$. Assume that $\det \hat{M}(x, \xi)$ is independent from ξ and different from zero for every $x \in \Omega$. Then there exists $N \in \mathscr{M}_{p \times p}(\mathfrak{A})$ corresponding to the multigrading (α_i, a_h) such that*

$$N \circ M = \text{identity}, \qquad M \circ N = \text{identity}.$$

Proof.   Assume that we have shown that every M satisfying the specified assumptions admits a left inverse N of type $(\alpha_i, a_h) : N \circ M = $ identity. We claim that we must also have $M \circ N = $ identity.

In fact, $\det \hat{N}(x, \xi)$ must be independent of ξ and different from zero and the total degree of N is zero. Thus by the above assumptions N admits a left inverse L of multigrading $(a_j, \alpha_i) : L \circ N = $ identity. From $N \circ M = $ identity we derive that $L \circ N \circ M = L$. Since $\mathscr{M}_{p \times p}(\mathfrak{A})$ is an associative algebra we deduce then, since $L \circ N = $ identity, that also $L \circ N \circ M = M$

and thus $L = M$. Thus M is a left inverse of N, i.e., N is a right inverse of M, i.e., $M \circ N =$ identity.

Therefore it will be sufficient to prove that under the specified assumptions M admits a left inverse.

The matrix $M(x, 0)$ can be viewed as an element of $\mathscr{M}_{p \times p}(\mathfrak{A})$ also of multigrading (a_j, α_i). Since $\det \hat{M}(x, 0) = \det M(x, \xi) \neq 0$, the matrix $(M(x, 0))^{-1}$ exists and can be viewed as an element of $\mathscr{M}_{p \times p}(\mathfrak{A})$ of multigrading (α_i, a_h). Indeed $1/(\det M(x, 0))$ can be considered as a differential operator of order zero. In the matrix $(M(x, 0))^{-1}$ the element of place (h, i) is

$$(1/\det M(x, 0)) \times \text{cofactor of } m_{ih}(x, 0) \quad \text{in } M(x, 0).$$

That cofactor can be considered as a differential operator of order

$$\leq \left(\sum_{j=1}^{p} a_j - a_h \right) - \left(\sum_{s=1}^{p} \alpha_s - \alpha_i \right) = \alpha_i - \alpha_h$$

because $\sum a_j = \sum \alpha_s$. Thus

$$S(x, D) = (M(x, 0))^{-1} M(x, D) \in \mathscr{M}_{p \times p}(\mathfrak{A})$$

corresponds to the multigrading (a_j, a_i). Moreover

$$\hat{S}(x, 0) = \text{identity}.$$

Let $(i_1, i_2, \ldots, i_p)$ be a permutation of $(1, 2, \ldots, p)$ such that $a_{i_1} \geq a_{i_2} \geq \cdots \geq a_{i_p}$ and let T denote the matrix $p \times p$ with 1 in the places $(1, i_1), (2, i_2), \ldots, (p, a_{i_p})$ and zero otherwise, and set $b_j = a_{i_j}$ for $j = 1, 2, \ldots, p$. We can consider the matrix T as an element of $\mathscr{M}_{p \times p}(\mathfrak{A})$ corresponding to the multigrading (a_j, b_i) [indeed, when (i, j) equals $(1, i_1), (2, i_2), \ldots, (p, i_p)$ we get $a_j - b_i = 0$]. We have ${}^tT = T^{-1}$ and T^{-1} can be considered as an element of $\mathscr{M}_{p \times p}(\mathfrak{A})$ corresponding to the multigrading (b_j, a_h). [Indeed, for (h, i) equal to $(i_1, 1)$, $(i_2, 2), \ldots, (i_p, p)$, where ${}^tT = T^{-1}$ has an element equal to 1, we have $b_j - a_h = 0$.] Then

$$N = T \circ S \circ T^{-1}$$

is an element of $\mathscr{M}_{p \times p}(\mathfrak{A})$ corresponding to the multigradation (b_j, b_i). In particular, in $N = (n_{ij}(x, D))$ the elements of place (i, j) with $i \geq j$ have an order $b_j - b_i \leq 0$. But $\hat{N}(x, 0) =$ identity and therefore the elements $n_{ij}(x, D)$ with $i > j$ must be zero, while all elements on the diagonal $i = j$ must be 1. We can thus write.

$$N(x, D) = I + U(x, D)$$

where I denotes the identity matrix and $U(x, D)$ is a matrix of differential operators with zeros on and above the diagonal $i = j$ [corresponding to the multigrading (b_j, b_i)]. Hence $U(x, D)$ is nilpotent: we have $U^p = 0$. Therefore

$$(I - U + U^2 - \cdots + (-1)^{p-1} U^{p-1})(I + U) = I,$$

and this proves that N admits a left inverse. But then S and therefore M admits a left inverse, as we wanted to prove.

An element $M(x, D) \in \mathcal{M}_{p \times p}(\mathfrak{A})$ of a given multigrading and of total degree zero, which is invertible, i.e., satisfies the assumptions of the previous proposition will also be called a *unit* of the algebra $\mathfrak{A}_{p \times p}(\mathfrak{A})$. ∎

(b) We give now some lemmas that we will need later.

By $C_0[t]$ we denote the graded ring of homogeneous polynomials in one variable t, with complex coefficients. It is a principal ideal ring. Let $A(t) = (a_{ij}(t))$ be a $p \times q$ matrix with elements in $C_0[t]$; $A(t) \in \mathcal{M}_{p \times q}(C_0[t])$.

The matrix $A(t)$ is homogeneous if integers a_j, $1 \le j \le q$, α_i, $1 \le i \le p$, are given such that

$$a_{ij}(t) = c_{ij} t^{a_j - \alpha_i}$$

with $c_{ij} \in C$ and $c_{ij} = 0$ if $a_j - \alpha_i < 0$. When $p = q$ the integer $\sum_{i,j} a_j - \alpha_i$ is called the total degree of $A(t)$.

LEMMA 2. *Let $A(t)$ be a $p \times q$ homogeneous matrix of multigrading* (a_j, α_i). *There exist homogeneous matrices*

$$R(t) \in \mathcal{M}_{q \times q}(C_0[t]) \qquad and \qquad L(t) \in \mathcal{M}_{p \times p}(C_0[t])$$

with the following properties:

(i) *$R(t)$ and $L(t)$ have total degree zero and for determinant a constant different from zero [i.e., $R(t)$ and $L(t)$ are units].*

(ii) *$R(t)$ has a multigrading of type (b_h, a_j) $(1 \le h, j \le q)$; $L(t)$ has a multigrading of type (α_i, β_s) $(1 \le i, s \le p)$.*

(iii) *We have*

$$L(t)A(t)R(t) = \begin{pmatrix} t^{k_1} & \cdots & 0 & \cdots & 0 \\ \vdots & \ddots & \vdots & & \vdots \\ 0 & \cdots & t^{k_r} & \cdots & 0 \\ \vdots & & \vdots & & \vdots \\ 0 & & 0 & \cdots & 0 \end{pmatrix}$$

with $k_s = b_s - \beta_s$ integers with $0 \le k_1 \le k_2 \le \cdots \le k_r$ and $r = \mathrm{rank}_{C(t)} A(t)$.

Proof. The proof goes by induction on the integer $p + q$, as for $p + q = 1$ the lemma is obvious. First one realizes that a permutation of rows (or columns) in the matrix $A(t)$ is obtained by multiplication on the left (or right) by a homogeneous matrix $L(R)$ with constant elements satisfying conditions (i) and (ii). Second, if $r = \text{rank}_{C(t)} A(t) = 0$, the lemma is also obvious. We can therefore assume that

$$a_{11}(t) = c_{11} t^{a_1 - \alpha_1} \neq 0; \qquad (\alpha)$$

$$\text{among the } a_{ij}(t) \neq 0 \quad a_{11}(t) \text{ has lowest degree.} \qquad (\beta)$$

Set

$$R(t) = \begin{pmatrix} c_{11}^{-1} & -c_{11}^{-1} c_{12} t^{a_2 - a_1} & \cdots & -c_{11}^{-1} c_{1q} t^{a_q - a_1} \\ 0 & 1 & \cdots & 0 \\ & & \vdots & \\ 0 & 0 & \cdots & 1 \end{pmatrix}.$$

It is a homogeneous matrix of multigrading (a_h, a_j). Also set

$$L(t) = \begin{pmatrix} 1 & 0 & \cdots & 0 \\ -c_{11}^{-1} c_{21} t^{\alpha_1 - \alpha_2} & 1 & \cdots & 0 \\ \vdots & & & \vdots \\ -c_{11}^{-1} c_{p1} t^{\alpha_1 - \alpha_p} & 0 & \cdots & 0 \end{pmatrix}.$$

This is a homogeneous matrix of multigrading (α_i, α_h). Then, with $k_1 = a_1 - \alpha_1$ we have

$$L(t)A(t)R(t) = \begin{pmatrix} t^{k_1} & 0 \\ 0 & C(t) \end{pmatrix}.$$

We have thus reduced the proof to the reduction into "canonical form" of the matrix $C(t)$ to which we can apply our induction. Note that every nonzero element in $C(t)$ has degree $\geq k_1$. Note that the canonical form of $A(t)$ is uniquely determined as t^{k_1} is the greatest common divisor of the elements of $A(t)$, $t^{k_1 + k_2}$ is the greatest common denominator (g.c.d.) of all second-order subdeterminants of $A(t)$, etc. ■

(c)   We separate the last from the first $n - 1$ variables in $\mathbf{R}^n$ and set $x = (y_1, \ldots, y_{n-1}, t) = (y, t)$, while we set for the variables ξ of the symbols $\xi = (\eta_1, \ldots, \eta_{n-1}, \tau) = (\eta, \tau)$.

LEMMA 3.   *Let $x_0 \in \Omega$ be given and let $A(x, D) \in \mathcal{M}_{p \times p}(\mathfrak{A})$ be of multigrading (a_j, b_i) with $b_1 \geq b_2 \geq \cdots \geq b_p$ and with $a_j - b_j = k_j$ for $1 \leq j \leq p$. We assume that $A(x_0, 0, \tau) = \text{diag}\langle t^{k_1}, \ldots, t^{k_p}\rangle.^\dagger$ We can find an open neighborhood*

---

† We denote in this way the matrix with the elements in angle brackets on the diagonal and zero elsewhere.

$U = U(x_0)$ of x_0 in Ω and a unit matrix $M(x, D) \in \mathcal{M}_{p \times p}(\mathfrak{A}(U))$, defined on U, of multigrading (b_i, b_h) such that

$$Q(x, D) = M(x, D)A(x, D)$$

has the properties:

(i) $Q(x_0, 0, \tau) = \mathrm{diag}\langle t^{k_1}, \ldots, t^{k_p}\rangle$;

(ii) for $i \neq j$, the element $q_{ij}(x, D)$ of $Q(x, D)$ in the place (i, j) has order in $\partial/\partial t$ strictly less than k_j: order in $\partial/\partial t$ of $q_{ij}(x, D) < k_j$.

Proof.   The proof is based on the following induction. For every integer $k \geq 0$ we choose an integer $h \geq 0$ and an integer l with $0 \leq l \leq p$ so that $k = l + hp$. Note that $p + hp$ can also be written as $(h + 1)p$, but, except for multiples of p, l and h are uniquely determined. For a matrix of differential operators Q defined in some open neighborhood U of x_0 in Ω and of multi-grading (a_j, b_i): $Q \in \mathcal{M}_{p \times p}(\mathfrak{A}(U))$, and for any integer $k \geq 0$ we consider the following property $P(k)$. We have

(i) $Q(x_0, 0, \tau) = \mathrm{diag}\langle \tau^{k_1}, \ldots, \tau^{k_p}\rangle$;

(ii) for $j \leq l$ the elements $q_{ij}(x, D)$ of Q of the jth column outside of the diagonal (i.e., for $i \neq j$ and $j \leq l$) have the property: order in $\partial/\partial t$ of $q_{ij}(x, D) \leq \max(k_j - 1, a_j - b_i - h - 1)$;

(iii) for $j > l$ the elements $q_{ij}(x, D)$ of Q in the jth column and outside of the diagonal (i.e., for $i \neq j$ and $j > l$) have the property: order in $\partial/\partial t$ of $q_{ij}(x, D) \leq \max(k_j - 1, a_j - b_i - h)$.

We note that $P((h + 1)p) = P(p + hp)$ and thus the property $P(k)$ is well defined for every integer $k \geq 0$. Also we note that the matrix A satisfies $P(0)$. We will show that if a matrix $Q(x, D)$ defined in some open neighborhood U of x_0 in Ω satisfies $P(k - 1)$ then we can find a smaller open neighborhood U of x_0 in Ω and a unit M in $\mathcal{M}_{p \times p}(\mathfrak{A}(U))$ of multigrading (b_j, b_h) such that $M(x, D)Q(x, D)$ satisfies $P(k)$. Let $h_0 = \sup\{a_j - b_i$ for $1 \leq i, j \leq p\}$. By iteration of the process indicated above we will reach after $h_0 p$ steps a matrix which verifies $P(h_0 p)$ and thus the conclusion of the lemma.

Let us assume that $Q(x, D)$ in some open neighborhood U of x_0 satisfies $P(k - 1)$ with $k = l + hp$ with $0 \leq l < p$. Because of property (i) of $P(k - 1)$ we have that in $q_{ll}(x, D)$ the coefficient of $(\partial/\partial t)^{k_l}$ is different from zero in an open neighborhood U of x_0 in Ω. In that neighborhood U we can write (in $\mathfrak{A}|U$), for $i \neq l$:

$$q_{il}(x, D) = -\mu_i(x, D)q_{ll}(x, D) + \lambda_i(x, D)$$

with order in $\partial/\partial t$ of $\lambda_i < k$.

We note that, if $i \neq l$, the global order of μ_i is $b - b_i$, and order of μ_i in $\partial/\partial t \leq b - b_i - h$ (since $i \neq l$). We set

$$
M(x, D) = \begin{pmatrix}
1 & \cdots & 0 & \mu_1 & 0 & \cdots & 0 \\
& \vdots & & \vdots & \vdots & \vdots & \\
0 & \cdots & 1 & \mu_{l-1} & 0 & \cdots & 0 \\
0 & \cdots & 0 & 1 & 0 & \cdots & 0 \\
0 & \cdots & 0 & \mu_{l+1} & 1 & \cdots & 0 \\
& \vdots & & \vdots & & \vdots & \\
0 & \cdots & 0 & \mu_p & 0 & \cdots & 1
\end{pmatrix}.
$$

We have $\det \hat{M}(x, \xi) = 1$. Thus M is a unit and of multigrading (b_i, b_h). We claim that $M(x, D)Q(x, D)$ satisfies $P(k)$.

Set $Q'(x, D) = M(x, D)Q(x, D)$ with $Q'(x, D) = (q'_{ij}(x, D))$. Then Q' is a matrix defined in a neighborhood U of x_0 and of multigrading (a_j, b_i). Moreover from $\hat{\mu}_i(x_0, 0, \tau) = 0$ one deduces that $\hat{M}(x_0, 0, \tau) = $ identity and therefore $\hat{Q}'(x_0, 0, \tau) = \hat{Q}(x_0, 0, \tau)$. Thus property (i) is satisfied. It remains to verify properties (ii) and (iii). Let $j < l$. Then for $i \neq j$

$$ q'_{ij}(x, D) = q_{ij}(x, D) + \mu_i(x, D)q_{lj}(x, D) $$

and order in $\partial/\partial t$ of $q_{ij} \leq \max(k_j - 1, a_j - b_i - h - 1)$, order in $\partial/\partial t$ of $\mu_i \leq b_l - b_i - h$, and order in $\partial/\partial t$ of $q_{lj} \leq \max(k_j - 1, a_j - b_l - h - 1)$. Now

$$ b_l - b_i - h + k_j - 1 = b_l - b_i - h + a_j - b_j - 1 \leq a_j - b_i - h - 1 $$

Since $b_l - b_j \leq 0$ by the assumption on the grading of A. Also

$$ b_l - b_i - h + a_j - b_l - h - 1 = a_j - b_i - 2h - 1 \leq a_j - b_i - h - 1 $$

since $h \geq 0$. For $j = l$ and $i \neq j$ we have $q'_{il} = q_{il} + \mu_i q_{ll} = \lambda_i$ and thus condition (ii) is satisfied by construction. This shows that condition (ii) of $P(k)$ is satisfied by Q'.

Let $j > l$ and $i \neq j$. Then

$$ q'_{ij} = q_{ij} + \mu_i q_{lj} $$

and order in $\partial/\partial t$ of $q_{ij} \leq \max(k_j - 1, a_j - b_i - h)$, order in $\partial/\partial t$ of $\mu_i \leq b_l - b_i - h$, and order in $\partial/\partial t$ of $q_{lj} \leq a_j - b_l = $ total order of q_j. Then

$$ b_l - b_i - h + a_j - b_l = a_j - b_i - h. $$

Therefore Q' satisfies also condition (iii) of $P(k)$. This completes the proof.

■

Remark.   We have seen that permutations of rows or columns in a multi-graded matrix $A(x, D) \in \mathcal{M}_{p \times p}(\mathfrak{A})$ is achieved by multiplications to the left or right by a multigraded matrix with constant coefficients which is a unit (cf. the proof of Proposition 2). By a permutation of rows and the corresponding permutation of columns we can always achieve that a matrix A of multigrading (a_j, b_i) with $a_j - b_j = k_j \geq 0$ for $1 \leq j \leq p$ and $\hat{A}(x, 0, \tau) = \text{diag}\langle \tau^{k_1}, \ldots, \tau^{k_p}\rangle$ is such that $b_1 \geq b_2 \geq \cdots \geq b_p \geq 0$.

(d)   Let now $x_0 = 0 \in \mathbf{R}^n$ and let $\mathcal{A}_0$ denote the ring of germs of complex-valued real analytic functions at the origin in $\mathbf{R}^n$. We consider for $\mathfrak{A}$ the noncommutative algebra

$$\mathfrak{A} = \mathcal{A}_0[\partial/\partial y_1, \ldots, \partial/\partial y_{n-1}, \partial/\partial t]$$

of differential operators with coefficients in $\mathcal{A}_0$.

Let $A(y, t; \partial/\partial y, \partial/\partial t) \in \mathcal{M}_{p \times p}(\mathfrak{A})$ be a matrix of multigrading (a_j, b_i) with $k_j = a_j - b_j \geq 0$ for $1 \leq j \leq p$, and let us assume that

(i) $\hat{A}(0, 0; 0, \tau) = \text{diag}\langle \tau^{k_1}, \ldots, \tau^{k_p}\rangle$,

(ii)   the element $a_{ij}(y, t; \partial/\partial y, \partial/\partial t)$ at place (i, j) in A has order in $\partial/\partial t$ strictly less than k_j if $i \neq j$.

Let $_{(n-1)}\mathcal{A}_0$ denote the ring of germs of complex-valued real analytic functions at the origin of $\mathbf{R}^{n-1}$, where $y = (y_1, \ldots, y_{n-1})$ are Cartesian coordinates. We consider the following Cauchy problem:

Given $v_{js} \in {}_{(n-1)}\mathcal{A}_0$, $1 \leq j \leq p$, $0 \leq s \leq k_j - 1$, *given* $f_i \in \mathcal{A}_0$, $1 \leq i \leq p$, *to find* $u \in \mathcal{A}_0$ *such that*

$$\sum_{j=1}^{p} a_{ij}(y, t; \partial/\partial y, \partial/\partial t) u_j = f_i, \qquad 1 \leq i \leq p, \qquad (*)$$

with the initial conditions:

$$\partial^s u_j / \partial t^s|_{t=0} = v_{js}, \qquad 1 \leq j \leq p, \quad 0 \leq s \leq k_j - 1. \qquad (**)$$

We claim that *under the above specified assumptions for the operator $A(x, D)$ the Cauchy problem $(*)$, $(**)$ admits a unique solution $u \in \mathcal{A}_0^p$.*

Proof.   We first remark that there exists a unique formal power series in t, with coefficients in $_{(n-1)}\mathcal{A}_0^p$, say, $u = (u_1, \ldots, u_p)$ with

$$u_j = \sum_0^\infty u_{js}(y) t^s, \qquad 1 \leq j \leq p, \quad u_{js} \in {}_{(n-1)}\mathcal{A}_0,$$

which solves the Cauchy problem (∗), (∗∗). Indeed, because of assumption (i) the coefficient of $\partial^{k_l}/\partial t^{k_l}$ in $a_{ll}(y, t; \partial/\partial y, \partial/\partial t)$ is different from zero and therefore the system (∗) permits us to obtain recursively all the u_{l_s} in terms of the $v_{j0}, \ldots, v_{j, k_j - 1}$ $(1 \leq j, l \leq p)$.

We now differentiate b_1 times the first equation of (∗), b_2 times the second one, b_3 times the third and so on. Then we consider the corresponding Cauchy problem:

$$\sum (\partial/\partial t)^{b_i}(a_{ij}(y, t; \partial/\partial y, \partial/\partial t)g_j) = (\partial/\partial t)^{b_i}f_i, \qquad 1 \leq i \leq p, \qquad (\alpha)$$

with initial conditions:

$$(\partial/\partial t)^s g_j|_{t=0} = u_{js}, \qquad 1 \leq j \leq p, \quad 0 \leq s \leq a_j - 1. \qquad (\beta)$$

This new system (α) corresponds to the multigrading $(a_j, 0)$. To it we apply Lemma 3, i.e., we can find a unit matrix M such that $M \circ A'$ [A' denoting the matrix of the system (α)] satisfies the conditions of the lemma. This shows that

$$M \circ A'u = M'((\partial/\partial t)^{b_1}f_1, \ldots, (\partial/\partial t)^{b_p}f_p)$$

with the initial conditions (β) is in Cauchy–Kowalewska form (cf. [9, Vol. II, p. 656] and [13, p. 14]). Therefore this system admits a unique solution $g \in \mathscr{A}_0^p$. Because of the choice of the initial conditions $g = u$ as formal power series in t. But this proves that the formal power series u given above is convergent and thus proves our claim. ∎

(e)   We set $\mathscr{P} = \mathbf{C}[\eta_1, \ldots, \eta_{n-1}, \tau]$, $_{(n-1)}\mathscr{P} = \mathbf{C}[\eta_1, \ldots, \eta_{n-1}]$ to denote the rings of polynomials in n or $n - 1$ variables, respectively.

We consider a matrix $F \in \mathscr{M}_{p \times p}(\mathscr{P})$ of type $p \times p$ with entries in $\mathscr{P}$ of the following form:

$$F(\eta, \tau) = \text{diag}\langle \tau^{k_1}, \ldots, \tau^{k_p} \rangle + R(\eta, \tau),$$

where the entries $r_{ij}(\eta, \tau)$ of $R(\eta, \tau)$ have the property:

$$\text{degree in } \tau \text{ of } r_{ij}(\eta, \tau) < k_j$$

and where $k_1 \geq 0, \ldots, k_p \geq 0$. If we agree that

$$\tau^{k_j - h} = \begin{cases} \tau^{k_j - h} & \text{if } k_j - h > 0, \\ 1 & \text{if } k_j - h = 0, \\ 0 & \text{if } k_j - h < 0, \end{cases}$$

then we can write

$$F(\eta, \tau) = \text{diag}\langle \tau^{k_1}, \ldots, \tau^{k_p} \rangle + \sum_{h > 0} c_h(\eta) \, \text{diag}\langle \tau^{k_1 - h}, \ldots, \tau^{k_p - h} \rangle,$$

where c_h is a matrix of $\mathscr{M}_{p \times p}(_{(n-1)}\mathscr{P})$.

LEMMA 4.   *Given $X \in \mathscr{P}^p$ we can find $Q \in \mathscr{P}^p$ and $R = {}^t(r_1, \ldots, r_p) \in \mathscr{P}^p$ such that*

(i)   *$X = FQ + R$ and*
(ii) *degree in τ of $r_j < k_j$.*

Moreover Q and R are uniquely determined by these properties.

Proof.   Existence of Q and R. Let $X = {}^t(x_1, \ldots, x_p)$. If $\sup_{1 \le j \le p} \deg$ in τ of $x_j < \inf(k_1, \ldots, k_p)$ we can take $Q = 0$, $R = X$. Then we can proceed by induction on the integer $l = \sup_j \deg$ in τ of x_j. We can find polynomials q_j, r_j with

$$x_j = \tau^{k_j} q_j + r_j, \qquad \text{degree in } \tau \text{ of } r_j < k_j.$$

Then

$$X = \left(F - \sum_{h > 0} c_h \operatorname{diag}\langle \tau^{k_1 - h}, \ldots, \tau^{k_p - h} \rangle \right) Q + R,$$

where $Q = {}^t(q_1, \ldots, q_p)$ and $R = {}^t(r_1, \ldots, r_p)$. Hence

$$X - FQ = \sum_{h > 0} c_h \operatorname{diag}\langle \tau^{k_1 - h}, \ldots, \tau^{k_p - h} \rangle Q + R. \tag{*}$$

Now, degree in τ of q_j = degree in τ of $x_j - k_j$, so that for $h > 0$ degree in τ of $\tau^{k_j - h} q_j <$ degree in τ of $x_j \le l$. Therefore the right-hand side of (*) has each component of a degree in τ less than l. We can thus apply to it the inductive hypothesis and we conclude with the existence of Q and R.

Uniqueness of Q and R.   We have to show that, if $FQ + R = 0$ and R satisfies assumption (ii) then $Q = R = 0$.

We may assume that $k_1 \ge k_2 \ge \cdots \ge k_p$, and assume, if possible, that $Q \ne 0$. Let $Q = {}^t(q_1, \ldots, q_p)$ with $q_1 = \cdots = q_{j-1} = 0$ and $q_j \ne 0$, and set

$$q_j = q_{j0} + q_{j1}\tau + \cdots + q_{js_j}\tau^{s_j}, \qquad \text{with} \quad q_{jh} \in {}_{(n-1)}\mathscr{P}, \quad q_{js_j} \ne 0.$$

In the jth component of $FQ + R$ the coefficient of $\tau^{k_j + s_j}$ is q_{js_j}. But then from $FQ + R = 0$ one deduces $q_{js_j} = 0$, getting a contradiction, that shows that necessarily $Q = 0$. Hence $R = 0$.

COROLLARY 1.   *With $F \in \mathscr{M}_{p \times p}(\mathscr{P})$ of the specified form we have*

$$\frac{\mathscr{P}^p}{F\mathscr{P}^p} \simeq {}_{(n-1)}\mathscr{P}^l \qquad \text{where} \quad l = k_1 + \cdots + k_p.$$

Proof.   From the previous lemma we derive that

$$\mathscr{P}^p \simeq F\mathscr{P}^p + \left(\bigoplus_{j=1}^{p} \bigoplus_{s=0}^{k_j - 1} {}_{(n-1)}\mathscr{P}\tau^s \right). \quad \blacksquare$$

COROLLARY 2.   *Under the same assumptions for F, given $M \in \mathcal{M}_{1 \times p}(\mathscr{P})$ there exists a unique $Q \in \mathcal{M}_{1 \times p}(\mathscr{P})$ and $R = (r_{ij}) \in \mathcal{M}_{1 \times p}(\mathscr{P})$ such that*

(i) $M = QF + R$ *and*
(ii)   *degree in τ of $r_{ij} < k_j$.*

Moreover, if F is a homogeneous matrix of multigrading (a_j, α_i) with $k_j = a_j - \alpha_j \geq 0$ (for $1 \leq j \leq p$) and if M is homogeneous of multigrading (a_j, β_h), then Q is homogeneous of multigrading (α_j, β_h) and R is homogeneous of multigrading (a_j, β_h).

Proof.   If we write

$$M = \begin{pmatrix} m_1 \\ \vdots \\ m_l \end{pmatrix}, \qquad Q = \begin{pmatrix} q_1 \\ \vdots \\ q_l \end{pmatrix}, \qquad R = \begin{pmatrix} r_1 \\ \vdots \\ r_l \end{pmatrix},$$

we must have $m_i = q_i F + r_i$ with degree in τ or $r_{ij} < k_j$. Thus Q and R are determined by Lemma 4. Moreover, if we write

$$m_{ij} = q_{ij} \tau^{k_j} + r_{ij}, \qquad \text{with degree in } \tau \text{ of } r_{ij} < k_j,$$

and m_{ij} is homogeneous of degree $a_j - \beta_i$, then r_{ij} is homogeneous of degree $a_j - \beta_i$, while q_{ij} is homogeneous of degree $\alpha_j - \beta_i$. From the proof of the existence of Q and R it follows that Q and R are homogeneous of the specified multigrading.   ∎

(f)   Let Ω be open in $\mathbf{R}^n$. With $\mathfrak{A}$ we denote the noncommutative algebra $\mathscr{E}(\Omega)[\partial/\partial y_1, \ldots, \partial/\partial y_{n-1}, \partial/\partial t]$ or $\mathscr{A}(\Omega)[\partial/\partial y_1, \ldots, \partial/\partial y_{n-1}, \partial/\partial t]$. We consider a matrix $F(y, t; \partial/\partial y, \partial/\partial t) \in \mathcal{M}_{p \times p}(\mathfrak{A})$ with the following property:

$$F(y, t; \partial/\partial y, \partial/\partial t) = \text{diag}\langle (\partial/\partial t)^{k_1}, \ldots, (\partial/\partial t)^{k_p} \rangle + S(y, t; \partial/\partial y, \partial/\partial t),$$

where $S = (s_{ij}(y, t; \partial/\partial y, \partial/\partial t))$ is such that

$$\text{order of } s_{ij}(y, t; \partial/\partial y, \partial/\partial t) \text{ in } \partial/\partial t < k_j.$$

We will also assume that F corresponds to a multigrading (a_j, α_i) with $k_j = a_j - \alpha_j \geq 0$ for $1 \leq j \leq p$.

LEMMA 5.   *Let $F(y, t; \partial/\partial y, \partial/\partial t)$ be as above. For every $M(y, t; \partial/\partial y, \partial/\partial t)$ in $\mathcal{M}_{1 \times p}(\mathfrak{A})$ of multigrading (a_j, β_h) we can find $Q(y, t; \partial/\partial y, \partial/\partial t)$ in $\mathcal{M}_{1 \times p}(\mathfrak{A})$ of multigrading (α_i, β_h) and $R(y, t; \partial/\partial y, \partial/\partial t)$ in $\mathcal{M}_{1 \times p}(\mathfrak{A})$ of multigrading (a_j, β_h) such that*

(i) $M(y, t; \partial/\partial y, \partial/\partial t) = Q(y, t; \partial/\partial y, \partial/\partial t)F(y, t; \partial/\partial y, \partial/\partial t)$
      $\qquad + R(y, t; \partial/\partial y, \partial/\partial t).$

(ii) *For every entry* $r_{ij}(y,t; \partial/\partial y, \partial/\partial t)$ *of* R *we have* order of $r_{ij}(y,t;$ $\partial/\partial y,\partial/\partial t)$ in $\partial/\partial t < k_j$.

Moreover Q and R are uniquely determined by properties (i) *and* (ii).

Proof. We fix a basic multigrading (a_j, β_i) and set

$$\mathcal{M}_{l \times p}(\mathfrak{A})_h = \{M = (m_{ij}(y,t; \partial/\partial y, \partial/\partial t) \in \mathcal{M}_{l \times p}(\mathfrak{A}) | \operatorname{ord} m_{ij} \le a_j - \beta_i - h\}.$$

The elements of $\mathcal{M}_{l \times p}(\mathfrak{A})_h$ can be viewed as corresponding to the multigrading $(a_j, \beta_i + h)$. For large h we have $\mathcal{M}_{l \times p}(\mathfrak{A})_h = 0$.

We will prove the existence of Q and R by a descending induction on h. Changing if need be the integers β, we may assume that the existence of Q and R is established for any $M \in \mathcal{M}_{l \times p}(\mathfrak{A})_h$ with $h \ge 1$. We will prove the existence of Q and R for any $M \in \mathcal{M}_{l \times p}(\mathfrak{A})_0$. From Corollary 2 of Lemma 4 we deduce the existence of Q_1 and R_1 of the prescribed multigrading such that

$$\hat{M}(y,t; \eta, \tau) = \hat{Q}_1(y,t; \eta, \tau)\hat{F}(y,t; \eta, \tau) + \hat{R}_1(y,t; \eta, \tau)$$

with the property:

entry $r_{ij}^{(1)}(y,t; \partial/\partial y, \partial/\partial t)$ of R_1 has order in $\partial/\partial t < k_j$.

Then

$$N(y,t; \partial/\partial y, \partial/\partial t) = M(y,t; \partial/\partial y, \partial/\partial t) - R_1(y,t; \partial/\partial y, \partial/\partial t)$$
$$- Q_1(y,t; \partial/\partial y, \partial/\partial t)F(y,t; \partial/\partial y, \partial/\partial t)$$

is such that its symbol of type (a_j, β_i) is 0. Therefore we can consider N as an element of $\mathcal{M}_{l \times p}(\mathfrak{A})_1$ to which we can apply the inductive assumption. From this we conclude the existence of Q and R.

To prove uniqueness, we need to show that, if

$$0 = Q(y,t; \partial/\partial y, \partial/\partial t)F(y,t; \partial/\partial y, \partial/\partial t) + R(y,t; \partial/\partial y, \partial/\partial t)$$

and R satisfies condition (ii) then necessarily $Q = R = 0$.

Let us assume, if possible, that $Q \ne 0$. Then there is a maximum $h \ge 0$ such that, replacing multigrading (a_j, β_i) by multigrading (α_j, β_i), we have

$$Q \in \mathcal{M}_{l \times p}(\mathfrak{A})_h, \quad \text{but} \quad Q \notin \mathcal{M}_{l \times p}(\mathfrak{A})_{h+1}. \tag{*}$$

Then we can consider Q of multigrading $(\alpha_j, \beta_i + h)$ and therefore R of multigrading $(a_j, \beta_i + h)$. Taking symbols we get

$$0 = \hat{Q}\hat{F} + \hat{R}.$$

Because of (*) we must have $\hat{Q} \ne 0$. But this contradicts Corollary 2 to Lemma 4.

4. Proof of Theorem 1

(a)   Let

$$\mathscr{E}^p(\Omega) \xrightarrow{A(x,D)} \mathscr{E}^q(\Omega) \xrightarrow{B(x,D)} \mathscr{E}^r(\Omega) \tag{1}$$

be the given complex of differential operators with real analytic coefficients on the open set $\Omega \subset \mathbf{R}^n$. Let $x_0 \in \Omega$ be given. We will assume $x_0 = 0$ at the origin of the coordinates in $\mathbf{R}^n$. We assume that the corresponding symbolic complex

$$\mathscr{E}^p(\mathbf{R}^n) \xrightarrow{\hat{A}(0,D)} \mathscr{E}^q(\mathbf{R}^n) \xrightarrow{\hat{B}(0,D)} \mathscr{E}^r(\mathbf{R}^n) \tag{2}$$

is exact. Here $\hat{A}(0,\xi)$ and $\hat{B}(0,\xi)$ denote the symbols for a multigrading (a_j, b_i) for A and (b_i, c_h) for B. Exactness of (2) is equivalent to the exactness of the sequence of $\mathscr{P}$-homomorphisms:

$$\mathscr{P}^p \xleftarrow{{}^t\hat{A}(0,\xi)} \mathscr{P}^q \xleftarrow{{}^t\hat{B}(0,\xi)} \mathscr{P}^r. \tag{3}$$

From Hilbert's syzygies theorem we deduce that we can find a third homogeneous matrix ${}^tC(\xi)$ of multigrading $(c_h, d_s)_{1 \le h \le r, \, 1 \le s \le l}$, such that we have an exact sequence of graded $\mathscr{P}$-homomorphisms:

$$\mathscr{P}^p \xleftarrow{{}^t\hat{A}(0,\xi)} \mathscr{P}^q \xleftarrow{\hat{B}(0,\xi)} \mathscr{P}^r \xleftarrow{{}^tC(\xi)} \mathscr{P}^l. \tag{3'}$$

Let $\mathscr{R}$ be the field of rational functions in the variables ξ and let $\rho_0 = \operatorname{rank}_{\mathscr{R}} \hat{A}(0,\xi)$, $\rho_1 = \operatorname{rank}_{\mathscr{R}} \hat{B}(0,\xi)$, $\rho_2 = \operatorname{rank}_{\mathscr{R}} C(\xi)$, so that

$$q = \rho_0 + \rho_1 \quad \text{and} \quad r = \rho_1 + \rho_2 \tag{4}$$

(cf. the proof of Proposition 1).

Let V denote the characteristic variety of the Hilbert resolution (3'):

$$V = \{\xi \in \mathbf{C}^n \,|\, \operatorname{rank}_{\mathbf{C}} \hat{A}(0,\xi) < \rho_0\}.$$

Let us choose $\xi^0 \in \mathbf{R}^n$ with $\xi^0 \notin V$ so that, according to Proposition 1, we get an exact sequence:

$$\mathbf{C}^p \xrightarrow{\hat{A}(0,\xi^0)} \mathbf{C}^q \xrightarrow{\hat{B}(0,\xi^0)} \mathbf{C}^r \xrightarrow{C(\xi^0)} \mathbf{C}^l. \tag{5}$$

Since V is a cone with vertex in 0 [because $A(0,\xi)$ is a homogeneous matrix] we necessarily have $\xi^0 \ne 0$. Thus we may assume $\xi^0 = (0, \ldots, 0, 1)$. Accordingly we distinguish the variables x as $(y_1, \ldots, y_{n-1}, t)$ and the variables ξ as $(\eta_1, \ldots, \eta_{n-1}, \tau)$.

(b)   *Reduction to Canonical Form of the Complex* (3')

(α)   We can find homogeneous matrices $R(\tau) \in \mathscr{M}_{p \times p}(\mathbf{C}_0[\tau])$ and $L(\tau) \in \mathscr{M}_{q \times q}(\mathbf{C}_0[\tau])$ of total degree 0 and determinant $\ne 0$ such that according to

Lemma 2 we have

$$L(\tau)\hat{A}(0,0;0,\tau)R(\tau) = \begin{pmatrix} & & 0 & & 0 \\ \tau^{k_1} & & & & \\ & \ddots & & 0 & \\ & & \tau^{k_{\rho_0}} & & \end{pmatrix}.$$

We apply the fiber transformation $R(\partial/\partial t)$ to $\mathscr{E}^p(\Omega)$ and $\mathscr{E}^p(\mathbf{R}^n)$ and the fiber transformation $L^{-1}(\partial/\partial t)$ to $\mathscr{E}^q(\Omega)$ and $\mathscr{E}^q(\mathbf{R}^n)$. Then A is replaced by $L(\partial/\partial t)A(y,t;\partial/\partial y,\partial/\partial t)R(\partial/\partial t)$ and B is replaced by $B(y,t;\partial/\partial y,\partial/\partial t) \times L^{-1}(\partial/\partial t)$.

Thus we can assume that

$$\hat{A}(0,0;0,\tau) = \begin{pmatrix} & & 0 & & 0 \\ \tau^{k_1} & & & & \\ & \ddots & & 0 & \\ & & \tau^{k_{\rho_0}} & & \end{pmatrix}.$$

and also that $b_1 \geq b_2 \geq b_3 \geq \cdots \geq b_{\rho_0}$. Then the last ρ_0 columns of $\hat{B}(0,0; 0,\tau)$ are zero since $\hat{B}(0,0; 0,\tau)\hat{A}(0,0; 0,\tau) = 0$.

(β) We can find homogeneous matrices $R(\tau) \in \mathscr{M}_{\rho_1 \times \rho_1}(\mathbf{C}_0[\tau])$ and $L(\tau) \in \mathscr{M}_{r \times r}(\mathbf{C}_0[\tau])$ of total degree zero and determinant different from zero such that, if we replace

$$\hat{A}(0,0;0,\tau) \quad \text{by} \quad \begin{pmatrix} R^{-1}(\tau) & 0 \\ 0 & I \end{pmatrix}\hat{A}(0,0;0,\tau)$$

and

$$\hat{B}(0,0;0,\tau) \quad \text{by} \quad L(\tau)\hat{B}(0,0;0,\tau)\begin{pmatrix} R(\tau) & 0 \\ 0 & I \end{pmatrix},$$

we have

$$\hat{A}(0,0;0,\tau) = \begin{pmatrix} & & 0 & & 0 \\ \tau^{k_1} & & & & \\ & \ddots & & 0 & \\ & & \tau^{k_{\rho_0}} & & \end{pmatrix},$$

$$\hat{B}(0,0;0,\tau) = \begin{pmatrix} & & 0 & & 0 \\ \tau^{h_1} & & & & \\ & \ddots & & 0 & \\ & & \tau^{h_{\rho_1}} & & \end{pmatrix},$$

and $b_1 \geq b_2 \geq \cdots \geq b_{\rho_0}, c_1 \geq c_2 \geq \cdots \geq c_{\rho_1}$.

We apply the fiber transformation $\begin{pmatrix} R(\partial/\partial t) & 0 \\ 0 & I \end{pmatrix}$ to $\mathscr{E}^q(\Omega)$ and to $\mathscr{E}^q(\mathbf{R}^n)$ and the fiber transformation $L^{-1}(\partial/\partial t)$ to $\mathscr{E}^r(\Omega)$ and $\mathscr{E}^r(\mathbf{R}^n)$. Then A and B are transformed in the way indicated above and C is replaced by $C(\partial/\partial y, \partial/\partial t)L^{-1}(\partial/\partial t)$. We can therefore assume that $\hat{A}$ and $\hat{B}$ have the form indicated above. Moreover the last ρ_1 columns in $\hat{C}(0,\tau) = C(0,\tau)$ will be zero.

(γ) We can find homogeneous matrices $R(\tau) \in \mathscr{M}_{\rho_2 \times \rho_2}(\mathbf{C}_0[\tau])$ and $L(\tau) \in \mathscr{M}_{l \times l}(\mathbf{C}_0[\tau])$ of total degree zero and determinant different from zero such that, if we replace

$$\hat{B}(0,0;0,\tau) \quad \text{by} \quad \begin{pmatrix} R^{-1}(\tau) & 0 \\ 0 & I \end{pmatrix} \hat{B}(0,0;0,\tau)$$

$$C(0,\tau) \quad \text{by} \quad L(\tau)C(0,\tau)\begin{pmatrix} R(\tau) & 0 \\ 0 & I \end{pmatrix},$$

we have

$$\hat{A}(0,0;0,\tau) = \begin{pmatrix} & & 0 & & 0 \\ \tau^{k_1} & & \ddots & & 0 \\ & & & \tau^{k_{\rho_0}} & \end{pmatrix},$$

$$\hat{B}(0,0;0,\tau) = \begin{pmatrix} & & 0 & & 0 \\ \tau^{h_1} & & & & 0 \\ & & \ddots & & \\ & & & \tau^{h_{\rho_1}} & \end{pmatrix},$$

$$C(0,\tau) = \begin{pmatrix} & & 0 & & 0 \\ \tau^{\lambda_1} & & & & 0 \\ & & \ddots & & \\ & & & \tau^{\lambda_{\rho_2}} & \end{pmatrix},$$

and moreover $b_1 \geq b_2 \geq \cdots \geq b_{\rho_0}$, $c_1 \geq c_2 \geq \cdots \geq c_{\rho_1}$, and $d_1 \geq d_2 \geq \cdots \geq d_{\rho_2}$.

We apply the fiber transformation $\begin{pmatrix} R(\partial/\partial t) & 0 \\ 0 & I \end{pmatrix}$ to $\mathscr{E}^r(\Omega)$ and $\mathscr{E}^r(\mathbf{R}^n)$ and the fiber transformation $L^{-1}(\partial/\partial t)$ to $\mathscr{E}^l(\mathbf{R}^n)$. Then we can assume that $\hat{A}$, $\hat{B}$, C have the form indicated above.

(δ) According to Lemma 3 we can find an open neighborhood U of the origin in $\mathbf{R}^n$ and unit matrices

$$M \in \mathscr{M}_{\rho_0 \times \rho_0}(\mathfrak{A}(U)), \qquad N \in \mathscr{M}_{\rho_1 \times \rho_1}(\mathfrak{A}(U)), \qquad P \in \mathscr{M}_{\rho_2 \times \rho_2}(\mathfrak{A}(U)),$$

where $\mathfrak{A}(U) = \mathscr{A}(U)[\partial/\partial y_1, \ldots, \partial/\partial y_{n-1}, \partial/\partial t]$, such that, if we write

$$A = \begin{pmatrix} A^{00} & A^{01} \\ A^{10} & A^{11} \end{pmatrix} \quad \text{with} \quad A^{10} \text{ of type } \rho_0 \times \rho_0,$$

$$B = \begin{pmatrix} B^{00} & B^{01} \\ B^{10} & B^{11} \end{pmatrix} \quad \text{with} \quad B^{10} \text{ of type } \rho_1 \times \rho_1,$$

$$C = \begin{pmatrix} C^{00} & C^{01} \\ C^{10} & C^{11} \end{pmatrix} \quad \text{with} \quad C^{10} \text{ of type } \rho_2 \times \rho_2,$$

the matrices MA^{10}, NB^{10}, PC^{10} are of the form

$$MA^{10} = \text{diag}\langle(\partial/\partial t)^{k_1}, \ldots, (\partial/\partial t)^{k_{\rho_0}}\rangle + R(y, t; \partial/\partial y, \partial/\partial t),$$
$$NB^{10} = \text{diag}\langle(\partial/\partial t)^{h_1}, \ldots, (\partial/\partial t)^{h_{\rho_1}}\rangle + S(y, t; \partial/\partial y, \partial/\partial t),$$
$$PC^{10} = \text{diag}\langle(\partial/\partial t)^{\lambda_1}, \ldots, (\partial/\partial t)^{\lambda_{\rho_2}}\rangle + T(y, t; \partial/\partial y, \partial/\partial t),$$

with $R = (r_{ij})$, $S = (s_{ij})$, $T = (t_{ij})$, and

$$\text{order of } r_{ij} \text{ in } \partial/\partial t < k_j \quad 1 \leq j \leq \rho_0,$$
$$\text{order of } s_{ij} \text{ in } \partial/\partial t < h_j \quad 1 \leq j \leq \rho_1,$$
$$\text{order of } t_{ij} \text{ in } \partial/\partial t < \lambda_j \quad 1 \leq j \leq \rho_2.$$

Because of the remark (d) of Section 3 MA^{10}, NB^{10}, PC^{10} lead to well-posed Cauchy–Kowalewska problems on $t = 0$. We assume $\Omega = U$ and we apply the fiber transformations:

$$\begin{pmatrix} I & 0 \\ 0 & M \end{pmatrix} \quad \text{to} \quad \mathscr{E}^q(\Omega), \qquad \begin{pmatrix} I & 0 \\ 0 & N \end{pmatrix} \quad \text{to} \quad \mathscr{E}^r(\Omega), \qquad \begin{pmatrix} I & 0 \\ 0 & P \end{pmatrix} \quad \text{to} \quad \mathscr{E}^l(\Omega).$$

This has the effect of replacing

$$A \quad \text{by} \quad \begin{pmatrix} I & 0 \\ 0 & M \end{pmatrix} A, \qquad B \quad \text{by} \quad \begin{pmatrix} I & 0 \\ 0 & N \end{pmatrix} B \begin{pmatrix} I & 0 \\ 0 & M^{-1} \end{pmatrix},$$

$$C \quad \text{by} \quad \begin{pmatrix} I & 0 \\ 0 & P \end{pmatrix} C \begin{pmatrix} I & 0 \\ 0 & N^{-1} \end{pmatrix}.$$

This will not affect the canonical form obtained for $\hat{A}(0, 0; 0, \tau)$, $\hat{B}(0, 0; 0, \tau)$, and $\hat{C}(0, 0; 0, \tau)$. Moreover we have achieved in $U = \Omega$ having the part A^{10} of A, B^{10} of B, and C^{10} of C in Cauchy–Kowalewska form. Note that C may be no longer with constant coefficients.

(ε)   *Remark.*   Let A, B, C have the form described above. By Lemma 5 we can write in a unique way

$$A^{00}(x, D) = Q_1(x, D)A^{10}(x, D) + R_1(x, D),$$
$$B^{00}(x, D) = Q_2(x, D)B^{10}(x, D) + R_2(x, D),$$
$$C^{00}(x, D) = Q_3(x, D)C^{10}(x, D) + R_3(x, D),$$

where the entries $r_{ij}^{(s)}(x, D)$ of $R_s(x, D)$ have order in $\partial/\partial t$ less than k_j if $s = 1$, h_j if $s = 2$, λ_j if $s = 3$. If we substitute

$$A \quad \text{by} \quad \begin{pmatrix} I & -Q_1 \\ 0 & I \end{pmatrix} A, \qquad B \quad \text{by} \quad \begin{pmatrix} I & -Q_2 \\ 0 & I \end{pmatrix} B \begin{pmatrix} I & Q_1 \\ 0 & I \end{pmatrix},$$

and

$$C \quad \text{by} \quad \begin{pmatrix} I & -Q_3 \\ 0 & I \end{pmatrix} C \begin{pmatrix} I & Q_2 \\ 0 & I \end{pmatrix},$$

then the entries A^{10}, B^{10}, and C^{10} remain unchanged, while A^{00}, B^{00}, and C^{00} are substituted by R_1, R_2, and R_3, respectively. Therefore one can also assume, if need be, that the entries in the jth column of A^{00} (resp., B^{00}, C^{00}) have order in $\partial/\partial t$ less than k_j (resp., h_j, λ_j).

(c)   A Complex of Cauchy Data

(α)   We set $\omega = U \cap \{t = 0\}$ so that ω is an open neighborhood of the origin in $\mathbf{R}^{n-1}$. We set $v_0 = k_1 + \cdots + k_{\rho_0}$, $v_1 = h_1 + \cdots + h_{\rho_1}$, and $v_2 = \lambda_1 + \cdots + \lambda_{\rho_2}$. We identify the space $\mathscr{E}^{v_0}(\omega)$ with the space of v_0-uples of functions $U = (u_{js_j})_{1 \le j \le \rho_0, \, 0 \le s_j \le k_j - 1}$ defined on ω. It is the space of "Cauchy data for the operator A^{10}." We identify the space $\mathscr{E}^{v_1}(\omega)$ with the space of v_1-uples of functions $V = (v_{js_j})_{1 \le j \le \rho_1, \, 0 \le s_j \le h_j - 1}$ defined on ω. It is the space of "Cauchy data for the operator B^{10}." We identify the space $\mathscr{E}^{v_2}(\omega)$ with the space of v_2-uples of functions $W = (w_{js_j})_{1 \le j \le \rho_2, \, 0 \le s_j \le \lambda_j - 1}$ defined on ω. It is a candidate space of "Cauchy data for the operator C^{10}."

We want to define two differential operators with coefficients in $\mathscr{A}(\omega)$: $R(y, \partial/\partial y) : \mathscr{E}^{v_0}(\omega) \to \mathscr{E}^{v_1}(\omega)$, $T(y, \partial/\partial y) : \mathscr{E}(\omega)^{v_1} \to \mathscr{E}^{v_2}(\omega)$ defining a complex:

$$\mathscr{E}^{v_0}(\omega) \xrightarrow{R} \mathscr{E}^{v_1}(\omega) \xrightarrow{T} \mathscr{E}^{v_2}(\omega). \qquad (6)$$

(β)   DEFINITION OF R.   Given $U = (u_{js})$ we can find a unique $u = {}^t(u_1, \ldots, u_{\rho_0})$ with u_j formal power series in t with coefficients in $\mathscr{E}(\omega)$ such that

$$A^{10}(y, t; \partial/\partial y, \partial/\partial t)u = 0,$$
$$(\partial/\partial t)^s u_j \big|_{t=0} = u_{js} \qquad \text{for} \quad 1 \le j \le \rho_0, \quad 0 \le s \le k_j - 1. \qquad (*)$$

We set $v = {}^t(v_1, \ldots, v_{\rho_1})$ with v_j formal power series in t with coefficients in $\mathscr{E}(\omega)$ and let

$$v = A^{00}(y, t; \partial/\partial y, \partial/\partial t)u. \qquad (**)$$

Then we define $V = {}^t(v_{js})$ by setting $v_{js} = (\partial/\partial t)^s v_j\big|_{t=0}$ for $1 \leq j \leq \rho_1$ and $0 \leq s \leq h_j - 1$. Because of $(*)$ every derivative in t of u can be expressed in terms of combinations with real analytic coefficients of y derivatives of the u_{js}. Thus we can find a differential operator $R(y, \partial/\partial y)$ with coefficients in $A(\omega)$ such that

$$V = R(y, \partial/\partial y)U.$$

Note that, because of $(*)$ and $(**)$ we have $\binom{v}{0} = A\binom{u}{0}$. Thus in particular we must have $B\binom{v}{0} = 0$, i.e.,

$$B^{00}v = 0 \qquad \text{and} \qquad B^{10}v = 0.$$

Also we note that $V = RU$ can be written in the form

$$v_{i\sigma} = \sum r_{i\sigma, js}(y, \partial/\partial y)u_{js}$$

with $r_{i\sigma, js}$ or order $\leq a_j - s - b_i + \sigma$, so that the matrix R can be considered of multigrading $(a_j - s, b_i - \sigma)$ with $1 \leq i \leq \rho_0$, $0 \leq s \leq k_j - 1$, $1 \leq i \leq \rho_1$, $0 \leq \sigma \leq h_i - 1$.

(γ) DEFINITION OF T.   Given $V = (v_{js})$ we can find a unique $v = {}^t(v_1, \ldots, v_{\rho_1})$ with v_j formal power series in t with coefficients in $\mathscr{E}(\omega)$ such that

$$B^{10}(y, t; \partial/\partial y, \partial/\partial t)v = 0,$$
$$(\partial/\partial t)^s v_j\big|_{t=0} = v_{js} \qquad \text{for} \quad 1 \leq j \leq \rho_1, \quad 0 \leq s \leq h_j - 1. \qquad (*)$$

We set $w = {}^t(w_1, \ldots, w_{\rho_2})$ with w_j formal power series in t with coefficients in $\mathscr{E}(\omega)$ and let

$$w = B^{00}(y, t; \partial/\partial y, \partial/\partial t)v. \qquad (**)$$

We define $W = (w_{js})$ by setting $w_{js} = (\partial/\partial t)^s w_j\big|_{t=0}$ for $1 \leq j \leq \rho_2$, $0 \leq s \leq \lambda_j - 1$. Because of $(*)$ every derivative in t of v can be expressed in terms of combinations with real analytic coefficients of y derivatives of the v_{js}. It follows that we can find a differential operator $T(y, \partial/\partial y)$ with coefficients in $\mathscr{A}(\omega)$ such that $W = T(y, \partial/\partial y)V$. Writing this as

$$w_{i\sigma} = \sum t_{i\sigma, js}(y, \partial/\partial y)v_{js},$$

then $t_{i\sigma, js}$ has order $\leq b_j - s - c_i + \sigma$, so that the matrix T can be considered of multigrading $(b_j - s, c_i - \sigma)$ with $1 \leq j \leq \rho_1$, $0 \leq s \leq h_j - 1$, $1 \leq i \leq \rho_2$, $0 \leq \sigma \leq \lambda_i - 1$.

Also note that, if $V = RU$, we must have, because of (∗) and the uniqueness of the solution of the Cauchy problem, that $v = A^{00}u$ by the remark at the end of point (β). It follows that $B^{00}v = 0$ and thus $W = TRU = 0$. This shows that (6) is a complex.

(d)   We prove now the following statement:

If the complex (6) admits the analytic Poincaré lemma at the origin $0 \in \mathbf{R}^{n-1}$, then the given complex (1) admits the analytic Poincaré lemma at the origin $0 \in \mathbf{R}^{n}$.

Let $\binom{f}{g} \in \mathscr{A}_0^q$ with $f \in \mathscr{A}_0^{\rho_1}$ and $g \in \mathscr{A}_0^{\rho_0}$, $q = \rho_0 + \rho_1$, be such that

$$B(y, t; \partial/\partial y, \partial/\partial t)\binom{f}{g} = 0.$$

We want to show that there exists a vector $\binom{u}{v} \in \mathscr{A}_0^p$, $u \in \mathscr{A}_0^{\rho_0}$, $v \in \mathscr{A}_0^{p-\rho_0}$, with

$$A(y, t; \partial/\partial y, \partial/\partial t)\binom{u}{v} = \binom{f}{g}.$$

We will seek a solution of this problem with $v = 0$. Then we must have $A^{10}u = g$. Let $w \in \mathscr{A}_0^{\rho_0}$ be such that $A^{10}w = g$. This is possible because A^{10} leads to a well-posed Cauchy problem. Let $\mu = u - w \in \mathscr{A}_0^{\rho_0}$. Then μ must satisfy

$$A^{00}\mu = \varphi = f - A^{00}w, \qquad A^{10}\mu = 0.$$

As a consequence of the condition $B\binom{f}{g} = 0$ we get

$$B^{10}\varphi = 0, \qquad B^{00}\varphi = 0.$$

Let now $V = (v_{js})$ with $v_{js} = (\partial/\partial t)^s \varphi_j|_{t=0}$ for $1 \le j \le \rho_1$, $0 \le s \le h_j - 1$, where $\varphi = {}^t(\varphi_1, \ldots, \varphi_{\rho_1})$. From $B^{10}\varphi = 0$ and $B^{00}\varphi = 0$ we derive that $T(y, \partial/\partial y)V = 0$. By assumption we can find $U = (u_{js})_{1 \le j \le \rho_0,\ 0 \le s \le k_j - 1}$ with $u_{js} \in {}_{(n-1)}\mathscr{A}_0$ (equal to germs of real analytic functions at the origin of $\mathbf{R}^{n-1}$), such that $R(y, \partial/\partial y)U = V$. We construct now the unique $\mu = {}^t(\mu_1, \ldots, \mu_{\rho_0}) \in \mathscr{A}_0^{\rho_0}$ such that

$$A^{10}\mu = 0, \qquad (\partial/\partial t)^s\mu_j = u_{js} \qquad \text{for}\quad 1 \le j \le \rho_0, \quad 0 \le s \le k_j - 1.$$

We claim that we also have $A^{00}\mu = \varphi$. This will prove the statement. To this purpose, set $\eta = A^{00}\mu - \varphi$. Since $RU = V$, we must have for $\eta = {}^t(\eta_1, \ldots, \eta_{\rho_1})$:

$$(\partial/\partial t)^s\eta_j|_{t=0} = 0 \qquad \text{for}\quad 1 \le j \le \rho_1, \quad 0 \le s \le h_j - 1. \qquad (∗∗)$$

Moreover we do have

$$B^{10}\eta = 0. \qquad (∗)$$

Indeed $B^{10}\eta = B^{10}A^{00}\mu - B^{10}\varphi = B^{10}A^{00}\mu = -B^{11}A^{10}\mu = 0$. From the uniqueness of the Cauchy problem $(*)$, $(**)$ we deduce that $\eta = 0$.

(e) We now prove the following statement:

The symbolic complex at the origin in $\mathbf{R}^{n-1}$ associated with the complex (6) is exact.

Let $_{(n-1)}\mathscr{P} = \mathbf{C}[\eta_1, \ldots, \eta_{n-1}]$. We apply Corollary 1 to Lemma 4 to the polynomial matrices $\hat{A}^{10}(0,0; \eta, \tau)$, $\hat{B}^{10}(0,0; \eta, \tau)$, $\hat{C}^{10}(0,0; \eta, \tau)$. We get isomorphisms:

$$_{(n-1)}\mathscr{P}^{v_0} \simeq \frac{\mathscr{P}^{\rho_0}}{{}^t\hat{A}^{10}(0,0; \eta, \tau)\mathscr{P}^{\rho_0}}, \qquad _{(n-1)}\mathscr{P}^{v_1} \simeq \frac{\mathscr{P}^{\rho_1}}{{}^t\hat{B}^{10}(0,0; \eta, \tau)\mathscr{P}^{\rho_1}}$$

$$_{(n-1)}\mathscr{P}^{v_2} \simeq \frac{\mathscr{P}^{\rho_2}}{{}^t\hat{C}^{10}(0,0; \eta, \tau)\mathscr{P}^{\rho_2}}.$$

We want to show that

$$_{(n-1)}\mathscr{P}^{v_0} \xleftarrow{\ {}^t\hat{R}(0,\eta)\ } {}_{(n-1)}\mathscr{P}^{v_1} \xleftarrow{\ {}^t\hat{T}(0,\eta)\ } {}_{(n-1)}\mathscr{P}^{v_2} \tag{7}$$

is an exact sequence. Taking into account the isomorphisms established above the complex (7) can be written also as

$$\frac{\mathscr{P}^{\rho_0}}{{}^t\hat{A}^{10}(0,0; \eta, \tau)\mathscr{P}^{\rho_0}} \xleftarrow{\ {}^tA^{00}(0,0;\eta,\tau)\ } \frac{\mathscr{P}^{\rho_1}}{{}^t\hat{B}^{10}(0,0; \eta, \tau)\mathscr{P}^{\rho_1}}$$

$$\xleftarrow{\ {}^tB^{00}(0,0;\eta,\tau)\ } \frac{\mathscr{P}^{\rho_2}}{{}^t\hat{C}^{10}(0,0; \eta, \tau)\mathscr{P}^{\rho_2}}. \tag{8}$$

(It is clear that this is a complex also by direct verification, as ${}^t\hat{A}^{00}\,{}^t\hat{B}^{00} = -{}^t\hat{A}^{10}\,{}^t\hat{B}^{01}$.) We admit this statement for the moment. The proof of this fact will be given at the end of the whole argument as an appendix [Section (g)]. Let us show now that this last complex (8) is exact. Let $X \in \mathscr{P}^{\rho_1}$ be such that ${}^t\hat{A}^{00}(0,0; \eta, \tau)X = {}^t\hat{A}^{10}(0,0; \eta, \tau)Z$ for some $Z \in \mathscr{P}^{\rho_0}$. Now we remark that $\mathrm{rank}_{\mathscr{R}}\,\hat{A}^{10}(0,0; \eta, \tau) = \mathrm{rank}_{\mathscr{R}}\,\hat{A}(0,0; \eta, \tau)$. It follows that ${}^t\hat{A}(0,0; \eta, \tau)\binom{X}{-Z} = 0$. Since the sequence $(3')$ is exact, we can find $\binom{Y}{L}$ belonging to $\mathscr{P}^q = \mathscr{P}^{\rho_1} \oplus \mathscr{P}^{\rho_0}$ with $Y \in \mathscr{P}^{\rho_1}$ and $L \in \mathscr{P}^{\rho_0}$ such that $\binom{X}{-Z} = {}^t\hat{B}(0,0; \eta, \tau)\binom{Y}{L}$. In particular we deduce that

$$X = {}^t\hat{B}^{00}(0,0; \eta, \tau)Y + {}^t\hat{B}^{10}(0,0; \eta, \tau)L,$$

which is what we wanted to prove.

(f) As a consequence of the statements (d) and (e) we derive that if Theorem 1 is true in $(n-1)$ variables, then it is also true in n variables. We have thus established an inductive procedure on the number of variables. Then Theorem 1 will be completely proved if we show that *Theorem 1 is true in the case of one variable* ($n = 1$).

Let (1) be given with Ω open in $\mathbf{R}$. Let $0 \in \Omega$ and let us assume that (3) is exact. Let $\mathscr{R}$ be the field of rational functions in ξ and let $\rho = \mathrm{rank}_{\mathscr{R}} \hat{A}(0, \xi)$. We can suppress $p - \rho$ columns in $A(x, D)$ to obtain a matrix $C(x, D)$ with the property $\mathrm{rank}_{\mathscr{R}} \hat{C}(0, \xi) = \rho$. We consider the new complex

$$\mathscr{E}^p(\omega) \xrightarrow{\;C(x,D)\;} \mathscr{E}^q(\omega) \xrightarrow{\;B(x,D)\;} \mathscr{E}^r(\omega). \qquad (1^*)$$

We remark that because of the condition $\mathrm{rank}_{\mathscr{R}} \hat{C}(0, \xi) = \mathrm{rank}_{\mathscr{R}} \hat{A}(0, \xi)$ the associated complex

$$\mathscr{P}^p \xleftarrow{\;{}^t\hat{C}(0,\xi)\;} \mathscr{P}^q \xleftarrow{\;{}^t\hat{B}(0,\xi)\;} \mathscr{P}^r \qquad (2^*)$$

is also exact. Let $f \in \mathscr{A}_0^q$ be such that $B(x, D)f = 0$. If we can find $v \in \mathscr{A}_0^p$ such that $C(x, D)v = f$, then if we set $u = \binom{v}{0} \in \mathscr{A}_0^p$ we derive that also $A(x, D)u = f$. It is therefore enough to show that the complex (1^*) admits the analytic Poincaré lemma. Now, given $f \in \mathscr{A}_0^q$ with $B(x, D)f = 0$ by virtue of Theorem 0 we can find $v = {}^t(v_1, \ldots, v_\rho)$ with v_j formal power series at the origin such that $C(x, D)v = f$. We can suppress $q - \rho$ rows in the matrix $C(x, D)$ to obtain a $\rho \times \rho$ matrix $E(x, D)$ with the property that $\det \hat{E}(0, \xi) \neq 0$. We do have $E(x, D)v \in \mathscr{A}_0^\rho$. Let $\det \hat{E}(0, \xi) = c_0 \xi^m$, with $c_0 \neq 0$ and let ${}^{co}\hat{E}(x, \xi)$ denote the matrix, adjoint to $\hat{E}(x, \xi)$, which in the place (r, s) has the cofactor of the element of place (s, r) of the matrix $\hat{E}(x, \xi)$. We do have ${}^{co}\hat{E}(0, \xi)\hat{E}(0, \xi) = c_0 \xi^m I$, where I is the identity $\rho \times \rho$ matrix. For x in a small open neighborhood U of 0 in $\mathbf{R}$ we do have

$$L(x, D) = {}^{co}\hat{E}(x, D)E(x, D) = c_0(x)I(d/dx)^m + \text{lower order terms}$$

with $c_0(x) \neq 0$ in U. From $E(x, D)v = g$ one deduces that

$$L(x, D)v = {}^{co}\hat{E}(x, D)g \in \mathscr{A}_0^\rho.$$

But this last system is of Cauchy–Kowalewska because of the above remark. Since we are in one variable all formal solutions of this system must be convergent (as the Cauchy data are constants and thus convergent). Therefore $v_j \in \mathscr{A}_0$ for $1 \leq j \leq \rho$. This proves our contention.

(g) *Appendix*

(α) For every integer $s \geq 0$ we have, according to Lemma 5:

$$\begin{aligned}
(\partial/\partial t)^s A^{00}(y, t; \partial/\partial y, \partial/\partial t) &= Q_s(y, t; \partial/\partial y, \partial/\partial t)A^{10}(y, t; \partial/\partial y, \partial/\partial t) \\
&\quad + R^{(s)}(y, t; \partial/\partial y, \partial/\partial t), \\
(\partial/\partial t)^s B^{00}(y, t; \partial/\partial y, \partial/\partial t) &= P_s(y, t; \partial/\partial y, \partial/\partial t)B^{10}(y, t; \partial/\partial y, \partial/\partial t) \\
&\quad + T^{(s)}(y, t; \partial/\partial y, \partial/\partial t),
\end{aligned} \qquad (*)$$

where $R^{(s)} = (R^{(s)}_{ij}(y, t; \partial/\partial y, \partial/\partial t)$ is of multigrading $(a_j, b_i - s)$ and $R^{(s)}_{ij}(x, D)$ has order in $(\partial/\partial t) < k_j$, $1 \leq j \leq \rho_0$; $T^{(s)} = (T^{(s)}_{ij}(y, t; \partial/\partial y, \partial/\partial t)$ is of multigrading $(b_j, c_i - s)$ and order of $T^{(s)}_{ij}(y, t; \partial/\partial y, \partial/\partial t)$ with respect to $(\partial/\partial t)$ is $< h_j$ for $1 \leq j \leq \rho_1$.

Let us write:

$$R^{(s)}_{ij} = \sum_{\sigma=0}^{k_j-1} R^{(s,\sigma)}_{ij}(y, \partial/\partial y)(\partial/\partial t)^\sigma,$$

$$T^{(s)}_{ij} = \sum_{\sigma=0}^{h_j-1} T^{(s,\sigma)}_{ij}(y, \partial/\partial y)(\partial/\partial t)^\sigma, \qquad \text{for} \quad t = 0.$$

(β)   Let

$$U = (u_{js})_{1 \leq j \leq \rho_0,\, 0 \leq s \leq k_j-1},$$
$$V = (v_{js})_{1 \leq j \leq \rho_1,\, 0 \leq s \leq h_j-1},$$
$$W = (w_{js})_{1 \leq j \leq \rho_2,\, 0 \leq s \leq \lambda_j-1}$$

represent elements in

$$\mathscr{E}^{\nu_0}(\omega), \quad \mathscr{E}^{\nu_1}(\omega), \quad \mathscr{E}^{\nu_2}(\omega), \text{ respectively.}$$

Suppose that $RU = V$. We will verify that

$$v_{is} = \sum_{j=1}^{\rho_0} \sum_{\sigma=0}^{k_j-1} R^{(s,\sigma)}_{ij}(y, \partial/\partial y)u_{j\sigma}. \qquad (**)$$

Similarly, if $TV = W$, we have that

$$w_{is} = \sum_{j=1}^{\rho_1} \sum_{\sigma=0}^{h_j-1} T^{(s,\sigma)}_{ij}(y, \partial/\partial y)v_{j\sigma}. \qquad (***)$$

We remark that

$$R^{(s,\sigma)}_{ij}(y, \partial/\partial y) \text{ is an operator of order } \leq a_j - \sigma - (b_i - s),$$
$$T^{(s,\sigma)}_{ij}(y, \partial/\partial y) \text{ is an operator of order } \leq b_j - \sigma - (c_i - s).$$

To do the direct verification we denote by $u = \rho_A(U)$ the formal power series in t with coefficients in $\mathscr{E}^{\rho_0}(\omega)$ such that

$$A^{10}(y, t; \partial/\partial y, \partial/\partial t)u = 0,$$
$$(\partial/\partial t)^s u_j|_{t=0} = u_{js} \qquad \text{for } 1 \leq j \leq \rho_0, \quad 0 \leq s \leq k_j - 1.$$

Conversely, given $u \in \mathscr{E}^{\rho_0}(\omega)\{\{t\}\}$ we set $U = \tau_A(u)$ with $U = (u_{js})$ and $u_{js} = (\partial/\partial t)^s u_j|_{t=0}$ for $1 \leq j \leq \rho_0$ and $0 \leq s \leq k_j - 1$. Similarly one defines ρ_B, τ_B, and τ_C. We want to show that

$$R(y, \partial/\partial y) = \tau_B A^{00}(y, t; \partial/\partial y, \partial/\partial t)\rho_A.$$

In fact, from relations (∗) we deduce that

$$(\partial/\partial t)^s A^{00}(y, t; \partial/\partial y, \partial/\partial t)(\rho_A(U))$$
$$= Q_s(y, t; \partial/\partial y, \partial/\partial t) A^{10}(y, t; \partial/\partial y, \partial/\partial t)(\rho_A(U))$$
$$+ R^{(s)}(y, t; \partial/\partial y, \partial/\partial t)(\rho_A(U))$$
$$= R^{(s)}(y, t; \partial/\partial y, \partial/\partial t)(\rho_A(U)).$$

Therefore by restriction to $t = 0$, if $V = RU$, we deduce (∗∗). Similarly one proves that $T(y, \partial/\partial y) = \tau_C B^{00}(y, t; \partial/\partial y, \partial/\partial t)\rho_B$ and hence (∗∗∗).

(γ)   We make explicit the isomorphism

$$\lambda_A : \frac{\mathscr{P}^{\rho_0}}{{}^t\hat{A}^{10}(0, 0; \eta, \tau)\,\mathscr{P}^{\rho_0}} \xrightarrow{\sim} {}_{(n-1)}\mathscr{P}^{\nu_0}.$$

Given $X \in \mathscr{P}^{\rho_0}$ we can find a unique $Y = {}^t(Y_1, \ldots, Y_{\rho_0})$ with

$$Y_j = \sum_0^{k_j - 1} Y_{js}(\eta)\tau^s$$

such that, for some $Z \in \mathscr{P}^{\rho_0}$, we have (by Lemma 4)

$$X = {}^t\hat{A}^{10}(0, 0; \eta, \tau)Z + Y.$$

Then $\lambda_A(X) = Y$. Similarly one describes the isomorphisms

$$\lambda_B : \frac{\mathscr{P}^{\rho_1}}{{}^t\hat{B}^{10}(0, 0; \eta, \tau)\mathscr{P}^{\rho_1}} \xrightarrow{\sim} {}_{(n-1)}\mathscr{P}^{\nu_1},$$

$$\lambda_C : \frac{\mathscr{P}^{\rho_2}}{{}^t\hat{C}^{10}(0, 0; \eta, \tau)\mathscr{P}^{\rho_2}} \xrightarrow{\sim} {}_{(n-1)}\mathscr{P}^{\nu_2}.$$

Now given $U(\eta) = (u_{js}(\eta))$ in ${}_{(n-1)}\mathscr{P}^{\nu_1}$ we construct $Z(\eta, \tau) = {}^t(Z_1(\eta, \tau), \ldots, Z_{\rho_1}(\eta, \tau))$ with

$$Z_j(\eta, \tau) = \sum_{s=0}^{h_j - 1} u_{js}(\eta)\tau^s.$$

This is a representative in $\mathscr{P}^{\rho_1}$ of the given element $U(\eta) \in {}_{(n-1)}\mathscr{P}^{\nu_1}$. We then set $X(\eta, \tau) = {}^t\hat{A}^{00}(0, 0; \eta, \tau)Z(\eta, \tau)$. From relations (∗) we deduce

$$ {}^t\hat{A}^{00}(0, 0; \eta, \tau)\tau^s = {}^t\hat{A}^{10}(0, 0; \eta, \tau){}^t\hat{Q}_s(0, 0; \eta, \tau) + {}^t\hat{R}^{(s)}(0, 0; \eta, \tau) $$

and therefore, for the jth component of $X(\eta, \tau)$ we have

$$X_j(\eta, \tau) = \sum_{i=1}^{\rho_1} \sum_{s=0}^{h_i - 1} {}^t\hat{R}_{ij}^{(s)}(0, 0; \eta, \tau)u_{is}(\eta)$$

$$+ \sum_{i=1}^{\rho_1} \sum_{s=0}^{h_i - 1} \sum_{k=1}^{\rho_0} {}^t\hat{A}_{ik}^{10}(0, 0; \eta, \tau){}^t\hat{Q}_{s,kj}(0, 0; \eta, \tau)u_{is}(\eta).$$

But this shows that the following diagram is commutative:

$$
\begin{array}{ccc}
(n-1)\mathscr{P}^{\nu_0} & \xleftarrow{\quad {}^t\hat{R}(0,\eta) \quad} & (n-1)\mathscr{P}^{\nu_1} \\[2pt]
\uparrow \lambda_A & & \uparrow \lambda_B \\[6pt]
\mathscr{P}^{\rho_0} & \xleftarrow{\quad {}^t\hat{A}^{00}(0,0;\eta,\tau) \quad} & \mathscr{P}^{\rho_1} \\[2pt]
{}^t\hat{A}^{10}(0,0;\eta,\tau)\mathscr{P}^{\rho_0} & & {}^t\hat{B}^{10}(0,0;\eta,\tau)\mathscr{P}^{\rho_1}
\end{array}
$$

A similar computation yields the commutativity of the other diagram

$$
\begin{array}{ccc}
(n-1)\mathscr{P}^{\nu_1} & \xleftarrow{\quad {}^t\hat{T}(0,\eta) \quad} & (n-1)\mathscr{P}^{\nu_2} \\[2pt]
\uparrow \lambda_B & & \uparrow \lambda_C \\[6pt]
\mathscr{P}^{\rho_1} & \xleftarrow{\quad {}^t\hat{B}^{00}(0,0;\eta,\tau) \quad} & \mathscr{P}^{\rho_2} \\[2pt]
{}^t\hat{B}^{10}(0,0;\eta,\tau)\mathscr{P}^{\rho_1} & & {}^t\hat{C}^{10}(0,0;\eta,\tau)\mathscr{P}^{\rho_2}
\end{array}
$$

This gives the proof of the isomorphism of the complex (7) with the complex (8).

5. Proof of Theorem 2

(a)   Given a differential operator in an open set $\Omega \subset \mathbf{R}^n$

$$L(x,D):\mathscr{E}^h(\Omega) \to \mathscr{E}^k(\Omega)$$

with coefficients in $\mathscr{E}(\Omega)$ or $\mathscr{A}(\Omega)$, we can consider its formal adjoint (with respect to the Lebesgue's measure dx in $\mathbf{R}^n$):

$$L^*(x,D):\mathscr{E}^k(\Omega) \to \mathscr{E}^h(\Omega)$$

defined by the property:

$$\int^t (L(x,D)u)\bar{v}\,dx = \int^t u\overline{(L^*(x,D)v)}\,dx$$

for any $u \in \mathscr{D}^h(\Omega)$, and $v \in \mathscr{D}^k(\Omega)$. The formal adjoint operator $L^*(x,D)$ is a differential operator with coefficients in $\mathscr{E}(\Omega)$ or $\mathscr{A}(\Omega)$ according to whether the coefficients of $L(x,D)$ are so. If L corresponds to a multigrading (a_j, b_i) then L^* corresponds to a multigrading $(-b_i, -a_h)$, i.e., if $L^* = (L^*_{hi})$ we have order of $L^*_{hi} \leq a_h - b_i$.

(b)   We consider the complex

$$\mathscr{E}^p(\Omega) \xrightarrow{A(x,D)} \mathscr{E}^q(\Omega) \xrightarrow{B(x,D)} \mathscr{E}^r(\Omega), \tag{1}$$

where A corresponds to a multigrading (a_j, b_i) and B corresponds to a multigrading (b_i, c_h). We may assume, adding to the a's, b's, and c's a same constant that all integers $a_1, \ldots, a_p, b_1, \ldots, b_q, c_1, \ldots, c_r$ are > 0 and let m be an upper bound for them.

We set $\Delta = -(\partial/\partial x_1)^2 - \cdots - (\partial/\partial x_n)^2$ to denote the Laplace operator. Then we define

$$\Delta^{2m-a}:\mathscr{E}^p(\Omega) \to \mathscr{E}^p(\Omega) \quad \text{by} \quad \Delta^{2m-a} = \operatorname{diag}\langle\Delta^{2m-a_1}, \ldots, \Delta^{2m-a_p}\rangle,$$
$$\Delta^{m-b}:\mathscr{E}^q(\Omega) \to \mathscr{E}^q(\Omega) \quad \text{by} \quad \Delta^{m-b} = \operatorname{diag}\langle\Delta^{m-b_1}, \ldots, \Delta^{m-b_q}\rangle,$$
$$\Delta^c:\mathscr{E}^r(\Omega) \to \mathscr{E}^r(\Omega) \quad \text{by} \quad \Delta^c = \operatorname{diag}\langle\Delta^{c_1}, \ldots, \Delta^{c_r}\rangle.$$

We then consider the following differential operator:

$$\square(x, D): \mathscr{E}^q(\Omega) \to \mathscr{E}^q(\Omega)$$

defined by

$$\square(x, D) = A(x, D)\Delta^{2m-a}A^*(x, D) + \Delta^{m-b}B^*(x, D)\Delta^c B(x, D)\Delta^{m-b}.$$

The element in the place (i, j) in $A\Delta^{2m-a}A^*$ has the form

$$\sum_s a_{is}(x, D)\Delta^{2m-a_s}a_{sj}^*(x, D)$$

and thus has order

$$\leq a_s - b_i + 4m - 2a_s + a_s - b_j = 4m - b_i - b_j.$$

The element in the place (i, j) in $\Delta^{m-b}B^*\Delta^c B\Delta^{m-b}$ has the form

$$\sum_s \Delta^{m-b_i}b_{is}^*(x, D)\Delta^{c_s}b_{sj}(x, D)\Delta^{m-b_j}$$

and therefore has order

$$\leq 2m - 2b_i + b_i - c_s + 2c_s + b_j - c_s + 2m - 2b_j = 4m - b_i - b_j.$$

Therefore the operator $\square(x, D)$ can be considered as corresponding to the multigrading $(2m - b_j, b_i - 2m)$.

(c)   We want to compute the symbol of $\square$. We set $|\xi|^2 = \sum_1^n \xi_i^2$ and we denote by $\hat{\square}_{rs}$, $\hat{A}_{rs}$, $\hat{A}_{rs}^*$, $\hat{B}_{rs}$, $\hat{B}_{rs}^*$ the elements in place (r, s) in the matrices $\hat{\square}(x, \xi)$, $\hat{A}(x, \xi)$, $\hat{A}^*(x, \xi)$, $\hat{B}(x, \xi)$, $\hat{B}^*(x, \xi)$. We have

$$\hat{A}_{rs}^*(x, \xi) = (-1)^{a_r - b_s}\hat{A}_{sr}(x, \xi),$$
$$\hat{B}_{rs}^*(x, \xi) = (-1)^{b_r - c_s}\overline{\hat{B}}_{sr}(x, \xi).$$

We have thus

$$\hat{\square}_{ik}(x, \xi) = |\xi|^{4m}\Bigg\{(-1)^{b_k} \sum_j |\xi|^{-2a_j}\hat{A}_{ij}(x, \xi)\overline{\hat{A}}_{kj}(x, \xi)$$

$$+ (-1)^{b_k} \sum_\lambda |\xi|^{2(-b_i - b_k + c_\lambda)}\overline{\hat{B}}_{\lambda i}(x, \xi)\hat{B}_{\lambda k}(x, \xi)\Bigg\}.$$

If we set $\xi = \lambda \xi^0$ with $|\xi|^0 = 1$ and $\lambda \in \mathbf{R}$, $\lambda \neq 0$, and if we set $w_k = (-\lambda)^{-b_k} v_k$ for $1 \leq k \leq q$, we get

$$
{}^t\overline{v}\,\widehat{\square}(x, \xi)v = \lambda^{4m}\left\{ \sum_j \widehat{A}_{ij}(x, \xi^0)\overline{\widehat{A}}_{kj}(x, \xi^0)w_i\overline{w}_k \right.
$$

$$
\left. + \sum_l \overline{\widehat{B}}_{li}(x, \xi^0)\widehat{B}_{lk}(x, \xi^0)w_i\overline{w}_k \right\}.
$$

Thus for $\xi \in \mathbf{R}^n - \{0\}$ the matrix $\widehat{\square}(x, \xi)$ is positive semidefinite. By assumption the sequence

$$
\mathbf{C}^p \xrightarrow{\widehat{A}(x_0, \xi)} \mathbf{C}^q \xrightarrow{\widehat{B}(x_0, \xi)} \mathbf{C}^r
$$

is exact for every $\xi \in \mathbf{R}^n - \{0\}$. This shows that for ξ real and different from zero the matrix $\widehat{\square}(x_0, \xi)$ is positive definite since (with respect to the scalar product in $\mathbf{C}^q(u|v) = {}^t\overline{u}v$) we must have for those ξ,

$$
\ker \widehat{A}(x_0, \xi) = \operatorname{Im} \widehat{B}(x_0, \xi) = (\ker {}^t\overline{\widehat{B}}(x, \xi))^0.
$$

But then, in a neighborhood U of x_0 in Ω the matrix $\widehat{\square}(x, \xi)$ is also positive definite. This shows that for $x \in U$, $\xi \in \mathbf{R}^n - \{0\}$, we have $\det \widehat{\square}(x, \xi) \neq 0$. Thus the operator $\square$ in U is elliptic in the sense of Douglis and Nirenberg (see [12, p. 210]).

(d) We now show that the C^∞ Poincaré lemma holds for (1) at the point x_0. Let $f \in \mathscr{E}_{x_0}^q$ be such that $B(x, D)f = 0$. Since $\square(x, D)$ is an elliptic operator in a neighborhood of x_0, we can find $w \in \mathscr{E}_{x_0}^q$ such that $\square(x, D)w = f$. We are looking for a solution $u \in \mathscr{E}_{x_0}^p$ of the equation $A(x, D)u = f$. If we replace u by $u - \Delta^{2m-a}A^*(x, D)w = v$ we are led to find a solution $v \in \mathscr{E}_{x_0}^p$ of the equation

$$
A(x, D)v = \Delta^{m-b}\mu, \tag{*}
$$

where

$$
\mu = B^*(x, D)\Delta^c B(x, D)\Delta^{m-b}w.
$$

Indeed

$$
A(x, D)v = f - A(x, D)\Delta^{2m-a}A^*(x, D)w
$$
$$
= \Delta^{m-b}B^*(x, D)\Delta^c B(x, D)\Delta^{m-b}w.
$$

Now we have $\square(x, D)\mu = 0$. Indeed

$$
\square(x, D)\mu = A(x, D)\Delta^{2m-a}A^*(x, D)B^*(x, D)\Delta^c B(x, D)\Delta^{m-b}w
$$
$$
+ \Delta^{m-b}B^*(x, D)\Delta^c B(x, D)(f - A(x, D)\Delta^{2m-a}A^*(x, D)w),
$$

and this is zero because $A^*(x, D)B^*(x, D) = 0$ [since $B(x, D)A(x, D) = 0$] and because $B(x, D)f = 0$ and $B(x, D)A(x, D) = 0$. It follows that, since $\square(x, D)$ has real analytic coefficients, μ must be real analytic, $\mu \in \mathscr{A}_{x_0}^q$. Then $\Delta^{m-b}\mu \in \mathscr{A}_{x_0}^q$. Moreover $B(x, D)\Delta^{m-b}\mu = 0$. In fact, applying $B(x, D)$ to both sides of the equation $\square(x, D)w = f$, we get

$$B(x, D)\Delta^{m-b}(B^*(x, D)\Delta^c B(x, D)\Delta^{m-b}w) = 0.$$

It follows that Eq. (∗) can be solved by Theorem 1. Thus, setting $u = v + \Delta^{2m-a}A^*(x, D)w$ we get a solution of the equation $A(x, D)u = f$, as we wanted.

6. SOME GENERAL REMARKS ON COMPLEXES OF DIFFERENTIAL OPERATORS

(a)   Let Ω be an open set in $\mathbf{R}^n$ and let $\mathscr{F}(\Omega)$ denote a subring of the ring of C^∞ functions on Ω, stable by differentiation. By this we mean that, if $f \in \mathscr{F}(\Omega)$, then $(\partial/\partial x_1)f \in \mathscr{F}(\Omega), \ldots, (\partial/\partial x_n)f \in \mathscr{F}(\Omega)$. We can have $\mathscr{F}(\Omega) = \mathscr{E}(\Omega)$, the full ring of C^∞ functions on Ω, or $\mathscr{F}(\Omega) = \mathscr{A}(\Omega)$, the ring of real analytic (complex-valued) functions on Ω, or, if $n = 2m$ and $\mathbf{R}^n = \mathbf{C}^m$, $\mathscr{F}(\Omega) = \mathscr{O}(\Omega)$, the ring of holomorphic functions on Ω. We also assume that for every $x_0 \in \Omega$ there is $f \in \mathscr{F}(\Omega)$ with $f(x_0) \neq 0$. Let us consider a complex of differential operators with coefficients in $\mathscr{F}(\Omega)$:

$$\mathscr{F}^{p_0}(\Omega) \xrightarrow{A^0(x,D)} \mathscr{F}^{p_1}(\Omega) \xrightarrow{A^1(x,D)} \mathscr{F}^{p_2}(\Omega) \to \cdots, \tag{1}$$

and let us assume that multigradings $a_1, \ldots, a_{p_0}$ on $\mathscr{F}^{p_0}(\Omega)$, $b_1, \ldots, b_{p_1}$ on $\mathscr{F}^{p_1}$, $c_1, \ldots, c_{p_2}$ on $\mathscr{F}^{p_2}, \ldots$, are given such that $A^0(x, D)$ corresponds to a multigrading (a_j, b_h), $A^1(x, D)$ corresponds to a multigrading $(b_h, c_s), \ldots$.

We denote by $\mathscr{H}(\Omega)$ the graded ring of homogeneous polynomials in the variables $\xi = (\xi_1, \ldots, \xi_n)$ with coefficients in $\mathscr{F}(\Omega)$:

$$\mathscr{H}(\Omega) = \mathscr{F}(\Omega)_0[\xi_1, \ldots, \xi_n],$$

the subscript 0 indicating that the ring is graded. Taking symbols we can consider the complex of multigraded $\mathscr{H}(\Omega)$-homomorphisms "associated to the symbolic complex of (1)":

$$\mathscr{H}(\Omega)^{p_0} \xleftarrow{{}^t\hat{A}^0(x,\xi)} \mathscr{H}(\Omega)^{p_1} \xleftarrow{{}^t\hat{A}^1(x,\xi)} \mathscr{H}(\Omega)^{p_2} \leftarrow \cdots.$$

Let $N_\Omega = \operatorname{coker} {}^t\hat{A}^0(x, \xi)$, so that we have a complex of $\mathscr{H}(\Omega)$-homomorphisms with exactness at $\mathscr{H}(\Omega)^{p_0}$:

$$0 \leftarrow N_\Omega \leftarrow \mathscr{H}(\Omega)^{p_0} \xleftarrow{{}^t\hat{A}^0(x,\xi)} \mathscr{H}(\Omega)^{p_1} \xleftarrow{{}^t\hat{A}^1(x,\xi)} \mathscr{H}(\Omega)^{p_2} \leftarrow \cdots. \tag{2'}$$

One can consider (2') as a free $\mathcal{H}(\Omega)$-resolution of the $\mathcal{H}(\Omega)$-module N_Ω, provided (2') is an exact sequence. We will therefore make the following assumption.

ASSUMPTION.   *The sequence* (2') *is an exact sequence of* $\mathcal{H}(\Omega)$-*homomorphisms.*

We say then that the complex (1) with the given multigrading is a *correct complex.*

(b)   Let (1) be a correct complex and let $x_0 \in \Omega$ be given. Let $\mathfrak{M}_{x_0}(\Omega)$ be the ideal of functions in $\mathcal{F}(\Omega)$ vanishing at x_0. This is the kernel of the multiplicative linear functional:

$$\delta_{x_0} : \mathcal{F}(\Omega) \to \mathbf{C}$$

represented by Dirac's measure at x_0, i.e., evaluation at x_0. Because of the previous assumption δ_{x_0} is surjective and thus $\mathbf{C} = \mathcal{F}(\Omega)/\mathfrak{M}_{x_0}(\Omega)$ can be considered as a $\mathcal{F}(\Omega)$-module. Tensoring the sequence (2) by the above $\mathcal{F}(\Omega)$-module $\mathbf{C}$ we obtain the complex

$$\mathcal{H}^{p_0} \xleftarrow{\ {}^t\tilde{A}^0(x_0, \xi)\ } \mathcal{H}^{p_1} \xleftarrow{\ {}^t\tilde{A}^1(x_0, \xi)\ } \mathcal{H}^{p_2} \leftarrow \cdots, \tag{3}$$

where $\mathcal{H} = \mathbf{C}_0[\xi_1, \ldots, \xi_n]$, the ring of homogeneous polynomials in n variables $\xi = (\xi_1, \ldots, \xi_n)$, with coefficients in $\mathbf{C}$, is here identified to $\mathcal{H} = \mathcal{H}(\Omega) \otimes_{\mathcal{F}(\Omega)} (\mathcal{F}(\Omega)/\mathfrak{M}_{x_0}(\Omega))$. If the complex (1) is a correct complex, then the homology of (3) is given by the $\mathcal{F}(\Omega)$-modules

$$\mathrm{Tor}^j_{\mathcal{F}(\Omega)}(N_\Omega, \mathcal{F}(\Omega)/\mathfrak{M}_{x_0}(\Omega)) \qquad \text{for} \quad j \geq 1.$$

We have therefore the following.

PROPOSITION 3.   *Given a correct complex* (1) *of differential operators with coefficients in* $\mathcal{F}(\Omega)$, *the necessary and sufficient condition that its symbolic complex be exact at* x_0 *in the jth place* (*i.e., that*

$$\mathscr{E}^{p_j - 1}(\mathbf{R}^n) \xrightarrow{\tilde{A}^{j-1}(x_0, D)} \mathscr{E}^{p_j}(\mathbf{R}^n) \xrightarrow{\tilde{A}^j(x, D)} \mathscr{E}^{p_j + 1}(\mathbf{R}^n)$$

be exact) *is that*

$$\mathrm{Tor}^j_{\mathcal{F}(\Omega)}(N_\Omega, \mathcal{F}(\Omega)/\mathfrak{M}_{x_0}(\Omega)) = 0$$

(*note that* $j \geq 1$).

(c)   We will assume from now on that $\mathcal{F}(\Omega)$ is one of the rings $\mathscr{E}(\Omega)$, $\mathscr{A}(\Omega)$ (Ω open in $\mathbf{R}^n$), $\mathcal{O}(\Omega)$ (Ω an open set of holomorphy in $\mathbf{R}^n = \mathbf{C}^m$).

PROPOSITION 4.   *Under the above specified assumptions we have*

$$\text{Tor}^{j}_{\mathscr{F}(\Omega)}(N_{\Omega}, \mathscr{F}(\Omega)/\mathfrak{M}_{x_0}(\Omega)) = 0$$

for any choice of $x_0 \in \Omega$ *provided*

$j > n$ *if* $\mathscr{F}(\Omega) = \mathscr{E}(\Omega)$ *or* $\mathscr{F}(\Omega) = \mathscr{A}(\Omega)$,

$j > m = n/2$ *if* $\mathscr{F}(\Omega) = \mathcal{O}(\Omega)$, Ω *open set of homomorphy in* $\mathbf{C}^m = \mathbf{R}^n$.

Proof.   We give the proof in the case $\mathscr{F}(\Omega) = \mathscr{A}(\Omega)$. The proof in the other two cases is similar. Let $\mathscr{A}$ be the sheaf of germs of real analytic functions and let $\mathfrak{M}_{x_0}$ be the subsheaf of ideals given by the germs vanishing at $x_0 \in \Omega$. We have an exact sequence of sheaves (Koszul complex)

$$0 \to \mathscr{A} \to \mathscr{A}^{c_{n,n-1}} \to \mathscr{A}^{c_{n,n-2}} \to \cdots \to \mathscr{A}^{c_{n,2}} \to \mathscr{A}^{c_{n,1}} \to \mathfrak{M}_{x_0} \to 0,$$

where $c_{n,i}$ denotes the binomial coefficient $n(n-1)\cdots(n-i+1)/i!$. From this we deduce a free resolution of $\mathscr{A}(\Omega)/\mathfrak{M}_{x_0}(\Omega)$ of length n. This entails the stated result. ∎

PROPOSITION 5.   *Let Ω be open and connected and let $\mathscr{F}(\Omega)$ denote the ring $\mathscr{A}(\Omega)$ or $\mathcal{O}(\Omega)$ (with Ω of holomorphy and $\mathbf{R}^n = \mathbf{C}^m$). Let the complex (1) be a correct complex. Then for every $j \geq 1$ the set*

$$A_j = \{x_0 \in \Omega \mid \text{Tor}^{j}_{\mathscr{F}(\Omega)}(N_{\Omega}, \mathscr{F}(\Omega)/\mathfrak{M}_{x_0}(\Omega)) \neq 0\}$$

is a proper analytic subset of Ω.

Proof.   (α)   We will first prove that A_j is an analytic set. To that purpose we first consider the following situation:

Let $M(\xi): \mathscr{H}^p \to \mathscr{H}^q$ be a homogeneous $q \times p$ matrix of multigrading (a_j, b_i). By this we mean that $M(\xi) = (m_{ij}(\xi))$ has the entry $m_{ij}(\xi)$ homogeneous in ξ of degree $a_j - b_i$. (Here $\mathscr{H} = \mathbf{C}_0[\xi_1, \ldots, \xi_n]$ denotes the graded ring of homogeneous polynomials in the indeterminates $\xi = (\xi_1, \ldots, \xi_n)$.)

LEMMA 6.   *Given n, p, q and the integers $a_1, \ldots, a_p, b_1, \ldots, b_q$, we can find an integer $\sigma \geq 0$ with the property: For any homogeneous matrix $M(\xi)$ of multigrading (a_j, b_i), $\ker M(\xi)$ has a set of generators with components homogeneous polynomials of degree $\leq \sigma$.*

Proof of Lemma 6.   For h an integer we set

$$(\mathscr{H}^p)_h = \{v = {}^t(v_1, \ldots, v_p) \in \mathscr{H}^p \mid v_j \text{ is homogeneous of degree } h - a_j$$
$$\text{for } 1 \leq j \leq p\},$$

$$(\mathcal{H}^q)_h = \{w = {}^t(w_1, \ldots, w_p) \in \mathcal{H}^q | w_i \text{ is homogeneous of degree } h - b_i$$
$$\text{for } 1 \leq i \leq q\},$$

so that $M(\xi)$ sends, for every h, $(\mathcal{H}^p)_h$ into $(\mathcal{H}^q)_h$. We assume first that $n = 1$. We will show that, if $Y \in \mathcal{H}^p$ belongs to a minimal set of generators of ker $M(\xi)$, then $Y \in (\mathcal{H}^p)_h$ with $h \leq \mu = \sup_j a_j$.

Indeed, if $Y \in (\mathcal{H}^p)_{\mu+k}$ with $k > 0$, each component Y_i of Y has a degree $\mu + k - a_j \geq k > 0$. Thus we can write $Y = \xi Z$ with $Z \in (\mathcal{H}^p)_{\mu+k-1}$. If $Y \in$ ker $M(\xi)$, then also $Z \in$ ker $M(\xi)$, but then Y cannot belong to a minimal basis of ker $M(\xi)$. In fact, expressing Z in terms of a basis of ker $M(\xi)$ containing Y, we see that Z is a combination of the other basis vectors except Y, by reasons of degree. Then Y is redundant in the basis and the basis is not minimal.

Thus we can proceed by induction on the number of variables ξ. We can assume that $M(\xi) \neq 0$, since, for $M(\xi) = 0$, ker $M(\xi) = \mathcal{H}^p$ and the lemma is trivial. Let $\rho \geq 1$ be the rank or $M(\xi)$ and, after a suitable linear change of the variable ξ we may assume that

(i)   the minor determinant $D(\xi)$ of the first ρ rows and columns of $M(\xi)$ is a nonzero polynomial of a certain degree $\mu > 0$, monic with respect to ξ_1 (the lemma is trivial also when the matrix $M(\xi)$ is constant);

(ii)   the degree μ is maximal among the degrees of all minor determinants of order ρ extracted from $M(\xi)$.

Certainly we have $\mu \leq \sum_1^q \sum_1^p |a_j - b_i| = \sigma_1$. Let

$$M(\xi) = \begin{pmatrix} R_1 \\ \vdots \\ R_q \end{pmatrix}.$$

We have

$$\ker M(\xi) = \{X \in \mathcal{H}^p | R_1 X = 0, \ldots, R_q X = 0\}.$$

Moreover all minors of order $\rho + 1$ of the matrix

$$\begin{pmatrix} R_i \\ R_1 \\ \vdots \\ R_\rho \end{pmatrix}, \qquad \text{for} \quad 1 \leq i \leq p,$$

have zero determinant; in particular the minors obtained by bordering the minor of $D(\xi)$ with one row and one column. This provides us with a system of

$p - \rho$ homogeneous vectors in ker $M(\xi)$ of the form

$$H_1(\xi) = \begin{pmatrix} h_{11}(\xi) \\ \vdots \\ h_{\rho 1}(\xi) \\ D(\xi) \\ 0 \\ \vdots \\ 0 \end{pmatrix}, \quad H_2(\xi) = \begin{pmatrix} h_{12}(\xi) \\ \vdots \\ h_{\rho 2}(\xi) \\ 0 \\ D(\xi) \\ \vdots \\ 0 \end{pmatrix}, \quad \ldots, \quad H_{p-\rho}(\xi) = \begin{pmatrix} h_{1\,p-\rho}(\xi) \\ \vdots \\ h_{\rho\,p-\rho}(\xi) \\ 0 \\ 0 \\ \vdots \\ D(\xi) \end{pmatrix}.$$

All components of these vectors have degree $\leq \mu$, because of assumption (ii) (cf. [10] and [3]).

Now every element of ker $M(\xi)$ is a combination over H of the vectors $H_1, \ldots, H_{p-\rho}$ and of a vector $L(\xi) \in \mathscr{H}^p$ having all components of degree in the variable $\xi_1 < \mu$. We write

$$L(\xi) = {}^t(L_1(\xi), \ldots, L_p(\xi))$$

with

$$L_j(\xi) = \sum_0^{\mu-1} \lambda_{jt}(\xi_2, \ldots, \xi_n)\xi_1^t, \qquad 1 \leq j \leq p,$$

$$M(\xi) = \sum_0^l M^{(h)}(\xi_2, \ldots, \xi_n)\xi_1^h, \qquad \text{where} \quad l \leq \sigma_1,$$

with $M^{(h)}$ homogeneous of multigrading $(a_j - h, b_i)$. We then have for

$$S(\xi) = {}^t(S_1(\xi), \ldots, S_q(\xi)) = M(\xi)L(\xi),$$

$$S_i(\xi) = \sum_0^{l+\mu-1} s_{i\eta}(\xi_2, \ldots, \xi_n)\xi_1^\eta, \qquad 1 \leq i \leq q,$$

that

$$s_{i\eta} = \sum_{j=1}^p \sum_{t=0}^{\mu-1} M_{ij}^{(\eta,t)}\lambda_{jt} \tag{*}$$

with $M_{ij}^{(\eta,t)} = M_{ij}^{(\eta-1)}$ homogeneous of degree $(a_j - \eta) - (b_i - t)$. Let $_{(n-1)}\mathscr{H} = C_0[\xi_2, \ldots, \xi_n]$ be the ring of homogeneous polynomials in the $n - 1$ indeterminates $\xi_2, \ldots, \xi_n$. Then (*) defines a homogeneous map $\tau: {}_{(n-1)}\mathscr{H}^{p\mu} \to {}_{(n-1)}\mathscr{H}^{(l+\mu)q}$. By the inductive assumption ker τ has a set of generators with all components of degree $\leq \sigma_2(n-1, p\mu, (l+\mu)q, \ldots, a_j - \eta, \ldots, b_i - t, \ldots)$, and we can take the supremum of these numbers for $l \leq \sigma_1, \mu \leq \sigma_1$. Then we can take $\sigma = (\sigma_2 + \sigma_1 - 1)$.

(β) Suppose that we have a homogeneous complex

$$\mathscr{H}^p \xrightarrow{\;M(\xi)\;} \mathscr{H}^q \xrightarrow{\;N(\xi)\;} \mathscr{H}^r \qquad (*)$$

with M homogeneous of multigrading (a_j, b_i) and N homogeneous of multigrading (b_i, c_h). For every integer h set

$$(\mathscr{H}^p)_h = \left\{ v = \begin{pmatrix} v_1 \\ \vdots \\ v_p \end{pmatrix} \in \mathscr{H}^p \,\middle|\, v_j \text{ is homogeneous of degree } h - a_j \text{ for } 1 \le j \le p \right\},$$

$$(\mathscr{H}^q)_h = \left\{ w = \begin{pmatrix} w_1 \\ \vdots \\ w_q \end{pmatrix} \in \mathscr{H}^q \,\middle|\, w_i \text{ is homogeneous of degree } h - b_i \text{ for } 1 \le i \le q \right\},$$

$$(\mathscr{H}^r)_h = \left\{ z = \begin{pmatrix} z_1 \\ \vdots \\ z_r \end{pmatrix} \in \mathscr{H}^r \,\middle|\, z_h \text{ is homogeneous of degree } h - c_h \text{ for } 1 \le h \le r \right\}.$$

Then for every h we have a complex

$$(\mathscr{H}^p)_h \xrightarrow{\;M(\xi)\;} (\mathscr{H}^q)_h \xrightarrow{\;N(\xi)\;} (\mathscr{H}^r)_h. \qquad (*)_h$$

From the previous lemma we deduce the following:

COROLLARY. *There exists an integer h_0, depending only on n, q, r and the integers $b_1, \ldots, b_q, c_1, \ldots, c_r$ with the following property: The complex $(*)$ is exact if and only if the complexes $(*)_h$ are exact for $h \le h_0$.*

Proof. Indeed exactness of $(*)$ implies exactness of $(*)_h$ for every h. Conversely there is an integer $\sigma = \sigma(n, q, r, \ldots, b_i, \ldots, c_h, \ldots)$ such that $\ker N(\xi)$ has a set of generators with all components of degree $\le \sigma$. Set $h_0 = \sup \sigma + b_i \,(1 \le i \le q)$. Then the exactness of $(*)_h$ for $h \le h_0$ implies the exactness of $(*)$ because then Im $M(\xi)$ contains a set of generators for $\ker N(\xi)$.

(γ) We apply the above considerations to the "symbolic complex" at the jth step:

$$\mathscr{H}(\Omega)^{p_{j-1}} \xleftarrow{\;{}^t\hat{A}^{j-1}(x,\xi)\;} \mathscr{H}(\Omega)^{p_j} \xleftarrow{\;{}^t\hat{A}^{j}(x,\xi)\;} \mathscr{H}(\Omega)^{p_{j+1}}. \qquad (\circ)$$

We have $\mathrm{Tor}^j_{\mathscr{F}(\Omega)}(N_\Omega, \mathscr{F}(\Omega))/\mathfrak{M}_{x_0}(\Omega)) = 0$ if and only if the complex

$$\mathscr{H}^{p_{j-1}} \xleftarrow{\;{}^t\hat{A}^{j-1}(x_0,\xi)\;} \mathscr{H}^{p_j} \xleftarrow{\;{}^t\hat{A}^{j}(x_0,\xi)\;} \mathscr{H}^{p_{j+1}} \qquad (*)$$

is exact. With obvious notations, in view of the above corollary there is an integer h_0, independent of x_0, such that $(*)$ is exact if and only if

$$(\mathscr{H}^{p_{j-1}})_h \xleftarrow{\;{}^t\hat{A}^{j-1}(x_0,\xi)\;} (\mathscr{H}^{p_j})_h \xleftarrow{\;{}^t\hat{A}^{j}(x_0,\xi)\;} (\mathscr{H}^{p_{j+1}})_h \qquad (*)_h$$

is exact for all $h \leq h_0$. We are thus reduced to prove a statement of the following type:

Consider a sequence of linear maps

$$\mathbf{C}^p \xrightarrow{A(x)} \mathbf{C}^q \xrightarrow{B(x)} \mathbf{C}^r, \qquad x \in \Omega, \qquad\qquad (**)$$

where $A(x)$ is a $q \times p$ matrix with real analytic (resp. holomorphic) entries for $x \in \Omega$ and similarly $B(x)$ an $r \times q$ matrix with real analytic (resp., holomorphic) entries for $x \in \Omega$. Assume that for every $x \in \Omega$ $(**)$ is a complex, i.e., $B(x)A(x) = 0 \; \forall x \in \Omega$. Then the set A of points $x \in \Omega$ where $(**)$ is not exact is analytic.

Let $\mathcal{K}(\Omega)$ denote the quotient field of $\mathcal{A}(\Omega)$ [resp., $\mathcal{O}(\Omega)$] and let $\mathrm{rank}_{\mathcal{K}(\Omega)} A(x) = \rho$, $\mathrm{rank}_{\mathcal{K}(\Omega)} B(x) = \sigma$. For every $x_0 \in \Omega$ we have

$$\mathrm{rank}_{\mathbf{C}} A(x_0) \leq \rho, \qquad \mathrm{rank}_{\mathbf{C}} B(x_0) \leq \sigma.$$

Moreover, since $(**)$ is a complex, we have $\rho + \sigma \leq q$. Exactness of $(**)$ is equivalent to the conditions:

$$\mathrm{rank}_{\mathbf{C}} A(x) = \rho, \qquad \mathrm{rank}_{\mathbf{C}} B(x) = \sigma, \qquad \rho + \sigma = q.$$

Therefore, either $\rho + \sigma < q$, and then $A = \Omega$, or $\rho + \sigma = q$ and then

$$A = \{x \in \Omega \,|\, \mathrm{rank}_{\mathbf{C}} A(x) < \rho\} \cup \{x \in \Omega \,|\, \mathrm{rank}_{\mathbf{C}} B(x) < \sigma\}.$$

In both cases A is analytic. We have thus proved that A_j is analytic for every $j \geq 1$.

(δ) It remains to show that $A_j \subsetneqq \Omega$. For this we will use the assumption that (1) is a correct complex. It will be enough to show that if ($\circ$) is exact then there exists a proper analytic subset $C \subsetneqq \Omega$ such that, for $x_0 \in \Omega - C$, the sequence ($*$) is exact. We will make use of the following considerations.

Let $A(x, \xi) = (a_{ij}(x, \xi))_{1 \leq i \leq q, \, 1 \leq j \leq p}$ be a $q \times p$ matrix with entries homogeneous polynomials in the indeterminates $\xi = (\xi_1, \ldots, \xi_l)$ with coefficients in $\mathcal{F}(\Omega)$. We will assume the matrix homogeneous of multigrading (a_j, b_i), i.e., degree in ξ of $a_{ij}(x, \xi) = a_j - b_i$. The desired statement is a consequence of the following lemma for the case $l = n$.

LEMMA 7. *We can find a homogeneous matrix $B(x, \xi) = (b_{rs}(x, \xi))$ with entries homogeneous polynomials in ξ with coefficients in $\mathcal{F}(\Omega)$, of type $p \times r$ (for some r) and of some multigrading (c_s, a_r) such that:*

(i) *Setting $\mathcal{H}_\Omega = \mathcal{F}(\Omega)_0[\xi_1, \ldots, \xi_l]$ the sequence of maps*

$$\mathcal{H}_\Omega^r \xrightarrow{B(x, \xi)} \mathcal{H}_\Omega^p \xrightarrow{A(x, \xi)} \mathcal{H}_\Omega^q$$

is a complex, i.e., $A(x, \xi)B(x, \xi) = 0$.

Furthermore we can find a proper analytic subset $C \subsetneqq \Omega$ such that

(ii)   *for any $x_0 \in \Omega - C$ and for any vector $S(\xi) = {}^t(s_1(\xi), \ldots, s_p(\xi))$ with $s_j(\xi)$ homogeneous polynomial of degree $h - a_j$ (some h integer) and with $S(\xi) \in \ker A(x_0, \xi)$ we can find homogeneous polynomials $q_1(\xi), \ldots, q_r(\xi)$ of suitable degrees such that*

$$S(\xi) = B(x_0, \xi) \begin{pmatrix} q_1(\xi) \\ \vdots \\ q_r(\xi) \end{pmatrix}.$$

Proof of Lemma 7.   We treat first the case $l = 0$. Then $A = A(x)$ is a matrix $(a_{ij}(x))$ with entries in $\mathscr{F}(\Omega)$. Let $\mathscr{K}(\Omega)$ be the quotient field of $\mathscr{F}(\Omega)$ and let $\rho = \operatorname{rank}_{\mathscr{K}(\Omega)} A(x)$. We may assume that the minor determinant $D(x)$ of the first ρ rows and columns of A is not identically zero. Set

$$A(x) = \begin{pmatrix} R_1 \\ \vdots \\ R_q \end{pmatrix}.$$

We have $\ker A(x) = \{X \in \mathscr{F}(\Omega)^p \,|\, R_1 X = 0, \ldots, R_\rho X = 0\}$. Moreover, all minors of order $\rho + 1$ of the matrix

$$\begin{pmatrix} R_i \\ R_1 \\ \vdots \\ R_\rho \end{pmatrix}, \qquad \text{for} \quad 1 \leq i \leq \rho,$$

have zero determinant; in particular, those obtained by bordering the minor of $D(x)$. This provides us with $p - \rho$ elements of $\ker A(x)$ that we write as columns of a matrix $B(x)$ of type

$$\begin{pmatrix} h_{11}(x) & h_{12}(x) & \cdots & h_{1\,p-\rho}(x) \\ \vdots & \vdots & \cdots & \vdots \\ h_{\rho 1}(x) & h_{\rho 2}(x) & \cdots & h_{\rho\,p-\rho}(x) \\ D(x) & 0 & \cdots & 0 \\ 0 & D(x) & \cdots & 0 \\ \vdots & \vdots & \cdots & \vdots \\ 0 & 0 & \cdots & D(x) \end{pmatrix}.$$

By construction we have $A(x)B(x) = 0$. We can thus consider the complex

$$\mathscr{F}(\Omega)^{p-\rho} \xrightarrow{\ B(x)\ } \mathscr{F}(\Omega)^p \xrightarrow{\ A(x)\ } \mathscr{F}(\Omega)^q$$

and we set $C = \{x \in \Omega \,|\, D(x) = 0\}$, so that $C \subsetneqq \Omega$ is a proper analytic subset of Ω. If $x_0 \in \Omega - C$ we have $\operatorname{rank}_{\mathbf{C}} A(x_0) = \rho$ and $\operatorname{rank}_{\mathbf{C}} B(x_0) = p - \rho$. Thus

$\text{rank}_{\mathbf{C}} A(x_0) + \text{rank}_{\mathbf{C}} B(x_0) = p$ and therefore the sequence

$$\mathbf{C}^{p-\rho} \xrightarrow{\; B(x_0) \;} \mathbf{C}^p \xrightarrow{\; A(x_0) \;} \mathbf{C}^q$$

is exact. This shows that also the conclusion (ii) of the lemma holds. The lemma is thus established when $l = 0$. We can proceed now by induction on the number l of indeterminates ξ. Let $\rho = \text{rank}_{\mathscr{H}(\Omega)(\xi)} A(x, \xi)$, and set

$$A(x, \xi) = \begin{pmatrix} R_1 \\ \vdots \\ R_q \end{pmatrix}.$$

We can assume that the matrix of the first ρ rows has rank ρ over $\mathscr{H}(\Omega)(\xi)$. Set

$$A'(x, \xi) = \begin{pmatrix} R_1 \\ \vdots \\ R_\rho \end{pmatrix} = (S_1, \ldots, S_p),$$

and let $D(x, \xi) = \det(S_1, \ldots, S_\rho)$. We can assume that (i) $D(x, \xi) \neq 0$; (ii) $r = \text{degree in } \xi_1 \text{ of } D(x, \xi) \geq \text{degree in } \xi_1 \text{ of } \det(S_{i_1}, \ldots, S_{i_\rho})$ for every choice of $\{i_1, \ldots, i_\rho\}$ in $\{1, \ldots, p\}$; (iii) $D(x, \xi) = a_0(x)\xi_1^r + \text{lower order terms in } \xi_1$, with $a_0(x) \neq 0$.

We set $C_1 = \{x \in \Omega \,|\, a_0(x) = 0\}$, so that $C_1 \subsetneqq \Omega$ is a proper analytic subset of Ω. As in the case $l = 0$ we construct a homogeneous $p \times (p - \rho)$ matrix of suitable multigrading (c_s, a_r) of the form

$$B_1(x, \xi) = \begin{pmatrix} * & * & \cdots & * \\ \vdots & \vdots & \cdots & \vdots \\ * & * & \cdots & * \\ D & 0 & \cdots & 0 \\ 0 & D & \cdots & 0 \\ \vdots & \vdots & \cdots & \vdots \\ 0 & 0 & \cdots & D \end{pmatrix}$$

and such that $A(x, \xi)B_1(x, \xi) = 0$. We have to establish property (ii). Let $x_0 \in \Omega - C_1$ be fixed and let $X(\xi) = {}^t(X_1(\xi), \ldots, X_p(\xi)) \in \ker A(x_0, \xi)$. Since $a_0(x_0) \neq 0$ we can divide $X_{\rho+s}(\xi)$ by $D(x_0, \xi)$ as polynomials in ξ_1, so that

$$X_{\rho+s}(\xi) = D(x_0, \xi)L_{\rho+s}(\xi) + Y_{\rho+s}(\xi)$$

for $1 \leq s \leq p - \rho$ with $Y_{\rho+s}(\xi)$ of degree $< r$ in ξ_1. Setting then

$$\begin{pmatrix} Y_1(\xi) \\ \vdots \\ Y_p(\xi) \end{pmatrix} = \begin{pmatrix} X_1(\xi) \\ \vdots \\ X_p(\xi) \end{pmatrix} - B_1(x_0, \xi) \begin{pmatrix} L_{1+\rho}(\xi) \\ \vdots \\ L_p(\xi) \end{pmatrix}$$

one realizes (see [10], see also [3]) that for every s with $1 \le s \le p$ degree in ξ_1 of $Y_s(\xi) < r$. Set

$$Y(\xi) = \theta_1(\xi_2, \ldots, \xi_l)\xi_1^{r-1} + \cdots + \theta_r(\xi_2, \ldots, \xi_l),$$
$$A(x_0, \xi) = \sigma_1(x_0; \xi_2, \ldots, \xi_l)\xi_1^t + \cdots + \sigma_{t+1}(x_0; \xi_2, \ldots, \xi_l).$$

From the condition $A(x_0, \xi)Y(\xi) = 0$ we deduce then a homogeneous system in $l - 1$ indeterminates $\xi_2, \ldots, \xi_l$

$$\sigma_1\theta_1 = 0,$$
$$\sigma_2\theta_1 + \sigma_1\theta_2 = 0$$
$$\vdots$$

with a homogeneous matrix

$$\sigma(x_0; \xi_2, \ldots, \xi_l) = \begin{pmatrix} \sigma_1 & & & & 0 \\ & \ddots & & & \\ & & \sigma_1 & & \\ \sigma_{t+1} & & & & \\ & \ddots & & & \\ 0 & & \sigma_{t+1} & & \end{pmatrix}.$$

By the inductive assumption we can find a set of elements with coefficients in $\mathscr{F}(\Omega)$,

$$\theta^{(1)}(x_0; \xi_2, \ldots, \xi_l), \quad \ldots, \quad \theta^{(k)}(x_0; \xi_2, \ldots, \xi_l)$$

with $\theta^{(j)} \in \ker \sigma\ \forall x_0 \in \Omega$ and a proper analytic set $C_2 \subsetneqq \Omega$ such that, if we set

$$Y^{(\alpha)}(x_0, \xi) = \theta_1^{(\alpha)}(x_0; \xi_2, \ldots, \xi_l)\xi_1^{r-1} + \cdots + \theta_r^{(\alpha)}(x_0; \xi_2, \ldots, \xi_l), \quad 1 \le \alpha \le \mu,$$

for fixed $x_0 \notin C_2$ every syzygy of the type $Y(\xi)$ is a combination of the $Y^{(\alpha)}(x_0, \xi)$ for $1 \le \alpha \le \mu$. We have then to take $C = C_1 \cup C_2$ and for $B(x, \xi) = (B_1(x, \xi), \Phi(x, \xi))$ where

$$\Phi(x, \xi) = (Y^{(1)}(x, \xi), \ldots, Y^{(\mu)}(x, \xi))$$

to be in the condition of property (ii) for every $x_0 \notin C$. This establishes the lemma and therefore also our statement.

Remark. In the case $\mathscr{F}(\Omega) = \mathscr{E}(\Omega)$ one would obtain by a similar argument that A_j is the zero set of a finite set of functions of $\mathscr{E}(\Omega)$. In particular A_j will be always a closed set. It follows then that if the complex (1) is correct and if at a point $x_0 \in \Omega$ the associated symbolic complex is exact, we can find a neighborhood U of x_0 in Ω such that the associated symbolic complex is still exact for any fixed $y \in U$.

(d)   Let $\mathscr{F}(\Omega)$ denote the ring $\mathscr{A}(\Omega)$ or $\mathcal{O}(\Omega)$ (for Ω open in $\mathbf{C}^m = \mathbf{R}^n$) and let K be a compact set. We define

$$\mathscr{F}(K) = \varinjlim_{K \subset \Omega} \mathscr{F}(\Omega)$$

the direct limit being taken under the natural restriction maps.

Let the differential operator $A^i(x, D)$ be defined in some open neighborhood U_i of K with coefficients in $\mathscr{F}(U_i)$ $(i = 0, 1, 2, \ldots)$. Instead of the complex (1) we can then consider the complex

$$\mathscr{F}(K)^{p_0} \xrightarrow{A^0(x,D)} \mathscr{F}(K)^{p_1} \xrightarrow{A^1(x,D)} \mathscr{F}(K)^{p_2} \to \cdots. \qquad (1)_K$$

Correspondingly we consider the graded ring $\mathscr{H}(K) = \mathscr{F}(K)_0[\xi_1, \ldots, \xi_n]$ of homogeneous polynomials in the $\xi = (\xi_1, \ldots, \xi_n)$ with coefficients in $\mathscr{F}(K)$. We can then consider the "associated symbolic complex"

$$0 \leftarrow N_K \leftarrow \mathscr{H}(K)^{p_0} \xleftarrow{{}^t\hat{A}^0(x,\xi)} \mathscr{H}(K)^{p_1} \xleftarrow{{}^t\hat{A}^1(x,\xi)} \mathscr{H}(K)^{p_2} \leftarrow \cdots, \qquad (2)_K$$

where $N_K = \mathrm{coker}({}^t\hat{A}^0(x, \xi) : \mathscr{H}(K)^{p_1} \to \mathscr{H}(K)^{p_0})$. We say that the complex $(1)_K$ is correct if $(2)_K$ is an exact sequence. We have then analogous statements to Propositions 3, 4, and 5. In particular, assuming $(1)_K$ to be a correct complex, the set

$$A_j = \{x_0 \in K \,|\, \mathrm{Tor}^j_{\mathscr{F}(K)}(N_K, \mathscr{F}(K)/\mathfrak{M}_{x_0}(K)) \neq 0,$$

where the sequence (3) fails to be exact at the jth place is a proper analytic subset defined in some neighborhood of K. Here $\mathfrak{M}_{x_0}(K)$ denotes the ideal of $\mathscr{F}(K)$ of functions vanishing at x_0. The advantage to deal with $\mathscr{F}(K)$ rather than with $\mathscr{F}(\Omega)$ is given by a theorem of Frish ([8], see also [6]) from which we deduce the following statements:

Let K be a compact semianalytic subset of $\mathbf{R}^n$. Then the ring $\mathscr{A}(K)$ is a Noetherian ring.

Let $K \subset \mathbf{C}^m$ be a semianalytic compact set of holomorphy (i.e., having a fundamental system of open neighborhood of holomorphy). Then the ring $\mathcal{O}(K)$ is Noetherian.

In particular, for K semianalytic of this sort, the graded rings $\mathscr{H}(K)$ are Noetherian rings.

7. EXISTENCE OF CORRECT RESOLUTIONS; INVOLUTIVE OPERATORS

(a)   By K we denote a semianalytic compact subset of $\mathbf{R}^n$, and when $\mathbf{R}^n = \mathbf{C}^m$ we will also assume that K is a compact of holomorphy. By $\mathscr{F}(K)$ we denote the ring $\mathscr{A}(\Omega)$ or, if $\mathbf{R}^n = \mathbf{C}^m$, the ring $\mathcal{O}(K)$, so that according to the theorem of Frish $\mathscr{F}(K)$ is a Noetherian ring. By $\mathfrak{D}(K) = \mathscr{F}(K)[D_1, \ldots, D_n]$ we will denote the ring of differential operators with coefficients in $\mathscr{F}(K)$. Setting

$$\mathfrak{D}(K)_h = \{a(x,D) \in D(K) | \text{order of } a(x,D) \leq h\}$$

we get a filtration of the ring

$$\mathfrak{D}(K): \mathfrak{D}(K)_0 \subset \mathfrak{D}(K)_1 \subset \mathfrak{D}(K)_2 \subset \cdots \mathfrak{D}(K) = \bigcup \mathfrak{D}(K)_h,$$

$$\mathfrak{D}(K)_r \mathfrak{D}(K)_s \subset \mathfrak{D}(K)_{r+s}.$$

The associated graded ring to this filtration

$$\bigotimes_{h=0}^{\infty} \mathfrak{D}(K)_h / \mathfrak{D}(K)_{h-1} \simeq \mathscr{F}(K)_0[\xi_1, \ldots, \xi_n]$$

is isomorphic to the graded ring $\mathscr{H}(K)$ of homogeneous polynomials in n indeterminates $\xi = (\xi_1, \ldots, \xi_n)$. The isomorphism is obtained by associating to each element $a(x,D) = \sum_{|\sigma| \leq h} a_\sigma(x) D^\sigma$ of $\mathfrak{D}(K)_h$ its symbol

$$\hat{a}(x, \xi) = \sum_{|\sigma| = h} a_\sigma(x) \xi^\sigma.$$

It follows then that also the ring $\mathfrak{D}(K)$ (as a left and right module over itself) is itself a Noetherian ring.

(b) Given a matrix $A^0(x, D)$ of type $p_1 \times p_0$ with entries in $\mathfrak{D}(K)$ we associate to it the $\mathfrak{D}(K)$-homomorphism

$$\mathfrak{D}(K)^{p_1} \xrightarrow{A^0(x,D)} \mathfrak{D}(K)^{p_0}$$

by sending each row vector $v(x, D) = (v_1(x, D), \ldots, v_{p_1}(x, D)) \in \mathfrak{D}(K)^{p_1}$ into the row vector $v(x, D) A^0(x, D)$. In this way we see that the set of "integrability conditions" for $A^0(x, D)$ is the $\mathfrak{D}(K)$-module that appears as the kernel of the $\mathfrak{D}(K)$-homomorphism $A^0(x, D)$. Suppose that we have assigned a multi-grading (a_j, b_i) to the operator $A^0(x, D)$. By this we mean that $A^0(x, D) = (a_{ij}^0(x, D))$ with order of $a_{ij}^0(x, D) \leq a_j - b_i$. We set for every integer h:

$$(\mathfrak{D}(K)^{p_0})_h = \{(r_1, \ldots, r_{p_0}) \in \mathfrak{D}(K)^{p_0} | \text{order of } r_j \leq a_j + h \text{ for } 1 \leq j \leq p_0\},$$

$$(\mathfrak{D}(K)^{p_1})_h = \{(s_1, \ldots, s_{p_1}) \in \mathfrak{D}(K)^{p_1} | \text{order of } s_i \leq b_i + h \text{ for } 1 \leq i \leq p_1\},$$

so that $A^0(x, D)$ sends $(\mathfrak{D}(K)^{p_1})_h$ into $(\mathfrak{D}(K)^{p_0})_h$ for every h.

Using the multigrading we can take symbols and we obtain an $\mathscr{H}(K)$-homomorphism given by the homogeneous matrix ${}^t\hat{A}^0(x, \xi)$:

$$\mathscr{H}(K)^{p_1} \xrightarrow{{}^t\hat{A}^0(x,\xi)} \mathscr{H}(K)^{p_0}$$

each column vector $v = {}^t(v_1, \ldots, v_{p_1})$ being sent into the column vector ${}^tA^0(x, \xi)v = w \in \mathscr{H}(K)^{p_0}$.

Given $v(x, D) = (v_1(x, D), \ldots, v_{p_1}(x, D)) \in (\mathfrak{D}(K)^{p_1})_h$ we can consider $w(x, D) = v(x, D) A^0(x, D)$ as an element of $(\mathfrak{D}(K)^{p_0})_h$ and we can take the corresponding symbol $\hat{w}(x, \xi) = {}^t(\hat{w}_1(x, \xi), \ldots, \hat{w}_{p_0}(x, \xi))$ belonging to $\mathscr{H}(K)^{p_0}$, where $\hat{w}_j(x, \xi)$ is the homogeneous polynomial symbol of the operator

$$v_1(x, D)a_{1j}(x, D) + \cdots + v_{p_1}(x, D)a_{p_1j}(x, D).$$

We note that ${}^t\hat{A}^0(x, \xi)$ is a homogeneous matrix of multigrading $(-b_j, -a_i)$ and that setting

$(\mathscr{H}(K)^{p_1})_h$

$$= \left\{ v = \begin{pmatrix} v_1 \\ \vdots \\ v_{p_1} \end{pmatrix} \in \mathscr{H}(K)^{p_1} \,\middle|\, v_i \text{ homogeneous of degree } b_i + h \text{ for } 1 \leq i \leq p_1 \right\},$$

$(\mathscr{H}(K)^{p_0})_h$

$$= \left\{ w = \begin{pmatrix} w_1 \\ \vdots \\ w_{p_0} \end{pmatrix} \in \mathscr{H}(K)^{p_0} \,\middle|\, w_j \text{ homogeneous of degree } a_j + h \text{ for } 1 \leq j \leq p_0 \right\},$$

the $\mathscr{H}(K)$-homomorphism ${}^t\hat{A}^0(x, \xi)$ sends $(\mathscr{H}(K)^{p_1})_h$ into $(\mathscr{H}(K)^{p_0})_h$. We have $\hat{w}(x, \xi) \in (\mathscr{H}(K)^{p_0})_h$. The elements $\hat{w}(x, \xi)$ describe a graded submodule of $\mathscr{H}(K)^{p_0}$, the submodule ${}^t\widehat{\operatorname{Im} A^0(x, D)}$. We have obviously

$$\widehat{{}^t\operatorname{Im} A^0(x, D)} \supset \operatorname{Im} {}^t\hat{A}^0(x, \xi).$$

We will say that the differential operator $A^0(x, D)$ defined in a neighborhood of K and with given multigrading (a_j, b_i) is *involutive* if

$$\widehat{{}^t\operatorname{Im} A^0(x, D)} = \operatorname{Im} {}^t\hat{A}^0(x, \xi).$$

Because of the Noetherianity of the ring $\mathscr{H}(K)$ we deduce that given any differential operator $A^0(x, D)$ on a neighborhood of K and given a multigrading of it (a_j, b_i) we can always extend the given operator by adding rows of graded operators $w(x, D) = v(x, D)A^0(x, D)$ to obtain a new multigraded operator which is involutive on K. An important property of involutive operators is the following:

Let $A^0(x, D)$ be involutive over K and let

$$w(x, D) \in (\operatorname{Im} A^0(x, D) \cap (\mathfrak{D}(K)^{p_0})_h).$$

Then we can find $v(x, D) \in (\mathfrak{D}(K)^{p_1})_h$ *such that*

$$w(x, D) = v(x, D)A^0(x, D).$$

Indeed, if h is very small, $(\mathfrak{D}(K)^{p_0})_h = 0$, thus $w = 0$ and we can take $v = 0$. By induction on h, because A^0 is involutive we can find a symbol ${}^t\hat{v}(x, \xi) \in (\mathscr{H}(K)^{p_1})_h$ such that $\hat{w}(x, \xi) = \hat{v}(x, \xi)\hat{A}^0(x, \xi)$. If $v(x, D) \in (\mathfrak{D}(K)^{p_1})_h$ is such that

$\widehat{v(x, D)} = \hat{v}(x, \xi)$, we have that

$$w(x, D) - v(x, D)A^0(x, D) \in (\mathfrak{D}(K)^{p_0})_{h-1}$$

since the symbol vanishes. We can thus apply to this operator the inductive hypothesis.

(c)   We have the following

LEMMA.   *Let $A^0(x, D)$ be a $p_1 \times p_0$ matrix of differential operators in $\mathfrak{D}(K)$ of some multigrading (a_j, b_i). Set*

$$\ker A^0(x, D) = \ker(A^0(x, D): \mathfrak{D}(K)^{p_1} \to \mathfrak{D}(K)^{p_0}),$$
$$\ker {}^t\hat{A}^0(x, \xi) = \ker({}^t\hat{A}^0(x, \xi): \mathscr{H}(K)^{p_1} \to \mathscr{H}(K)^{p_0}).$$

We have

$$\widehat{{}^t\ker A^0(x, D)} \subset \ker {}^t\hat{A}^0(x, \xi)$$

and if A^0 is involutive on K, then

$$\widehat{{}^t\ker A^0(x, D)} = \ker {}^t\hat{A}^0(x, \xi).$$

Proof.   Set $N = \ker A^0(x, D)$, ${}^tM = \ker {}^t\hat{A}^0(x, \xi)$ and let

$$v(x, D) = (v_1(x, D), \ldots, v_{p_1}(x, D)) \in N.$$

This means that $v(x, D)A^0(x, D) = 0$. Let $v(x, D) \in (\mathfrak{D}(K)^{p_1})_h$, so that ${}^t\hat{v}(x, \xi) \in (\mathscr{H}(K)^{p_1})_h$. Taking symbols we obtain $\hat{v}(x, \xi)\hat{A}^0(x, \xi) = 0$, and therefore ${}^t\hat{A}^0(x, \xi){}^t\hat{v}(x, \xi) = 0$, so that ${}^t\hat{v}(x, \xi) \in \ker {}^t\hat{A}^0(x, \xi)$. This shows that ${}^t\hat{N} \subset {}^tM$, which is the first part of the statement of the lemma.

Let now ${}^t\hat{S}(x, \xi) \in \ker {}^t\hat{A}^0(x, \xi)) \cap (\mathscr{H}(K)^{p_1})_h$ be given. We can find $S(x, D) \in (\mathfrak{D}(K)^{p_1})_h$ such that $\widehat{S(x, D)} = \hat{S}(x, \xi)$. Let $L(x, D) = S(x, D)A^0(x, D)$. We have $L(x, D) \in (\mathfrak{D}(K)^{p_0})_h$, but by the choice of $S(x, D)$ its symbol is zero and therefore $L(x, D) \in (\mathfrak{D}(K)^{p_1})_{h-1}$. Because A^0 is involutive on K we can find an operator $C(x, D)$ in $(\mathfrak{D}(K)^{p_1})_{h-1}$ such that $L(x, D) = C(x, D)A^0(x, D)$. Thus we obtain

$$(S(x, D) - C(x, D))A^0(x, D) = 0,$$

i.e., $S(x, D) - C(x, D) \in \ker A^0(x, D)$. Moreover

$$S(x, D) - C(x, D) \in (\mathfrak{D}(K)^{p_1})_h$$

and

$$\widehat{{}^tS(x, D) - C(x, D)} = \widehat{{}^tS(x, D)} = {}^t\hat{S}(x, \xi).$$

This proves the desired statement.

(d) *Existence of Involutive Correct Resolutions*
   of an Involutive Operator over K

Let $A^0(x, D)$ be an involutive differential operator over K of a given multigrading (a_j, b_i): $A^0(x, D): \mathfrak{D}(K)^{p_1} \to \mathfrak{D}(K)^{p_0}$. By Lemma 8 we can find a differential operator $A^1(x, D)$ which is involutive and corresponds to a multigrading (b_i, c_h) such that we have the exact sequences:

$$\mathfrak{D}(K)^{p_2} \xrightarrow{A^1(x,D)} \mathfrak{D}(K)^{p_1} \xrightarrow{A^0(x,D)} \mathfrak{D}(K)^{p_0},$$
$$\mathscr{H}(K)^{p_2} \xrightarrow{{}^t\hat{A}^1(x,\xi)} \mathscr{H}(K)^{p_1} \xrightarrow{{}^t\hat{A}^0(x,\xi)} \mathscr{H}(K)^{p_0}.$$

Proceeding with the involutive operator A^1 as with A^0 and so on one obtains a correct resolution of the operator $A^0(x, D)$ by involutive operators, i.e., a complex

$$\mathscr{F}(K)^{p_0} \xrightarrow{A^0(x,D)} \mathscr{F}(K)^{p_1} \xrightarrow{A^1(x,D)} \mathscr{F}(K)^{p_2} \to \cdots \qquad (1)_K$$

such that each operator $A^i(x, D)$ is involutive multigraded and such that the complex

$$\mathscr{H}(K)^{p_0} \xleftarrow{{}^t\hat{A}^0(x,\xi)} \mathscr{H}(K)^{p_1} \xleftarrow{{}^t\hat{A}^1(x,\xi)} \mathscr{H}(K)^{p_2} \leftarrow \cdots \qquad (2)_K$$

is exact. This establishes the existence of correct involutive resolutions.

We now make the following observation. Let $v^{(\alpha)}(x, D) \in (\mathfrak{D}(K)^{p_1})_h$ for $1 \le \alpha \le l$ be elements of $\ker A^0(x, D)$. We assume that their symbols ${}^t\hat{v}^{(\alpha)}(x, \xi) \in (\mathscr{H}(K)^{p_1})_h$, $1 \le \alpha \le l$, generate $\ker {}^t\hat{A}^0(x, \xi)$. Then it follows that the operators $v^{(\alpha)}(x, D)$ generate $\ker A^0(x, D)$ in $\mathfrak{D}(K)^{p_1}$. To construct the resolution $(1)_K$ then one proceeds as follows. We construct ${}^t\hat{A}^1(x, \xi)$ so that $(2)_K$ is exact and then choose an involutive $A^1(x, D)$ correspondingly, so that

$$\cdots \to \mathfrak{D}(K)^{p_2} \xrightarrow{A^1(x,D)} \mathfrak{D}(K)^{p_1} \xrightarrow{A^0(x,D)} \mathfrak{D}(K)^{p_0} \qquad (3)_K$$

is exact at $\mathfrak{D}(K)^{p_1}$. We then construct ${}^t\hat{A}^2(x, \xi)$ so that $(2)_K$ is exact and choose an involutive $A^2(x, D)$ so that $(3)_K$ is exact at $\mathfrak{D}(K)^{p_2}$, etc.

We then deduce that if the exact sequence $(2)_K$ stops with an injective $\mathscr{H}(K)$-homomorphism at the dth place

$$\mathscr{H}(K)^{p_{d-1}} \xrightarrow{{}^t\hat{A}^{d-1}(x,\xi)} \mathscr{H}(K)^{p_d} \to 0$$

then the involutive correct resolution $(1)_K$ has also length d, i.e., we have in $(1)_K$

$$\mathscr{F}(K)^{p_{d-1}} \xrightarrow{A^{d-1}(x,D)} \mathscr{F}(K)^{p_d} \to 0$$

with

$$\mathfrak{D}(K)^{p_{d-1}} \xrightarrow{A^{d-1}(x,D)} \mathfrak{D}(K)^{p_d} \to 0$$

exact.

Set for any integer l, $\xi = (\xi_1, \ldots, \xi_l)$ as a set of indeterminates and let $_l\mathcal{H}(K) = \mathcal{F}(K)_0[\xi_1, \ldots, \xi_l]$ be the graded ring of homogeneous polynomials in l variables ξ with coefficients in $\mathcal{F}(K)$.

For $l = 0$ we get $_0\mathcal{H}(K) = \mathcal{F}(K)$. Let a matrix $p_0 \times p_1$ $^tA^0(x)$ with elements in $\mathcal{F}(K)$ be given; we can consider $^tA^0(x)$ as an $\mathcal{F}(K)$-homomorphism $^tA^0(x): \mathcal{F}(K)^{p_1} \to \mathcal{F}(K)^{p_0}$.

We will assume that K is connected and that, when $\mathbf{R}^n = \mathbf{C}^m$, K has a fundamental system of topologically contractible neighborhoods in $\mathbf{C}^m$. From [1, Proposition 1] we deduce then that we can construct an exact sequence

$$0 \to \mathcal{F}(K)^{p_d} \xrightarrow{\,^tA^{d-1}(x)\,} \mathcal{F}(K)^{p_{d-1}} \to \cdots \to \mathcal{F}(K)^{p_2} \xrightarrow{\,^tA^1(x)\,} \mathcal{F}(K)^{p_1}$$
$$\xrightarrow{\,^tA^0(x)\,} \mathcal{F}(K)^{p_0}$$

of length $d \leq n + 1$ ($d \leq m + 1$ if $\mathbf{R}^n = \mathbf{C}^m$).

We now apply the inductive procedure of Hilbert on the number l of variables (see [10], see also [3]). We deduce then the following.

Given a homogeneous matrix $^t\hat{A}^0(x, \xi)$ of some multigrading (a_j, b_i) with elements in $_l\mathcal{H}(K)$ we can consider it as an $_l\mathcal{H}(K)$-homomorphism $^t\hat{A}^0(x, \xi): {}_l\mathcal{H}(K)^{p_1} \to {}_l\mathcal{H}(K)^{p_0}$. Under the above assumptions on K, we can find an exact sequence of homogeneous homomorphisms

$$0 \to {}_l\mathcal{H}(K)^{p_d} \xrightarrow{\,^t\hat{A}^{d-1}(x,\xi)\,} {}_l\mathcal{H}(K)^{p_{d-1}} \to \cdots \to {}_l\mathcal{H}(K)^{p_2} \xrightarrow{\,^tA^1(x,\xi)\,} {}_l\mathcal{H}(K)^{p_1}$$
$$\xrightarrow{\,^tA^0(x,\xi)\,} {}_l\mathcal{H}(K)^{p_0}$$

of length $d \leq n + l + 1$. For $l = n$ we obtain the following.

PROPOSITION 6. *Let K be a semianalytic connected compact set which in the case $\mathbf{R}^n = \mathbf{C}^m$ is a compact set of holomorphy, having a fundamental sequence of topologically contractible open neighborhoods. Let $\mathfrak{D}(K)^{p_1} \xrightarrow{A^0(x,D)} \mathfrak{D}(K)^{p_0}$ be a multigraded involutive operator with coefficients in $\mathcal{F}(K)$. Then $A^0(x,D)$ admits a correct involutive resolution*

$$\mathcal{F}(K)^{p_0} \xrightarrow{A^0(x,D)} \mathcal{F}(K)^{p_1} \xrightarrow{A^1(x,D)} \mathcal{F}(K)^{p_2} \to \cdots \xrightarrow{A^{d-1}(x,D)} \mathcal{F}(K)^{p_d} \to 0,$$
$$(1)_K$$

i.e., such that the sequences of multigraded morphisms

$$\mathfrak{D}(K)^{p_0} \xleftarrow{A^0(x,D)} \mathfrak{D}(K)^{p_1} \xleftarrow{A^1(x,D)} \mathfrak{D}(K)^{p_2} \leftarrow \cdots \xleftarrow{A^{d-1}(x,D)} \mathfrak{D}(K)^{p_d} \leftarrow 0 \quad (3)_K$$
$$\mathcal{H}(K)^{p_0} \xleftarrow{\,^t\hat{A}^0(x,\xi)\,} \mathcal{H}(K)^{p_1} \xleftarrow{\,^t\hat{A}^1(x,\xi)\,} \mathcal{H}(K)^{p_2} \leftarrow \cdots \xleftarrow{\,^t\hat{A}^{d-1}(x,\xi)\,} \mathcal{H}(K)^{p_d} \leftarrow 0$$
$$(2)_K$$

are exact sequences of length $d \leq 2n + 1$ ($d \leq n + 1$ if $\mathbf{R}^n = \mathbf{C}^m$).

Note that in the case $\mathbf{R}^n = \mathbf{C}^m$ when A^0 has holomorphic coefficients the operator A^0 can be considered as a matrix $A^0 = (a_{ij}(z, \partial/\partial z))$ where a_{ij} is a polynomial in $\partial/\partial z_1, \ldots, \partial/\partial z_m$ with holomorphic coefficients in a neighborhood of K. Here $z_1, \ldots, z_m$ denote the complex coordinates of $\mathbf{C}^m$. But then in the symbol the number of variables l can be reduced to $l = m$ (i.e., reduced by the half).

Remark. One may ask whether the correct minimal length for an involutive resolution could be $d \leq n + 1$ (resp., $d \leq m + 1$). If the number of variables is $n \geq 2$ one could also expect $d \leq n$.

(e)   We end up this section with some remarks on the nth operator $A^{n-1}(x, D)$ of the resolution. These we formulate in the following statement for the case $\mathscr{F}(K) = \mathscr{A}(K)$.

STATEMENT.   *We can suppress a certain number of rows in $A^{n-1}(x, D)$ to obtain a matrix $B^{n-1}(x, D)$ of type $\rho \times p_{n-1}$ and we can find a thin analytic subset C of some open neighborhood of K such that*

(i)   *for every $x_0 \in K - C$ all columns of ${}^t\hat{A}^{n-1}(x_0, \xi)$ are linear combinations over $\mathscr{H}$ of the columns left in the symbol ${}^t\hat{B}^{n-1}(x_0, \xi)$ of B^{n-1}.*
(ii)   *For every $x_0 \in K - C$ the columns of the symbol of the new operator B^{n-1} are linearly independent over $\mathscr{H}$.*

Then for every $x_0 \in K - C$ we have an exact sequence

$$0 \to \mathscr{H}^\rho \xrightarrow{{}^t\hat{B}^{n-1}(x_0, \xi)} \mathscr{H}^{p_{n-1}}. \tag{$*$}$$

Therefore by Theorem 1 (and Theorem 2) we have the exact sequence

$$\mathscr{A}^{p_{n-1}}_{x_0} \xrightarrow{B^{n-1}(x, D)} \mathscr{A}^\rho_{x_0} \to 0$$

$\left[\text{or if } B^{n-1}(x, D) \text{ is elliptic at } x_0, \mathscr{E}^{p_{n-1}}_{x_0} \xrightarrow{B^{n-1}(x, D)} \mathscr{E}^\rho_{x_0} \to 0\right].$

Sketch of the proof.   The conditions (i) and (ii) are translated into the Rouché–Capelli rule for a linear system with coefficients made up by the coefficients of the entries of ${}^t\hat{A}^{n-1}(x, \xi)$ as polynomials in ξ. This shows that there exists a thin analytic set C, defined in a neighborhood of K, with properties (i) and (ii). The exactness of the sequence $(*)$ is then a consequence of a lemma of Hilbert and Gröbner (see [1, Lemma, p. 195]) applied to the ideal generated by $\xi_1, \ldots, \xi_n$ and of the condition (ii).

Remark.   We have a similar statement in the case of holomorphic operators, i.e., in the case $\mathscr{F}(K) = \mathcal{O}(K)$.

REFERENCES

1. A. ANDREOTTI AND H. GRAUERT, Theorème de finitude pour la cohomologie des espaces complexes, *Bull. Soc. Math. France* **90** (1962), 193–259.
2. A. ANDREOTTI, C. D. HILL, S. ŁOJASIEWICZ, AND B. MAC KICHAN, Complexes of differential operators, the Mayer–Vietoris sequence, *Invent. Math.* **35** (1976), 43–86.
3. A. ANDREOTTI AND M. NACINOVICH, Complexes of partial differential operators, *Ann. Scuola Norm. Sup. Pisa* **3** (1976), 553–621.
4. A. ANDREOTTI AND M. NACINOVICH, Some remarks on formal Poincaré lemma, *in* "Complex Analysis and Algebraic Geometry, A Collection of Papers dedicated to K. KODAIRA" (W. BAILY AND T. SHIODA, eds.), pp. 295–305, Cambridge Univ. Press, London and New York, 1977.
5. A. ANDREOTTI AND M. NACINOVICH, Analytic convexity. *Ann. Scuola Norm. Sup. Pisa*, **7** (1980), 287–372.
6. C. BĂNICĂ AND O. STĂNĂSILĂ, "Methodes algebriques dans la theorie globale des espaces complexes," Vols. I and II, Gauthier-Villars, Paris, 1977.
7. A. DOUGLIS AND L. NIRENBERG, Interior estimates for elliptic systems of partial differential equations, *Comm. Pure Appl. Math.* **8** (1955), 503–538.
8. J. FRISH, Points de platitude d'un morphisme d'espaces analytiques, *Invent. Math.* **2** (1970), 118–138.
9. E. GOURSAT, "Cours d'Analyse Mathematique," Gauthier-Villars, Paris, 1942.
10. D. HILBERT, "Gesammelte Abhandlungen," Vol. 2, Chelsea, New York, 1965.
11. H. LEWY, An example of a smooth partial differential equation without solutions, *Ann. Math.* **66** (1957), 155–158.
12. C. B. MORREY, "Multiple Integrals in the Calculus of Variations," Springer-Verlag, Berlin and New York, 1966.
13. I. G. PETROWSKI, "Lectures on Partial Differential Equations," Wiley (Interscience), New York, 1957.

Weighted Inequalities in L^2
and Lifting Properties

RODRIGO AROCENA,
MISCHA COTLAR, AND
CORA SADOSKY[†]

Facultad de Ciencias
Universidad Central de Venezuela
Caracas, Venezuela

TO LAURENT SCHWARTZ IN ADMIRATION
FOR HIS OUTSTANDING CONTRIBUTIONS
TO SCIENCE AND TO JUSTICE AND SOCIAL
PROGRESS

INTRODUCTION

This paper is a self-contained continuation of the study begun in [10, 11, 3].

Let X and Y be two locally compact spaces, Γ a linear set of functions defined in X, and T a linear operator in Γ, assigning to each $f \in \Gamma$ a function Tf defined on Y. The L^2 weighted problem for T is to determine the pairs of positive measures (μ, ν) acting on Y and X, respectively, which satisfy the inequality

$$\int |Tf|^2 \, d\mu \leq M \int |f|^2 \, d\nu \quad \forall f \in \Gamma \tag{1}$$

for some constant M. We write $(\mu, \nu) \in [T, \Gamma]_M = [T]_M$ if (1) holds.

The L^2 weighted problems arise in several contexts [8, 14–17, 18, 22, 24, 25]. The more special problem of finding pairs (μ, ν) satisfying (1) with a prescribed M is important in works such, as [4, 22, 3], and in some questions of prediction theory.

Natural basic questions arising in the study of the L^2 weighted problem are (i) characterization of $[T]_M$; (ii) characterization of $\{\mu : (\mu, \mu) \in [T]_M\}$; (iii) characterization of the ν such that $(\mu, \nu) \in [T]_M$ for some $\mu \neq 0$; (iv) characterization of the extremal elements of $[T]_M$ and of $\{\mu : (\mu, \mu) \in [T]_M\}$; and (v) characterization of the Fourier transforms $\{\hat{\mu} : (\mu, \mu) \in [T]_M\}$, when this makes sense.

[†] Present address: Department of Mathematics, Howard University, Washington, D.C.

95

In this paper answers to these questions are given in the one-dimensional case by a unified procedure through the application of a general lifting theorem on quadratic forms (see Section 2). This is done for a class of operators T acting in $\mathbb{R}$ or in the unit circle $\mathbb{T}$ when the domain Γ of T is a certain subspace of $C_\infty(\mathbb{R})$ (the class of continuous functions vanishing at ∞) or of $C(\mathbb{T})$. The pattern for applying lifting properties to weighted problems is as follows.

We impose on T a condition, which is satisfied in several classical cases:

ASSUMPTION. *There exists a system* $\{\Gamma_1, \ldots, \Gamma_N\}$ *of N linear subspaces of $C_\infty(X)$, such that* (a) $\Gamma \subset \Gamma_1 \oplus \cdots \oplus \Gamma_N$, *and* (b) *for each* $j, k = 1, \ldots, N$, *and for each measure μ in Y, there is a measure v_{jk} in X, $v_{jk} = v_{jk}(\mu)$, such that*

$$\int_Y (Tf_j)(\overline{Tf_k})\, d\mu = \int_X f_j \bar{f}_k\, dv_{jk} \quad \forall f_j \in \Gamma_j, \quad f_k \in \Gamma_k. \tag{2}$$

EXAMPLES.   (1)   If $X = Y$, the assumption is satisfied whenever

$$Tf_j = \lambda_j f_j \quad \forall f_j \in \Gamma_j, \quad j = 1, \ldots, N, \quad \text{with} \quad \lambda_j \in C_\infty(X) \tag{2a}$$

and, in particular, if the λ_j are constants, so that the Γ_j are eigenspaces of T. In this case

$$v_{jk} = \lambda_j \bar{\lambda}_k \mu. \tag{2b}$$

(2)   The assumption is satisfied if

$$(Tf_j)(\overline{Tf_k}) = T_{jk}(f_j \bar{f}_k), \quad f_j \in \Gamma_j, \quad j = 1, \ldots, N, \tag{2c}$$

because then (2) holds for $v_{jk} = T^*_{jk}\mu$.

By (a) of the Assumption, (1) can be rewritten as

$$\sum_{j, k=1}^N \int (Tf_j)(\overline{Tf_k})\, d\mu \leq M \sum_{j, k=1}^N \int f_j \bar{f}_k\, dv, \tag{1a}$$

$$\forall \underline{f} = (f_1, \ldots, f_N) \in \underline{\Gamma} = \Gamma_1 \times \cdots \times \Gamma_N. \tag{1b}$$

From (1a) and (b) of the Assumption, we get the following equivalence that will be used throughout the paper:

$$(\mu, v) \in [T]_M \quad \text{iff} \quad \sum_{j, k=1}^N \int_X f_j \bar{f}_k\, d\rho_{jk} \geq 0 \quad \forall \underline{f} \in \underline{\Gamma}, \tag{3}$$

where

$$\rho_{jk} = Mv - v_{jk}(\mu). \tag{3a}$$

Inequality (1a) suggests the following definition: $(\mu, v) \in ((T))_M = ((T, \Gamma))_M = $ weak class $[T]_M$, with respect to Γ, if the weaker condition

$$\sum_{j \neq k} \int (Tf_j)(\overline{Tf_k}) \, d\mu \leq M \sum_{j,k} \int f_j \overline{f_k} \, dv \quad \forall \underline{f} \in \Gamma \tag{4}$$

holds. In (4) μ may be complex, $v \geq 0$. If (2) is valid only for $j \neq k$, we still have, instead of (3) with (3a),

$$(\mu, v) \in ((T))_M \text{ iff (3) holds with } \quad \rho_{jk} = Mv - (1 - \delta_{jk})v_{jk}(\mu). \tag{4a}$$

Each $N \times N$ matrix of complex measures in X, $\rho = (\rho_{jk})$, with $\rho_{kj} = \overline{\rho_{jk}}$, gives rise to a quadratic form

$$\rho(\underline{f}, \underline{f}) = \sum_{j,k=1}^{N} \int f_j \overline{f_k} \, d\rho_{jk}$$

on $\underline{C}_0 = C_0(X) \times \cdots \times C_0(X)$ $(C_0 = \{f \in C(X) : \operatorname{supp} f \text{ compact}\})$. We write

$$\rho \geq 0 \quad \text{if} \quad \rho(\underline{f}, \underline{f}) \geq 0 \quad \forall \underline{f} \in \underline{C}_0, \tag{5}$$

and

$$\rho \succ 0 \text{ (with respect to } \Gamma) \quad \text{if} \quad \rho(\underline{f}, \underline{f}) \geq 0 \quad \forall \underline{f} \in \Gamma, \tag{5a}$$

that is, if the restriction of the quadratic form ρ to Γ is positive. Thus, by (3) and (4),

$$(\mu, v) \in [T]_M \quad \text{iff} \quad \rho \succ 0 \text{ for } \rho \text{ given by (3a)}, \tag{5b}$$

$$(\mu, v) \in ((T))_M \quad \text{iff} \quad \rho \succ 0 \text{ for } \rho \text{ given by (4a)}. \tag{5c}$$

From (5b) and (5c) it is clear that the knowledge of the set $\{\rho : \rho \succ 0$ with respect to $\Gamma\}$ will be of great help for the solution of the L^2 weighted problems for T. Note that this set depends only on Γ and not on T. Now, the smaller set $\{\rho : \rho \geq 0\}$ is quite familiar, since its extremal elements are the elementary positive definite (p.d.) matrices of the form

$$\delta_{z,a} = (a_{jk}\delta_z), \tag{6}$$

where δ_z is the Dirac measure at $z \in X$ and $\mathbf{a} = (a_{jk})$ is an $N \times N$ p.d. numerical matrix. In several cases it can be proved, under suitable hypothesis on the Γ_j, that the extremal elements of the set $\{\rho : \rho \succ 0\}$ are the same elementary p.d. matrices, and that each $\rho \succ 0$ is the limit of convex combinations of elementary p.d. matrices in the weak topology induced by Γ. This result can be restated as the *lifting property* [11]: For each $\rho \succ 0$ (with respect to Γ) there is a $\sigma \geq 0$ such that $\rho(\underline{f}, \underline{f}) = \sigma(\underline{f}, \underline{f})$ for $\underline{f} \in \Gamma$. In other words, if the restriction of the quadratic form ρ to $\underline{\Gamma} \cap \underline{C}_0$ is positive, then it can be lifted to a positive form σ on the whole of $\underline{C}_0$.

As already noted, here ρ need not be associated with an operator as in (3a) or (4a). However if T satisfies condition (2) for the system Γ_j (for all j, k or only for $j \neq k$), and if the lifting property holds for that $\underline{\Gamma}$, then it entails a general characterization of the classes $[T]_M$, or $((T))_M$, with control over the norm M.

In this paper we concentrate on the one dimensional case: $X = \mathbb{R}$ or $X = \mathbb{T}$, and $N = 2$. Here the lifting property and some of its variants hold for natural choices of Γ_j and furnish, by a unified procedure, refinements of classical results in prediction theory and the duality $H^1 - BMO$, as well as new results for weighted inequalities.

In Part I the lifting property in $\mathbb{T}$ and $\mathbb{R}$ is proved by an argument which gives explicit recurrent relations for the lifting. These are applied in Part III to the nonconstructive part of Fefferman's duality theorem. Then the connection of the lifting property with a modified Bochner–Schwartz theorem is discussed.

In Part II the matrices $\rho \succ 0$ are described for the basic cases and some abstract variants of the lifting property are considered.

In Part III applications are given to $T = H$, the Hilbert transform, and $T = P$, the Poisson transform. First, the Riesz classes $\mathscr{R}_M^{(n)}$ of pairs (μ, v) that satisfy the Riesz inequality $\int |Hf|^2 \, d\mu \leq M \int |f|^2 \, dv$ for all f with n vanishing moments are characterized. For $\mu = v$ in the case of $\mathbb{T}$, and for $\mu = v$ and $n = 0$ in the case of $\mathbb{R}$, this result gives a refinement of classical theorems of Helson–Szegö [15] and Helson–Sarason [16], with control over M. This control leads in a natural way to refinements of results of Helson–Sarason [16] and of Ibragimov–Rozanov [18] in prediction theory, and to characterization of the extremal elements and of the Fourier transform of the set $\{\mu : (\mu, \mu) \in \mathscr{R}_M^{(n)}\}$. Finally, the interest of the weak classes $((T))_M$ is shown. It is first observed that if $dv = dx$ is the Lebesgue measure, then

$$\{\mu : (\mu, dx) \in \mathscr{R}_M^{(0)} = [H, \Gamma^0]_M\} = \{\mu : d\mu = \phi \, dx, \phi \in L^\infty, \|\phi\|_\infty \leq M\}$$

and

$$\{\mu : (\mu, dx) \in ((H, \Gamma^0))_M\} = \{\mu : d\mu = \phi \, dx, \phi \in BMO, \|\phi\|_{BMO} \leq M\}. \quad (7)$$

Since $\{\mu : (\mu, dx) \in [P, \Gamma^0]_M\} = \{\mu \text{ Carleson measure}\}$, (7) suggests then the definition of weak M-Carleson measures as the elements of $\{\mu : (\mu, dx) \in ((P, \Gamma^0))_M\}$. The lifting properties furnish then basic facts of the Fefferman–Stein duality theory and the identification of BMO with the balyage of Carleson measures [2, 7, 20] with some refinements.

Remark. In order to get to the applications to weighted problems given in Part III, for which only a particular case of the lifting theorems is required it is enough to read the simpler abstract versions given in Part II and skip the second statement of Lemma 1, Lemma 2, and Theorem 1.

I. LIFTING AND BOCHNER–SCHWARTZ THEOREMS

1. *Preliminaries*

In this chapter we shall concentrate on the one-dimensional case and use the following notations and basic properties. $\mathbb{T}$ will denote the unit circle; dt Lebesgue measure; $e_n(t) = e^{int}$; $\mathscr{P} = \mathscr{P}_+ + \mathscr{P}_- = \{$trigonometric polynomials$\}$, $\mathscr{P}_+ = \{\sum_{n \geq 0} c_n e_n(t)\}$, $\mathscr{P}_- = \{\sum_{n < 0} c_n e_n(t)\}$; $\hat{f}$ the Fourier transform of f; Hf the Hilbert transform defined by $Hf = H(f_+ + f_-) = -if_+ + if_-$ if $f_\pm \in \mathscr{P}_\pm$; $\tilde{f} = Hf + i\hat{f}(0)$, the conjugated function; and H^p = the closure of $\mathscr{P}_+$ in $L^p(\mathbb{T}, dt) = \{f \in L^p : \hat{f}(n) = 0 \text{ for } n < 0\}$, $1 \leq p < \infty$.

We shall use often the following property of $\mathscr{P}_+$, which can be easily verified by observing that $f \in \mathscr{P}$ implies $e_m f \in \mathscr{P}_+$ for m large enough, and $|e_m f| = |f|$ (cf. Lemma 1).

WEAK FEJÉR–RIESZ PROPERTY. If $0 \leq f \in C(\mathbb{T})$ then there exist $\phi_n \in \mathscr{P}_+$ with $\|f - |\phi_n|^2\|_\infty \to 0$ such that each $1/\phi_n$ is in the closure of $\mathscr{P}_+$ in $C(\mathbb{T})$.

From this property it follows that if μ is a complex measure in $\mathbb{T}$ then

$$\mu(\phi\bar{\phi}) \geq 0 \text{ (or } = 0) \text{ for all } \phi \in \mathscr{P}_+ \text{ implies } \mu \geq 0 \text{ (or } = 0). \qquad (8a)$$

On the other hand, by the F. and M. Riesz theorem [31, 35],

$$\mu(\phi\bar{\psi}) \geq 0 \text{ for all } (\phi, \psi) \in \mathscr{P}_+ \times \mathscr{P}_- \text{ implies } d\mu = h\,dt, h \in H^1 \qquad (8b)$$

In the case of the real line $\mathbb{R}$, we have to deal with tempered measures, and therefore instead of $\mathscr{P}_\pm$, consider, for k a positive integer, the classes

$$\begin{aligned} \Gamma^k &= \{f \in C(\mathbb{R}) : \|f\|_{(k)} = \sup(1 + |x|)^k |f(x)| < \infty\}, \\ \Gamma^k_\pm &= \{f \in \Gamma^k : \hat{f}(x) = 0 \text{ for } x \gtrless 0\}. \end{aligned} \qquad (9)$$

If $f = f_+ + f_-, f_\pm \in \Gamma^1_\pm$, then Hf is defined by

$$Hf = H(f_+ + f_-) = -if_+ + if_-.$$

LEMMA 1 (Weak Fejér–Riesz property in $\mathbb{R}$). *If $0 \leq f \in \Gamma^{2k}$ then there exist $f_n \in \Gamma^k_+$ such that $f = \lim|f_n|^2 (in \ \Gamma^{2k})$. Furthermore, $f < |f_n|^2$ and each $1/f_n$ is in $C(\mathbb{R})$ and in the closure of Γ^k_+ in $L^2(dt)$. (See Remark at the end of the Introduction.)*

Proof. Since $\mathscr{D}$, the space of indefinitely differentiable functions with compact support, is dense in Γ^{2k}, we may assume $0 \leq f \in \mathscr{D}$.

It is easy then to get a sequence g_n such that $|g_n|^2 \to f$ in $\mathscr{S}$ and $\hat{g}_n \in \mathscr{D}$. Setting $f_n = g_n \exp(iN_n t)$, $\hat{f}_n(\xi) = \hat{g}_n(\xi + N_n)$ will have support in $(0, \infty)$ for large N_n, hence $f_n \in \Gamma^k_+$ and $|f_n|^2 = |g_n|^2 \to f$ in S. Applying this result to $g = f + 2\varepsilon(1 + |x|)^{-2k}$ and using $\||g - |f_n|^2\|| < \varepsilon$, we get $|f_n(x)|^2 \geq f(x) + 2\varepsilon(1 + |x|)^{-2k} - \varepsilon(1 + |x|)^{-2k} > f(x)$. Since $0 < f_n \in H^2(\mathbb{R})$, by known facts on factorization through Blaschke products [35] we may get $1/f_n \in H^2$ without changing the value of $|f_n|$. ∎

From Lemma 1 it follows that if μ is a 2k-tempered measure, that is, if $dv = (1 + |x|)^{-2k} d\mu$ is a finite measure (so that $\mu(|\phi|^2)$ is defined if $\phi \in \Gamma^{2k}$), then

$$\mu(\phi\bar{\phi}) \geq 0 \ (\text{or} = 0) \quad \text{for all} \quad \phi \in \Gamma^k_+ \cap C_0 \text{ implies } \mu \geq 0 \ (\text{or} = 0). \quad (10)$$

Instead, if such a μ satisfies

$$\mu(\phi\bar{\psi}) = 0 \quad \text{for all} \quad \phi \in \Gamma^k_+ \cap C_0, \quad \psi \in \Gamma^k \cap C_0,$$

then

$$d\mu = (i + x)^{2k}h(x)\,dx, \qquad h \in H^1. \quad (10a)$$

In fact, taking $\phi = (i + x)^{-k}e^{isx}$, $s \geq 0$, $\bar{\psi} = (i + x)^{-k}$,

$$\mu(\phi\bar{\psi}) = \int e^{isx}(i + x)^{-2x}\,d\mu = \int e^{isx}\,dv = 0 \quad \forall s \geq 0,$$

and v being finite, (10a) follows.

From the above properties of the classes $\mathscr{P}_+, \Gamma^k_+$ we deduce (see Remark at the end of the Introduction):

LEMMA 2. *Let $\mathbf{f} = (f_{jk})$ be a 2×2 matrix of functions such that the numerical matrix $(f_{jk}(t))$ is p.d. for each $t \in \mathbb{T}$ (respectively, $t \in \mathbb{R}$), $f_{jk} \in \mathscr{P}$ (resp., $\in \Gamma^{2k}$) and $f_{12} \in \mathscr{P}_+$ (resp., $\in \Gamma^k_+$). Then $\mathbf{f}$ is the limit in $C(\mathbb{T})$ (resp., in Γ^{2k}) of matrices of the form*

$$\psi = (\psi_j\bar{\psi}_k) + (\gamma_j\bar{\gamma}_k), \qquad \text{with} \quad (\psi_1, \psi_2) \in \mathscr{P}_+ \times \mathscr{P}_-, \quad (11)$$
$$(\gamma_1, \gamma_2) \in \mathscr{P}_+ \times \mathscr{P}_-, \quad \gamma_2 = 0 \quad (\text{resp., } \in \Gamma^k_+ \times \Gamma^k_-).$$

Proof. Consider first the case of $\mathbb{T}$. Since $\mathbf{f}(t)$ is p.d. we have $f_{11}(t) \geq 0$, $f_{22}(t) \geq 0$, and $|f_{12}(t)|^2 \leq f_{11}(t)f_{22}(t)$. By the weak Fejér–Riesz property we can approximate $\mathbf{f}$ by matrices ϕ of the form $\phi_{11} = |\phi|^2$, $\phi_{22} = |\psi|^2$, $(\phi, \psi) \in \mathscr{P}_+ \times \mathscr{P}_-$ with $\phi_{12} = f_{12} \in \mathscr{P}_+$, $|\phi_{12}|^2 \leq |\phi|^2|\psi|^2$. Hence $\phi_{12} = \theta\phi\bar{\psi}$, where $|\theta| \leq 1$ and since $\phi_{12} \in \mathscr{P}_+$, and by (8), $1/\phi$ and $1/\bar{\psi}$ are limits of elements in $\mathscr{P}_+$, θ is also the limit of elements in $\mathscr{P}_+$. Setting $\psi_1 = \theta\phi$, $\psi_2 = \psi$, $|\phi|^2(1 - |\theta|^2) = |\gamma_1|^2$, $\gamma_1 \in \mathscr{P}_+$, $\gamma_2 = 0$, we have that $\mathbf{f}$ is limit of matrices ψ of the form (11).

In the case of $\mathbb{R}$ the proof follows the same pattern, using Lemma 1 (letting $f = f_{11}$ and $f = f_{22}$), which asserts that $f < |f_n|^2$, so that the approximating matrix $(|f_n|^2, f_{12}, f_{21}, |g_n|^2)$, with $f_n = \phi$, $g_n = \psi \in \Gamma_+^k$, will also be p.d. As above we shall have $\phi_{12} = f_{12} = \theta\phi\bar{\psi}$ with $|\theta| \le 1$, $\theta \in H^2 \cap C$, which together with $\phi \in \Gamma_+^k$ gives $\phi\theta \in \Gamma_+^k$. ∎

The properties above shall be used for the study of matrices of measures, for which the notations of the Introduction shall be maintained. But now the ρ will be 2×2 matrices of measures ρ_{jk}, $j, k = 1, 2$, and we shall take $\Gamma_1 \times \Gamma_2 = \mathcal{P}_+ \times \mathcal{P}_-$ in the case of $\mathbb{T}$, and $\Gamma_1 \times \Gamma_2 = \Gamma_+^k \times \Gamma_+^k$, with fixed k, in the case of $\mathbb{R}$. These choices of Γ come naturally in the study of the L^2 weighted problem for the Hilbert transform H. In fact, as pointed out above, these Γ_1, Γ_2 are the eigenspaces of H, so that $T = H$ satisfies (2a) with $\lambda_1 = -i$, $\lambda_2 = i$, and as will be shown in Section 2, the lifting property holds in these situations.

Though we consider mainly the matrix ρ as a quadratic form $\rho(\underline{f}, \underline{f})$ on $\underline{f} = (f_1, f_2)$, it will be convenient to consider also ρ as a *linear form* which assigns to each quadruple of functions $\mathbf{f} = (f_{jk})$, $f_{jk} \in C_0(X)$, the number $\rho(\mathbf{f}) = \sum_{j,k} \int f_{jk} d\rho_{jk}$. If K_0 is the cone generated by the quadruples of the form $(f_{jk} = \phi_j\bar{\phi}_k)$, where $(\phi_1, \phi_2) \in \Gamma$, then $\rho \succ 0$ (with respect to Γ) means that the form ρ is positive with respect to the cone K_0:

$$\rho(\mathbf{f}) = \rho(\underline{f}, \underline{f}) = \sum_{j,k=1}^{2} \int f_j\bar{f}_k \, d\rho_{jk} \ge 0 \; \forall \underline{f} \in \Gamma \cap C_0. \tag{12}$$

Letting $f_1 = \lambda_1\phi_1$, $f_2 = \lambda_2\phi_2$ in (12), we get for fixed $(\phi_1, \phi_2) \in \Gamma \cap C_0$ a positive quadratic form in λ_1, λ_2. From this and the above considerations it follows easily that $\rho \succ 0$ with respect to Γ iff

$$\rho_{11} \ge 0, \qquad \rho_{22} \ge 0, \qquad \rho_{21} = \bar{\rho}_{12},$$

and

$$\left| \int \phi_1\bar{\phi}_2 \, d\rho_{12} \right|^2 \le \left(\int |\phi_1|^2 \, d\rho_{11} \right)\left(\int |\phi_2|^2 \, d\rho_{22} \right) \quad \forall(\phi_1, \phi_2) \in \Gamma \cap C_0. \tag{12a}$$

In the case of $\mathbb{T}$, where $\underline{\Gamma} = \mathcal{P}_+ \times \mathcal{P}_-$, letting $\phi_2 = e_{-1}$, (12a) gives

$$\left| \int \phi \, d\rho_{12} \right| \le (\|\rho_{11}\| \|\rho_{22}\|)^{1/2} \|\phi\|_\infty \quad \forall\phi \in \mathcal{P}_1 = e_1 \mathcal{P}_+. \tag{12b}$$

In the case of $\mathbb{R}$, if the ρ_{jk} are $2k$-tempered measures, $dv_{jk} = d\rho_{jk}(1 + |x|)^{-2k}$, (12a) is valid for $\phi_1 = (1 + |x|)^{-k}\phi$, $\phi \in C_\infty(\mathbb{R})$, $\bar{\phi}_2 = (1 + |x|)^{-k}$ and we get

$$\left| \int \phi \, dv_{12} \right| \le (\|v_{11}\| \|v_{22}\|)^{1/2} \|\phi\|_\infty \quad \forall\phi \in C_\infty(\mathbb{R}). \tag{12c}$$

These considerations show that if ρ_{jj} and ρ_{12} are merely linear functionals in $\mathscr{P}$ and $\mathscr{P}_+$ (or Γ^{2k} and Γ^{2k}_+) satisfying (12), then ρ_{jj} (or $\rho_{jj}(1 + |x|)^{-2k}$) extend to finite positive measures, and ρ_{12} (or $\rho_{12}(1 + |x|)^{-2k}$) extend to a finite complex measure in $\mathbb{T}$ (or $\mathbb{R}$).

On the other hand, $\rho \geq 0$ means that (12) is satisfied for all $(f_1, f_2) \in \mathscr{P} \times \mathscr{P}$ (or $\Gamma^k \times \Gamma^k$). Letting $f_1 = f_2 \to 1_E$ we obtain that

$$\rho \geq 0 \quad \text{iff} \quad \text{numerical matrix } (\rho_{jk}(E)) \text{ is p.d. for all compacts sets } E. \quad (13)$$

If $d\rho_{jk}/dt = a_{jk}(t)$ then $a_{jj} \geq 0$ and $|a_{12}(t)|^2 \leq a_{11}(t)a_{22}(t)$ a.e. (13a)

We shall write $\rho \sim \sigma$ [with respect to (w.r.t.)Γ] if $\rho(f, f) = \sigma(f, f)$ for $f \in \Gamma \cap C_0$, or equivalently, if $\rho = \sigma + \tau$ with $\tau \sim 0$, i.e., $\tau = 0$ in $\Gamma \cap C_0$. In this case, if $\rho' =$ restriction of ρ to Γ, then $\sigma \sim \rho$ is a lifting of the quadratic form ρ' from $\Gamma = \Gamma_1 \times \Gamma_2$ to $(\Gamma_1 + \Gamma_2) \times (\Gamma_1 + \Gamma_2)$. In the case of $\mathbb{T}$, it follows (8a) and (8b) (setting $f_1 = 0$ or $f_2 = 0$ in (12)) that

$$\tau \sim 0 \quad \text{w.r.t.} \quad \mathscr{P}_+ \times \mathscr{P}_- \quad \text{iff} \quad \tau_{11} = \tau_{22} = 0$$

and   (14)

$$d\tau_{12} = h \, dt, \quad h \in H^1.$$

Similarly, in case of the $\mathbb{R}$, it follows from (10) and (10a) that (if τ is $2k$-tempered) then

$$\tau \sim 0 \quad \text{w.r.t.} \quad \Gamma^k_+ \times \Gamma^k_- \quad \text{iff} \quad \tau_{11} = \tau_{22} = 0$$

and   (14a)

$$d\tau_{12} = (i + x)^{2k} h(x) \, dx, \quad h \in H^1.$$

Therefore, every lifting σ of $\rho | \Gamma$, $\sigma = \rho + \tau$, $\tau \sim 0$, is determined by a function $h \in H^1$.

2. Lifting Theorem

We now prove the lifting property for $X = Y = \mathbb{T}$ or $\mathbb{R}$ and for $N = 2$, $\Gamma = \mathscr{P}_+ \times \mathscr{P}_-$ or $\Gamma = \Gamma^k_+ \times \Gamma^k$, $\rho = (\rho_{jk})$ 2×2 matrices of measures.

The existence of this lifting was already proved in [11] (in detail only for $\mathbb{T}$) by a nonconstructive argument. In view of applications in Part III and other questions (see Remark 1 to follow), we give here a variant of that proof, for both $\mathbb{T}$ and $\mathbb{R}$, with an explicit recurrent relation for the lifting. (See Remark at the end of the Introduction.)

THEOREM 1 [11]. *Let $X = Y = \mathbb{T}$ and $\Gamma = \mathscr{P}_+ \times \mathscr{P}_-$ or $X = Y = \mathbb{R}$ and $\Gamma = \Gamma^k_+ \times \Gamma^k_-$, k fixed positive integer. If $\rho \succ 0$ with respect to Γ (and the ρ_{jk} are $2k$-tempered in the case of $\mathbb{R}$), then there exist $\sigma \geq 0$ with $\sigma = \rho$ on Γ, i.e., $\sigma = \rho + \tau$, $\tau \sim 0$.*

Proof. (i) Consider first the case of $\mathbb{T}$. According to the remark preceding (12) we shall reduce the thesis to an extension of positive linear forms associated with ρ. Since we need to work with real forms while ρ_{12} is in general a complex measure (but ρ_{jj} are positive real), we introduce the space $\mathscr{E} = \{\mathbf{f} = (f_{jk}), f_{jj} \in Re\,\mathscr{P}, f_{21} = \bar{f}_{12} \in \mathscr{P}\}$ and the subspaces $\mathscr{E}_n = \{\mathbf{f} \in \mathscr{E}, f_{12} \in e_{-n}\mathscr{P}_+\}$, $n \geq 0$, so that $\mathscr{E}_1 \subset \mathscr{E}_2 \subset, \ldots, \subset \mathscr{E} = \bigcup_n \mathscr{E}_n, \mathscr{E}_{n+1} = \mathscr{E}_n + \{ce_{-n}\}$, where $\mathbf{e}_{-n} = (0, e_{-n}, e_n, 0) = (e_{jk})$ with $e_{jj} = 0$, $e_{12} = e_{-n}$. Let $\mathscr{M} = \{\boldsymbol{\mu} = (\mu_{jk}) : \mu_{jj} \text{ positive measures}, \mu_{21} = \overline{\mu_{12}}\}$, so that each $\boldsymbol{\mu} \in \mathscr{M}$ acts as a real linear form on $\mathscr{E}$, $\boldsymbol{\mu}(\mathbf{f}) = \sum_{jk}/\int f_{jk} d\mu_{jk}$ and our matrix ρ is in $\mathscr{M}$. Let $K \subset \mathscr{E}$ be the cone generated by the elements of the form $(f_{jk} = f_j \bar{f}_k)$, with $f_1, f_2 \in \mathscr{P}$, so that $\mathbf{f} \in K$ iff the numerical matrix $(f_{jk}(t))$ is p.d. for all $t \in \mathbb{T}$, equivalently.

$$\mathbf{f} \in K \quad \text{iff} \quad f_{jj} \geq 0 \quad \text{and} \quad |f_{12}|^2 \leq f_{11} f_{22}. \tag{15}$$

Let $K_n \subset \mathscr{E}_n$, $n \geq 0$, be the cone generated by the elements of the form $(f_{jk} = \phi_j \bar{\phi}_k)$, with $\phi_1 \in e_{-n}\mathscr{P}_+$, $\phi_2 \in \mathscr{P}_-$. Then

$$\boldsymbol{\mu} \geq 0 \quad \text{iff} \quad \boldsymbol{\mu}(f) \geq 0 \quad \forall \mathbf{f} \in K$$

and $\hspace{7cm}$ (16)

$$\boldsymbol{\mu} > 0 \quad \text{iff} \quad \boldsymbol{\mu}(\phi) \geq 0 \quad \forall \phi \in K_0,$$

so that our ρ is a positive linear form with respect to the cone K_0. To prove the theorem it is enough to prove now that the restriction of $\mathbf{f}$ to $\mathscr{E}_0$ can be extended to a linear functional in $\mathscr{E}$, positive with respect to K [see the remark following (12c)]. Now, by Lemma 2, $K \cap \mathscr{E}_n$ coincides with the closure of K_n (in the topology of $C(\mathbb{T})$) and, in particular, $K \cap \mathscr{E}_0$ is the closure of K_0. Therefore the restriction of ρ to $\mathscr{E}_0$ is a positive linear form with respect to $K \cap \mathscr{E}_0$ and, by a known theorem of M. Riesz and M. Krein (see [9]), it extends to a form in $\mathscr{E}$, positive with respect to K. Moreover, using the fact that $K \cap \mathscr{E}_n$ is the closure of K_n, the extension can be carried out by induction with explicit recurrent expressions, as follows. We first extend the restriction σ^0 of ρ to $\mathscr{E}_0$ to σ^1 in $\mathscr{E}_1$ (positive with respect to K), by setting $\sigma^1(\mathbf{f} + c e_{-1}) = \sigma^0(\mathbf{f}) + c\sigma^1(\mathbf{e}_{-1})$ for $f \in \mathscr{E}_0$, where

$$\sigma^1(\mathbf{e}_{-1}) = \inf\{\sigma^0(\mathbf{f}) + \rho(\mathbf{e}_{-1}), f \in K \cap \mathscr{E}_1 = \bar{K}_1\}. \tag{17}$$

It is easy to see that $\sigma^1(\mathbf{e}_{-1})$ is finite and, that since σ^0 is positive in $\mathscr{E}_0$, σ^1 is positive in $\mathscr{E}_1$. Then we extend σ^1 to a form σ^2, positive in $\mathscr{E}_2 = \mathscr{E}_1 + c e_{-2}$, and so on, and define σ^n by the recurrent formula

$$\sigma^n_{12}(e_{-n}) = \sigma^n(\mathbf{e}_{-n}) = \inf\{\sigma^{n-1}(\mathbf{f}) + \rho(\mathbf{e}_{-n}), \mathbf{f} \in \bar{K}_n\}$$

$$= \rho_{12}(e_{-n}) + \inf\left\{\sum_{j,k=1}^{2} \int \phi_j \bar{\phi}_k \, d\sigma^{n-1}_{jk}, \phi_1 \in e_{-n+1}\mathscr{P}_+, \phi_2 \in \mathscr{P}_-\right\}. \tag{18}$$

(ii)   The case of $\mathbb{R}$ is proved in exactly the same way, by using Lemma 2 for $\mathbb{R}$, (10), and (12c), taking, instead of e_{-n}, the function $(i - x)^{-n}$, and working in $\Gamma^{2k} = \Gamma_{+}^{2k} + \Gamma_{-}^{2k}$. ■

REMARK 1.   Note that, if $\sigma = \rho + \tau$, $\tau \sim 0$, $d\tau_{12} = h \, d\tau$ then in the case of $\mathbb{T}$, (18) gives a recurrent relation for the coefficients $\hat{h}(n) = \sigma_{12}^{n}(e_{-n}) - \rho_{12}(e_{-n})$, and similarly for $\mathbb{R}$. Since $\sigma \sim \rho$ implies that σ and ρ define the same functional on $\mathscr{E}_0$, it follows from Theorem 1 that the sets $\{\rho \succ 0\}$ and $\{\sigma \geq 0\}$ have the same extremal elements, given by (6), and each $\rho \succ 0$ is the limit of convex combinations of such elements (6), in the topology induced by $\mathscr{E}_0$.

COROLLARY 1.   *If $\rho \succ 0$ and the elements of ρ are bounded functions, $d\rho_{jk} = a_{jk}(x) \, dx$, $a_{jk} \in L^{\infty}$, then there exists a $\sigma \sim \rho$ such that $\sigma \geq 0$ and the elements of σ are also bounded functions, $d\sigma_{jk} = b_{jk}(x) \, dx$, with $b_{12} = a_{12} + h$, $h \in H^1$ [or $(i + x)^{2k} h \in H^1$].*

Proof.   In fact, since $\rho_{jj} = \sigma_{jj}$ and $\sigma_{12} = \rho_{12} + h \, dx$, it follows that $d\sigma_{jk} = b_{jk} \, dx$ and $b_{jj} = a_{jj} \in L^{\infty}$. From $\sigma \geq 0$ and (13a) we get then that also $b_{12} \in L^{\infty}$. ■

If $(d\rho_{jk} = a_{jk} \, dx)$ with $a_{jk} \in L^{\infty}$, $a_{21} = \overline{a_{12}}$, then ρ can be identified with the self-adjoint operator A in the Hilbert space $\mathscr{K} = L^2(\mathbb{T}) \times L^2(\mathbb{T})$, defined by $A\underline{f} = \underline{g}$, with $g_j = \sum_k a_{jk} f_k$ if $\underline{f} = (f_1, f_2)$. In this case

$$\rho = (a_{jk}) \succ 0 \quad \text{iff} \quad \text{compression } PA \text{ of } A \text{ to } H_{+}^2 \times H_{-}^2 \text{ is positive,} \quad (19)$$

where P is the orthogonal projection of $\mathscr{K}$ onto $\mathscr{H} = H_{+}^2 \times H_{-}^2$. Moreover, if U is the unitary operator in $\mathscr{K}$ defined by $U\underline{f} = \underline{g}$, with $g_j = e_1 f_j$ if $\underline{f} = (f_1, f_2)$, then the operators A given by such (a_{jk}) are exactly the operators A satisfying $UA = AU$. We may relate now Corollary 1 to classical lifting theorems considered by Sz. Nagy and Foias [26], Sarason [29], Nehari [27], and others. If $\mathscr{H}$ is a subspace of a Hilbert space $\mathscr{K}$, $U : \mathscr{K} \to \mathscr{K}$ an isometry and $X : \mathscr{H} \to \mathscr{H}$ a bounded operator in $\mathscr{H}$, we say that $Y : \mathscr{K} \to \mathscr{K}$ is a U-lifting of X if $UY = YU$ and $XP = PYP$, where P is the projection of $\mathscr{K}$ on $\mathscr{H}$. If in addition $\|Y\| = \|X\|$ we say that Y is a *norm-preserving U-lifting* of X. If X and Y are self-adjoint operators then we may consider the two following generalizations of the condition $\|Y\| = \|X\|$:

(i) $\|Y\|_{-} = \|X\|_{-}$, where $\|Y\|_{-} = \sup\{a \in \mathbb{R} : aI \leq Y\}$
$$= \inf\{(Yf, f) : \|f\| = 1\};$$

(ii) $\|Y - cI\|_{\mathscr{K}} = \|X - cI\|_{\mathscr{H}}$ for large enough $c > 0$.

LEMMA 3.   *If X and Y are self-adjoint, Y a U-lifting of X, then Y satisfies (i) iff Y satisfies (ii).*

Proof. If (i) holds and $a = \|X\|_-$, then by hypothesis, $0 \le Y - a$. If $c > 0$ is large enough, then $-c \le X - a - c \le 0$, $-c \le Y - a - c \le 0$, and, by definition of a, we must have $\|X - a - c\| = c$, hence $\|Y - a - c\| = c$ also. Conversely, if (ii) holds, then for c large enough we have $\|Y - c\| = \|X - c\|$, $Y - c \le 0$, $X - c \le 0$. If $a = \|X\|_-$ then $a - c = \|X - c\| = \|Y - c\|$, $a \le Y$, $\|Y\|_- \ge a = \|X\|_-$, hence $\|Y\|_- = \|X\|_-$. ∎

Let us say that Y *is a weak norm-preserving U-lifting of* X if Y is a U-lifting satisfying one of the equivalent conditions (i) or (ii). From Corollary 1 we obtain immediately:

COROLLARY 2. *Let* $\mathscr{K} = L^2 \times L^2$, $\mathscr{H} = H_+^2 \times H_-^2$. *For every operator* $A \sim (a_{jk})$, $a_{jk} \in L^\infty$, *the operator* $X = PAP : \mathscr{H} \to \mathscr{H}$ *has a weak norm-preserving U-lifting, where* Uf, $P = $ *projection on* $\mathscr{H}$.

Proof. Let $a \le \|X\|_-$, so that $0 \le X - a = P(A - a)P$ and $A - a \succ 0$. By Corollary 1, there exists an operator Y of the form $Y = A - a + \tau = (a_{jk})$, where $a_{jk} \in L^\infty$, $\tau_{jj} = 0$, $d\tau_{12} = h\,dt$, $h \in H^1$, such that $Y \ge 0$ and $UY = YU$. Hence $P(Y + a)P = X$, so that $Y + a$ is a U-lifting of X and, since $Y + a \ge a = \|X\|_-$, we have $\|Y + a\| = \|X\|_-$. ∎

Therefore, Corollary 2 expresses a weak norm-preserving U-lifting property for the subspace $\mathscr{H} = H_+^2 \times H_-^2$ and the isometry $Uf = e_{-1}f$. Let us remark that Corollary 2 holds also if $\mathscr{H} = H^2 \times L^2$ or if $\mathscr{H} = \bar{H}^2 \times H^2$, but in these cases the proof is straightforward.

REMARK 2. By Corollary 2, Corollary 1 is similar in character to the Nehari extension for the Hankel operator [27]. Adamjan *et al.* [1] have studied in great detail and by spectral theory methods several questions concerning the possible extension of Nehari's theorem. The recurrent relations in the proof of Theorem 1 may be of use in considering similar questions for our case, but we will not go into this problem here.

3. *Modified Bochner–Schwartz Theorem*

A kernel $K(j, n)$ [respectively, $K(x, y)$] defined in $(j, n) \in \mathbb{Z} \times \mathbb{Z}$ (in $\mathbb{R} \times \mathbb{R}$) is called a Toeplitz kernel if there exists a function $K(n)$, $n \in \mathbb{Z}$ ($K(x)$, $x \in \mathbb{R}$) such that $K(j, n) = K(j - n)$ ($K(x, y) = K(x - y)$). The Herglotz–Bochner theorem says that a Toeplitz kernel is p.d. iff $K(n) = \hat{\mu}(n)$, where μ is a positive measure in $\mathbb{T}$. The theorem of Bochner–Schwartz extends this fact to continuous and distribution Toeplitz kernels in $\mathbb{R}^d$. In other terms, each p.d. Toeplitz kernel K determines a unitary representation $n \mapsto U^n$ in a Hilbert space $\mathscr{H}$ such that $K(j, n) = (U^{j-n}\eta, \eta)$, for a fixed $\eta \in \mathscr{H}$, and if $P(t)$ is the spectral measure of U, then $d(P(t)\eta, \eta) = d\mu$, with $\hat{\mu}(n) = K(n)$.

Now, in $\mathbb{Z}$ there are two natural cones, $\mathbb{Z}_1 = \mathbb{Z}_+$ and $\mathbb{Z}_2 = \mathbb{Z}_-$, and the corresponding splitting $\mathbb{Z} \times \mathbb{Z} = \bigcup_{\alpha,\beta=1}^{2} \mathbb{Z}_{\alpha\beta}$, $\mathbb{Z}_{\alpha\beta} = \mathbb{Z}_\alpha \times \mathbb{Z}_\beta$, $\alpha, \beta = 1, 2$. We say that $K(j,n)$ is a *modified Toeplitz kernel* if there are functions $K_{\alpha\beta}(n)$, $n \in \mathbb{Z}$, such that $K(j,n) = K_{\alpha\beta}(j - n)$ wherever $(j,n) \in \mathbb{Z}_{\alpha\beta}$. This is equivalent to $K(j,n) = \sum_{\alpha,\beta} K_{\alpha\beta}(j - n)1_\alpha(j)1_\beta(n)$, $1_\alpha = $ characteristic function of $\mathbb{Z}_\alpha$. Similar definitions can be given for $\mathbb{R}$.

PROPOSITION 1. *If K is a modified Toeplitz kernel in $\mathbb{Z} \times \mathbb{Z}$, then the following properties are equivalent:* (a) *K is p.d.;* (b) *there exists a positive matrix measure* $\mu = (\mu_{\alpha\beta}) \geq 0$ *with* $K(j,n) = \hat{\mu}_{\alpha\beta}(j - n)$ *for* $(j,n) \in \mathbb{Z}_{\alpha\beta}$, $\alpha, \beta = 1, 2$; (c) *there exists a unitary representation* $n \mapsto U^n$ *in a Hilbert space* $\mathcal{H}$, *and two elements* $\eta_\alpha \in \mathcal{H}$, $\alpha = 1, 2$, *such that* $K(j,n) = (U^{j-n}\eta_\alpha, \eta_\beta)$ *for* $(j,n) \in \mathbb{Z}_{\alpha\beta}$.
Similar results hold for tempered measures in $\mathbb{R}$.

Proof. (a) implies (b): We have $K(j,n) = K_{\alpha\beta}(j - n)$ for $(j,n) \in \mathbb{Z}_{\alpha\beta}$ and $\sum_{j,n} K(j,n)\lambda(j)\overline{\lambda(n)} \geq 0$ for $\lambda(n)$ of compact support. Therefore, if $\lambda(n) = 0$ for $n < 0$, then $\sum K_{11}(j - n)\lambda(j)\overline{\lambda(n)} \geq 0$, and since $K(j - n) = K((j + N) - (n + N))$, we see that this is true if $\lambda(n) = 0$ for $n < -N$. Thus $K_{11}(j - n)$ is p.d., and similarly K_{22} is p.d., so $K_{jj}(n) = \hat{\rho}_{jj}(n)$ with $\rho_{jj} \geq 0$. We define now in $\mathscr{P}_+$ a linear functional ρ_{12}, by setting $\rho_{12}(e_n) = K_{12}(n)$ for $n > 0$. Then the hypothesis $\sum_{\alpha\beta} \sum_{jn} K_{\alpha\beta}(j - n)1_\alpha(j)1_\beta(n)\lambda(m)\lambda(n) \geq 0$ becomes the inequality (12). By the remark following (12c), this implies that ρ_{12} and $\rho_{21} = \overline{\rho_{12}}$ can be extended to finite measures satisfying (12). By the lifting theorem (Theorem 1), there is a $\mu \sim \rho$ with $\mu \geq 0$. Since then $\mu_{jj} = \rho_{jj}$ and $\mu_{12} = \rho_{12} + h\,dt$, $h \in H^1$, we obtain that $\hat{\mu}_{12}(n) = \hat{\rho}_{12}(n)$ for $n > 0$. Hence, $\hat{\mu}_{\alpha\beta}(j - n) = K(j,n)$ for $(j,n)\mathbb{Z}_{\alpha\beta}$. (b) implies (c): it is enough to set $\mathcal{H} = L^2 \times L^2$, $(f,g) = \sum_{\alpha\beta} \int f_\alpha \overline{g}_\beta \, d\mu_{\alpha\beta}$, $Uf = e_1 f$, and take $\eta^1 = (\eta_\alpha^1)$ with $\eta_1^1 \equiv 1$, $\eta_2^1 \equiv 0$, $\eta^2 = (\eta_\alpha^2)$ with $\eta_1^2 \equiv 0$, $\eta_2^2 \equiv 1$. (c) implies (a): If $dP(t)$ is the spectral measure of U we set $\mu_{\alpha\beta} = d(P(t)\eta_\alpha, \eta_\beta)$. Then $K_{\alpha\beta} = \hat{\mu}_{\alpha\beta}$ in $\mathbb{Z}_{\alpha\beta}$ and $(\mu_{\alpha\beta}) \geq 0$, hence $K = \sum_{\alpha\beta} K_{\alpha\beta}(j - n)1_\alpha(j)1_\beta(n)$ is p.d. ∎

From the proof of Proposition 1 it is clear that the *lifting theorem (Theorem 1) is equivalent to the Herglotz–Bochner–Schwartz theorem for modified Toeplitz kernels.*

In the following remarks we mention some other interpretations of these two theorems, which cannot be discussed here.

REMARK 3. If K is a kernel in $\mathbb{R}$ and D a differential operator, then K is said to be *D-symmetric* if $D_x K(x, y) = \bar{D}_y K(x, y)$, and *D-elementary* if $D_x K = \bar{D}_y K = \lambda K$, $\lambda \in \mathbb{C}$. If $D = id/dx$, then K is *D*-symmetric iff K is Toeplitz, and K is *D*-elementary iff $K = \exp i\lambda(x - y)$. The Bochner–Schwartz theorem is then a particular case of a theory, initiated by M. Krein and developed in great detail by Berezanskii [5], of representation of p.d. D-symmetric kernels in terms of integrals of D-elementary ones. We say that K is a *modified D-symmetric kernel* if the equality $D_x K(x, y) = \bar{D}_y K(x, y)$ holds in the set $\{x \neq 0, y \neq 0\} \subset \mathbb{R} \times \mathbb{R}$. Then K is *modified Toeplitz iff it is modified $i\,d/dx$-symmetric*, and the first part of Proposition 1 is equivalent to the Krein–Berezanskii integral representation in $\{x \neq 0, y \neq 0\}$ for $D = i\,d/dx$ (see [11]). We say that $(x, y) \mapsto U_{xy} \in L(\mathcal{H})$ is a *D-representation in the Hilbert space* $\mathcal{H}$ if $D_x U_{xy} = B U_{xy}$, B a fixed operator. The modified D-symmetric kernels K have then the representation $K(x, y) = (U_{xy}\eta_\alpha, \eta_\beta)$ for $(x, y) \in \mathbb{R}_{\alpha\beta}$, where U_{xy} is a D-representation of $\mathbb{R}$.

Similarly we can generalize the Naimark dilation theorem which says that if K is a p.d. Toeplitz kernel with values in $L(\mathcal{H})$, then there is a Hilbert space $\mathcal{H}_1 \supset \mathcal{H}$ and a unitary representation $x \mapsto U_x$ in $\mathcal{H}$, with $K(x) = $ the compression of U_x to $\mathcal{H}$.

The consideration of second-order operators D leads to new extensions of Proposition 1, Theorem 1, and related questions.

REMARK 4.   The p.d. Toeplitz kernels are closely related to the Cara-théodory and Nevanlinna–Pick interpolation theorems, and the considera-tion of operator-valued functions or kernels (as in the Naimark theorem) leads to dilation theorems of Nagy and Sarason. All these questions arise also for modified Toeplitz kernels and D-representations. We finally remark that the functions $K_{\alpha\beta}(n)$ in the definition of a modified Toeplitz kernel K are not uniquely determined and questions of the type considered by Adamjan *et al.* (see Remark 2) arise also here.

II.   OTHER DESCRIPTIONS OF Γ-POSITIVE MATRICES

4.   *Analytic Description*

Consider first the case of $\mathbb{T}$, where we take $\Gamma = \mathscr{P}_+ \times \mathscr{P}_-$ and let us describe the matrices $\mu \succ 0$ (with respect to Γ). If $\mu_{jj} \geq 0$ and $|\mu_{12}(E)|^2 \leq \mu_{11}(E)\mu_{22}(E)$ for all sets $E \subset \mathbb{T}$, then it is trivial that $\mu \succ 0$. Therefore we only consider here the interesting case, which generally occurs in applications, where

$$\mu_{11} = \mu_{22} \quad \text{and} \quad |\mu_{12}(E)| \geq \mu_{11}(E) \quad \text{for all} \quad E \subset \mathbb{T}. \quad (20)$$

LEMMA 4.   *Let $\mu_{jk} = w_{jk}\,dt + s_{jk}$, s_{jk} the singular part of μ_{jk}, $j, k = 1, 2$, be such that (20) is satisfied. Then $\mu \succ 0$ iff $(w_{jk}) \succ 0$ and $|s_{12}(E)| = s_{11}(E)$ for all $E \subset \mathbb{T}$.*

Proof. If $\mu > 0$ then by the lifting theorem, $\mu_{22} = \mu_{11} \geq 0$ and there exists $h \in H^1$ with $|\mu_{12}(E) + h(E)| \leq \mu_{11}(E)$ for all $E \subset \mathbb{T}$ [see (13)], where $h(E) = \int_E h\, dt$. Dividing by $|E|$, the Lebesgue measure of the set, and letting $|E| \to 0$, $t \in E$, we get $|w_{12}(t) + h(t)| \leq w_{11}(t)$, so that $(w_{jk}) > 0$. For $|E| = 0$ we have $|\mu_{12}(E)| \leq \mu_{11}(E)$, which together with (20), gives $|\mu_{12}(E)| = \mu_{11}(E)$ for all $|E| = 0$, and hence $|s_{12}| = s_{11}$. The converse is also clearly true. ∎

By Lemma 4 it is enough to consider the matrices of functions $\mathbf{w} = (w_{jk})$ satisfying $w_{11} = w_{22}$ and $|w_{12}| \geq w_{11}$. Consider first the case where $w_{12} \geq 0$. If $\mathbf{w} > 0$ then, by the lifting theorem, there exists $h \in H^1$ with $|w_{12} - h| \leq w_{11}$, so that there exists a function η, $|\eta| \leq 1$, such that

$$h = w_{11}(w_{12}/w_{11} - \eta), \quad w_{12}/w_{11} \geq 1, \quad w_{11} \geq 0. \tag{21}$$

From (21) it follows that $|\arg h| \leq \pi/2$, $\operatorname{Re} h \geq 0$. Moreover, $\operatorname{Re} h > 0$ a.e., because $\operatorname{Re} h(t) = 0$ implies $\eta(t) = 1$, $w_{12}(t)/w_{11}(t) = 1$ and hence $h(t) = 0$. But $h \in H^1$ and can vanish only on a set of measure zero. Setting $w_{12}/w_{11} - \eta = u$, $h = w_{11}u$, $\arg h = \arg u$. Since u is the vector of origin η in the unit disk and vertex w_{12}/w_{11} on the positive real axis, it is clear that if $|\arg u|$ is close to $\pi/2$, then w_{12}/w_{11} has to be close to 1 and $|u|$ must be small. Similarly, if the value of $|\arg u|$ is fixed, then $|u|$ must be in a fixed interval $(0, a)$, and once the values of $|\arg u|$ and $|u|$ are chosen, then w_{12}/w_{11} must be in a fixed interval. Since $\operatorname{Re} h > 0$, $h \in H^1$ implies $h = C\exp(\tilde{v} - iv)$, with $|v| < \pi/2$ and $\exp \tilde{v} = |h| = w_{11}|u|$, we get

PROPOSITION 2. *There are fixed functions Φ, assigning to each $s \in (-\pi/2, \pi/2)$ an interval $\Phi(s) \subset \mathbb{R}_+$, and Ψ, assigning to each pair (s, u), $u \in \Phi(s)$, an interval $\Psi(s, u) \subset (1, \infty)$, such that $\mathbf{w}$ satisfies $\mathbf{w} > 0$ with $w_{12} \geq 0$, $w_{12} \geq w_{11} = w_{22}$ iff*

$$w_{11} = |u|^{-1}e^{\tilde{v}} \quad \text{with} \quad |u(t)| \in \Phi(v(t)), \quad |v| < \pi/2, \quad e^{\tilde{v}} \in L^1, \tag{22}$$

$$w_{12} = |u|^{-1}e^{\tilde{v}} \cdot w' \quad \text{with} \quad w' \in \Psi(v, |u|). \tag{22a}$$

In particular, if the value $w_{12}/w_{11} = \alpha > 1$ is given, then

$$|\cos v|^2 \geq (\alpha^2 - 1)/\alpha. \tag{22b}$$

In order to extend this result for nonpositive w_{12} we need some auxiliary lemmas.

LEMMA 5. (i) *Assume that on $\mathbb{T}$ we have $\phi = e_{-n}h$, $\operatorname{Re} \phi > 0$, where h is holomorphic in $|z| \leq 1$. Then*

$$|\phi| = e^{\tilde{v}}|Q_n|^2 \quad \text{with} \quad |v| < \pi/2 \text{ a.e.}, \tag{23}$$

Q_n analytic polynomial of degree $\leq n$ and moreover,

$$Q_n = \prod_{k=1}^{n} (1 + a_k e^{ikt}) \quad with \quad |a_k| < 1. \tag{23a}$$

(ii) If for each p, $\phi_p = e_{-n}h_p$ with $\mathrm{Re}\,\phi_p > 0$ on $\mathbb{T}$ and h_p holomorphic in $|z| \leq 1$, and if $\phi_p \to \phi$ pointwise, $\phi \neq 0$ a.e., then (23a) holds with $|a_k| \leq 1$.

Proof. (i) Since h is holomorphic in $|z| \leq 1$, $\mathrm{var\,arg}\, h = \mathrm{var\,arg}\,\phi - \mathrm{var\,arg}$ $e_{-n} = n$ and $e_{-n}h = \phi$ is $\neq 0$ in $\mathbb{T}$, it follows that h has exactly n zeros in $|z| < 1$, so that $h(z) = B(z)g(z)$, $B(z) = \prod_1^n (z - \alpha_k)/(1 - \overline{\alpha}_k z)$, $|\alpha_k| < 1$, $g \neq 0$ in $|z| \leq 1$. Hence $z^n \phi(z) = B(z)g(z)$ and, all these functions being continuous, we may choose the argument so that, setting $v(e^{it}) = \arg \phi(e^{it})$, we have

$$\arg g(t) = v(t) + nt - \arg B(t) = v(t) + 2\sum_1^n \arg(1 - \overline{\alpha}_k e^{it}), \quad e^{it} \sim t.$$

Therefore,

$$(\arg g(t))^{\tilde{}} = \tilde{v}(t) + \sum_1^n \log|1 - \overline{\alpha}_k e^{it}|^2,$$

and since

$$|\phi(t)| = |g(t)| = \exp(\log|g|) = |g(0)| \exp(-\arg g(t))^{\tilde{}},$$

we get (23) and (23a).

(ii) By (i), each $|\phi_p|$ has representation (23), (23a) with $v = v_p = \arg \phi_p$, $|v_p| < \pi/2$, $|\alpha_k| = |\alpha_{k,p}| < 1$. Since $\phi \neq 0$, we have $|\phi_p| \to |\phi|$, $v_p \to v = \arg \phi$ a.e., and since $|v_p| < \pi/2$, $v_p \to v$ also in L^2. Therefore, $\tilde{v}_p \to \tilde{v}$ in L^2, and pointwise for a subsequence of $\{v_p\}$, as well as $\alpha_{k,p} \to \alpha_k$, $|\alpha_k| \leq 1$. Passing to the limit we obtain (23) for ϕ. ∎

LEMMA 6.   (i) If $\phi = e_{-n}h$, $h \in H^1(\mathbb{T})$, $\mathrm{Re}\,\phi > 0$, then (23a) holds with $|a_k| \leq 1$. (ii) If h is holomorphic in $|z| < 1$, $\mathrm{Re}(z^{-n}h(z)) > 0$ in $0 < |z| < 1$ and if $z^{-n}h(z) \to \phi(t)$ pointwise for $z \to e^{it}$, $\phi \neq 0$ a.e., then (23) holds.

Proof. (i) Since $h = \sum_0^\infty c_k e_k(t)$, $\phi = c_0 e_{-n} + \cdots + c_{n-1}e_{-1} + \sum_0^\infty c_{k+n}e_k$, we have, for $0 < r < 1$, (P_r the Poisson integral) that

$$P_r \phi = \phi_r(t) = c_0 r^n e_{-n} + \cdots + c_{n-1} r e_{-1} + \sum_0^\infty c_{k+n} r^k e_{k+n}, \quad |c_{k+n}| \leq c, \ |r| < 1,$$

$h'_r = e_n \phi_r$ holomorphic in $|z| \leq 1$. Since $\phi_r = e_{-n}h'_r$, $\phi_r \to \phi$, it is enough to apply Lemma 5(ii). (ii) This also follows from Lemma 5(ii) applied to $\phi_r(t) = r^{-n}e_{-n}h(re^{it})$. ∎

LEMMA 7.   *Assume that for each p we have, on $\mathbb{T}$, $\phi_p = e_{-n}h_p$, $h_p \in H^1$, $\mathrm{Re}\,\phi_p > 0$, and that the measures $\phi_p\,dt$ satisfy $\|\phi_p dt\| \leq c$, $\phi_p\,dt \to \phi\,dt$ in the weak-* topology, for $\phi \neq 0$ a.e. Then (23) holds with $|a_k| \leq 1$.*

Proof. We have $h_p \to h$ in the weak-* topology with $\phi = e_{-n}h$, and for each $r < 1$, $P_r\phi_p(t) \to P_r\phi(t)$ uniformly in each domain $|z| \le r_0 < 1$. In the proof of Lemma 6 (i), we have seen that then $\{P_r\phi_p\}$ fulfills the same assumptions as $\{\phi_p\}$, so that $P_r\phi_p = (\exp \tilde{v}_{rp})|P_{rp}|^2$, $|v_{rp}| < \pi/2$. Passing to the limit in p and then in $r \to 1$, and applying Lemma 5(ii), we get (23) for ϕ. ∎

We extend now Proposition 2 to nonpositive w_{12}, assuming that $e_{-n}w_{12}$ is positive for some $n > 0$, or that w_{12} is a weak limit of functions w_{12}^p with $e_{-n}w_{12}^p \ge 0$.

THEOREM 2. (a) *Assume that for each p we have matrix of functions* $\mathbf{w}^p \succ 0$ *such that* $w_{11}^p = w_{22}^p$, $|w_{12}^p| \ge w_{11}^p$, $e_{-n}w_{12}^p \ge 0$, *and that* $w_{11} = \lim w_{11}^p$ *pointwise and* $w_{12}^p \, dt \to w_{12}dt$ *in the weak-* topology. Then* $\mathbf{w} \succ 0$ *and*

$$
\begin{aligned}
w_{11} &= e^{\tilde{v}}|u|^{-1}|P_n|^2 \quad \text{with} \quad |v| < \pi/2, \quad u \in \Phi(v), \\
w_{12} &= w_{11} \cdot w' \quad \text{with} \quad w' \in \Psi(v, |u|),
\end{aligned}
\tag{24}
$$

where Φ, Ψ are the fixed functions of Proposition 2 and P_n is an analytic polynomial of degree $\le n$.

(b) *If* $w_{11} = w_{22}$, $|w_{12}| > w_{11}$, $e_{-n}w_{12} > 0$ *then* $\mathbf{w} \succ 0$ *iff* (24) *holds with* $e^{\tilde{v}} \in L^1$. *In particular, if the value* $|w_{12}|/w_{11} = \alpha > 1$ *is given then* (22b) *holds for v in* (24).

Proof. (a) As above, for each p, there are functions $|\eta_p| < 1$ and $h_p \in H^1$ with

$$
h_p = w_{11}^p\left(\frac{|w_{12}^p|e_n}{w_{11}^p} - \eta_p\right), \qquad e_{-n}h^p = w_{11}^p\left(\frac{|w_{12}^p|}{w_{11}^p} - \eta_p'\right) = w_{11}^p u_p,
$$

where $\text{Re}(e_{-n}h_p) \equiv \text{Re}(\phi_p) > 0$, $h_p \in H^1$. By passing to a subsequence, we may assume that $|w_{12}^p| \, dt$ and $w_{11}^p\eta \, dt$ converge in the weak-* topology so that $e_{-n}h = w_{11}(|w_{12}|/w_{11} - \eta)$ where $|\eta| < 1$ and $h \, dt$ is the limit of $e_{-n} \cdot h_p \, dt$. Applying Proposition 2 and Lemma 7, we get (24). (b) is included in (a) and Proposition 2. ∎

THEOREM 2A. *Let $\mathbf{w}$ be a matrix of functions defined in $\mathbb{R}$ such that* $(1 + |x|)^{-2k}w_{jn} \in L^1$, k *fixed positive integer*, $w_{11} = w_{22}$, $w_{12} \ge 0$, $w_{12} \ge w_{11}$. *Then* $\mathbf{w} \succ 0$ *with respect to* $\Gamma_+^k \times \Gamma_-^k$ *iff*

$$
w_{11} = \prod_{k=1}^m |(1 - \alpha_k)x + i(1 + \alpha_k)|^2|u|^{-1}e^{\tilde{v}},
\tag{24a}
$$

u, v, *and* w_{12} *as in* (24), $|\alpha_k| \le 1$, $0 \le m \le k - 1$.

Proof. By the lifting theorem for $\mathbb{R}$ we have that $\mathbf{w} \succ 0$ iff $w_{11} \ge 0$ and there exists $h = (i + x)^{2k}h_1$, with $h_1 \in H^1$ and $|w_{12} - h| \le w_{11}$, so that there is a function η, $|\eta| < 1$, with $h = (i + x)^{2k}h_1 = w_{11}(w_{12}/w_{11} - \eta)$, $|\arg h| < \pi/2$. By the change of variables $z = (x - i)/(x + i)$, the functions $h(x)$, $h_1(x)$ are transformed into holomorphic functions $\phi(z)$, $\phi_1(z)$ such that $\phi(z) = c(1 - z)^{-2k}\phi_1(z) = (1 - z)^{-2k+2}h'(z)$, with $h' \in H^1$, $|\arg \phi| < \pi/2$. Setting $\mathscr{H}(z) = (1 - z)^{k-1}(1 - 1/z)^{k-1}\phi = (-1)^{k-1}z^{-k+1}h'$, we have, for $z \in \mathbb{T}$, $\mathscr{H} = ((1 - z)(1 - \bar{z}))^{k-1}h$. Therefore $\mathscr{H} \in L^1$, $\arg \mathscr{H} = \arg \phi$, $h' \in H^1$ and we may apply Lemma 6(i) to get $|\phi| = |\mathscr{H}| = e^{\tilde{v}}\prod_k |1 + a_k e^{it}|^2$, $|a_k| \leq 1$. Changing variables to $\mathbb{R}$ we get (24a). ∎

COROLLARY 3. *If* $w_{12} = |w_{12}|e_n$ *and* $|w_{12}|/w_{11} > \lambda > 1$, $w_{11} \geq 0$, *then* $\mathbf{w} \succ 0$ *implies*

$$\left| \int_{\mathbb{T}} P(t)|w_{12}| \, dt \right| \leq \frac{1}{\lambda} \int_{\mathbb{T}} |P| \, |w_{12}| \, dt \quad \forall P \in e_{n+1}\mathscr{P}_+.$$

Proof. We have $e_{-n}h = |w_{12}| + \eta w_{11}$, $w_{11} \geq 0$, $|\eta| \leq 1$, $h \in H^1$. Hence $P \in e_{n+1}\mathscr{P}_+$ implies

$$\int P|w_{12}| \, dt + \int P\eta w_{11} \, dt = 0,$$

$$\left| \int P|w_{12}| \, dt \right| \leq \int (w_{11}/|w_{12}|)|P| \, |w_{12}| \, dt \leq (1/\lambda) \int |P| \, |w_{12}| \, dt. \quad ∎$$

5. Abstract Variants of the Lifting Theorem

In most of the following applications we encounter only 2×2 matrices $\boldsymbol{\mu}$ satisfying $\mu_{11} = \mu_{22} = \mu_1 \geq 0$ and $\mu_{12} = \bar{\mu}_{21} = \mu_2$. Therefore the quadruple (μ_{jk}) reduces to a pair μ_1, μ_2, and setting $p(f) = \int|f| \, d\mu_1 = \int|f| \, d\mu_{11}$, $l(f) = \int f \, d\mu_2 = \int f \, d\mu_{12}$, the condition $\boldsymbol{\mu} \succ 0$ becomes

$$p(|f_+|^2) + 2\operatorname{Re} l(f_+\bar{f}) + p(|f_-|^2) \geq 0, \quad f_+ \in \Gamma_1, \quad f_- \in \Gamma_2. \tag{25}$$

In this case, when $\mu_{11} = \mu_{22} = \mu_1$, the proof of the lifting theorem (Theorem 1) is much easier and can be given in an abstract form with variants, which deals with triples (p, l, p) satisfying inequalities of type (25), as follows.

Let X be an abstract space, Γ_+, Γ_-, W, V vector spaces of functions defined on X, such that $\Gamma_+ \cdot \bar{\Gamma}_-$ and $\Gamma_- \cdot \bar{\Gamma}_+ \subset W \subset V$ (here $\Gamma_\alpha \cdot \bar{\Gamma}_\beta = \{f_\alpha \bar{f}_\beta : f_\alpha \in \Gamma_\alpha, f_\beta \in \Gamma_\beta\}$) and that f, $g \in V$ implies $\operatorname{Re} f \in V$, $|f| \in V$, and $fg \in V$, where V is a topological vector space.

Let p and q be two (semi)norms on V acting on the scalars (real or complex), satisfying some of the following:

(a) $f_\pm \in \Gamma_\pm \Rightarrow p(|f_\pm|^2) \leq c_1 q(|f_\pm|^2)$,
(a_1) $f \in W \Rightarrow p(|f|) \leq c_1' q(\operatorname{Re} f)$,
(b) $\forall f \in V$, $q(|f|^2) \leq c_2 p(|f|^2)$,

(c) $0 \leq f \leq g \Rightarrow q(f) \leq q(g)$,

(d)   (*factorization property*)

$$\forall f \in W, \exists g^n_\pm \in \Gamma_\pm \text{ such that } f = \lim_n g^n_+ \overline{g^n_-} \text{ and } \lim_n q(|g^n_\pm|^2) \leq q(f),$$

(d_1) the same as (d) for p instead of q.

If l is a continuous linear functional on V we write

$$(p, l, p) \geq 0 \qquad \text{if } p(|f_1|^2) + 2 \operatorname{Re} l(f_1 \overline{f_2}) + p(|f_2|^2) \geq 0 \;\; \forall f_1, f_2 \in V, \qquad (25a)$$

$$(p, l, p) > 0(\Gamma) \quad \text{if } p(|f_+|^2) + 2 \operatorname{Re} l(f_+ \overline{f_-}) + p(|f_-|^2) \geq 0 \;\; \forall f_\pm \in \Gamma_\pm, \qquad (25b)$$

$$(p, l, p) > 0(|\Gamma|) \quad \text{if } p(|f_+|^2) - 2|l(|f_+| |f_-|)| + p(|f_-|^2) \geq 0 \;\; \forall f_\pm \in \Gamma_\pm. \qquad (25c)$$

PROPOSITION 3.   (i)   *If p, q are (semi)norms on $\mathbb{C}$ that satisfy (a), (b), (c), (d) with $c_1 = c_2 = 1$, then $(p, l, p) > 0(\Gamma)$ iff $\exists l_1$ continuous linear functional on V, such that $l_1 = l$ on W and $(p, l_1, p) \geq 0$, iff $|l_1(f)| \leq q(|f|) \; \forall f \in V$, and iff $|\operatorname{Re} l_1(f)| \leq q(|f|) \; \forall f \in W$.*

(ii)   *If p, q are (semi)norms on $\mathbb{R}$ that satisfy $(a_1), (b_1), (c)$, and (d_1), and if l is a real functional (on real functions), then $(p, l, p) > 0(\Gamma)$ iff $\exists l_1$, a continuous linear functional on V, and $c = c(c'_1, c'_2) > 0$, such that $l_1 = l$ on $\operatorname{Re} W$ and $(cp, l_1, cp) > 0(|\Gamma|)$. Furthermore, $l_1 = l_1^+ - l_1^-$, where $l_1^\pm$ are positive forms and $(cp, l_1^\pm, cp) > 0(|\Gamma|)$.*

Proof.   (i) From the hypotheses if follows easily that

$$|l(f_+ \overline{f_-})| \leq \tfrac{1}{2}(p(|f_+|^2) + p(|f_-|^2)).$$

For every $f \in W$, conditions (a) and (d) imply

$$|l(f)| \leq \lim_n \tfrac{1}{2}(p(|g^n_+|^2) + p(|g^n_-|^2)) \leq \lim_n \tfrac{1}{2}(q(|g^n_+|^2) + q(|g^n_-|^2)) \leq q(|f|).$$

Since it is therefore $|l(f)| \leq q(|f|) \; \forall f \in W$, by the Hahn–Banach theorem, $\exists l_1$ on V such that $l_1 = l$ on W and $|l_1(f)| \leq q(|f|) \forall f \in V$. Then, by (a) and (c),

$$|l_1(f_1 \overline{f_2})| \leq q(|f_1 \overline{f_2}|) \leq \tfrac{1}{2}(q(|f_1|^2) + q(|f_2|^2))$$
$$\leq \tfrac{1}{2}(p(|f_1|^2) + p(|f_2|^2)),$$

which means that $(p, l_1, p) \geq 0$, and the proposition follows. (ii) From the hypotheses follows

$$|l(\operatorname{Re} f_+ \overline{f_-})| \leq \tfrac{1}{2}(p(|f_+|^2) + p(|f_-|^2)).$$

For every $f \in W$, conditions (a_1) and (d_1) imply

$$|l(\operatorname{Re} f)| \leq \lim_n \tfrac{1}{2}(p(|g^n_+|^2) + p(|g^n_-|^2)) \leq p(|f|) \leq c'_1 q(\operatorname{Re} f).$$

Then, $|l(f)| \leq c_1' q(f)$ for all $f \in \text{Re } W$ and $\exists l_1$ on V such that $l_1 = l$ on Re W and $|l_1(f)| \leq c_1' q(f)$ for all $f \in V$. Therefore,

$$2|l_1(|f_+||f_-|)| \leq 2c_1' q(|f_+||f_-|) \leq c_1'(q(|f_+|^2) + q(|f_-|^2))$$
$$\leq c_1' c_2'(p(|f_+|^2) + p(|f_-|^2)),$$

which is $(cp, l_1, cp) > 0(|\Gamma|)$ with $c = c_1' c_2'$.

Furthermore, condition (c) tells that the (semi)norm q is normal (with respect to the usual cone of positive functions) and, by a well-known theorem of M. Krein, $l_1 = l_1^+ - l_1^-$, $l_1^\pm$ positive functionals, with $|l_1^\pm(f)| \leq c_1' q(f)$, and the proposition follows. ∎

Let us deduce from Proposition 3(i) a simplified proof of Theorem 1 for the case $\mu_{11} = \mu_{22} \geq 0$, $\mu_{21} = \bar{\mu}_{12}$. Consider first the case of $\mathbb{T}$ and let $\mu \succ 0$ with respect to $\mathcal{P}_+ \times \mathcal{P}_-$. Set $X = \mathbb{T}$, $\Gamma_\pm = \mathcal{P}_+ \cdot \bar{\mathcal{P}}_- = \{f \in \mathcal{P} : \hat{f}(n) = 0$ for $n \leq 0\} = e_1 \mathcal{P}_+$, $V = C(\mathbb{T})$, $p(f) = q(f) = \int |f| \, d\mu_{11}$, $l(f) = \int f \, d\mu_{12}$, $\forall f \in V$. Then conditions (a), (b), and (c) are clearly satisfied with $c_1 = c_2 = 1$ and (d) holds in a stronger form: For all $f \in W = e_1 \mathcal{P}_+$, there exist $g_\pm \in$ closure in C of $\Gamma_\pm$ with $f = g_+ \bar{g}_-$ and $|g_\pm|^2 = |f|$ (in fact, by the factorization theorem for analytic functions, $f = e_1 h^2 B$, $h \in$ closure of $\mathcal{P}_+$, $|B| = 1$ on $\mathbb{T}$, and it is enough to set $g_+ = e_1 hB$, $g_- = \bar{h}$). Thus Proposition 3(i) applies, and since $\mu \succ 0$ means now $(p, l, p) > 0(\Gamma)$, there exists a continuous functional l_1 on V such that $l_1 = l$ on W and $(p, l_1, p) \geq 0$. If v_{12} is the measure determined by l_1, then we have $v_{12} = \mu_{12}$ on $W = e_1 \mathcal{P}_+ = \mathcal{P}_+ \cdot \bar{\mathcal{P}}_-$, and

$$\int |f_1|^2 \, d\mu_{11} + 2\,\text{Re} \int f_1 \bar{f}_2 \, dv_{12} + \int |f_2|^2 \, d\mu_{11} \geq 0.$$

Thus, setting $v_{11} = v_{22} = \mu_{11}$, $\overline{v_{21}} = v_{12}$, it is $v = (v_{jk}) \geq 0$ and $v \sim \mu$, which proves Theorem 1 for $\mathbb{T}$ and $\mu_{11} = \mu_{22}$.

In the case of $\mathbb{R}$, set $X = \mathbb{R}$, $V = B(\mathbb{R}) = \{f$ Borel measurable: $(1 + |x|)^{2k+2} f(x)$ is bounded$\}$, k a fixed positive integer, $H_\pm = \{f \in L^1 : \hat{f}(x) = 0$ for $x \gtrless 0\}$, $W = B \cap H_\pm = \Gamma_\pm^{k+1}$ [for the definition, see (9)].

Given $\mu \succ 0$ with respect to $\Gamma_+ \times \Gamma_-$, $\mu_{11} = \mu_{22}$, we set $p(f) = q(f) = \int |f| \, d\mu_{11}$, $l(f) = \int f \, d\mu_{12}$, $\forall f \in V$. The norms p, q clearly satisfy conditions (a), (b), (c) with $c_1 = c_2 = 1$. The factorization condition (d) can be easily deduced as above, from the factorization $f = g^2 B$ of H^2 functions. Again, $\mu \succ 0$ means $(p, l, p) > 0(\Gamma)$. Therefore, by Proposition 3(i), there exists a measure $v_{12} \sim l_1$, such that $v_{12} - \mu_{12}$ vanishes on $\Gamma_+ \bar{\Gamma}_-$ and $(p, l_1, p) \geq 0$, i.e.,

$$\int |f_1|^2 \, d\mu_{11} + 2\,\text{Re} \int f_1 \bar{f}_2 \, dv_{12} + \int |f_2|^2 \, d\mu_{11} \geq 0 \quad \forall f_1, f_2 \in V.$$

Thus, setting $v_{11} = \mu_{11} = v_{22}$, $(v_{jk}) = v \geq 0$, and $v \sim \mu$. This proves Theorem 1 for $\mathbb{R}$ and $\mu_{11} = \mu_{22}$.

III. PREDICTION PROBLEMS AND WEAK CARLESON MEASURES

6. *Weighted Problems for the Hilbert Transform with*
    Vanishing Moments

Let H be the Hilbert transform in $\mathbb{R}$, $\mu \geq 0$ a measure in $\mathbb{R}$, $L^1((1 + |x|)^n dx) = \{f \in L^1(\mathbb{R}); f(x)(1 + |x|)^n \in L^1(dx)\}$. We define the Riesz classes by

$$(\mu, v) \in \mathscr{R}_M^{(n)}(\mathbb{R}) \quad \text{if} \quad \int_{-\infty}^{\infty} |Hf|^2 d\mu \leq M \int_{-\infty}^{\infty} |f|^2 dv \quad \forall f \in \mathscr{S} \cap \mathscr{L}_n, \quad (26)$$

where $\mathscr{S}$ is the Schwartz space and

$$\mathscr{L}_n = \left\{ f \in L^1((1 + |x|)^n dx) : \int x^k f(x) dx = 0 \text{ for } k = 0, \dots, n - 1 \right\} \quad (26a)$$

if $n \geq 1$ and $\mathscr{L}_0 = \{\text{all measurable functions}\}$. We also write $\mu \in \mathscr{R}_M^{(n)}$ for $(\mu, \mu) \in \mathscr{R}_M^{(n)}$, and

$$\alpha(M) = \pi/2 - \arccos(M - 1)/(M + 1),$$
$$\beta(M, \phi) = \text{arch}(\cos \phi (1 - (M - 1)^2/(M + 1)^2)^{-1/2}). \quad (27)$$

THEOREM 3. *For a given measure $\mu \geq 0$ in $\mathbb{R}$, the following assertions are equivalent*: (a) $\mu \in \mathscr{R}_M^{(n)}(\mathbb{R})$; (b) μ *is a tempered measure of order* $2(n + 1)$, *such that $\boldsymbol{\mu} \succ 0$ with respect to* $\Gamma_+^{n+1} \times \Gamma_-^{n+1}$ [*see* (9)] *with $\mu_{11} = \mu_{22} = (M - 1)\mu$, $\mu_{12} = \mu_{21} = (M + 1)\mu$;* (c) $d\mu = w(x) dx = e^u |h| dx$, *where $h(x) = (i + x)^{2(n+1)} h_1$, $h_1 \in H^1$, $\|\arg h\|_\infty \leq \alpha(M) < \pi/2$, $|u| \leq \beta(M, \arg h)$* [$\alpha$ *and β as in* (27)]; (d) μ *is tempered of order* $2(n + 1)$, $d\mu = w(x) dx$ *with*

$$w(x) = \prod_{k=1}^{m} |(1 - \alpha_k)x + i(1 + \alpha_k)|^2 e^{u+v}, \quad |\alpha_k| \leq 1, \quad k = 1, \dots m, \quad 0 \leq m \leq n,$$

$$\|v\|_\infty \leq \alpha(M) < \pi/2, \quad |u| \leq \beta(M, v).$$

Proof. (a) *is equivalent to* (b): Assume that (a), i.e., (26) for $\mu = v$, holds. It is easy to see that $\mu \in \mathscr{R}_M^{(n)}$ implies μ is $(2n + 1)$-tempered. If $f_\pm \in \Gamma_\pm^{n+1}$ then $(1 + |x|)^{n+1} f_\pm$ are bounded, the derivatives $D^k \hat{f}_\pm$ are continuous for $k \leq n - 1$ and therefore $D^k \hat{f}_\pm(0) = 0$ for $k \leq n - 1$ [since $\hat{f}_\pm(\xi) = 0$ for $\xi < 0$]. Hence the first $n - 1$ moments of $f_\pm$ vanish, i.e., $f_\pm \in \mathscr{L}_n$, and $f_\pm$ are in the closure of $\mathscr{S} \cap \mathscr{L}_n$ so that (26) holds for $f = f_+ + f_-$. Since $Hf_\pm = \mp if_\pm$, we obtain that for every $(f_+, f_-) \in \Gamma_+^{n+1} \times \Gamma_-^{n+1}$,

$$\int f_+ \bar{f}_+ d\mu_{11} + 2 \operatorname{Re} \int f_+ \bar{f}_- d\mu_{12} + \int f_- \bar{f}_- d\mu_{22} \geq 0, \quad (26b)$$

$\mu_{11} = \mu_{22} = (M - 1)\mu$, $\mu_{12} = (M + 1)\mu$. Thus $\boldsymbol{\mu} \succ 0$ and (b) holds. Conversely, assume $\boldsymbol{\mu} \succ 0$, i.e., (26) holds for $\mu = v$ and $f = f_+ + f_-, f_\pm \in \Gamma_\pm^{n+1}$, and let us prove that (a) holds. For this it is enough to show that $f \in \mathscr{S} \cap \mathscr{L}_n$ implies $f = f_+ + f_-$ with $f_+ \in \Gamma_{\pm}^{n+1}$ and $\hat{f}_{\pm}(\xi) = \hat{f}(\xi)$ for $\xi \lessgtr 0$ and zero otherwise. Since $f \in \mathscr{S} \cap \mathscr{L}_n$ implies that $\hat{f} \in \mathscr{S}$ and $D^k\hat{f}(0) = 0$ for $k \leq n - 1$, we have that $\hat{f}_+ \in C^{n-1}(\mathbb{R})$ and $D^k\hat{f}_+(0) = 0$ for $k \leq n - 1$. Therefore, the derivative of $D^{n-1}\hat{f}_+$ in the sense of distributions coincides then with $D^n\hat{f}(\xi)$ for $\xi > 0$ and is zero for $\xi < 0$, and its Fourier antitransform is $x^n f_+(x)$. Since $g = D^n\hat{f}_+$ coincides with $D^n\hat{f} \in \mathscr{S}$ on $\xi > 0$ and is zero on $\xi < 0$, we have that $\int |g(\xi + t) - g(\xi)| \, d\xi \leq c|t|$, and so $|x^n f_+(x)| \leq c(1 + |x|)^{-1}$, $f_+(x)(1 + |x|)^{n+1}$ is bounded, and $f_+ \in \Gamma_+^{n+1}$. (b) *is equivalent to* (c) *and* (d): We saw that (b) is equivalent to (26b), i.e., that $\boldsymbol{\mu} \succ 0$. Since $\mu_{11} = \mu_{22} = (M - 1)\mu$, $\mu_{12} = (M + 1)\mu$ and, by Lemma 4, singular part of μ_{12} = singular part of μ_{11}, it follows that $d\mu = w(x) \, dx$ and $(\mu_{jk}) = (w_{jk} \, dx)$. It is enough now to apply Theorem 2A (with e^u in place of $|u|^{-1}$) taking into account that $w_{12}/w_{11} = (M + 1)w/(M + 1)w = (M + 1)/(M - 1)$. ∎

Consider now the case of $\mathbb{T}$ and the Hilbert transform H in $\mathbb{T}$. We write $(\mu, v) \in \mathscr{R}_M^{(n)}(\mathbb{T}) = [H, e_n\mathscr{P}_+ + \mathscr{P}_-]_M$ if

$$\int_{\mathbb{T}} |Hf|^2 \, d\mu \leq M \int_{\mathbb{T}} |f|^2 \, dv \quad \forall f \in e_n\mathscr{P}_+ + \mathscr{P}_- \tag{28}$$

is satisfied, or equivalently if the preceding inequality is satisfied for all $f \in \mathscr{P}$ such that $\hat{f}(k) = 0$ for $0 \leq k \leq n - 1$. Since (28) holds for all f of the form $f = e_n f_+ + f_-$, $f_{\pm} \in \mathscr{P}_{\pm}$, it can be rewritten as $\boldsymbol{\mu} \succ 0$ with respect to $\mathscr{P}_+ \times \mathscr{P}_-$, with

$$\mu_{11} = \mu_{22} = Mv - \mu, \qquad \mu_{12} = (Mv + \mu)e_n. \tag{29}$$

In the case $\mu = v$, from (29) and Lemma 4 we obtain $d\mu = w(t) \, dt$ and the equivalence of (29) with

$$\mathbf{w} \succ 0 \quad \text{with } w_{11} = w_{22} = (M - 1)w, \ w_{12} = (M + 1)we_n, \ w \geq 0. \tag{29a}$$

Since here $|w_{21}|/w_{11} = (M + 1)/(M - 1)$, we get the following from Theorem 2.

THEOREM 3A. *For a positive measure μ in $\mathbb{T}$ the following are equivalent:* (a) $\mu \in \mathscr{R}_M^{(n)}(\mathbb{T})$; (b) $d\mu = e^u|h| \, dt$, *where* $h = e_{-n}h_1$, $h_1 \in H^1(\mathbb{T})$, $\|\arg h\|_{\infty} \leq \alpha(M) < \pi/2$, $|u| \leq \beta(M, \arg h)$; (c) $d\mu = w(t) \, dt$, *with* $w(t) = Q(t)R(t)e^{u + \tilde{v}}$, *with*

$$\|v\|_{\infty} \leq \alpha(M) < \pi/2, \qquad |u| \leq \beta(M, v), \tag{29b}$$

$$Q(t) = c \prod_{k=1}^{m} |\beta_k - e^{it}|^2, \qquad |\beta_k| = 1,$$

$$R(t) = \prod_{k=1}^{n-m} |\gamma_k - e^{it}|^2, \qquad |\gamma_k| < 1, \tag{29c}$$

$0 \leq m \leq n$, $\alpha(M)$, $\beta(M, \gamma)$ as in (27). [*Note that $R(t)$ can be incorporated in the factor* $\exp(u + \tilde{v})$.]

Theorem 3A is a refined version of a theorem of Helson–Sarason [16] (of Helson–Szegö [15], for $n = 0$). However, Theorem 3 corresponds to a new situation not considered by those authors and solves a problem posed by Stein and Muckenhoupt. Applying Proposition 1 we obtain the characterization of the Fourier transform of $\mathscr{R}_M^{(n)}$.

THEOREM 3B. *Given a sequence of numbers* (c_m), $c_{-m} = \overline{c_m}$, *there exists a measure* $\mu \in \mathscr{R}_M^{(n)}(\mathbb{T})$ *with* $\hat{\mu}(m) = c_m$ *iff the kernel* $K_n(m, k) = \sum_{\alpha\beta}^2 K_{\alpha\beta}(m - k)1_\alpha(m)1_\beta(k)$ *is p.d., where* $K_{11}(m) = (M - 1)c_{m-n} = K_{22}(m)$, $K_{12}(m) = (M + 1)c_{m+n} = \overline{K_{21}(m)}$. *A similar result holds for* $\mathscr{R}_M^{(n)}(\mathbb{R})$.

Let us characterize now the pairs $(\mu, v) \in \mathscr{R}_M^{(n)}(\mathbb{T})$. By the lifting theorem, (29) is equivalent to the existence of a function $h \in H^1(\mathbb{T})$ such that

$$\mu(E) \leq Mv(E) \quad \text{and} \quad |((Mv + \mu)e_n - h)(E)| \leq |(Mv - \mu)(E)| \quad (30)$$

for all $E \subset \mathbb{T}$. Since $|E| = 0$ implies $h(E) = 0$ and $\mu \leq Mv$, we have

$$d\mu = w(t)\, dt, \qquad 0 \leq w \in L^1. \quad (30a)$$

Writing $dv = \rho\, dt + v_s$, it follows from Lemma 4 that (29) is equivalent to

$$\left(\frac{d\mu_{jk}}{dt}\right) \succ 0, \quad \frac{d\mu_{11}}{dt} = \frac{d\mu_{22}}{dt} = M\rho - w, \quad \frac{d\mu_{12}}{dt} = (M\rho + w)e_n. \quad (30b)$$

Since (30b) entails

$$e_{-n}h = M\rho\left(\left(1 + \frac{w}{M\rho}\right) + \left(1 - \frac{w}{M\rho}\right)\eta\right), \quad |\eta| \leq 1, \quad w \leq M\rho, \quad (30c)$$

we get with the same argument as in Theorem 2,

THEOREM 3C. *The following conditions are equivalent*:

(a) $(\mu, v) \in \mathscr{R}_M^{(n)}(\mathbb{T})$;

(b) $d\mu = w\, dt$, $dv = \rho\, dt + v_s$ *and there exists* $h \in H^1(\mathbb{T})$ *such that* $\mu \leq Mv$, $M\rho - w \geq 0$, *and*

$$|(M\rho + w)e_n - h| \leq M\rho - w \quad \text{a.e.} \quad (31)$$

(c) $d\mu = w\, dt$, $M\rho - w = M\, dv/dt - w \geq 0$ *and*

$$M\rho = e^{\tilde{v}}|u|^{-1}|Q_n|^2, \qquad |v(t)| < \pi/2 \quad \text{a.e.},$$

$$0 < |\mu(t)| < \Phi(v(t)), \quad w = M\rho w', \quad 0 \leq |w'(t)| \leq \Psi(v(t)), \quad e^{\tilde{v}} \in L^1, \quad (31a)$$

where Q_n is an analytic polynomial of degree $\leq n$ and Φ and Ψ are fixed functions (as in Theorem 2).

In particular, given $0 \leq \rho \in L^1$, there is some $w > 0$ with $(w, \rho) \in \mathcal{R}_M(\mathbb{T})$ iff

$$M\rho = e^{\tilde{v}}|u|^{-1}, \qquad |v| < \pi/2 \quad a.e. \qquad and \qquad 0 < u(t) < \Phi(v(t)) \quad \forall t. \quad (31b)$$

A similar result (of the type of Theorem 2A) characterizes $(\mu, v) \in \mathcal{R}_M^{(n)}(\mathbb{R})$.

Observe that if $d\mu = (1 + x^2)^{-1} dx$, $x \in \mathbb{R}$, and if

$$Tf(x) = Hf(x) - c(f) = \int_{-\infty}^{\infty} \frac{f(t)}{x - t} dt - \int_{-\infty}^{\infty} \frac{f(t)}{i - t} dt \quad (32)$$

then $\int |Tf|^2 d\mu \leq M \int |f|^2 d\mu$, for some M and all f in a dense subset of $L^2(\mu)$. Let us characterize the class $\mathcal{R}_M^{(-1)}(\mathbb{R}) = [T]_M$ for $Tf = Hf(x) - c(f)$. Since $Tf(x) = (i - x)Hg(x)$ for $g(x) = (i - x)^{-1}f(x)$, this is equivalent to characterizing the μ for which

$$\int |Hg(x)|^2 (1 + x^2) d\mu \leq M \int |g(x)|^2 (1 + x^2) d\mu \quad \forall g(x) = \frac{f(x)}{i - x}.$$

Thus $dv = (1 + x^2) d\mu$ satisfies $\int |Hg|^2 dv \leq M \int |g|^2 dv$ but only for $g(x) = (i - x)^{-1}f(x)$, i.e., for $g = g_+ + g_-$ with $g_\pm \in \Gamma_\pm^{-1} = \{f : (1 + |x|)^{-1}f(x)$ bounded and $\hat{f}(\xi) = 0$ for $\xi \gtrless 0\}$. The same argument as in Theorem 3 yields:

THEOREM 3D. *A positive measure* $\mu \in \mathcal{R}_M^{(-n)}(\mathbb{R})$ *iff* $d\mu = e^u|h| dx$, *where* $h(x) = (i + x)^{-2n}h_1(x)$, $h_1 \in H^1(\mathbb{R})$, $\|\arg h\|_\infty \leq \alpha(M) < \pi/2$, *and iff* $d\mu = w(x) dx, w(x) = (1 + x^2)^{-m}e^{u + \tilde{v}}, 0 \leq m \leq n, u \in L^\infty, \|v\|_\infty < \pi/2, (1 + x^2)^n e^{\tilde{v}} \in L^1$.

Here $\mu \in \mathcal{R}_M^{(-n)}(\mathbb{R})$ *means that* $\int |Tf|^2 d\mu \leq M \int |f|^2 d\mu$ *for* $Tf(x) = Hf(x) = Hf(x) - c_0(f) - c_1(f)x - \cdots - c_{n-1}(f)x^{n-1}$, *where* $c_k(f) = (-1)^k \int_{-\infty}^{\infty} f(t) \times (i - t)^{-k-1} dt, 0 \leq k \leq n - 1$.

Remark 5. In Theorem 3, (26) is required to hold for functions f satisfying $\int x^k f(x) dx = 0$ for $k = 0, \ldots, n - 1$. Our method allows us also to obtain the characterization of the classes $\mathcal{R}_M^\alpha(\mathbb{R})$ of pairs of finite measures (μ, v) satisfying (26) for functions f of the form $f = \sum c_k \exp(i\alpha_k x)$, with $c_k = 0$ for $0 \leq \alpha_k \leq \alpha$, α fixed. The measures such that $(\mu, \mu) \in \mathcal{R}_M^\alpha(\mathbb{R})$ for some M were studied in Ibragimov–Rozanov [18], and a deep study of the class $\mathcal{R}_1^\alpha$ was done by Koosis in [22]. Thus the lifting method yields also a refinement of those results of Ibragimov–Rozanov and partially of those of Koosis for $M > 1$.

7. Applications to Past and Future

We work here in $\mathbb{T}$ and write $e_n \mathscr{P}_+ = \mathscr{F}_n$, $n \geq 0$, $\mathscr{M}(\mathbb{T})$ for the class of positive measures in $\mathbb{T}$. For each $\mu \in \mathscr{M}(\mathbb{T})$, set

$$\rho_n = \rho_n(\mu) = \sup\{|\textstyle\int f\bar{g}\,d\mu| : f \in \mathscr{F}_n, g \in \mathscr{P}_-, \textstyle\int |f|^2\,d\mu = \int |g|^2\,d\mu = 1\}. \quad (33)$$

The number $\alpha_n = \arccos \rho_n(\mu)$ is called the angle between the past $\mathscr{P}_-$ and the future $\mathscr{F}_n$, with respect to μ. We set

$$B_n(\rho) = \{\mu \in \mathscr{M} : \rho_n(\mu) \leq \rho\}, 0 \leq \rho \leq 1; \qquad B_n = \bigcup\{B_n(\rho) : 0 \leq \rho \leq 1\}.$$
$$(33a)$$

The characterization of B_0 was first given by Helson–Szegö [15], that of B_n, by Helson–Sarason [16]. The measures μ satisfying $\rho_n(\mu) \to 0$ for $n \to \infty$ were also studied in [16], and those for which $\rho_n(\mu)$ decreases "rapidly" to zero, were studied by Ibragimov–Rozanov [18]. A thorough study of prediction from finite sections of time has been done by Krein [23], the case of $\rho = 0$ has been done by Koosis [22], and all of these considerations have been developed by Dym–McKean [12]. All the characterizations mentioned give complete descriptions of the corresponding classes, but in practice do not allow us to decide whether a given μ belongs to the required class; on the other hand, they do not characterize $B_n(\rho)$ for fixed ρ, nor do they treat the case of pairs of measures (μ, ν). An important characterization of B_0, which allows us to decide whether a given μ belongs to B_0 and which leads to significant extensions is given by Hunt et $al.$ [17] (see also [24]). Theorems 3A and 3C give characterizations of $B_n(\rho)$ for fixed ρ and for pairs of measures. In fact, setting $\rho = (M - 1)/(M + 1)$, or $M = (1 + \rho)/(1 - \rho)$, then, by (33), $\mu \in B_n(\rho)$ iff

$$\left|\int f\bar{g}(M + 1)\,d\mu\right|^2 \leq \left(\int |f|^2(M - 1)\,d\mu\right)\left(\int |g|^2(M - 1)\,d\mu\right) \quad \forall f \in \mathscr{F}_n, g \in \mathscr{P}_-$$

and by (12a) this is equivalent to $\mu > 0$ with $\mu_{11} = \mu_{22} = (M - 1)\mu$, $\mu_{12} = \bar{\mu}_{21} = e_n(M + 1)\mu$. Thus,

$$\mu \in B_n(\rho) \text{ iff } \mu \in \mathscr{R}_M^{(n)}(\mathbb{T}) \text{ for } M = (1 + \rho)/(1 - \rho) \quad (33b)$$

and such μ are characterized by Theorem 3A. Moreover, Theorem 3B gives a characterization of the Fourier transforms $\{\hat{\mu} : \mu \in B_n(\rho)\}$, and in particular, these theorems give characterizations of the classes B_n and $\{\hat{\mu} : \mu \in B_n\}$. (For the case $n = 0$, these characterizations of $B_0(\rho)$ and $\{\hat{\mu} : \mu \in B_0(\rho)\}$ were given in [11]). We shall indicate now briefly how Theorems 3–3C lead naturally to results of Helson–Sarason and Ibragimov–Rozanov concerning the behaviors of $\rho_n(\mu)$ for $n \to \infty$, with some refinements.

By (33b), $w \in B_n(\rho)$ if $w = |P|^2 w_0$, where $P \in \mathscr{P}_+$ is of degree $\leq n$ and

$$w_0 = \exp(u + \tilde{v}) \in B_0(\rho) \tag{33c}$$

with u, v as in (29b), so that $B_n(\rho) = B_n(0)B_0(\rho)$. Let $B = \bigcup_n B_n$, $B_n \uparrow B$, and for each $\mu \in B$ set $N(\mu) = \min\{n : \mu \in B_n\} = \min\{n : \rho_n(\mu) < 1\}$. For each $n \geq N(w\,dt)$ we have then

$$w = |Q|^2 |R|^2 w_n, \qquad \text{where} \quad w_n \in B_0(\rho_n(w)) \tag{33d}$$

and

$$Q_n = c \prod_{k=1}^{m} |\beta_k - e^{it}|^2, \qquad |\beta_k| = 1,$$

$$R_n = \gamma \prod_{k=1}^{n-m} |\gamma_k - e^{it}|^2, \qquad |\gamma_k| < 1, \tag{33e}$$

and moreover

$$0 \leq m = N(w) \leq n. \tag{34}$$

In fact, if n_1, $n_2 \geq N(w)$ then $|Q_{n_1}|/|Q_{n_2}| = (|R_{n_1}|/|R_{n_2}|)(w_{n_1}/w_{n_2})^{1/2}$ with Q_n, R_n as in (33e) and $(w_{n_1}/w_{n_2})^{1/2}$ integrable; so that every zero of Q_{n_2} is a zero of Q_{n_1} and conversely. And if it were $\deg Q = l < N(w)$, we would have $w \in B_l$, since the R_n are bounded below and $|R|^2 w \in B_0$. Thus,

$$w = |Q|^2 |R_k|^2 w_{N(w)+k} \qquad \forall k \geq 0$$

with

$$\deg Q = N(w), \qquad \deg R_k \leq k, \qquad w_{N(w)+k} \in B_0(\rho_{N(w)+k}). \tag{34a}$$

Hence

$$w = |Q|^2 w_{N(w)}, \qquad \deg Q = N(w), \qquad w_{N(w)} \in B_0(\rho_{N(w)}). \tag{34b}$$

Similarly, $w_{N(w)} = |R_k|^2 w_{N(w)+k}$, $w_{N(w)+k} \in B_0(\rho_{N(w)+k})$, $|R_k|^2 \in B_k(0)$ [with the notations of (34a)].

Since $\rho_n(\mu)$ is nonincreasing, we set for $\mu \in B$,

$$\rho_\infty(\mu) = \lim_n \rho_n(\mu) \in (0, 1). \tag{35}$$

Now, by (34b) and (33c) (with $n = 0$), we have that

$$\log w_{N(w)} \in L^\infty + \widetilde{L^\infty} = \{u + \tilde{v} : u, v \in L^\infty\}.$$

For each $\phi \in L^\infty + \widetilde{L^\infty}$, we denote the distance of ϕ to $C(\mathbb{T}) \subset L^\infty + \widetilde{L^\infty}$ by

$$d(\phi, C(\mathbb{T})) = \inf\{\max(\|u\|_\infty, \|v\|_\infty) : \phi = f + u + \tilde{v}, \qquad f \in C(\mathbb{T})\}.$$

PROPOSITION 4

$$d(\log w_{N(w)}, C(\mathbb{T})) \leq \max\{(\pi/2\text{-arcos } \phi_\infty(w)), \text{arch}(1 - \rho_\infty^2(w))^{-1/2}\}.$$

Proof. From what was shown above it follows that $\log w_{N(w)} =$ constant $+ \log|R_k|^2 + u_{N(w)+k} + \tilde{v}_{N(w)+k}$ for $k \geq 0$, where $f =$ constant $+ \log|R_k|^2 \in C(\mathbb{T})$ and $v_{N(w)+k}, u_{N(w)+k}$ are as in (29b).

PROPOSITION 5. $\rho_\infty(w) = 0$ *iff for every* $\varepsilon > 0$ *w has a representation* $w = |Q|^2|R|^2 e^{u+\tilde{v}}$, *where Q and R are of type* (33e) *and* $\|u\| + \|v\|_\infty < \varepsilon$.

Proof. Let $\rho_\infty(w) = 0$ and take k so that $\rho_{N(w)+k} < \varepsilon$. Since the $w_{N(w)+k}$ in (34a) is of the form constant times $\exp(u + \tilde{v})$ with u, v satisfying (29b) with $(M - 1)/(M + 1) = \rho_{N(w)+k}$, it is clear that for ε small enough we get the described representations for w. Conversely, assume that, for each $\varepsilon > 0$, w has the representation and let us show that, given $\delta > 0$, there exists an n with $w \in B_n(\delta)$. If $\varepsilon > 0$ is small enough, we shall have $w = |Q|^2|R|^2 \exp(u + \tilde{v})$ with

$$\|v\|_\infty < \pi/2 - \text{arcos } \delta, \|u\|_\infty \leq \text{arch}(\cos \varepsilon(1 - \delta^2)^{-1/2}) \leq \text{arch}(\cos(1 - \delta^2)^{-1/2}),$$

and hence $c \exp(u + \tilde{v}) \in B_0(\delta)$. Setting $n = \deg Q + \deg R$, we have $w \in B_n(\delta)$. ■

Let $C + \tilde{C}$ be the closure of $C(\mathbb{T})$ in $L^\infty + \widetilde{L^\infty}$. From Propositions 4 and 5 we obtain

PROPOSITION 6. $\rho_\infty(w) = 0$ *iff* $\log w_{N(w)} \in C + \tilde{C}$.

Therefore we get the theorem equivalent to Sarason [30]:

$$\rho_\infty(w) = 0 \quad \text{iff} \quad w = |P|^2 e^\phi, \; P \in \mathscr{P}_+, \; \phi \in C + \tilde{C}. \tag{35a}$$

From (33c) we have immediately

PROPOSITION 7.   *Let* $d\mu = w\,dt \in B$, $\{c_n\} \subset (0, 1)$, $c_n \to 0$. *Then for* $n \to \infty$, *we have* $\rho_n(\mu) \leq c_n$ *iff for each n, w has a representation*

$$w = |P_n|^2 \exp(u_n + \tilde{v}_n), \quad \text{with } P_n \in \mathscr{P}_+, \; \deg P_n \leq n, \; \|v\|_\infty \leq \pi/2 - \text{arcos } c_n,$$
$$|u_n| \leq \text{arch}(\cos v_n(1 - c_n^2)^{-1/2}).$$

We give now two conditions, one necessary and other sufficient, where only the norms of u and v appear.

PROPOSITION 8.   *Let c_n be as in Proposition 7*, $b_n = c_n^2$.

(i)   *If* $\rho_n(\mu) \leq c_n$ *then for each n*, $w = |P_n|^2 \exp(u_n + \tilde{v}_n)$, P_n *and v_n as in*

Proposition 7,

$$\|u_n\|_\infty \leq \tfrac{1}{2}\log(1 + c_n)/(1 - c_n).$$

(ii) *For each n, $w = |P_n|^2 \exp(u_n + \tilde{v}_n)$ with*

$$\|v_n\|_\infty \leq \pi/2 - \arccos b_n, \quad \|u_n\|_\infty \leq \operatorname{arch}(1 + b_n)^{1/2}, \quad P_n \text{ as above,}$$

then $\rho_n(u) \leq c_n$.

Proof. (i) $\operatorname{arch}(\cos v_n(1 - c_n)^{-1/2}) \leq \operatorname{arch}(1 - c_n^2)^{-1/2} \leq \tfrac{1}{2}\log(1 + c_n)/(1 - c_n)$.
(ii) We have

$$\|v_n\|_\infty \leq \pi/2 - \arccos c_n, \qquad \cos v_n \geq \cos(\pi/2 - \arccos b_n) = (1 - b_n^2)^{1/2},$$

hence

$$\operatorname{arch}(\cos v_n(1 - c_n^2)^{-1/2} \geq \operatorname{arch}((1 - b_n^2)/(1 - c_n^2))^{1/2} = \operatorname{arch}(1 + b_n)^{1/2} \geq \mu$$

and it is enough to apply Proposition 7. ∎

PROPOSITION 9. *Let $\{c_n\} \subset (0, 1)$, $c_n \to 0$. If $w = |P_n|^2 \exp(u_n + \tilde{v}_n)$ is as in Proposition 7, with $\|v_n\|_\infty = O(c_n^2)$, $\|u_n\|_\infty = O(c_n)$, then $\rho_n(w) = O(c_n)$.*

Proof. We have

$$\lim_{x \to 0+} \frac{\pi/2 - \arccos x}{x} = \lim_{x \to 0+} \frac{\sin(\pi/2 - \arccos x)}{x} = 1,$$

and similarly for $\operatorname{arch}(1 + x^2)^{1/2}/x$. Hence, there is $\lambda > 0$ such that $\lambda x \leq \pi/2 - \arccos x$, $\lambda x \leq \operatorname{arch}(1 + x^2)^{1/2}$, for all $0 \leq x \leq 1$. By hypothesis, $\|v_n\|_\infty \leq Kc_n^2$, $\|u_n\|_\infty \leq Kc_n$, $K > \lambda$, hence if $c_n' = Kc_n/\lambda$ then $\|v_n\|_\infty \leq \lambda(Kc_n/\lambda)^2 \leq \pi/2 - \arccos c_n'^2$, $\|u\|_\infty \leq \operatorname{arch}(1 + c_n'^2)^{1/2}$. From Proposition 8 it follows then that $\rho_n(\mu) \leq c_n' = Kc_n/\lambda$. ∎

Finally, observe that since $\mu \in B_n(\rho)$ is equivalent to $\boldsymbol{\mu} \succ 0$, $\mu_{11} = \mu_{22} = (M - 1)\mu$, $\mu_{12} = (M + 1)e_n\mu$, we obtain from Corollary 3 (Section 4) that

$$\mu \in B_n(\rho) \qquad \text{iff} \qquad \left|\int_{\mathbb{T}} P \, d\mu\right| \leq \rho \int_{\mathbb{T}} |P| \, d\mu \quad \forall P \in \mathscr{F}_{n+1} = e_{n+1}\mathscr{P}_+$$

and

$$\text{iff} \left|\int_{\mathbb{T}} e_{n+1} P \, d\mu\right| \leq \rho \int_{\mathbb{T}} |P| \, d\mu \quad \forall P \in \mathscr{P}_+. \tag{36}$$

If $P = \sum_n^{n+k} a_j e_j, \mu(e_j) = \hat{w}(-j)$, then $\int P \, d\mu = \sum_n^{n+k} \hat{w}(j)a_j$, and letting $a_n = e^{int}$, we get

$$\left|\sum_{2^k}^{2^{k+1}} \hat{w}(n)e^{int}\right| \leq \rho_{2^k} \int_0^{2^k} \left|\sum^{2^k} e^{in(t-s)}\right| w(s) \, ds.$$

From this last relation it follows easily that

$$\text{if} \quad \rho_n(\mu) = O(n^{-r-\beta}) \quad \text{then} \quad w \in C^r \quad \text{and} \quad w^{(r)} \in \text{Lip}(\beta - \varepsilon) \quad (36a)$$

(see [18]).

8. *Characterization of the Extremal Elements of $B_n(\rho)$*

By (33c), $w \in B_n(\rho)$ iff

$$w = |Q|^2 |R|^2 w_0, \qquad \deg Q + \deg R = q + r \leq n, \qquad (37)$$

Q, R as in (33e).

LEMMA 8. (i) *If $q < n$ then $w \notin \text{Ext } B_n(\rho) = $ the set of extremal elements of $B_n(\rho)$. (ii) If $w = Q^2 w_0$, Q as in (33e) and $w_0 \in \text{Ext } B_0(\rho)$ then $w \in \text{Ext } B_n(\rho)$.*

Proof. (i) Consider two cases: (a) $r = -q + n$, therefore $r > 0$, $|R|^2 \neq$ constant, $|R|^2 \geq m > 0$. Let w_1, w_2 be given by $w_j = |Q|^2(|R|^2 + (-1)^j m)w_0$, $j = 1, 2$. Then $w_1, w_2 \in B_n(\rho)$, $w = (w_1 + w_2)/2$, $w_1/w \neq$ constant. (b) $r < -q + n$. Let $1 = |S_1|^2 + |S_2|^2$, $S_1, S_2 \in \mathscr{P}_+$, $\deg S_1 = \deg S_2 = 1$, $S_1/S_2 \neq$ constant. Then, $w = |Q|^2|R|^2|S_1|^2 w_0 + |Q|^2|R|^2|S_2|^2 w_0 = w_1 + w_2$ with $w_1, w_2, B_n(\rho)$, noncolinear.

(ii) Let $w = w^1 + w^2 = |P_1|^2 w_0^1 + |P_2|^2 w_0^2$, as in (33c). If β_j is not a zero of Q and of P_1, then $|P_1|/|Q|$ is not integrable and $(w_0/w_0^1)^{1/2} - |P_1|/|Q|$ is not integrable either, contradicting the fact that w_0 and $1/w_0^1 \in B_0$. ■

In [3] one of the authors gave a characterization of $\text{Ext } B_0(\rho)$. From that characterization and Lemma 8 we obtain

THEOREM 4. *The following conditions are equivalent:*

(a) $w \in \text{Ext } B_n(\rho)$.
(b) $w = \prod_{k=1}^n |\beta_k - e^{it}|^2 w_0$, $|\beta_k| = 1$, $k = 1, \ldots m$, $w_0 \in \text{Ext } B_0(\rho)$.
(c) *there exists a unique pair (u, v) such that:*

    (i) $\|v\|_\infty \leq \arccos \rho$,
    (ii) $|u| \leq \text{arch}[\cos v(1 - \rho^2)^{-1/2}]$,
    (iii) $w = c \prod_{k=1}^n |\beta_k - e^{it}|^2 e^{u+\tilde{v}}$, $c = $ constant, $|\beta_k| = 1$, $k = 1, \ldots n$.

In this case $|u| = \text{arch}[\cos v(1 - \rho^2)^{-1/2}]$, a.e.

9. *Weak Classes $((T))_M$ and Carleson Measures*

In this section we illustrate how the weak classes $((T))_M$, defined in (4) of the Introduction, appear naturally and become of interest in the study of the space $BMO \sim L^\infty + \widetilde{L^\infty}$ and of the Carleson measures. In Section 6 we saw that $(\mu, \mu) \in \mathscr{R}_M(\mathbb{T}) = \mathscr{R}_M^{(0)}(\mathbb{T}) = [H, \mathscr{P}_+ \times \mathscr{P}_-]_M$, for some M, iff $d\mu = w\,dt$ with $\log w \in L^\infty + \widetilde{L^\infty}$, and in Section 7 that $(\mu, \mu) \in \mathscr{R}_{M_n}^{(n)}(\mathbb{T})$ with $M_n \to 1$ iff $d\mu = w\,dt$ with $\log w \in C + \tilde{C} = $ closure of C in $L^\infty + \widetilde{L^\infty}$. The following proposition shows that the classes $L^\infty + \widetilde{L^\infty}$ appear in a more natural way without logarithms, by considering weak classes of pairs (μ, dt), for μ a real measure and dt the Lebesgue measure.

PROPOSITION 10. (i) $(\mu, dt) \in [H]_M = \mathscr{R}_M^{(0)}$ iff $d\mu = \phi\,dt$, $0 \le \phi \in L^\infty$, $\|\phi\|_\infty \le M$.

(ii) $(\mu, dt) \in ((H))_M = $ weak $\mathscr{R}_M^{(0)}(\mathbb{T})$ iff $d\mu = \phi\,dt$, $\phi = u + \tilde{v} + C \in L^\infty + \widetilde{L^\infty}$, $\|u^2 + v^2\|_\infty^{1/2} \le M$, C constant.

(iii) $(\mu, dt) \in $ weak $\mathscr{R}_M^{(n)}(\mathbb{T})$ iff $d\mu = \phi\,dt$, $\phi = Q_n + u + \tilde{v} \in \mathscr{P}_n + L^\infty + \widetilde{L^\infty}$, $\|u^2 + v^2\|_\infty^{1/2} \le M$. In particular $(\mu, dt) \in $ weak $\mathscr{R}_{M_n}^{(n)}(\mathbb{T})$ for all $n = 1, 2, \ldots$, with $M_n \to 1$ iff $d\mu = \phi\,dt$, $\phi \in C + \tilde{C}$.

Proof. (i) By definition, $(\mu, dt) \in \mathscr{R}_M^{(0)}(\mathbb{T})$ iff $(\mu_{jk}) \succ 0$ with respect to $\mathscr{P}_+ \times \mathscr{P}_-$ with $\mu_{11} = \mu_{22} = M\,dt - \mu$, $\mu_{12} = \bar{\mu}_{21} = M\,dt + \mu$, which is equivalent (cf. Sections 4 and 6) to $d\mu = \phi\,dt$ with $M - \phi \ge 0$ and $|M + \phi - h| \le M - \phi$ a.e., $h \in H^1(\mathbb{T})$. But if $\phi \le M$ it is enough to take $h = 0$, then condition $0 \le \phi \le M$ alone is necessary and sufficient.

(ii) $(\mu, dt) \in ((H))_M$ iff $(\mu_{jk}) \succ 0$ with respect to $\mathscr{P}_+ \times \mathscr{P}_-$, with $\mu_{11} = \mu_{22} = M\,dt$, $\mu_{12} = \bar{\mu}_{21} = M\,dt + \mu$ [see (4a)]. This is equivalent to $d\mu = \phi\,dt$, $|M + \phi - h| \le M$ a.e., $h \in H^1(\mathbb{T})$. Write $h = h_1 + ih_2$, $M + \phi - h = M\eta$, $|\eta| \le 1$; we get $h_2 = \text{Im } M\eta = v$ and $M + \phi - h_1 = M + \phi - \tilde{v} + C = \text{Re } M\eta$. Thus, $\phi = u + \tilde{v} + C$, where $|u| = |\text{Re } M\eta|$, $|u|^2 + |v|^2 \le |\text{Re } M\eta|^2 + |\text{Im } M\eta|^2 \le M^2$.

(iii) Now, $(\mu, dt) \in \mathscr{R}_M^{(n)}(\mathbb{T})$ is equivalent to $(\mu_{jk}) \succ 0$ with respect to $e_n \mathscr{P}_+ \times \mathscr{P}_-$, with $\mu_{11} = \mu_{22} = M\,dt$, $\mu_{12} = \bar{\mu}_{21} = e_n(M\,dt + \mu)$. This is now equivalent to $d\mu = e_n \phi\,dt$, $(M + \phi)e_n - h = M\eta$, $|\eta| \le 1$, $h \in H^1$ or $M + \phi - e_{-n}h = M\eta_1$, $|\eta_1| \le 1$. Writing $-e_{-n}h = P^n + h'$ with $P^n = \sum_{-n}^{-1} c_k e_k$, $h' \in H^1(\mathbb{T})$, we have $\phi + M + P_1^n + h_1' + i(P_2^n + h_2') = M\eta_1$ and $h_1' = -\tilde{h}_2' + $ const. Thus, $P_2^n + h_2' = v + \text{Im } M\eta_1$, $\phi + M + P_1^n + \tilde{P}_2^n - (P_2^n + h_2')^{\sim} = u = \text{Re } M\eta_1$. Hence $\phi = Q_n + u + \tilde{v}$ where $|u|^2 + |v|^2 \le M^2$. ∎

Consider now the classes $[T]_M$ and $((T))_M$ for $T = P = $ the Poisson operator that transforms $f \in C(\mathbb{T})$ into its harmonic extension to the disk $\bar{\Delta} = \{z : |z| \le 1\}$, instead of $T = H$.

A measure $v \ge 0$ defined in Δ is said to be M-Carleson if $(v, dt) \in [P]_M$, i.e.,

$$\int_\Delta |Pf|^2 \, dv \le M \int_\mathbb{T} |f|^2 \, dt \quad \forall f \in C(\mathbb{T}). \tag{38}$$

A complex measure v is M-Carleson if $|v|$ is M-Carleson.

We define a real measure v in $\bar\Delta$ to be *weak M-Carleson* if $(v, dt) \in ((P))_M$ with respect to $\mathscr{P}_+ \times \mathscr{P}_-$. Since $P(f_+ \bar f_-) = P(f_+)\overline{P(f_-)}$ and $\int f_+ \bar f_- \, dt = 0$ for all $f_\pm \in \mathscr{P}_\pm$, by (4), condition (38) can be rewritten as

$$M \int_{\mathbb{T}} |f_+|^2 \, dt - 2 \operatorname{Re} \int_\Delta P(f_+ \bar f_-) \, dv + M \int_{\mathbb{T}} |f_-|^2 \, dt \geq 0. \tag{39}$$

If v^b = balyage of $v = P^*v$ is the measure on $\mathbb{T}$ defined by

$$\int_{\mathbb{T}} f \, dv^b = \iint_\Delta Pf \, dv,$$

(39) takes the form $\sum_{j,k=1}^2 \int_{\mathbb{T}} f_j \bar f_k \, dv_{jk} \geq 0$ for $f_1 = f_+$, $f_2 = f_-$, i.e., $(v_{jk}) \succ 0$, with respect to $\mathscr{P}_+ \times \mathscr{P}_-$, for $v_{11} = v_{22} = M \, dt$, $v_{12} = v_{21} = v^b \geq 0$. By the lifting theorem, this is equivalent to $dv^b = \phi \, dt$ with $|v^b - h| \leq M$, for $h \in H^1(\mathbb{T})$ By the same argument as in Proposition 10(ii), we get (a) and (b) of the following theorem.

THEOREM 5. *For a measure v defined in $\bar\Delta$, the following are as equivalent*: (a) v *is weak M-Carleson*; (b) *balyage of* $v = v^b = \phi \, dt$ *with* $\phi = u + \tilde v + \text{const}$, $\|u^2 + v^2\|_\infty^{1/2} \leq M$; (c) *balyage of* $v = \phi \, dt$ *with* $\phi \in$ *dual of* H_1, *with norm* $\leq M$, *where* $H_1 = \{f = \operatorname{Re} F, F \in H^1, \|f\|_{H_1} = \int |F| \, dt\}$.

Proof. The equivalence of (a) and (b) was already established. We also saw that (a) is equivalent to $p(|f_+|^2) + 2 \operatorname{Re} l(f_+ \bar f_-) + p(|f_-|^2) \geq 0$ for all $f_\pm \in \mathscr{P}_\pm$, with $p(f) = M\int |f| \, dt$, $l(f) = \int f\phi \, dt$. By Proposition 3(i), this is equivalent to

$$\left| \int (\operatorname{Re} F)\phi \, dt \right| \equiv |\operatorname{Re} l(F)| \leq M \int |F| \, dt$$

$$= M\|\operatorname{Re} F\|_{H_1} \qquad \text{for all} \quad F \in W \sim H^1.$$

Hence (a) is equivalent to (c). ∎

Defining v to be M, n-*Carleson* if $(v, dt) \in ((P, e_n \mathscr{P}_+ \times \mathscr{P}_-))_M$, we get similarly that v is M, n-Carleson iff $dv^b = (Q_n + u + \tilde v) \, dt$, $Q_n \in \mathscr{P}_n$, $\|u^2 + v^2\|_\infty^{1/2} \leq M$.

THEOREM 6. *For every weak M-Carleson measure v there exists a cM-Carleson measure μ, such that balyage μ = balyage v, with control over the constant c.*

Proof. We apply Proposition 3(ii) to $X = \bar\Delta$, $\Gamma_+(\Delta) = \{F_\pm = Pf_\pm : f_\pm \in \mathscr{P}_\pm\}$,

$$p(F) = \int_{\mathbb{T}} |F| \, dt, \qquad q(F) = \int_{\mathbb{T}} F^* \, dt, \qquad F^*(t) = \sup\{|F(re^{it})|, 0 < r < 1\},$$

$W = \Gamma_+(\Delta)$, $V = \{F \in B(\bar\Delta) \text{ (bounded Borel measureable function)}: q(F) <$

$\infty\}$, $l(F) = \iint_{\bar{\Delta}} F \, dv$. Then, by (39), v is weak M-Carleson iff $(p, l, p) > 0(\Gamma)$. That the hypotheses (a_1) and (b_1) of Proposition 3 are satisfied is a well-known theorem of Burkholder *et al.* [6], while hypothesis (c) is evident and hypothesis (d_1) follows from the classical factorization of analytic functions $F = G^2 B$, and the fact that p is concentrated on $\mathbb{T}$, where $|B| = 1$. Therefore, Proposition 3(ii) enables us to obtain the existence of $\mu = \mu^+ - \mu^-$, with $\mu = v$ on harmonic functions ($\subset$ closure in C of Re W), which entails $\mu^b = v^b$. The measures μ, $\mu^\pm$ satisfy inequalities such as $(cp, l_1, cp) > 0(|\Gamma|)$ for the corresponding functionals. Taking $f_+ = \bar{f}_- = F_+$, the last inequality gives

$$\iint_{\bar{\Delta}} |F_+|^2 \, d\mu \le cM \int_{\mathbb{T}} |F_+|^2 \, dt \quad \forall \text{ analytic functions } F_+. \tag{40}$$

This shows that μ is cM-Carleson (see Remark 6). Moreover, the support of μ is contained in Δ, since $q(1_{\mathbb{T}}) = 0$ and $|\mu^\pm(1_{\mathbb{T}})| = |\int_{\mathbb{T}} d\mu^\pm| \le q(1_{\mathbb{T}}) = 0$. ∎

Remark 6. Conditions (40) and (38) are equivalent because (38) clearly implies (40), and (40) implies

$$\iint |Pf|^2 \, d\mu \le \iint (|Pf|^2 + |P\tilde{f}|^2) \, d\mu \equiv \iint |F_+|^2 \, d\mu \le c \int |F_+|^2 \, dt$$

$$\le c \int (|f|^2 + |\tilde{f}|^2) \, dt \le 2c \int |f|^2 \, dt.$$

By the classical factorization theorem, (38) and (40) are equivalent to

$$\iint |F_+| \, d\mu \le M \int |F_+| \, dt \quad \text{for analytic } F_+. \tag{40a}$$

And by a theorem of Carleson (for a short proof, see [2]),

$$\mu \text{ is } M\text{-Carleson for some } M \quad \text{iff} \quad \mu(R(I)) \le M_1 |I|$$
$$\text{for all intervals } I \subset \mathbb{T}, \tag{40b}$$

where $R(I)$ is the "square" in Δ with basis at I (with "control over M and M_1").

From Theorem 5, the above remarks, and the inequality $\|\tilde{f}\|_2 \le \|f\|_2$, we obtain that for a measure $v \ge 0$ in $\bar{\Delta}$ the following conditions are equivalent (with $F = Pf$, $f \in H^1$):

(i) v is weak M-Carleson (for some M);

(ii) $|\iint F \, dv| \le M \int |F| \, dt$; $\qquad\qquad\qquad\qquad\qquad\qquad$ (41)

(iii) $|\iint \text{Re } F \, dv| \le M \int |F| \, dt$.

Let us point out that, even though their balyages coincide, *there are weak Carleson measures that are not Carleson*. In fact, it can be shown that the function h, given by $h(re^{i\theta}) = g(r)$ for $|\theta| \le \pi(1 - r)$ and zero otherwise, $g(r) = r^{-1}(\sin \pi(1 - r))^{-1}(1 - r)^{-1/2}$ for $\frac{1}{2} < r < 1$ and zero otherwise, is weak Carleson but not Carleson. The proof of this fact is lengthy and will not be given here.

We remark further that the modified Bochner–Schwartz theorem (Section 3) allows us to give a *characterization of the Fourier transform of* $\{v^b:v$ *weak Carleson*$\}$, i.e., of $\iint z^n \, dv$, for v weak Carleson.

A function ϕ is said to belong to BMO if

$$\sup\left\{\frac{1}{|I|}\int_I |\phi(t) - \phi_I|\, dt : I \subset \mathbb{T} \text{ intervals}\right\} \equiv \|\phi\|_* < \infty,$$

where $\phi_I = \int_I \phi \, dt$ [19]. From (40b) and from the properties of the Poisson integral it is easy to see (cf. [2]) that if v is Carleson then $v^b = \phi$ is in BMO with control over $\|\phi\|_*$.   (42)

From Theorems 5 and 6 and (41) we obtain the equivalence of (1) $\phi \in$ dual of H^1; (2) $\phi = u + \tilde{v}, u, v \in L^\infty$; (3) $\phi =$ balyage of v, v weak Carleson; (4) $\phi =$ balyage of v_1, v_1 Carleson.   (43)

From (42) and (43) it follows that each of these conditions implies $\phi \in BMO$. That (2) implies BMO is a theorem of Stein, Spanne, and Peetre. The equivalence of (1) and (2) is part of the duality theorem of C. Fefferman. The equivalence of (2) and (4) is a theorem of Carleson [7], Amar–Bonami [2] and Jones [20]. Thus, the lifting properties lead in a natural and unified way to some basic facts of the duality theory, with simplified proof and some refinements. Moreover, Fefferman's proof of the equivalence of (1) and (2) is nonconstructive and other authors [21] gave constructive procedures for the decomposition $\phi = u + \tilde{v}$. Since the proof of Theorem 1 gives explicit recurrent relations for the lifting, we obtain also an *explicit recurrent formula for the determination of u and $\tilde{v}$ in (2).*

Remark 7. A variant of Proposition 3 allows us to deduce also a simplified proof of the equivalence of (1) and (4) in $\mathbb{R}^d, d > 1$. The proof of Theorem 3 applies to Rosenbloom's weighted inequality (see [25]) as well as to the Hardy–Littlewood maximal operator, and shall be developed elsewhere.

ACKNOWLEDGMENTS

We are pleased to acknowledge our gratitude to Professor L. Nachbin for inviting us to contribute to this volume, and to Professors Iu. Berezanskii, B. Muckenhoupt and E. Stein for helpful comments.

REFERENCES

1. V. M. ADAMJAN, D. Z. AROV, AND M. G. KREIN, Infinite Hankel matrices and generalized Carathéodory–Fejér problems, *Funct. Anal. Appl.* **2** (1968), 1–19.
2. E. AMAR AND A. BONAMI, Mesures de Carleson d'ordre α et solutions au bord de l'equation $\bar{\partial}$, *Bull. Soc. Math. France* **107** (1979), 23–48.
3. R. AROCENA, A refinement of the Helson–Szegö theorem and determination of the extremal measures, *Studia Math.* **70**, in press; summary in *C. R. Acad. Sci. Paris Sér. A* **288** (1979), 721–724.
4. W. BECKNER, Inequalities in Fourier analysis, *Ann. Math.* **102** (1975), 159–182.
5. I. BEREZANSKII, "Eigenfunction expansions for Selfadjoint Operators," Translations of Mathematics, Vol. 17, Amer. Math. Soc., Providence, Rhode Island, 1968; Naukova Dumka, Kiev, 1965 (in Russian).
6. D. BURKHOLDER, R. GUNDY, AND M. SILVERSTEIN, A maximal function characterization of H^p, *Trans. Amer. Math. Soc.* **157** (1971), 137–153.
7. L. CARLESON, Two remarkes on H^1 and BMO, *Adv. in Math.* **22** (1976), 269–277.
8. R. R. COIFMAN AND C. FEFFERMAN, Weighted norm inequalities for maximal functions and singular integrals, *Studia Math.* **51** (1974), 241–250.
9. M. COTLAR AND R. CIGNOLI, "An Introduction to Functional Analysis," North-Holland Publ., Amsterdam, 1974.
10. M. COTLAR AND C. SADOSKY, A moment theory approach to the Riesz theorem on the conjugate function with general measures, *Studia Math.* **53** (1975), 75–101.
11. M. COTLAR AND C. SADOSKY, On the Helson–Szegö theorem and a related class of modified Toeplitz kernels, *Proc. Symp. Pure Math. AMS* **35**: I (1979), 383–407.
12. H. DYM AND H. P. MCKEAN, "Gaussian Processes, Function Theory, and the Inverse Spectral Method," Academic Press, New York, 1976.
13. C. FEFFERMAN AND E. M. STEIN, H^p spaces of several variables, *Acta Math.* **129** (1972), 137–193.
14. V. GAPOSHKIN, A generalization of a theorem of M. Riesz, *Mat. Sb.* **46** (1958), 359–372.
15. H. HELSON AND G. SZEGÖ, A problem in prediction theory, *Ann. Mat. Pura Appl.* **51** (1960), 107–138.
16. H. HELSON AND D. SARASON, Past and future, *Math. Scand.* **21** (1967), 5–16.
17. R. A. HUNT, B. MUCKENHOUPT, AND R. L. WHEEDEN, Weighted norm inequalities for the conjugate function and Hilbert transform, *Trans. Amer. Math. Soc.* **176** (1973), 227–252.
18. I. IBRAGIMOV AND Y. ROZANOV, "Random Gaussian Processes," Springer-Verlag, Berlin and New York, 1978; Mir, Moscow, 1974 (in Russian).
19. F. JOHN AND L. NIRENBERG, On functions of bounded mean oscillation, *Comm. Pure Appl. Math.* **14** (1961), 414–426.
20. P. W. JONES, Constructions with function of bounded mean oscillation, Ph.D. Thesis, U.C.L.A., 1978.
21. P. W. JONES, Carleson measures and the Fefferman–Stein decomposition of BMO(R), *Ann. Math.* **111** (1980), 197–208.
22. P. KOOSIS, Weighted quadratic means and Hilbert transforms, *Duke Math. J.* **38** (1971), 609–634.
23. M. KREIN, On a fundamental approximation problem, *Dokl. Akad. Nauk SSR* **84** (1954), 13–16.
24. B. MUCKENHOUPT, Weighted norm inequalities for the Hardy maximal function, *Trans. Amer. Math. Soc.* **165** (1972), 207–226.

25. B. MUCKENHOUPT, Weighted norm inequalities for classical operators, *Proc. Symp. Pure Math. AMS* **35**: I (1979), 69–83.
26. B. SZ. NAGY AND C. FOIAS, "Analyse harmonique des operateurs de l'espace de Hilbert," Akad. Kiadó, Budapest, 1967.
27. Z. NEHARI, On bounded bilinear forms, *Ann. of Math.* **65** (1957), 153–162.
28. J. PEETRE, On convolution operators leaving $L^{p,\lambda}$ spaces invariant, *Ann. Mat. Pura Appl.* **72** (1966), 295–304.
29. D. SARASON, Generalized interpolation in H, *Trans. Amer. Math. Soc.* **127** (1970), 179–203.
30. D. SARASON, An attendem to "Past and future," *Math. Scand.* **30** (1972), 62–64.
31. D. SARASON, Function theory in the unit circle, Lecture Notes, Virgina Polyt. Inst. & State Univ., 1978.
32. L. SCHWARTZ, "Théorie des Distributions," Hermann, Paris, 1951.
33. S. SPANNE, Some function spaces defined using mean oscillation, *Ann. Scuola Norm. Sup. Pisa* **19** (1965), 593–608.
34. E. M. STEIN, Singular integrals, harmonic functions and differentiability properties of functions of several variables, *Proc. Symp. Pure Math. AMS* **10** (1967), 316–335.
35. A. ZYGMUND, "Trigonometric Series," 2nd ed., Vols. I and II, Cambridge Univ. Press, Cambridge, 1959.

Green's Functions for Self-Dual Four-Manifolds

M. F. ATIYAH

Mathematical Institute
University of Oxford
Oxford, England

DEDICATED TO LAURENT SCHWARTZ

1. INTRODUCTION

The Penrose twistor theory converts problems of four-dimensional Riemannian geometry into complex analytic geometry of three dimensions. It applies whenever the Riemannian curvature is self-dual, in the sense that the Weyl tensor (the conformally invariant part of the Riemannian curvature) is self-dual. For a general exposition of this theory (in the positive definite case) see [2]. Now on a Riemannian four-manifold there is a conformally invariant version of the Laplace operator (obtained by adding one-sixth of the scalar curvature to the usual Laplacian) and this has therefore a conformally invariant Green's function (under appropriate global hypotheses). The purpose of this chapter is to show what corresponds to this Green's function in the twistor theory. It turns out that we get a very well-known object in complex analytic geometry associated to the diagonal in $Z \times Z$, where Z is the twistor space.

We begin in Section 2 by reviewing some basic material of complex geometry involving the Ext-groups and the Serre–Grothendieck duality. We explain how to a complex submanifold Y of a complex manifold X one can, under suitable assumptions, associate a "fundamental class" which we call the Serre class of Y in X. Then in Section 3 we state and prove the main result (Theorem 1) which shows that the Green's function $G(x, y)$, regarded as a function of x with y held fixed, corresponds to the Serre class of the line Py in the twistor space Z. This result is then put into a two-variable form (Theorem 2) in Section 4, where we prove that $G(x, y)$ corresponds to the Serre class of the diagonal in $Z \times Z$.

In addition to dealing with the scalar Laplacian, it is also possible to deal with its covariant analog for a vector bundle with self-dual connection (a Yang–Mills instanton). This is explained in Section 5 and the particular case when $M = S^4$ is examined in detail. The explicit Green's function formula first found in [4, 5] is rederived by twistor methods. The formulas are

129

so simple that they seemed to call for some general explanation and this was the original motivation for the present investigation. A similar presentation for this case has been given by Drinfeld and Manin [6].

The last part of the chapter is an application of Theorem 1 to the explicit computation of the Green's function for the ALE spaces of Gibbons and Hawking [8]. These spaces have a finite cyclic fundamental group at ∞ and they have been investigated from the twistor point of view by Hitchin [11]. In Section 6 we review Hitchin's description of these manifolds and we extend his results by showing how to compactify conformally (at the expense of introducing some minor singularities at ∞). Hitchin's twistor space is algebraic and given by simple equations. In Section 7 we then proceed to apply the theory of Section 3 and construct the Serre class explicitly. This amounts, in fact, to constructing a rank 2 vector bundle over Z associated to the curve Py. Translating back to M, we obtain an explicit formula for the Green's function. In fact, Page [14] by direct methods has also found this Green's function and, since his formula looks rather different from ours, we spend some time in Section 8 in showing that they coincide.

As Hitchin has indicated in [11] there should be further ALE-manifolds corresponding to all the finite subgroups of $SU(2)$. In principle the methods of this paper give an algebraic method of constructing the corresponding Green's function, once the twistor space and its family of curves are explicitly known. In fact, Theorem 2 shows that the basic construction involves only the twistor space itself. Of course, the explicit solution of the algebraic problem may be difficult to carry out in practice. This is exemplified by another problem for the A_k-manifolds of Section 6 which was suggested by Hawking. On these manifolds, which have two-dimensional homology, there are nontrivial Yang–Mills $U(1)$-fields. One can ask for the Green's functions associated to these on the lines of Section 5. In principle, it is just a question of computing the Serre class of the diagonal in $Z \times Z$, twisted in a certain sense by a holomorphic line bundle. The line bundles are not hard to describe but their Serre classes are not so easy to identify.

In Section 3 we take some care to identify the absolute constant $(1/4\pi^2)$ occurring in the Green's function. When carrying out computations, however, in Section 7 it is simpler to ignore the exact constant (and even the sign). The constant is easily normalized at the end of the calculations.

2. REVIEW OF COMPLEX GEOMETRY

In this section we shall review certain parts of the theory of complex manifolds which go beyond sheaf cohomology and are not so widely known.

This will be essential for our study of Green's functions. A good general reference for most of what we need can be found in [9].

Let X be a compact complex manifold of complex dimension n and let $\mathcal{O}_X$ denote the sheaf of holomorphic functions on X. Then we can introduce the sheaf cohomology groups $H^q(X, \mathcal{O}_X)$ defined for $q = 0, 1, \ldots$. They are finite-dimensional vector spaces and vanish for $q > n$. More generally we can replace $\mathcal{O}_X$ here by the sheaf of holomorphic sections of any holomorphic vector bundle E over X or, more generally still, by any *coherent* sheaf S (i.e., a sheaf of $\mathcal{O}_X$-modules which is locally the cokernel of a homomorphism $\mathcal{O}_X^p \to \mathcal{O}_X^q$). A typical example of such an S is the ideal sheaf J_Y consisting of functions vanishing on some fixed complex analytic submanifold $Y \subset X$. The quotient $\mathcal{O}_X/J_Y$ is also a coherent sheaf on X: it is equal to $\mathcal{O}_Y$ (extended by zero on $X - Y$).

If E is a holomorphic vector bundle then Serre's duality theorem asserts that $H^q(X, E)$ and $H^{n-q}(X, E^* \otimes K)$ are dual vector spaces. Here E^* is the dual bundle of E and K is the canonical line-bundle of X (or equivalently the bundle Ω^n of holomorphic n-forms). If we replace E by an arbitrary coherent sheaf S there is a generalization of Serre's duality theorem due to Grothendieck. To formulate this we need to know that, for any two coherent sheaves A, B on X one can define groups $\text{Ext}_X^q(A, B)$. For $q = 0$ we have

$$\text{Ext}_X^0(A, B) = \text{Hom}_X(A, B)$$

the space of global $\mathcal{O}_X$-homomorphisms $A \to B$. If $A = \mathcal{O}_X$ then

$$\text{Ext}_X^q(A, B) \cong H^q(X, B) \tag{2.1}$$

so that the Ext-groups are a generalization of the cohomology groups. They have appropriate exact sequence properties and, like $\text{Hom}(A, B)$, they are covariant in B and contravariant in A. Grothendieck's duality theorem now asserts that $H^q(X, S)$ is dual to $\text{Ext}_X^{n-q}(S, K)$. If S is the sheaf of sections of a vector bundle E then, generalizing (2.1) we have

$$\text{Ext}_X^{n-q}(E, K) \cong H^{n-q}(X, E^* \otimes K) \tag{2.2}$$

so that we recover Serre's duality theorem.

If now $Y \subset X$ is a complex submanifold of codimension m we can apply the Serre duality theorem on Y and compare it with the Grothendieck duality theorem on X applied to the sheaf $\mathcal{O}_Y$. This shows that

$$H^q(Y, \mathcal{O}_Y) \cong \text{Ext}_X^{q+m}(K_Y, K_X) \tag{2.3}$$

both being dual to $H^{n-m-q}(Y, K_Y) = H^{n-m-q}(X, K_Y)$, where K_Y in the second factor is the extension by zero of the canonical bundle of Y.

In particular, taking $q = 0$ in (2.3) so that the left-hand side reduces to the constants, we get a canonical element

$$\mu(Y) \in \text{Ext}_X^m(K_Y, K_X) \tag{2.4}$$

which we shall refer to as the Grothendieck class of Y. If there is a holomorphic line-bundle L on X which extends the bundle $K_Y^* \otimes K_X$ on Y then (2.4) can be rewritten

$$\mu(Y) \in \text{Ext}_X^m(\mathcal{O}_Y, L). \tag{2.5}$$

For example when $X = P_3$, $Y = P_1$ we get

$$\mu(P_1) \in \text{Ext}_{P_3}^2(\mathcal{O}_{P_1}, \mathcal{O}(-2)). \tag{2.6}$$

Note that in (2.5) and (2.6) a definite isomorphism $L \cong K_Y^* \otimes K_X$ has to be chosen: this is just a question of normalization since the different isomorphisms differ by constant multiples (if Y is connected).

Using the exact sequence

$$0 \to J_Y \to \mathcal{O}_X \to \mathcal{O}_Y \to 0$$

we deduce the long exact sequence of Ext-groups

$$\to \text{Ext}_X^{m-1}(\mathcal{O}_X, L) \to \text{Ext}_X^{m-1}(J_Y, L) \xrightarrow{\delta} \text{Ext}_X^m(\mathcal{O}_Y, L) \to \text{Ext}_X^m(\mathcal{O}_X, L) \to \quad (2.7)$$

Assume now that

$$H^q(X, L) = 0 \qquad \text{for} \quad q = m, m-1, \tag{2.8}$$

then (2.7) shows that we have an isomorphism

$$\text{Ext}_X^{m-1}(J_Y, L) \xrightarrow[\cong]{\delta} \text{Ext}_X^m(\mathcal{O}_Y, L) \tag{2.9}$$

and hence a class

$$\lambda(Y) \in \text{Ext}_X^{m-1}(J_Y, L) \tag{2.10}$$

such that $\delta(\lambda(Y)) = \mu(Y)$.

As a special case we now take $m = 2$ so that $\lambda(Y) \in \text{Ext}_X^1(J_Y, L)$. Since Ext^1 classifies extension classes, i.e., short exact sequences:

$$0 \to L \to S \to J_Y \to 0, \tag{2.11}$$

we obtain a sheaf $S(Y)$ defined by $\lambda(Y)$. Since $J_Y = \mathcal{O}_X$ outside Y it is clear that S is locally free of rank 2 outside Y, whatever extension we have in (2.11). For the extension $\lambda(Y)$ one can show that S is everwhere locally free and so corresponds to a rank 2 vector bundle E. Then (2.11) is equivalent to saying that $E \otimes L^*$ has a section vanishing exactly on Y. In this way we obtain a correspondence between rank 2 vector bundles on X and certain submanifolds of codimension 2 in X. For example when $X = P_3$ we obtain a correspondence between rank 2 vector bundles on P_3 and certain algebraic curves Y in P_3. Since this correspondence was first studied by Serre we shall refer to the class $\lambda(Y)$ in (2.11) as the *Serre class*. In [3] this correspondence was used as a way of constructing $SU(2)$-instantons, and we shall discuss this in further detail in the next section.

Finally we note that the Ext-groups can be restricted to open sets. Thus if A, B are coherent sheaves on X and $U \subset X$ is open then we have a natural homomorphism

$$\text{Ext}_X^q(A, B) \to \text{Ext}_U^q(A, B).$$

In particular, restricting to $X - Y$ we see that the class $\lambda(Y)$ of (2.10) restricts to an element of $H^{m-1}(X - Y, L)$. This is the guise in which we shall meet and employ our fundamental classes.

3. One-Variable Green's Functions

We shall now show how to relate Green's functions for self-dual metrics to suitable Serre–Grothendieck classes in the twistor space. Let M be a four-dimensional manifold with a self-dual metric, in the sense of [2], i.e., one for which the Weyl tensor is self-dual. We denote by Δ the differential operator

$$\Delta = d*d + R/6, \tag{3.1}$$

where R is the scalar curvature. This operator depends only on the conformal structure of M when regarded as acting on $\frac{1}{4}$-densities with values in $\frac{3}{4}$-densities [12]. We assume now that M is *compact*, although we shall relax this condition somewhat later in order to include more interesting examples. Moreover we assume that $\Delta\phi = 0$ has $\phi = 0$ as its only global solution. Under these assumptions Δ is invertible and its inverse operator G has a kernel $G(x, y)$ which is the Green's function. Since Δ is conformally invariant so is $G(x, y)$ when properly interpreted, i.e as a $\frac{1}{4}$-density in both x and y. If we fix y then, as a function of x, $G(x, y)$ satisfies the differential equation

$$\Delta_x G(x, y) = 0 \qquad \text{for} \quad x \neq y. \tag{3.2}$$

At $x = y$ $G(x, y)$ is singular and its "leading term" is given by the Euclidean space formula

$$G(x, y) = \frac{1}{4\pi^2} \cdot \frac{1}{r(x, y)^2} + O\left(\frac{1}{r}\right), \tag{3.3}$$

where $r(x, y)$ is the distance from x to y. Properties (3.2) and (3.3) uniquely characterize $G(x, y)$.

We recall now that the Penrose twistor space of M is a complex three-dimensional manifold Z with a real fibration over M whose fibres are complex projective lines [2]. If M is a spin-manifold there is a natural holomorphic line-bundle H on Z such that $H^{-4} \cong K_Z$. If M is not a spin-manifold then H does not exist but H^2 always exists. When $M = S^4$ and $Z = P_3$ the bundle H is the standard Hopf bundle and following the notation of this example we shall write $\mathcal{O}(n)$ for the sections of H^n and $S(n)$ for $S \otimes \mathcal{O}(n)$. If M is not a spin-manifold then we should only use even values of n.

One of the basic facts of twistor theory (see [12]) is that, for any open set $U \subset M$ and the corresponding $\tilde{U} \subset Z$, there is a natural isomorphism:

$$H^1(\tilde{U}, \mathcal{O}(-2)) \cong \{\text{solutions of } \Delta\phi = 0 \text{ in } U\}. \tag{3.4}$$

Taking $U = M - \{y\}$ it follows from (3.2) that the Green's function $G(x, y)$, regarded as function of x, corresponds to a canonical element

$$\gamma_y \in H^1(Z - P_y, \mathcal{O}(-2)) \tag{3.5}$$

where P_y is the projective line over the point y. To find the Green's function it is therefore enough to find the element γ_y in (3.5).

Now in Section 2 we defined the Grothendieck class $\mu(Y)$ for any sub-manifold $Y \subset X$. For $Y = P_y \subset Z$ we recall that the normal bundle is isomorphic to $H \oplus H$ and so $K_Y^* \otimes K_X$ is isomorphic to the restriction of H^{-2}. Hence from (2.5) we have an element

$$\mu(P_y) \in \text{Ext}^2_Z(\mathcal{O}_{P_y}, \mathcal{O}(-2)) \tag{3.6}$$

generalizing the special case of (2.6) when $M = S^4$, $Z = P_3$. Moreover the sheaf $\mathcal{O}(-2)$ satisfies (2.8), namely,

$$H^1(Z, \mathcal{O}(-2)) = H^2(Z, \mathcal{O}(-2)) = 0. \tag{3.7}$$

To see this we use (3.4) with $U = M$, $\tilde{U} = Z$ and recall that we have assumed that $\phi = 0$ is the only global solution of $\Delta\phi = 0$. This gives the vanishing of $H^1(Z, \mathcal{O}(-2))$, and $H^2(Z, \mathcal{O}(-2))$ then vanishes by Serre duality, since $K_Z \cong H^{-4}$. Thus we are in the situation in which we have the isomorphism of (2.9)

$$\text{Ext}^1_Z(J_{P_y}, \mathcal{O}(-2)) \cong \text{Ext}^2_Z(\mathcal{O}_{P_y}, \mathcal{O}(-2)) \tag{3.8}$$

and hence a Serre class.

$$\lambda(P_y) \in \text{Ext}^1_Z(J_{P_y}, \mathcal{O}(-2)) \tag{3.9}$$

corresponding to the Grothendieck class $\mu(P_y)$. Our main result is then the following:

THEOREM 1. *The Green's function $G(x, y)$ of the conformally invariant Laplacian Δ on a self-dual compact four-manifold M corresponds, up to a factor $1/4\pi^2$, to the Serre class of the line P_y in the twistor space Z of M. More precisely the class $(1/4\pi^2)\lambda(P_y)$ of (3.9), when restricted to $Z - P_y$, coincides with the element γ_y of (3.5).*

Remarks. (1) As noted before $G(x, y)$ is a $\frac{1}{4}$-density in both x and y. Thus γ_y is only defined once we have picked a $\frac{1}{4}$-density at y. Similarly $\lambda(P_y)$ depends on a choice of isomorphism $H^{-2} \to K^*_{P_y} \otimes K_Z$ (along P_y), and these isomorphisms form a one-dimensional space which is naturally identified with the $\frac{1}{4}$-densities at y. In Theorem 1 the understanding is that we pick the same $\frac{1}{4}$-density at y to define both γ_y and $\lambda(P_y)$.

(2) Although we have formulated Theorem 1 globally in situations when we have a unique Green's function the result and proof are essentially local. If we work locally the Green's function is only unique modulo solutions of $\Delta\phi = 0$. Correspondingly from the exact sequence (2.7) with $m = 2$ we see that the Serre class is not uniquely determined by the Grothendieck class but can be modified by elements of $H^1(Z, \mathcal{O}(-2))$, and these correspond to the ambiguity in the Green's function. The essential local content of Theorem 1 therefore is that the Grothendieck class corresponds to the $1/r^2$-singularity.

In view of (3.4), the element $\lambda(P_y)$ corresponds to a $\frac{1}{4}$-density $\phi(x)$, defined in $M - y$, and satisfying $\Delta_x\phi(x) = 0$. To identify $(1/4\pi^2)\phi$ with the Green's function (the assertion of Theorem 1) all that we have to do is to check the local behavior (3.3) as $x \to y$. We shall carry this out in the following three steps:

(i) We shall first prove that $\phi(x)$ is a meromorphic function with a simple pole on the hypersurface C_y (in the complexification M^c of M) representing lines which intersect P_y. Now the local equation of C_y near y is of the form $\rho(x) = 0$, where $\rho = r^2 +$ higher order terms (since infinitesimally the C_y define the light-cones of the conformal structure). Hence

$$\phi(x) = \frac{a(x)}{\rho(x)} = \frac{a_0}{r^2} + O\left(\frac{1}{r}\right)$$

has the form (3.3) for a suitable constant a_0 (possibly dependent on y).

(ii) To identify the constant a_0 we shall consider the limit of $r^2\phi(x)$ along a fixed curve through y, and show that a_0 is a fixed universal constant.

(iii) The final step in calculating this constant is to take $M = R^4$ and compute carefully, making explicit the various conventions being adopted.

To carry out the first step we first note that we can restrict our element $\lambda(P_y)$ to any line P_x (for $x \in M^c$) which lies entirely inside $Z - P_y$, i.e., which does not intersect P_y. This shows that ϕ is holomorphic on $M^c - C_y$. To show that it has a simple pole on C_y we introduce the space $W \subset M^c \times Z$ which represents the correspondence between points x of M^c and lines P_x of Z. The complex dimension of W is five and there are holomorphic fibrations

$$\beta \diagup \overset{W}{} \diagdown \alpha$$
$$M^c \qquad\qquad Z$$

The fibers of β are projective lines and those of α have dimension 2. By definition of the cone C_y we have

$$C_y = \beta(\alpha^{-1}P_y)$$

and so if we put $\tilde{P}_y = \alpha^{-1}P_y$, $\tilde{C}_y = \beta^{-1}C_y$, we have $\tilde{P}_y \subset \tilde{C}_y$. Note that these subspaces of W have dimension 3 and 4, respectively. Now our class $\lambda(P_y)$ of (3.9) lifts naturally to a class

$$\lambda(\tilde{P}_y) \in \text{Ext}^1_W(J_{\tilde{P}_y}, \mathcal{O}(-2))$$

and the corresponding function ϕ on M^c is determined by restricting $\lambda(\tilde{P}_y)$ to $W - \tilde{C}_y$ where the ideal sheaf $J_{\tilde{P}_y}$ is the same as $\mathcal{O}_W$ so that Ext^1 can be replaced by H^1. Since $\tilde{P}_y \subset \tilde{C}_y$ we have

$$J_{\tilde{P}_y} \supset J_{\tilde{C}_y}$$

for their ideal sheaves and so we have a homomorphism

$$\text{Ext}^1_W(J_{\tilde{P}_y}, \mathcal{O}(-2)) \to \text{Ext}^1_W(J_{\tilde{C}_y}, \mathcal{O}(-2))$$

under which $\lambda(\tilde{P}_y)$ goes into an element λ'. Clearly, when we restrict to $W - \tilde{C}_y$, λ and λ' coincide so that λ' also determines ϕ. On the other hand $\tilde{C}_y$ is a divisor so that $J_{\tilde{C}_y}$ is locally free and hence

$$\text{Ext}^1_W(J_{\tilde{C}_y}, \mathcal{O}(-2)) \cong H^1(W, \text{Hom}(J_{\tilde{C}_y}, \mathcal{O}(-2))).$$

Since multiplication by a section of $J_{\tilde{C}_y}$ maps

$$H^1(W, \text{Hom}(J_{\tilde{C}_y}, \mathcal{O}(-2))) \to H^1(W, \mathcal{O}(-2))$$

it follows that $H^1(W, \text{Hom}(J_{\tilde{C}_y}, \mathcal{O}(-2)))$ corresponds to solutions of $\Delta\phi = 0$ with a simple pole on C_y. This completes step (i) of the proof.

For step (ii) we fix a nonsingular analytic curve Γ through y, then its inverse image $\tilde{\Gamma} \subset Z$ is a real three-manifold fibered by complex lines and its complexification $Q = \tilde{\Gamma}^c$ is a complex two-manifold in Z fibered by complex lines (parametrized by Γ^c). The standard example is when $M = S^4$,

Γ is a great circle, and Q is a quadric surface with P_y as a generator. Our aim is to examine the behavior of $\phi(x)$ as $x \to y$ along the curve Γ, and for this it is sufficient to restrict the Serre class $\lambda(P_y)$ to the surface Q. The best way to do this is to use the vector bundle interpretation of the Serre class in which we give a rank 2 vector bundle E over Z with a section s vanishing to first order on P_y. When we restrict E and s to the surface Q (which contains P_y) we have a section vanishing on a divisor. Since we can take the geodesic distance r as a local parameter on Γ, with $r = 0$ defining the point y, we can write $s = rs'$ where s' is a nonvanishing section of the bundle $E' = E|_Q \otimes L^*$, where L is the line-bundle defined by the divisor P_y on Q. Since $\det E' = \det E|_Q \otimes L^{-2}$ it follows that s' defines an extension on Q of the form

$$0 \to \mathcal{O} \to E' \to \mathcal{O}(2) \otimes L^{-2} \to 0 \tag{3.10}$$

and so an element of $H^1(Q, \mathcal{O}(-2) \otimes L^2)$. By its construction this element is the restriction to Q of $r^2\lambda(P_y)$. It follows that the constant term a_0 in the expansion of

$$\phi(x) = \frac{a_0}{r^2} + O\left(\frac{1}{r}\right)$$

is given by restricting (3.10) to P_y. This way of computing a_0 shows that it depends only on the first neighborhood of P_y in Z and thus is independent of y or Z. Thus a_0 is a universal constant.

To evaluate a_0 we may therefore take the standard case when $M = R^4$. It is also necessary to spell out rather carefully the details concerning densities and conformal factors. The twistor space Z is $P_3(C) - P_1(C)$ and we choose homogeneous coordinates (x, y, u, v) so that $u = v = 0$ is the line removed and $x = y = 0$ defines the line P_0 corresponding to the origin of R^4. The fibers of Z are the lines given by the equations

$$x = au + bv, \qquad y = -\bar{b}u + \bar{a}v \tag{3.11}$$

so that the two complex coordinates a, b give the four real coordinates of R^4. We pick $x \wedge y$, $u \wedge v$ as basic alternating forms. This gives $u\,dv - v\,du$ as the basic section of $K_{P_0}(2)$. Differentiating the Eqs. (3.11) at P_0 gives

$$dx \wedge dy = (|da|^2 + |db|^2)u \wedge v$$

showing that we get the standard metric on R^4. Thus $u\,dv - v\,du$ corresponds to the standard $\frac{1}{4}$-density of R^4.

If f is of degree -2 and holomorphic in $U_0 \cap U_1$ where $U_0 \cup U_1 = Z$ then f defines an element $(f) \in H^1(Z, \mathcal{O}(-2))$. To get the corresponding solution ϕ of the Laplace equation on R^4 we have to "evaluate" $\omega = f(u\,dv - v\,du)$ on each fiber of Z. This is done by integrating $(1/2\pi i)\omega$

in each fiber round a closed contour lying inside $U_0 \cap U_1$. For meromorphic f this can also be given by summing the residues inside U_0 (or, with a change of sign, inside U_1). In particular, taking $f = 1/uv$, U_0 the complement of $u = 0$, U_1 the complement of $v = 0$, we have

$$\omega = \frac{u \, dv - v \, du}{uv} = \frac{dv}{v}$$

if we put $u = 1$ and use v as an inhomogeneous coordinate. Since ω has total residue 1 inside U_0 we see that (f) corresponds to the constant function 1 on R^4.

Since P_0 is given by $x = y = 0$ its Serre class is represented by the cocycle $1/xy$, with U_0, U_1 being the complements of $x = 0$ and $y = 0$, respectively. To get the corresponding function ϕ on R^4 we substitute for x and y from (3.11) and compute the total residue inside U_0 of

$$dv/(a + bv)(-\bar{b} + \bar{a}v).$$

The only relevant residue arises from the second factor in the denominator and gives

$$\frac{1}{|a|^2 + |b|^2} = \frac{1}{r^2}.$$

This verifies Theorem 1 for R^4 and by identifying the universal constant a_0 it completes the proof of Theorem 1 in general.

Remarks. (1) According to Hadamard [10] the Green's function for any metric has the form

$$G(x, y) = A/r^2 + B \log r + C,$$

where A, B, C are analytic functions. Theorem 1 clearly implies

(a) $B = 0$, i.e., there is no logarithmic term;
(b) $r^2 = 0$ is the equation of the cone C_y, i.e., x is at zero distance from $y \Leftrightarrow P_x$ intersects P_y.

Note that these conclusions are purely local. McLenaghan [13] has studied Lorentzian metrics for which $B = 0$; these correspond according to Hadamard to having a Huygens principle. McLenaghan derives some necessary conditions (applicable also in the Riemannian case) and these are indeed satisfied by self-dual metrics.[†] Hadamard [10] gives an explicit formula for A. Theorem 1 contains implicitly a local twistorial construction for the principal part of the Green's function, i.e., for A. It would be instructive

[†] My thanks are due to R. S. Ward for drawing my attention to the paper of McLenaghan.

to compare this in detail with Hadamard's description. Concerning (b) the corresponding infinitesimal statement is essentially the definition of the conformal structure. Thus (b) is an integrated version.

(2) In Theorem 1 we have assumed that $\Delta\phi = 0 \Rightarrow \phi = 0$ so that $G(x, y)$ is well defined. However, if $G(x, y)$ can be constructed satisfying (3.2) and (3.3) for all y then it is the inverse of Δ and so $\Delta\phi = 0$ has only the trivial solution. More explicitly (3.2) and (3.3) imply that, distributionally, $\Delta_x G(x, y) = \delta_y$, where δ_y is the Dirac measure at y, and then

$$\Delta\phi = 0 \Rightarrow (\phi, \Delta_x G(x, y)) = 0 \Rightarrow \phi(y) = 0.$$

Thus if $G(x, y)$ exists for all y, or even for an open set of y (since ϕ is analytic), we deduce that $\phi = 0$. In the twistor description this means that, if we can find an element $g_y \in \operatorname{Ext}^1_Z(J_{P_y}, \mathcal{O}(-2))$ such that $\delta(g_y) = \mu(g_y) \in \operatorname{Ext}^2_Z(\mathcal{O}_{P_y}, \mathcal{O}(-2))$ for all y, then $\mathcal{O}(-2)$ has vanishing H^1 and H^2 and $g_y = (1/4\pi^2)\lambda(P_y)$ corresponds to the Green's function. In terms of the exact sequence (2.7), which is here

$$\to H^1(Z, \mathcal{O}(-2)) \to \operatorname{Ext}^1_Z(J_{P_y}, \mathcal{O}(-2)) \xrightarrow{\delta} \operatorname{Ext}^2_Z(\mathcal{O}_{P_y}, \mathcal{O}(-2))$$
$$\xrightarrow{j} H^2(Z, \mathcal{O}(-2)), \tag{3.12}$$

this can be seen as follows. The Grothendieck dual of the homomorphism j is the natural restriction

$$H^1(P_y, \mathcal{O}(-2)) \xrightarrow{j^*} H^1(Z, \mathcal{O}(-2)),$$

which corresponds to evaluating a solution ϕ of $\Delta\phi = 0$ at the point y. Thus δ is surjective if and only if all such ϕ vanish at y. Hence surjectivity of δ for all y implies that $\phi \equiv 0$ and so $\mathcal{O}(-2)$ has vanishing H^1 and H^2. These observations will be useful in computation of examples since they bypass the need for a direct proof that H^1 vanishes.

(3) The 't Hooft Ansatz for constructing self-dual Yang-Mills fields starts from a solution ϕ of the Laplace equation with a number of $1/r^2$-singularities. In the Penrose picture this corresponds to the rank 2 vector bundle E defined by a bundle of disjoint lines. Moreover, outside these lines, E can be given by an extension

$$0 \to H^{-1} \to E \to H \to 0$$

the class of which is the Serre class of the collection of lines. More generally, as explained in [3] bundles E can arise as extensions

$$0 \to H^{-l} \to E \to H^l \to 0$$

outside some algebraic curve $\Gamma \subset Z$. The Serre class corresponds to a solution ϕ of the zero Maxwell field equations for spin $(l - 1)$. By the same arguments as used to prove Theorem 1 one can show that near its singular surface $\hat{\Gamma}$ (the projection of Γ into M), ϕ has the behavior

$$\phi = \frac{a_0}{r^2} + \text{regular terms},$$

where r is the distance to $\hat{\Gamma}$ and a_0 is a certain canonical field defined along $\hat{\Gamma}$. Note that $\hat{\Gamma}$ is a very special type of surface in M: it satisfies a certain differential equation which can be expressed in terms of its second fundamental form, and depends only on the conformal structure of M.

4. Two-Variable Green's Functions

In the preceding section we treated the Green's function $G(x, y)$ as a function of x with y being kept fixed. This destroys the transparent symmetry between the two variables and we shall now show how to give a twistor description which preserves the symmetry.

Since $G(x, y)$ is a function on $M \times M$ it is natural to look at the complex manifold $Z \times Z$. On this we can consider the sheaves $\mathcal{O}(p, q)$ of sections of $H^p \boxtimes H^q$. In particular we shall be interested in $\mathcal{O}(-2, -2)$ and the cohomology group $H^2(U, \mathcal{O}(-2, -2))$ for any open set $U \subset Z \times Z$. If $(x, y) \in M \times M$ and $P_x \times P_y \subset U$ we can restrict this cohomology group to $P_x \times P_y$ where it becomes one dimensional. Thus any $\Phi \in H^2(U, \mathcal{O}(-2, -2))$ defines a scalar function ϕ [actually a $(\frac{1}{4}, \frac{1}{4})$ density] on a region of $M \times M$. This suggests that the Green's function should arise from such a Φ with U being an appropriate subspace of $Z \times Z$.

Now, as described in Section 2, the diagonal D in $Z \times Z$ has a Grothendieck class

$$\mu(D) \in \text{Ext}^3_{Z \times Z}(\mathcal{O}_D, \mathcal{O}(-2, -2)). \tag{4.1}$$

Since $H^q(Z, \mathcal{O}(-2)) = 0$ for $q = 0, 1, 2$ the Künneth formula implies that

$$H^q(Z \times Z, \mathcal{O}(-2, -2)) = 0 \quad \text{for} \quad q = 2, 3. \tag{4.2}$$

Using the exact sequence (2.7) with $m = 3$ we then see that there is a well-defined Serre class

$$\lambda(D) \in \text{Ext}^2_{Z \times Z}(J_D, \mathcal{O}(-2, -2)). \tag{4.3}$$

Restricting to the complement of the diagonal this gives an element of

$$H^2(Z \times Z - D, \mathcal{O}(-2, -2)). \tag{4.4}$$

If $(x, y) \in M \times M$ is a pair such that P_x and P_y do not intersect, then $P_x \times P_y$ does not meet D and so the Serre class $\lambda(D)$ can be "evaluated" on $P_x \times P_y$.

As might be surmised this gives the Green's function in symmetric form (up to $4\pi^2$). We state this as

THEOREM 2.   *Let z be the twistor space of a compact self-dual four-manifold M, and assume that the conformally invariant Laplacian equation $\Delta\phi = 0$ has no global solutions, so that the Green's function $G(x, y)$ is well defined. Then*

$$G(x, y) = (1/4\pi^2)\lambda(D)[P_x \times P_y],$$

where $\lambda(D) \in \text{Ext}^2_{Z \times Z}(J_D, \mathcal{O}(-2, -2))$ is the Serre class of the diagonal D in $Z \times Z$.

We shall prove Theorem 2 by reducing it to Theorem 1. Thus we shall fix y and we shall compare the restriction $\lambda(D)_y$ of $\lambda(D)$ to $Z \times P_y$ with the Serre class $\mu(P_y)$. Naturality properties of the Grothendieck class show that $\mu(D)$ restricted to $Z \times P_y$ coincides with the Grothendieck class $\mu(D_y)$, where D_y is the diagonal of $P_y \times P_y$ (considered as subvariety of $Z \times P_y$). Since

$$H^q(Z \times P_y, \mathcal{O}(-2, -2)) = 0 \qquad \text{for} \quad q = 2, 3 \qquad (4.5)$$

it follows that we have a Serre class

$$\lambda(D_y) \in \text{Ext}^2_{Z \times P_y}(J_{D_y}, \mathcal{O}(-2, -2)) \qquad (4.6)$$

and, by naturality, this coincides with $\lambda(D)_y$.

Next we introduce the quadric

$$Q_y = P_y \times P_y \subset Z \times P_y.$$

Since $D_y \subset Q_y$ we have a commutative diagram

$$\text{Ext}^2_{Z \times P_y}(J_{D_y}, \mathcal{O}(-2, -2)) \overset{\sim}{\to} \text{Ext}^3_{Z \times P_y}(\mathcal{O}_{D_y}, \mathcal{O}(-2, -2))$$

$$(4.7)$$

$$\text{Ext}^2_{Z \times P_y}(J_{Q_y}, \mathcal{O}(-2, -2)) \overset{\sim}{\to} \text{Ext}^3_{Z \times P_y}(\mathcal{O}_{Q_y}, \mathcal{O}(-2, -2)),$$

both horizontal isomorphisms being consequences of (4.5). In the top row $\lambda(D_y)$ goes to $\mu(D_y)$. To understand the bottom row we consider the following diagram

$$\text{Ext}^1_Z(J_{P_y}, \mathcal{O}(-2)) \otimes \text{Ext}^1_{P_y}(\mathcal{O}_{P_y}, \mathcal{O}(-2))$$

$$\downarrow \qquad \to \text{Ext}^2_Z(\mathcal{O}_{P_y}, \mathcal{O}(-2)) \otimes \text{Ext}^1_{P_y}(\mathcal{O}_{P_y}, \mathcal{O}(-2))$$

$$\downarrow \qquad (4.8)$$

$$\text{Ext}^2_{Z \times P_y}(J_{Q_y}, \mathcal{O}(-2, -2)) \to \qquad \text{Ext}^3_{Z \times P_y}(\mathcal{O}_{Q_y}, \mathcal{O}(-2, -2))$$

in which the vertical arrows are given by the natural multiplication in Ext-groups. Application of Grothendieck duality shows easily that all arrows in

(4.7) and (4.8) are isomorphisms and all vector spaces involved have dimension one. For example, the two Ext-groups on the right in (4.7) are dual to $H^1(D_y, \mathcal{O}(-2,0))$ and $H^1(Q_y, \mathcal{O}(-2,0))$ while the top right-hand Ext-group in (4.8) is dual to $H^1(P_y, \mathcal{O}(-2))$. All three of these spaces have a canonical generator [once we have chosen an isomorphism $\mathcal{O}(-2) = K_{P_y}$] and these generators all correspond by the maps in (4.7) and (4.8). Now the canonical generator in the top right-hand Ext-group of (4.7) is the Grothendieck class $\lambda(D_y)$, while the corresponding generator in (4.8) is $\mu(P_y) \otimes d_y$ where d_y is the chosen generator of $H^1(P_y, \mathcal{O}(-2))$ and $\mu(P_y)$ is the Grothendieck class determined by this choice. Passing to the left-hand sides of (4.7) and (4.8) it follows that the Serre class $\lambda(D_y)$ and $\lambda(P_y) \otimes d_y$ have the same image in $\mathrm{Ext}^2_{Z \times P_y}(J_{Q_y}, \mathcal{O}(-2,-2))$ and hence in $H^2((Z - P_y) \times P_y, \mathcal{O}(-2,-2))$. But, by Theorem 1, we have (for $P_x \cap P_y = \varnothing$):

$$G(x, y) = \lambda(P_y) \otimes d_y[P_x \times P_y] = \lambda(P_y)[P_x] \otimes d_y[P_y]$$

(the factor $d_y[P_y]$ is simply a piece of notation to describe the fact that $G(x, y)$ is a $\frac{1}{4}$-density). Since $P_x \times P_y \subset (Z - P_y) \times P_y$ whenever P_x and P_y are disjoint it follows that $G(x, y)$ is equally well given by evaluating $\lambda(D_y)$ on $P_x \times P_y$. Recalling that $\lambda(D_y)$ is equal to $\lambda(D)_y$ we have finally

$$G(x, y) = \lambda(D)[P_x \times P_y]$$

as stated in Theorem 2.

The class $\lambda(D)$ exhibits clearly the symmetry of the Green's function. By extracting the appropriate parts of the proof of Theorem 2 we could have produced a direct proof that the twistor description given in Section 3 was symmetrical. In other words (and omitting the notational factor $d_y[P_y]$)

$$\lambda(P_y)[P_x] = \lambda(P_x)[P_y] \tag{4.9}$$

for any two disjoint lines P_x and P_y. This symmetry formula can be viewed as formally analogous to the behavior of linking numbers in ordinary homology. Thus we should view $\lambda(P_y)[P_x]$ as a kind of *holomorphic linking number* for the two lines P_x and P_y in the three-manifold Z (we have symmetry rather than skew-symmetry because everything is complex and so even dimensional). Once the symmetry (4.9) has been established it follows that the term a_0, which entered in Section 3 and was independent of x, is also independent of y and so an absolute constant. This gives an alternative proof of the last part of Theorem 1, modulo the explicit determination of the constant—in applications the appropriate constant is in any case very easily found.

One interesting aspect of Theorem 2 is that the "singular set" of the Serre class $\lambda(D)$ is just D itself (where J_D differs from $\mathcal{O}_{Z \times Z}$). A direct look at $G(x, y)$

which is singular on the diagonal of $M \times M$ leads us to look for a sheaf cohomology class in the complete inverse image of $M \times M$–diagonal in $Z \times Z$. Thus we would expect the singular set to be the inverse image of this diagonal (and so to have real dimension 8) consisting of pairs $(z_1, z_2) \in Z \times Z$ lying on the same real fiber of Z. This space of course contains the diagonal of $Z \times Z$ but is much larger. We may paraphrase this by saying that the twistor singularities of the Green's function are smaller than might have been expected on naive grounds.

5.   GREEN'S FUNCTIONS FOR YANG–MILLS FIELDS

The conformal Laplacian has a natural covariant analog when we have a unitary vector bundle E with connection over M. This is defined by

$$\Delta_E = \nabla_E^* \nabla_E + R/6, \tag{5.1}$$

where ∇_E is the covariant derivative, ∇_E^* its adjoint and R the scalar curvature of M. If the connection on E is self-dual then E lifts to a holomorphic bundle $\tilde{E}$ on Z [2] and as before we have an isomorphism

$$H^1(\tilde{U}, \tilde{E}(-2)) \cong \{\text{space of solutions of } \Delta_E \phi = 0 \text{ on } U\}, \tag{5.2}$$

where U is any open set in M, and $\tilde{U}$ its inverse image in Z.

It is natural therefore to ask for a twistor interpretation of the Green's function $G_E(x, y)$ for this new Laplacian. We assume that $\Delta_E \phi = 0$ has no global solutions so that $G_E(x, y)$ is well defined. Note that this is not a scalar function but takes its values in $\text{Hom}(E_y, E_x)$. We start therefore by considering a matrix version of the Grothendieck class, namely the element

$$\mu_{\tilde{E}}(D) \in \text{Ext}^3_{Z \times Z}(\mathcal{O}_D, \tilde{E}^*(-2) \boxtimes E(-2)), \tag{5.3}$$

which corresponds, via the basic isomorphism (2.3), to the identity endomorphism of $\tilde{E}$. Using the assumed vanishing of cohomology we obtain a corresponding Serre class

$$\lambda_{\tilde{E}}(D) \in \text{Ext}^2_{Z \times Z}(J_D, \tilde{E}^*(-2) \boxtimes \tilde{E}(-2)). \tag{5.4}$$

The proofs of Theorems 1 and 2 go over with only trivial changes to give the corresponding matrix result, namely,

THEOREM 3.   *Let E be a unitary bundle with a self-dual connection on a compact self-dual four-manifold M, and assume that $\Delta_E \phi = 0$ has only $\phi = 0$*

as global solution where Δ_E is the conformally invariant Laplacian as in (4.1). Let $\tilde{E}$ be the corresponding holomorphic vector bundle on the twistor space Z and let $\lambda_{\tilde{E}}(D)$ be the Serre class of the diagonal D on $Z \times Z$ extended to endomorphisms of $\tilde{E}$ as in (4.4). Then the Green's function $G_E(x, y)$ is given by evaluating $\lambda_{\tilde{E}}(D)$ on $P_x \times P_y$ whenever P_x and P_y are disjoint.

As an illustration we shall use this Theorem when $M = S^4$ and E is an $SU(2)$-instanton [or, more generally, an $Sp(n)$-instanton] to rederive the simple form of the Green's function found empirically in [4, 5].

We recall the basic linear algebra construction of [1] for $Sp(n)$-instantons. This can be displayed in the diagram of exact sequences:

$$
\begin{array}{ccccccc}
& 0 & & 0 & & 0 & \\
& \downarrow & & \downarrow & & \downarrow & \\
0 \to & W(-1) & \to & Q^* & \to & \tilde{E} & \to 0 \\
& \| & & \downarrow & & \downarrow & \\
0 \to & W(-1) & \overset{A}{\to} & V & \to & Q & \to 0 \\
& \downarrow & & \downarrow A^* & & \downarrow & \\
& 0 & & \to W(1) & = & W(1) & \\
& & & \downarrow & & \downarrow & \\
& & & 0 & & 0 &
\end{array}
\qquad (5.5)
$$

Here W, V are fixed vector spaces of dimensions k, and $2k + 2n$, respectively (where k is the instanton number). In (5.5) we also regard W, V as trivial vector bundles on Z. The basic data is the homomorphism A, and the whole diagram is self-dual in a skew-symmetric manner, V itself having a fixed skew form.

We now fix a point $y \in M$ and apply $\mathrm{Ext}_Z^1(J_{P_y}, \ )$ to the bundles in the diagram (twisted by -2). A little use of Grothendieck duality and the vanishing of various sheaf cohomology groups then shows that we get

$$
\begin{array}{c}
0 \\
\downarrow \\
\mathrm{Ext}_Z^1(J_{P_y}, Q^*(-2)) \overset{\approx}{=} \mathrm{Ext}_Z^1(J_{P_y}, \tilde{E}(-2)) \\
\downarrow \beta \\
\mathrm{Ext}_Z^1(J_{P_y}, V(-2)) \\
\downarrow \\
\mathrm{Ext}^1(J_{P_y}, W^*(-1)) \\
\downarrow \\
0
\end{array}
\qquad (5.6)
$$

Now Theorem 3 implies that the Green's function $G_E(x, y)$ is given by the element

$$\lambda_y(E) \in \operatorname{Ext}_Z^1(J_{P_y}, \operatorname{Hom}(E_y, \tilde{E}(-2)))$$

corresponding to the identity endomorphism of E_y. Here we use the fact that $\tilde{E}|_{P_y}$ is trivial and naturally isomorphic to $P_y \times E_y$. Applying $\operatorname{Hom}(E_y, \quad)$ to (5.6) we see that $G_E(x, y)$ is determined by the element

$$\gamma_y = \beta\alpha^{-1}(\lambda_y(E)) \in \operatorname{Ext}_Z^1(J_{P_y}, \mathcal{O}(-2)) \otimes \operatorname{Hom}(E_y, V)$$

given by $\gamma_y = \lambda(P_y) \otimes i_y$, where i_y is the natural inclusion of E_y in V [obtained by restricting (5.5) to P_y and decomposing].

It remains to make explicit the way in which γ_y determines the Green's function. For this we pick a point $x \in M$ so that P_x and P_y are disjoint and restrict the groups in (5.6) [after applying $\operatorname{Hom}(E_y, \quad)$] to the line P_x. Thus (5.6) will map into the commutative diagram

$$0$$
$$\downarrow$$
$$H^1(P_x, \operatorname{Hom}(E_y, W(-3))) \to H^1(P_x, \operatorname{Hom}(E_y, Q^*(-2))) \to H^1(P_x, \operatorname{Hom}(E_y, \tilde{E}(-2)))$$

$$\| \wr$$
$$\operatorname{Hom}(E_y, E_x)$$
$$\uparrow \pi$$
$$H^1(P_x, \operatorname{Hom}(E_y, V(-2))) \cong \qquad \operatorname{Hom}(E_y, V)$$
$$\downarrow$$
$$0 \tag{5.7}$$

where π is given by the natural projection $\pi_x : V \to E_x$.

Hence finally we get the formula

$$G_E(x, y) = G(x, y)\pi_x i_y, \tag{5.8}$$

where $G(x, y)$ is the scalar Green's function of S^4. This is the simple form discovered in [4, 5].

6.   REVIEW OF THE A_k-MANIFOLDS

As an application of our methods we shall compute explicitly the scalar Green's function for the gravitational multi-instantons of Gibbons and Hawking [8]. These manifolds have been studied from the twistor viewpoint by Hitchin [11], who relates them to the singularities of type A_k, and we shall begin by reviewing his results. One point should be mentioned at the outset and that is the noncompactness of these manifolds. This can be overcome by a conformal compactification at the expense of allowing a mild singularity at infinity. This will necessitate the extension of our complex methods to allow for such singularities but this is a minor problem.

We begin by describing the twistor space as in [11]. We consider the equation

$$xy = \prod_{i=1}^{k} (z - p_i(u, v)) \tag{6.1}$$

in the five complex variables x, y, z, u, v. The p_i are homogeneous quadratics of the form

$$p_i(u, v) = a_i u^2 + 2b_i uv + c_i v^2$$

with $\bar{c}_i = -a_i$ and $\bar{b}_i = b_i$, and k is a fixed positive integer (the gravitational instanton number). If we give the five variables the weights $(k, k, 2, 1, 1)$ then Eq. (6.1) becomes homogeneous and it defines a three-dimensional compact complex algebraic variety $\bar{Z}$ as a subspace of the "weighted" projective space $[C^5$-0 modulo the C^*-action $(x, y, z, u, v) \rightarrow (\lambda^k x, \lambda^k y, \lambda^2 z, \lambda u, \lambda v)]$. Note that $\bar{Z}$ has certain singular points as follows:

(i) the two points $(1, 0, 0, 0, 0)$ and $(0, 1, 0, 0, 0)$ where we have a quotient singularity $(C^3$ modulo a cyclic group of order k);
(ii) points where $p_i(u, v) = p_j(u, v)$ for some i, j.

The p_i are supposed sufficiently generic so that we do not get multiple coincidences in (ii) and the points in (ii) are then ordinary double points.

Hitchin's twistor space is obtained from $\bar{Z}$ by omitting the curve $u = v = 0$ (containing the two singular points in (i)) giving a space Z, and then resolving the double points by adding curves. For many purposes it is more convenient to work with the singular model where we have the simple equation (6.1), and to make allowances for the singularities. The curve $u = v = 0$ should be viewed as the "line at ∞" and corresponds to the conformal compactification of the four-manifold.

Hitchin then defines the family of curves (projective lines) on Z given by the equations

$$x = A \prod_{i=1}^{k} (u - \alpha_i v),$$

$$y = B \prod_{i=1}^{k} (u - \beta_i v), \tag{6.2}$$

$$z = au^2 + 2buv - \bar{a}v^2,$$

with appropriate conditions on the α_i, β_i, A, B. The main point is that we factorize

$$(au^2 + 2buv - \bar{a}v^2) - p_i(u,v) = (a - a_i)(u - \alpha_i v)(u - \beta_i v)$$

and assign the α_i to x and the β_i to y. The parameter space of these curves is a four-manifold M with a self-dual conformal structure. Moreover there is a choice of metric which also satisfies Einstein's equations, i.e., the Ricci tensor is zero. This is the Gibbons–Hawking metric (for $k = 2$ it is due to Eguchi and Hanson [7]).

As mentioned before M is not compact and at ∞ it has a fundamental group which is cyclic of order k. The metric is asymptotically Euclidean on the k-fold cover and so the one-point compactification $\bar{M}$ of M has a singular point at ∞, but the conformal structure is well defined on the k-fold cover of a neighborhood of ∞. We can therefore form the twistor space of $\bar{M}$ by including the P_1-fiber over ∞. Locally, near ∞, the twistor space will look like the twistor space of R^4 modulo the standard action of Z_k [as subgroup of $SU(2) \subset SO(4)$]. This action has two fixed points in the fiber over the base point and so the twistor space of $\bar{M}$ has two quotient singularities over ∞. One can now check that $\bar{Z}$ (with its double points removed) is precisely the twistor space of $\bar{M}$, the curve $u = v = 0$ being the fiber at ∞. The main point is that this curve, when we lift to the k-fold cover, has normal bundle $H \oplus H$, and this follows from its defining equations being $u = v = 0$.

Although $\bar{M}$ is not quite a manifold there is no difficulty in doing differential geometry on $\bar{M}$: near ∞ we always use the k-fold cover and consider functions, etc. invariant under Z_k. In particular, the conformal Laplacian Δ is well defined. If the equation $\Delta\phi = 0$ has only $\phi = 0$ as global solution (on $\bar{M}$) then the Green's function $G(\xi, \eta)$ is defined and can be constructed by using Theorem 1. (We shall now use Greek letters for points on M to avoid confusion with x, y on Z.) This will be carried out in detail in the next section.

The Einstein metric on M is a complete metric and it can be characterized (up to a constant) as that metric in the conformal class whose density or volume element dm is $[G(\xi, \infty)]^4$. In particular

$$dm = \rho(\xi)\,d\bar{m} \qquad \text{with} \quad \rho(\xi) \to 0 \quad \text{as} \quad \xi \to \infty, \qquad (6.3)$$

where $\bar{m}$ is any metric in the conformal class which is defined on $\bar{M}$. Hence, for a $\frac{1}{4}$-density, square integrability on M (for the Einstein metric) implies square integrability on $\bar{M}$. Since the Green's function, as an operator, is characterized by the property

$$\Delta G f = f \qquad \text{for} \quad f \in L^2,$$

it follows that the Green's function for $\bar{M}$ restricts to give the Green's function for M. This justifies our use of the conformal compactification. Note incidentally that the assumption about $\Delta\phi = 0$ having no global solutions on $\bar{M}$ is equivalent to

$$\Delta\phi = 0 \quad \text{and} \quad \int_M \rho^{1/2}|\phi|^2 \, dm < \infty \Rightarrow \phi = 0, \tag{6.4}$$

where ρ is the factor in (6.3).

7. Green's Function for the A_k-Manifolds

In order to construct the Green's function for the manifold $\bar{M}$ what we have to do, in view of Theorem 1, is to construct the Serre class of the line P_η. Equivalently, as explained in Section 2, we have to find a rank 2 vector bundle on $\bar{Z}$ with a holomorphic section vanishing precisely on P_η. Before proceeding further we should comment on the singularities of $\bar{Z}$. For the points at ∞ the words vector bundle, section, etc., will always mean the corresponding object on the (local) k-fold cover invariant under the cyclic group Z_k. For the double points in the finite part of $\bar{Z}$ we shall assume that P_η does not pass through any of these. The rank 2 bundle we shall construct will be well defined at these singular points and is therefore trivial on the curves into which these points get blown up when we desingularize Z to get the genuine twistor space. The final formula to be derived for $G(\xi, \eta)$ will of course remain valid (by continuity) even for those η which have been excluded, i.e., for which P_η passes through a double point.

We start therefore with a fixed curve P_η of Hitchin's family given by the three equations $f = g = h = 0$ where

$$\begin{aligned}
f &\equiv x - A \prod (u - \alpha_i v), \\
g &\equiv y - B \prod (u - \beta_i v), \\
h &\equiv z - (au^2 + 2buv - \bar{a}v^2).
\end{aligned} \tag{7.1}$$

We note that the equations $f = h = 0$ will imply $g = 0$ in view of the identity

$$xy = \prod (z - p_i(u, v)) \tag{7.2}$$

unless $u = \alpha_i v$ for some i. This means that our curve is a complete intersection, with equations $f = h = 0$, in the open set U_A given by $\mathscr{A} \neq 0$ where we put

$$\mathscr{A} = A \prod (u - \alpha_i v), \qquad \mathscr{B} = B \prod (u - \beta_i v).$$

Similarly in U_B, given by $\mathscr{B} \neq 0$, it is the complete intersection $g = h = 0$. Now generically the parameters α_i are distinct from the β_i and so $\mathscr{A}$ and

$\mathscr{B}$ never vanish simultaneously except at ∞, i.e., when $u = v = 0$. Under this genericity assumption [which can be dropped when we arrive at our final formula for $G(x, y)$] U_A and U_B together cover the whole of Z. We shall therefore begin by constructing our rank 2 bundle E over Z by using an appropriate transition matrix over $U_A \cap U_B$.

First we define a polynomial θ by substituting for z from (7.1) into (7.2) and expanding:

$$
\begin{aligned}
xy &= \prod (z - p_i(u, v)) \\
&= \prod ((a - a_i)(u - \alpha_i v)(u - \beta_i v) + h) \qquad (7.3) \\
&= AB \prod (u - \alpha_i v)(u - \beta_i v) + \theta h.
\end{aligned}
$$

In other words θ is defined as

$$
\theta = (xy - \mathscr{A}\mathscr{B})/h \qquad (7.4)
$$

but the expansion (7.3) shows that the denominator in (7.4) divides the numerator.

We now consider the polynomial identities

$$
f = -\frac{x}{\mathscr{B}} g + \frac{\theta}{\mathscr{B}} h, \qquad h = -\frac{h}{\mathscr{B}} g + \frac{y}{\mathscr{B}} h. \qquad (7.5)
$$

The determinant of the 2×2 matrix in (7.5) is

$$
-(xy + \theta h)/\mathscr{B}^2 = -\mathscr{A}/\mathscr{B} \quad \text{by (7.4)},
$$

which is invertible and nonzero in $U_A \cap U_B$. Since f, g, h have "weights" $(k, k, 2)$ they should be viewed as sections of H^k, H^k, H^2, respectively, where H is the standard line-bundle over Z. The identities (7.5) then assert that, if we define a vector bundle E over Z by gluing $(H^k \oplus H^2)$ on U_A with $(H^k \oplus H^2)$ on U_B via the 2×2 matrix of (7.5), then we get a section s of E given by

$$
s = (f, h) \quad \text{in } U_A, \qquad s = (g, h) \quad \text{in } U_B.
$$

Thus $s = 0$ precisely on our curve P_η. Our assertion is that the pair (E, s) does indeed define the Serre class and that, when translated back into M, it gives the correct Green's function. Certainly the Green's function obtained in this way will satisfy the Laplace equation and will have the correct behavior as $x \to y$. What still needs to be checked is that it has the correct asymptotic properties at ∞. This could be done by explicitly looking at the final formula. However, by using our conformal compactification of M to $\bar{M}$, it is sufficient instead to check that the pair (E, s) extend from Z to $\bar{Z}$.

To cover $\bar{Z}$ we need two more open sets, namely U_x, defined by $x \neq 0$, and U_y defined by $y \neq 0$. The following identities then provide the necessary extension of (E, s) to $\bar{Z}$:

$$\left.\begin{aligned} f &= -\frac{\mathscr{A}}{y} g + \frac{\theta}{y} h \\[2mm] h &= \frac{h}{y} g + \frac{\mathscr{B}}{y} h \end{aligned}\right\} \text{ in } U_x \cap U_y,$$

$$\left.\begin{aligned} f &= -\frac{x}{\mathscr{B}} g + \frac{\theta}{\mathscr{B}} h \\[2mm] h &= \phantom{-\frac{x}{\mathscr{B}} g +} h \end{aligned}\right\} \text{ in } U_x \cap U_B,$$

$$\left.\begin{aligned} f &= -\frac{\mathscr{A}}{y} g + \frac{\theta}{y} h \\[2mm] h &= \phantom{-\frac{\mathscr{A}}{y} g +} h \end{aligned}\right\} \text{ in } U_A \cap U_y,$$

$$\left.\begin{aligned} f &= f \\[2mm] h &= \frac{h}{x} f + \frac{\mathscr{A}}{x} h \end{aligned}\right\} \text{ in } U_A \cap U_x,$$

$$\left.\begin{aligned} g &= g \\[2mm] h &= \frac{h}{y} g + \frac{\mathscr{B}}{y} h \end{aligned}\right\} \text{ in } U_B \cap U_y.$$

The important point to note is that each 2×2 matrix is nonsingular and invertible in the appropriate intersection. The section s in U_x and U_y is given by

$$s = (f, h) \quad \text{in } U_x, \qquad s = (g, h) \quad \text{in } U_y.$$

It remains to translate the Serre class of P_η, as described by the pair (E, s), into the explicit Green's function. We recall the general procedure. Outside the curve P_η the section s is nonvanishing and hence gives an extension

$$0 \to \mathscr{O} \to E \to H^2 \to 0 \tag{7.6}$$

and so an element of $H^1(\bar{Z} - P_\eta, \mathscr{O}(-2))$. Evaluating this element on the curve P_ξ then gives $G(\xi, \eta)$. The precise extension (7.6) depends on a choice of $\frac{1}{4}$-density at η, and $G(\xi, \eta)$ then becomes a $\frac{1}{4}$-density at ξ. If we use the Einstein metric of M then this fixes these $\frac{1}{4}$-densities. This corresponds (see [11]) to the definite choice of isomorphism $K_{P_\xi} \to H^{-2}$ for all $\xi \in M$, arising from the holomorphic projection $Z \to P_1(C)$ given by $(x, y, z, u, v) \mapsto (u, v)$. In fact the Einstein metric is only determined up to a constant factor and so we shall only work up to such a constant. The final normalization is best carried out in terms of the explicit coordinates of M.

The extension (7.6) is given by gluing together extensions over U_A and U_B as follows:

$$
\begin{array}{ccccc}
\mathcal{O} & \xrightarrow{(g,h)} & H^k \oplus H^2 & \xrightarrow{\mathscr{B}^{-1}(h,-g)} & H^2 \text{ in } U_B \\
\| & & \downarrow T & & \| \\
\mathcal{O} & \xrightarrow{(f,h)} & H^k \oplus H^2 & \xrightarrow{\mathscr{A}^{-1}(-h,f)} & H^2 \text{ in } U_A
\end{array}
\tag{7.7}
$$

when T is the 2×2 matrix of (7.5). Now let U_h, U_f denote the open sets given by $h \neq 0$, $f = 0$, respectively, and restrict the top part of (7.7) to $U_B \cap U_h$ and the bottom part to $U_A \cap U_f$. Both extensions can then be split by the maps $H^2 \to H^k \oplus H^2$ given by $(\mathscr{B}h^{-1}, 0)$ and $(0, \mathscr{A}f^{-1})$, respectively. Subtracting these splittings using (7.5) we get the map $H^2 \to H^k \oplus H^2$ given in the U_A description, by

$$(xh^{-1}, \mathscr{A}f^{-1} - 1).$$

Hence, as a map $H^2 \to \mathcal{O}$, the difference of the two splittings is given by

$$xh^{-1}f^{-1}. \tag{7.8}$$

Thus (7.8) is the explicit 1-cocycle for the extension (7.6) in the portion of Z covered by $(U_A \cap U_f) \cup (U_B \cap U_h)$. Now if we restrict to a line P_ξ, given by equations $x = \mathscr{A}'$, $y = \mathscr{B}'$, $z = h'$ corresponding to parameters a', b', A', B' the equations $h = 0$, $f = 0$, $\mathscr{A} = 0$, $\mathscr{B} = 0$ all correspond to finite sets of points on P_ξ which are disjoint for generic ξ. Thus generically the cocyle given by (7.8) can be used to compute the Green's function $G(\xi, \eta)$. Substituting the equations of P_ξ and putting $v = 1$ we have (cf. Section 3) to compute the sum of the residues of $xh^{-1}f^{-1}$ in $U_A \cap U_f$. Since $f \neq 0$ in U_f, only the two zeros of h contribute and so we get the formula

$$G(\xi, \eta) = C \sum_h \operatorname{Res}\left(\frac{\mathscr{A}'}{hf}\right) du, \tag{7.9}$$

where C is a constant and we sum over the two zeros of h. More explicitly in terms of the parameters of the lines P_ξ, P_η we recall that

$$
\begin{aligned}
\mathscr{A}' &= A' \prod (u - \alpha_i'), \\
f &= A' \prod (u - \alpha_i') - A \prod (u - \alpha_i), \\
h &= (a'u^2 + 2b'u - \bar{a}') - (au^2 + 2bu - \bar{a}).
\end{aligned}
\tag{7.10}
$$

Hence (7.9) becomes

$$G(\xi, \eta) = \frac{C}{(a' - a)(\gamma - \delta)} \left\{ \frac{A' \prod (\gamma - \alpha_i')}{g(\gamma)} - \frac{A' \prod (\delta - \alpha_i')}{g(\delta)} \right\}, \tag{7.11}$$

where γ, δ are the two zeros of h.

This completes our determination of the Green's function except for making explicit the value of the constant C. In the next section we shall compare (7.11) with the formula derived by Page [14], at the same time making explicit the normalization of the Einstein metric and the corresponding value for C.

Note that we have bypassed the problem of proving that $\Delta\phi = 0$ has only $\phi = 0$ as global solution on $\bar{M}$. Remark 2 following Theorem 1 showed that the existence of the Serre class implied the invertibility of Δ, and moreover we need only the Serre class of P_η for an open set of η. Since we constructed this explicitly in our case for generic η, the invertibility of Δ is assured. Alternatively, using the fact that $\bar{Z}$ is defined as a hypersurface in a weighted projective space, it is not difficult by standard algebraic geometry to prove that $H^1(\bar{Z}, \mathcal{O}(-2)) = 0$, which is the twistor equivalent of the invertibility of Δ.

We conclude with a couple of comments arising from our computations. The rank 2 vector bundle E which we have constructed, using the transition matrix T of (7.5) in $U_A \cap U_B$, restricts to yield the normal bundle of the curve P_η given by Eq. (7.1). But on P_η the matrix T reduces to

$$
\begin{pmatrix}
-\mathscr{A} & \theta \\
\dfrac{\mathscr{B}}{} & \mathscr{B} \\
0 & 1
\end{pmatrix},
$$

which represents $E\big|_{P_\eta}$ as an extension

$$0 \to \mathcal{O} \to E \to H^2 \to 0. \tag{7.12}$$

Computing the 1-cocycle of this extension, along the same lines as our previous computation, we find

$$\frac{\theta}{\mathscr{A}\mathscr{B}} = \sum \frac{du}{(a - a_i)(u - \alpha_i)(u - \beta_i)}.$$

Taking the sum of the residues in U_B we get

$$\sum \frac{1}{(a - a_i)(\alpha_i - \beta_i)} = \sum \frac{1}{\Delta_i} = \gamma > 0$$

in Hitchin's notation. Since this is nonzero, the extension (7.12) is nontrivial and so $E\big|_{P_\eta}$ must be isomorphic to $H \oplus H$, as required for the Penrose theory to apply. This verification was bypassed in [11]. If we go into the complexification, then the same calculation shows that the conformal structure will break-down when $\gamma = 0$. This is consistent with the Hawkings–Gibbons formula for the metric (see the next section).

Finally, we observe that the "line at ∞," P_∞ is given by $u = v = 0$, which implies that the Green's function $G(\infty, x)$ is everywhere constant, provided we identify all P_ξ by the projection onto (u, v). Put more invariantly this implies that the Einstein metric is that metric in the conformal class for which the Riemannian density is $\{G(\infty, x)\}^4$.

8. COMPARISON WITH RESULTS OF PAGE

As explained by Hitchin [11] the real curve given by Eqs. (7.1) can be parametrized by the four real variables (Re a, Im a, b, arg A) and the Einstein metric then takes the form

$$\gamma\, dx \cdot dx + \gamma^{-1}(dt + \boldsymbol{\omega} \cdot dx)^2, \tag{8.1}$$

where x stands for the 3-vector (Re a, Im a, b) with the standard Euclidean metric $|a|^2 + b^2$, $t = 2\arg A$, $\gamma = \sum(1/|x - x_i|)$ and curl $\omega = \operatorname{grad} \gamma$. In Hitchin's case, there is an explicit choice of ω so that

$$\boldsymbol{\omega} \cdot dx = -\operatorname{Im} \sum \frac{\alpha_i}{\Delta_i}\, da. \tag{8.2}$$

Now let us use polar coordinates in R^3 so that

$$a = r\sin\theta e^{i\phi}, \qquad b = r\cos\theta \tag{8.3}$$

with similar formulas for $a - a_i$, $b - b_i$ in terms of polar coordinates r_i, θ_i, ϕ_i centered at the point (a_i, b_i). Then converting Hitchin's notation into ours we have

$$\Delta = r, \qquad \alpha = \frac{1 - \cos\theta}{\sin\theta}\, e^{-i\phi} = \tan\frac{\theta}{2}\, e^{-i\phi}.$$

Substituting into (8.2) we find the following expression for the Einstein metric

$$\gamma\, dx \cdot dx + \gamma^{-1}(dt - \sum(1 - \cos\theta_i)\, d\phi_i)^2. \tag{8.4}$$

Page in [14] takes his metric in the form

$$\gamma\, dx \cdot dx + \gamma^{-1}(d\tau + \sum\cos\theta_i\, d\phi_i)^2 \tag{8.5}$$

which agrees with (8.4) under the change of variable

$$\tau = t - \sum\phi_i. \tag{8.6}$$

Recall now that $AB = \prod(a - a_i)$ so that

$$\arg(AB) = \sum\phi_i.$$

Since, by definition of t and τ,

$$\arg A = (\tau + \sum \phi_i)/2 \qquad (8.7)$$

it follows that

$$\arg A/B = \tau \qquad (8.8)$$

is Page's variable.

We return now to our formula (7.11) for the Green's function. We distinguish between the two roots γ, δ of

$$(a' - a)a^2 + 2(b' - b)u - \overline{(a' - a)} = 0$$

by taking

$$\gamma = \frac{-(b' - b) + \Delta}{(a' - a)}, \qquad \delta = \frac{-(b' - b) - \Delta}{(a' - a)} \qquad (8.9)$$

where $\Delta = \{|a' - a|^2 + (b' - b)^2\}^{1/2}$ is the three-space Euclidean distance between the points (a', b') and (a, b). Then we have

$$(a' - a)(\gamma - \delta) = 2\Delta.$$

If we now put

$$R_+ = \frac{A \prod (\gamma - \alpha_i)}{A' \prod (\gamma - \alpha_i')}, \qquad R_- = \frac{A \prod (\delta - \alpha_i)}{A' \prod (\delta - \alpha_i')},$$

$$S_\pm = 1 - R_\pm, \qquad (8.10)$$

then the Green's function formula (7.11) takes the form

$$G = \frac{C}{2\Delta}\left(\frac{1}{S_+} - \frac{1}{S_-}\right). \qquad (8.11)$$

Comparing this with Page's formulas [14, (17) and (18)] we see that they coincide provided we can identify our $S_\pm$ with his. We must therefore compute the products $R_\pm$ in (8.10) in terms of our polar coordinates. We shall proceed to evaluate the individual terms in these products.

The calculations which follow are elementary but tedious and to simplify them we introduce variables λ, μ by

$$\lambda = r^{1/2} \sin(\theta/2)e^{i\phi/2}, \qquad \mu = r^{1/2} \cos(\theta/2)e^{-i\phi/2}. \qquad (8.12)$$

so that $|\lambda|^2 + |\mu|^2 = r$ and the quadratic

$$au^2 + 2bu - \bar{a} = r \sin \theta e^{i\phi} u^2 + 2r \cos \theta - r \sin \theta e^{-i\phi}$$

factorizes as

$$2(\bar{\mu}u - \bar{\lambda})(\lambda u + \mu) \qquad (8.13)$$

with $\alpha = \bar{\lambda}/\bar{\mu}$ and $\bar{\beta} = -\mu/\lambda$. Corresponding variables λ', μ', λ_i, μ_i, λ'_i, μ'_i relate similarly to the quadratics given by (a', b'), $(a - a_i, b - b_i)$, $(a' - a_i, b' - b_i)$, respectively.

In terms of these new variables (8.9) gives the following expression for δ:

$$\delta = \frac{|\mu'|^2 - |\lambda'|^2 + |\lambda|^2 - |\mu|^2 + \Delta}{2(\lambda\bar{\mu} - \lambda'\bar{\mu}')}. \tag{8.14}$$

For brevity we now introduce three further quantities defined by

$$v = \bar{\lambda}\lambda' + \bar{\mu}\mu', \qquad \rho = r + r' + \Delta, \qquad \sigma = r + r' - \Delta, \tag{8.15}$$

where, as before, r and r' stand for $|\lambda|^2 + |\mu|^2$ and $|\lambda'|^2 + |\mu'|^2$. With all this notation we then have the following two algebraic lemmas

LEMMA 1.

$$4|v|^2 = \rho\sigma$$

LEMMA 2.

$$\frac{\bar{\mu}\delta - \bar{\lambda}}{\bar{\mu}'\delta - \bar{\lambda}'} = \frac{2v}{\sigma}.$$

If we grant, for the moment, these lemmas, we deduce that

$$\frac{\bar{\mu}\delta - \bar{\lambda}}{\bar{\mu}'\delta - \bar{\lambda}'} = \sqrt{\frac{\rho}{\sigma}}\,\frac{v}{|v|}. \tag{8.16}$$

We now apply this with $(\lambda, \mu, \lambda', \mu')$ replaced by $(\lambda_i, \mu_i, \lambda'_i, \mu'_i)$ and note that δ, which is a root of the difference of the two quadratics (a', b') and (a, b), is equally the root of the difference of $(a' - a_i, b' - b_i)$ and $(a - a_i, b - b_i)$. Hence δ is the same for all i and so (8.16) yields

$$\prod_i \frac{\bar{\mu}_i\delta - \bar{\lambda}_i}{\bar{\mu}'_i\delta - \bar{\lambda}'_i} = \prod\left(\sqrt{\frac{\rho_i}{\sigma_i}}\cdot\frac{v_i}{|v_i|}\right). \tag{8.17}$$

Comparing A with $\prod \mu_i$ we see that (using [10, (3.4)])

$$|A|^2 = \prod r_i(1 + \cos\theta_i), \qquad \arg A = \frac{\tau + \sum\phi_i}{2},$$

$$\prod|\mu_i| = \prod r_i^{1/2}\cos\theta_i/2, \qquad \arg\prod\bar{\mu}_i = \frac{\sum\phi_i}{2}.$$

Hence

$$A = 2^k e^{i\tau/2}\prod\bar{\mu}_i. \tag{8.18}$$

Using (8.18) we deduce from (8.17) that R_-, defined by (8.10), is given by

$$R_- = e^{i(\tau - \tau')/2} \prod \left(\sqrt{\frac{\rho_i}{\sigma_i}} \cdot \frac{v_i}{|v_i|} \right). \tag{8.19}$$

Quite similarly we find

$$R_+ = e^{i(\tau - \tau')/2} \prod \left(\sqrt{\frac{\sigma_i}{\rho_i}} \cdot \frac{v_i}{|v_i|} \right). \tag{8.20}$$

Finally, if we compute v, as defined by (8.15), in terms of the polar coordinates, we get

$$v = (rr')^{1/2} \left\{ \sin \frac{\theta}{2} \sin \frac{\theta'}{2} e^{i(\phi' - \phi)/2} + \cos \frac{\theta}{2} \cos \frac{\theta'}{2} e^{i(\phi - \phi')/2} \right\}$$

$$= (rr')^{1/2} \left\{ \cos \frac{\phi' - \phi}{2} \cos \frac{\theta - \theta'}{2} - i \sin \frac{\phi' - \phi}{2} \cos \frac{\theta + \theta'}{2} \right\}$$

So that

$$\arg v = \tan^{-1} \left\{ \frac{\cos[(\theta + \theta')/2]}{\cos[(\theta - \theta')/2]} \cdot \tan \frac{\phi - \phi'}{2} \right\}. \tag{8.21}$$

These formulas finally identify our $R_\pm$ with the corresponding expressions in [14, (13)–(16)]:

$$R_\pm = \exp(i(T \pm iU)),$$

so that our $S_\pm$ agree with those of Page. The constant C in (8.12) is therefore normalized to be $\frac{1}{8}\pi^2$.

It remains for us to establish Lemmas 1 and 2. We begin with Lemma 1. Since

$$\rho\sigma = (r + r')^2 - \Delta^2,$$
$$\Delta^2 = \{|\mu'|^2 - |\lambda'|^2 + |\lambda|^2 - |\mu|^2\}^2 + 4|\lambda\bar{\mu} - \lambda'\bar{\mu}'|^2.$$

Lemma 1 is equivalent to

$$(r + r')^2 - \{|\mu'|^2 - |\lambda'|^2 + |\lambda|^2 - |\mu|^2\} = 4|v|^2 + 4|\lambda\bar{\mu} - \lambda'\bar{\mu}'|^2. \tag{8.22}$$

Recalling that $\tau = |\lambda|^2 + |\mu|^2$ and similarly for τ', (8.22) is equivalent to

$$(|\lambda|^2 + |\mu'|^2)(|\lambda'|^2 + |\mu|^2) = |v|^2 + |\lambda\bar{\mu} - \lambda'\bar{\mu}'|^2. \tag{8.23}$$

But

$$|v|^2 = |\lambda\lambda'|^2 + |\mu\mu'|^2 + \bar{\lambda}\lambda'\mu\bar{\mu}' + \lambda\bar{\lambda}'\bar{\mu}\mu'$$

and

$$|\lambda\bar{\mu} - \lambda'\bar{\mu}'|^2 = |\lambda\mu|^2 + |\lambda'\mu'|^2 - \bar{\lambda}\mu\lambda'\bar{\mu}' - \lambda\bar{\mu}\bar{\lambda}'\mu'.$$

Adding these two we obtain (8.23) as required.

For the proof of Lemma 2 we have, from (8.14),

$$\bar{\mu}\delta - \bar{\lambda} = \frac{\bar{\mu}\{|\mu'|^2 - |\lambda'|^2 + |\lambda|^2 - |\mu|^2 + \Delta\}}{2(\lambda\bar{\mu} - \lambda'\bar{\mu}')} - \bar{\lambda}$$

$$= \frac{-\mu\sigma + 2\bar{\mu}'v}{2(\lambda\bar{\mu} - \lambda'\bar{\mu}')} \quad \text{using (8.15).}$$

Similarly,

$$\bar{\mu}'\delta - \bar{\lambda}' = \frac{\bar{\mu}\rho - 2\bar{\mu}v}{2(\lambda\bar{\mu} - \lambda'\bar{\mu}')}.$$

Taking the ratios we obtain

$$\frac{\bar{\mu}\delta - \bar{\lambda}}{\bar{\mu}'\delta - \bar{\lambda}'} = \frac{-\bar{\mu}\sigma + 2\bar{\mu}'v}{-2\bar{\mu}v + \bar{\mu}'\rho}.$$

In view of Lemma 1 the 2×2 matrix on the right has a vanishing determinant and so the ratio is equal to that given by either column. Hence

$$\frac{\bar{\mu}\delta - \bar{\lambda}}{\bar{\mu}'\delta - \bar{\lambda}'} = \frac{2v}{\rho}$$

as required.

ACKNOWLEDGMENTS

I am indebted to N. J. Hitchin, who first suggested that one should look for the twistor version of the Green's function, and to D. Page for providing me with his results whilst I was in the midst of my own computations. Comparison with his formulae has prevented many elementary errors, even though the identification in Section 8 turned out to be much harder than expected. I am also grateful to J–P. Serre for the invitation to lecture at the Collège de France when these results, in outline, were first presented.

REFERENCES

1. M. F. ATIYAH, N. J. HITCHIN, V. G. DRINFELD, AND Y. I. MANIN, Construction of instantons, *Phys. Lett. A* **61** (1977), 81–83.
2. M. F. ATIYAH, N. J. HITCHIN, AND I. M. SINGER, Self-duality in four-dimensional Riemannian geometry, *Proc. Roy. Soc. London Ser. A* **362** (1978), 425–461.
3. M. F. ATIYAH AND R. S. WARD, Instantons and algebraic geometry, *Comm. Math. Phys.* **55** (1977), 117–124.
4. N. H. CHRIST, E. J. WEINBERG, AND N. K. STANTON, General self-dual Yang-Mills solutions, *Phys. Rev. D* **18** (1978), 2013–2025.

5. E. J. CORRIGAN, D. B. FAIRLIE, S. TEMPLETON, AND P. GODDARD, A Green's function for the general self-dual guage field, *Nucl. Phys. B* **140** (1978), 31–44.

6. V. G. DRINFELD AND Y. I. MANIN, Description of Instantons II, *Proc. Int. Sem. High Energy Phys. Quantum Field Theory, Serpukhov*, Vol. 1, 1978.

7. T. EGUCHI AND A. HANSON, Asymptotically flat solutions to Euclidean gravity, *Phys. Lett. B* **74** (1978), 249–251.

8. G. W. GIBBONS AND S. W. HAWKING, Gravitational multi-instantons, *Phys. Lett. B* **78** (1978), 430.

9. P. A. GRIFFITHS AND J. HARRIS, "Principles of Algebraic Geometry," Wiley, New York, 1978.

10. J. HADAMARD, "Lectures on Cauchy's Problem in Linear Partial Differential Equations," Yale Univ. Press, New Haven, Connecticut, 1923.

11. N. J. HITCHIN, Polygons and gravitons, *Math. Proc. Cambridge Philos. Soc.* **85** (1979), 465–476.

12. N. J. HITCHIN, Linear field equations on self-dual spaces, *Proc. Roy. Soc. London, Ser. A* **370** (1980), 173–191.

13. R. G. MCLENAGHAN, An explicit determination of the empty spacetimes on which the wave equation satisfies Huygens' principle, *Proc. Cambridge Philos. Soc.* **65** (1969), 139–155.

14. D. PAGE, Green's function for gravitational multi-instantons, *Phys. Lett. B* **85** (1979), 369–372.

MATHEMATICAL ANALYSIS AND APPLICATIONS, PART A
ADVANCES IN MATHEMATICS SUPPLEMENTARY STUDIES, VOL. 7A

Contingent Derivatives of Set-Valued Maps and Existence of Solutions to Nonlinear Inclusions and Differential Inclusions[†]

JEAN PIERRE AUBIN

Centre de Recherche de Mathématiques de la Decision
Université de Paris–Dauphine
Paris, France

DEDICATED TO LAURENT SCHWARTZ

We use the Bouligand contingent cone to a subset K of a Hilbert space at $x \in K$ for defining contingent derivatives of a set-valued map, whose graphs are the contingent cones to the graph of this map, as well as the upper contingent derivatives of a real-valued function. We develop a calculus of these concepts and show how they are involved in optimization problems and in solving equations $f(x) = 0$ and/or inclusions $0 \in F(x)$. They also play a fundamental role for generalizing the Nagumo theorem on flow invariance and for generalizing the concept of Liapunov functions for differential equations and/or differential inclusions.

Contents

[†] Sponsored by the United States Army under Contract No. DAAG29-75-C-0024.

159

INTRODUCTION

Everyone knows the crucial importance in both pure and applied analysis of the concept of derivative of a function or a distribution discovered by Laurent Schwartz.

Let V be a locally integrable function defined on an open set $\Omega \subset \mathbb{R}^n$ and $v \in \mathbb{R}^n$. We form the differential quotients $\nabla_h V(\cdot, v) \doteq [V(\cdot + hv) - V(\cdot)]/h$. Instead of requiring that $\nabla_h V(\cdot, v)$ converges for the topology of the pointwise convergence, one is still content with the much weaker convergence of $\nabla_h V(\cdot, v)$, in the space of distributions. In other words, one can find a weak enough topology that allows the convergence of the differential quotients $\nabla_h V(\cdot, v)$.

However, many problems arising in nonlinear analysis, in optimization, and in differential equations still require the pointwise convergence of the differential quotients $\nabla_h V(\cdot, v)$, but allow to use lim sup or lim inf instead of the limit. This was already proposed by Dini when V is a locally Lipschitzean function. A few years ago, Clarke suggested using $\lim \sup_{h \to 0+, y \to x} \nabla_h V(y, v)$, whose charm lies in the fact that it is always convex and continuous with respect to v.

We propose in this paper to take in consideration another candidate, namely $\lim \inf_{h \to 0+, w \to v} \nabla_h V(x, w)$. The main justification for this is that it works well for solving the problems we were studying: we hope to convince the reader by presenting some results in the following pages.

Also, this concept can be defined not only for real-valued functions, but can be adapted for vector-valued as well as set-valued maps.

Indeed, one way to see this is to consider the graph of a map. If we can define a tangent space to this graph, then we know that it is the graph of its derivative. If not, we can still define a "tangent cone" to this graph and decide to look at it as the graph of some set-valued map that, hopefully, retains enough properties of a would be derivative to deserve to be presented to the public.

As a seducing candidate, we can think of the Clarke tangent cone, which is *always* closed and convex. But there is an older candidate, the *contingent cone*, introduced by Bouligand in the early 1930s, in connection with the theory of derivatives of functions of one or two variables. We claim that it would be unwise to bury and forget it. The contingent cone $D_K(x)$ is defined by

$$D_K(x) = \bigcap_{\substack{\alpha > 0 \\ \varepsilon > 0}} \bigcup_{h \in]0,\alpha[} \left(\frac{1}{h}(K - x) + \varepsilon B \right). \tag{1}$$

For this literature, see Saks [27] and Rockafellar [26].

We note that when Int $K \neq \emptyset$ and when $x \in$ Int K, then $D_K(x) = X$. So, conditions involving the contingent cones are boundary conditions, in the sense that they are trivial when $x \in$ Int K. In 1943, Nagumo [21] proved that if a continuous and bounded map f from K to $\mathbb{R}^n$ satisfy

$$\forall x \in K, \qquad f(x) \in D_K(x) \tag{2}$$

then there exists a trajectory $x(\cdot)$ of the differential equation

$$x' = f(x), \qquad x(0) = x_0 \qquad \text{where} \quad x_0 \text{ is given in } K \tag{3}$$

that remains in the closure of K. Moreover, if for all $x_0 \in K$, there exists a trajectory of the differential equation that remains in K, condition (2) is satisfied.

Analogous statements remain true for differential inclusions (see Haddad [18], Aubin *et al.* [9] when K is convex, and Aubin and Clarke [4]).

Also, we can use this contingent cone to solve nonlinear equations.

For instance, let f be a continuously differential map from a neighborhood of a compact subset $K \subset \mathbb{R}^n$ to $\mathbb{R}^p$. Assume that

$$\forall x \in K, \qquad \exists u \in D_K(x) \quad \text{such that} \quad \nabla f(x)u = -f(x). \tag{4}$$

Then there exists a solution $\bar{x} \in K$ to the equation $f(x) = 0$ (see Aubin and Clarke [5]). We shall extend this result to inclusions $0 \in F(x)$ as well as finding other results in this direction.

Finally, in optimization theory, contingent cones play a role. For instance, if $x_0 \in K$ minimizes a continuously differentiable function U defined on a neighborhood of K, then

$$\forall u \in D_K(x_0), \qquad \langle \nabla U(x_0), u \rangle \geq 0. \tag{5}$$

These results, among many other applications, justify a further study of contingent cones, despite the unfortunate fact that they can fail to be convex. If one needs convexity (for using duality correspondence between convex cones and their polars, for instance), one should use the Clarke tangent cone (see Clarke [12–14] and Rockafellar [24–26]. So, we face the dilemma of either using a convex tangent cone, which may be too small, or using the contingent cone which appears naturally in many instances, but which is not generally convex. Fortunately, when K is closed and convex or when K is a smooth manifold, these two cones coincide.

JEAN PIERRE AUBIN

We proceed as in elementary calculus, when the derivatives of real-valued functions are defined from the tangents to the graph. Actually, if F is a set-valued map and if (x_0, y_0) belongs to graph(F) of F, we can define its contingent cone $D_{\text{graph}(F)}(x_0, y_0)$, which is a closed cone (not necessarily convex). We define the *contingent derivative* $DF(x_0, y_0)$ of F at x_0 and $y_0 \in F(x_0)$ to be the set-valued map whose graph is $D_{\text{graph}(F)}(x_0, y_0)$ (see Fig. 1). We shall characterize the contingent derivative: $v_0 \in DF(x_0, y_0)(u_0)$ if and only if

$$\liminf_{\substack{h \to 0+ \\ u \to u_0}} d\left(v_0, \frac{F(x_0 + hu) - y_0}{h} \right) = 0. \tag{6}$$

This formula captures the idea of a derivative as a suitable limit of differential quotients.

If one desires to use a concept of derivatives, which would be a set-valued map whose graph is closed and convex (these are called convex processes by Rockafellar [23]), he may use the "Clarke derivative," whose graph is the Clarke tangent cone to the graph of F. (See Ioffé [19] for a another approach.)

Again, the advantages of convexity should be weighted against valuable properties of contingent derivatives in the field of nonlinear equations and differential equations.

What about real-valued functions? Since a real-valued function V is a particular case of a set-valued map, we can define its contingent derivative: $v_0 \in DV(x_0)(u_0)$ if and only if

$$\liminf_{\substack{h \to 0+ \\ u \to u_0}} \left| v_0 - \frac{V(x_0 + hu) - V(x_0)}{h} \right| = 0. \tag{7}$$

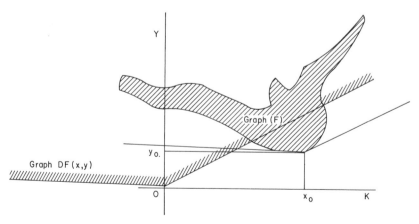

FIGURE 5.1

In many instances, the order relation of real numbers play an important role. This is the case in optimization theory, in differential inequalities, and in Liapunov stability of trajectories of differential equations or inclusions. In this point of view, it is natural to associate with a real-valued function $x \mapsto V(x)$ the set-valued map $V_+(x) \doteq V(x) + \mathbb{R}_+$ (whose graph is the epigraph of V). So, we check that the contingent derivative $DV_+(x_0, v(x_0))(u_0)$ is the half line $[D_+ V(x_0)(u_0), \infty[$ where

$$D_+ V(x_0)(u_0) = \liminf_{\substack{h \to 0+ \\ u \to u_0}} \frac{V(x_0 + hu) - V(x_0)}{h}. \tag{8}$$

We shall say that $D_+ V(x_0)$ is the *upper contingent derivative* of V. By using the Clarke derivative of V_+ at $(x, V(x))$, we obtain the Clarke generalized directional derivative (see Clarke [12, 13] and Rockafellar [24–26]). Let us mention that the variational principle holds true. When V is a function from K to $\mathbb{R}$ and when $x_0 \in K$ minimizes V on K, then,

$$\forall u \in X, \qquad 0 \le D_+ V(x_0)(u). \tag{9}$$

What makes this property useful is the calculus of upper contingent derivatives for computing $D_+ V(x)$ in terms of derivatives of other functions from which V is constructed. As an example, consider the case where $V = U|_K$ is the restriction to K of a continuously differentiable function U. One can prove that

$$D_+ V(x_0)(u_0) = \begin{cases} \langle \nabla U(x_0), u_0 \rangle & \text{when} \quad u_0 \in D_K(x_0), \\ +\infty & \text{when} \quad u_0 \notin D_K(x_0). \end{cases} \tag{10}$$

So, property (9) becomes

$$\forall u \in D_K(x_0), \qquad 0 \le \langle \nabla U(x_0), u \rangle, \tag{11}$$

or, equivalently, if $D_K(x_0)^-$ denotes the negative polar cone,

$$-\nabla U(x_0) \in D_K(x_0)^-. \tag{12}$$

Contingent derivatives do play an important role in sensibility analysis for optimization problems, which is of great relevance in economics, for instance. Let $F: K \subset \mathbb{R}^n \to \mathbb{R}^m$ be a compact-valued map and $U: F(K) \times K \to \mathbb{R}$ be a real-valued function $(x, y) \to U(x, y)$ which is lower semicontinuous with respect to x. We define the marginal function V by

$$V(y) \doteq \min\{U(x, y) | x \in F(y)\} \tag{13}$$

and the marginal (set-valued) map G by

$$G(y) = \{x \in F(y) | U(x, y) = V(y)\}. \tag{14}$$

We shall prove the following facts:

$$\forall v_0 \in \operatorname{Dom} DF(y_0, x_0), \qquad \forall u_0 \in DF(y_0, x_0)(v_0)$$

we have

$$D_+ V(y_0)(v_0) + D_+(-U)(x_0, y_0)(u_0, v_0) \leq 0 \tag{15}$$

and, for the contingent derivative of the marginal map G,

$$\forall v_0 \in \operatorname{Dom} DG(y_0, x_0), \qquad \forall u_0 \in DG(y_0, x_0)(v_0),$$

we have

$$D_+ U(x_0, y_0)(u_0, v_0) + D_+(-V)(y_0)(v_0) \leq 0. \tag{16}$$

In this line of thought, we can state a differential version of Ekeland's variational principle: Let K be a closed subset and $V: K \to [0, \infty[$ a lower semicontinuous function. Then we can associate with any $\varepsilon > 0$ and $x_\varepsilon \in K$ satisfying $V(x_\varepsilon) \leq \inf V(x) + \varepsilon^2$ an element $\bar{x}_\varepsilon \in K$ which satisfies

(i) $\|x_\varepsilon - \bar{x}_\varepsilon\| \leq \varepsilon,$

(ii) $\forall u \in X, 0 \leq D_+ V(\bar{x}_\varepsilon)(u) + \varepsilon\|u\|.$ $\tag{17}$

This result is as useful as the original version of Ekeland's theorem. It yields surjectivity theorems and inverse function theorems. For instance, we shall prove that when an upper semicontinuous map F from a closed subset $K \subset X$ to the compact subsets of Y satisfies

$$\exists c > 0 \quad \text{such that,} \quad \forall x \in K, \quad \forall y \in F(x), \quad \forall v \in Y,$$
$$\exists u \in X \quad \text{satisfying} \quad v \in DF(x, y)(u) \quad \text{and} \quad c\|u\| \leq \|v\|, \tag{18}$$

then F maps K onto Y.

We also shall use this kind of approach for solving inclusions $0 \in F(x_*)$. (We shall say that x_* is a stationary point of F.)

We introduce two functions

$$V: K \to \mathbb{R}_+ \quad \text{and} \quad W: K \times \overline{\operatorname{co}}(F(x)) \to \mathbb{R}_+ \tag{19}$$

and we shall say that V is a Liapunov function for F with respect to W if

$$\forall x \in K, \qquad \exists v \in F(x) \quad \text{such that} \quad D_+ V(x)(v) + W(x, v) \leq 0. \tag{20}$$

We shall observe that when V is lower semicontinuous and lower semi-compact (i.e., the subsets $\{x \in K \mid V(x) \leq \lambda\}$ are relatively compact for all $\lambda \in \mathbb{R}$),

$$\exists x_* \in K \quad \text{and} \quad v_* \in F(x_*) \quad \text{such that} \quad W(x_*, v_*) = 0. \tag{21}$$

We note that when we assume that $W(x_*, v_*) = 0$ if and only if $v_* = 0$, such an x_* is a stationary point of F. This is not all. Assume, for instance, that

(i) F is bounded, upper semicontinuous and has compact convex values,

(ii) V is continuous and lower semicompact,

(iii) W is continuous and is convex with respect to v. (22)

We shall prove that V is a Liapunov function for F with respect to W if and only if for all $x_0 \in K$, there exists a trajectory $x(\cdot) \in \mathscr{C}(0, \infty; \mathbb{R}^n)$ of the differential inclusion

$$x' \in F(x), \qquad x(0) = x_0, \tag{23}$$

which is monotone in the sense that

$$\forall s > t, \qquad V(x(s)) - V(x(t)) + \int_s^t W(x(\tau), x'(\tau))\, d\tau \leq 0. \tag{24}$$

In this case, subsequences $x(t_n)$ and $x'(t_n)$ have "almost cluster points" $x_* \in K$ and $v_* \in F(x_*)$ satisfying property (22), where almost cluster points are analogs for measurable classes of functions of cluster points for usual functions. Monotone trajectories yield information on the behavior of the nonincreasing function $t \to V(x(t))$ when $t \to \infty$. We also prove that under assumptions (22, i) and (22, iii), the function V_F defined on K by

$$V_F(x_0) = \inf\left\{ \int_0^\infty W(x(\tau), x'(\tau))\, d\tau \text{ when } x(\cdot) \text{ is a solution to (23)} \right\}, \tag{25}$$

is the smallest Liapunov function for F with respect to W when Liapunov functions do exist. This provides a bridge between Liapunov stability theory and Carathéodory–Bellman approach to optimal control theory.

1. Bouligand's Contingent Cone

Let K be a nonempty subset of a Hilbert space X.[†] We shall define the Bouligand contingent cone as follows.

Definition 1.   We say that the subset

$$D_K(x) = \bigcap_{\varepsilon > 0} \bigcap_{\alpha > 0} \bigcup_{0 < h \leq \alpha} \left(\frac{1}{h}(K - x) + \varepsilon B \right) \tag{1}$$

is the *contingent cone* to K at x.

[†] This is for simplicity. Several results of this paper are true for topological vector spaces.

In other words, $v \in D_K(x)$ if and only if

$$\forall \varepsilon > 0, \quad \forall \alpha > 0, \quad \exists u \in v + \varepsilon B, \quad \exists h \in \,]0, \alpha] \quad \text{such that} \quad x + hu \in K. \quad (2)$$

It is quite obvious that $D_K(x)$ is a *closed cone*, which is contained in the closed cone $T_K(x)$ defined by

$$T_K(x) \doteq \text{cl}\left(\bigcup_{h > 0} \frac{1}{h}(K - x) \right)$$

They coincide when K is a closed convex subset (see, e.g., Rockafellar [26]). We also note that

$$\text{if} \quad x \in \text{Int}(K), \quad \text{then} \quad D_K(x) = X. \quad (3)$$

We characterize the contingent cone by using the distance function $d_K(\cdot)$ to K defined by

$$d_K(x) \doteq \inf\{\|x - y\| \mid y \in K\}.$$

PROPOSITION 1

$$v \in D_K(x) \quad \text{if and only if} \quad \liminf_{h \to 0+} \frac{d_K(x + hv)}{h} = 0.$$

Proof. (a) Let $v \in D_K(x)$. For all $\varepsilon > 0$, $\alpha > 0$, there exist $h \in \,]0, \alpha]$ and $u \in v + \varepsilon B$ such that $x + hu \in K$. Hence

$$d_K(x + hv)/h \le (1/h)\|x + hv - (x + hu)\| \le \|u - v\| \le \varepsilon.$$

So, $\forall \varepsilon > 0, 0 \le \sup_\alpha \inf_{h \le \alpha} [d_K(x + hv)/h] \le \varepsilon$. This proves that

$$\liminf_{h \to 0+} [d_K(x + hv)/h] = 0.$$

(b) *Conversely*, if

$$\liminf_{h \to 0+} [d_K(x + hv)/h] = \sup_{\alpha > 0} \inf_{h < \alpha} [d_K(x + hv)/h] = 0,$$

we deduce that $\forall \varepsilon > 0, \forall \alpha > 0, \exists h \le \alpha$ such that $d_K(x + hv)/h \le \varepsilon/2$. Thus, there exists $y \in K$ such that $\|x + hv - y\|/h \le d_K(x + hv)/h + \varepsilon/2$. Hence $u = (y - u)/h \in v + \varepsilon B$ and satisfies $x + hu = y \in K$. ∎

Remark. We recognize the "Nagumo condition" implying the existence of trajectories remaining in a given subset K.

We can also characterize the contingent cone in terms of sequences.

PROPOSITION 2. $v \in D_K(x)$ *if and only if there exists a sequence of strictly positive numbers h_n and of elements $u_n \in X$ satisfying*

(i) $\lim\limits_{n \to \infty} u_n = v$, (ii) $\lim\limits_{n \to \infty} h_n = 0$, (iii) $\forall n \geq 0$, $x + h_n u_n \in K$. $\qquad$ (4)

Proof. It is left as an exercise.

Remark. For all $x \in X$, we have $D_X(x) = X$. We shall set $D_\emptyset(x) \doteq \emptyset$.

Remark. It is easy to see that the contingent cone to K and the contingent cone to the closure $\bar{K}$ of K coincide:

$$\forall x \in K, \qquad D_K(x) = D_{\bar{K}}(x).$$

Therefore, there is no danger in speaking of $D_K(x)$ even when $x \in \bar{K}$ and $x \notin K$.

PROPOSITION 3. *Let $K \subset X$ be a closed subset. We denote by $\pi_K(y)$ the subset of elements $x \in K$ such that $\|x - y\| = d_K(y)$. We obtain the following inequalities*

$$\forall y \notin K, \quad \forall x \in \pi_K(y), \quad \forall v \in \overline{\mathrm{co}}\, D_K(x), \qquad then \quad (y - x, v) \leq 0. \qquad (5)$$

Proof. Let $x \in \pi_K(y)$ and $v \in D_K(x)$. We deduce from the inequalities $\|y - x\| - d_K(x + hv) = d_K(y) - d_K(x + hv) \leq \|y - x - hv\|$ that

$$\frac{\langle y - x, v \rangle}{\|y - x\|} = \lim_{h \to 0+} \frac{\|y - x\| - \|y - x - hv\|}{h} \leq \liminf_{h \to 0+} \frac{d_K(x + hv)}{h} = 0$$

for $y \neq x$, $u \to \|u\|$ is differentiable at $u \neq 0$ and $v \in D_K(x)$. So $\langle y - x, v \rangle \leq 0$ for all $v \in D_K(x)$, and, consequently, for all $v \in \overline{\mathrm{co}}\, D_K(x)$. ∎

We deduce from this proposition a criterion of convexity of the contingent cone. Let us recall that a set-valued map F from M to N is lower semicontinuous at $x_0 \in M$ if for any $\varepsilon > 0$ and for any $y_0 \in F(x_0)$ there exists $\eta > 0$ such that, $F(x) \cap (y_0 + \varepsilon B) \neq \emptyset$ for all $x \in x_0 + \eta B$.

THEOREM 1. (B. Cornet). *Let us assume that*

$$x \in \bar{K} \mapsto \overline{\mathrm{co}}\, D_K(x) \text{ is lower semicontinuous at } x_0 \in \bar{K}. \qquad (6)$$

Then the contingent cone $D_K(x_0)$ to K at x_0 is a closed convex cone.

Proof. (a) Let $v_0 \in \overline{\mathrm{co}}\, D_K(x_0)$. For proving that $v_0 \in D_K(x_0)$, fix $\varepsilon > 0$ and let $\eta \doteq \eta(\varepsilon)$ such that, thanks to (6),

$$d_{\overline{\mathrm{co}}\, D_K(x)}(v_0) \leq d_{\overline{\mathrm{co}}\, D_K(x_0)}(v_0) + \varepsilon/2 \qquad \text{when} \quad \|x - x_0\| \leq \eta.$$

We take $h \doteq \eta/2\|v_0\|$ and, for $t \in \,]0, h[$, we set $y_t \doteq x_0 + tv_0$, we choose $x_t \in \pi_K(y_t)$ and $v_t \doteq \pi_{\overline{co}\, D_K(x_t)}(x_0)$. Hence Proposition 3 implies that

$$\langle y_t - x_t, v_t \rangle \leq 0. \tag{7}$$

We observe that $\|x_t - x_0\| \leq \|x_t - y_t\| + \|y_t - x_0\| \leq 2\|y_t - x_0\|$ [for $x_t \in \tau_K(y_t)$]. Consequently,

$$\|x_t - y_t\| \leq 2t\|v_0\| \leq \eta \qquad \text{when} \quad t \leq h \tag{8}$$

and thus,

$$\|v_t - v_0\| = d_{\overline{co}\, D_K(x_t)}(v_0) \leq \varepsilon/2. \tag{9}$$

Hence, inequalities (7), (8), and (9) imply

$$\begin{aligned}
\langle y_t - x_t, v_0 \rangle &\leq \langle y_t - x_t, v_0 - v_t \rangle + \langle y_t - x_t, v_t \rangle \\
&\leq \|y_t - x_t\|\|v_0 - v_t\| \leq 2t\varepsilon/2.
\end{aligned} \tag{10}$$

Let us set

$$f(t) \doteq \tfrac{1}{2} d_K(x_0 + tv)^2.$$

It is a locally Lipschitzean function, which is thus almost everywhere differentiable. If $t \in \,]0, h[$ is such that $f'(t)$ exists, we obtain

$$\begin{aligned}
f'(t) &= \lim_{\theta \to 0+} \frac{1}{2\theta} (d_K(y_t + \theta v_0)^2 - d_K(y_t)^2) \\
&\leq \lim_{\theta \to 0+} \frac{1}{2\theta} (\|y_t - x_t + \theta v_0\|^2 - \|y_t - x_t\|^2) \quad \text{for} \quad x_t \in \pi_K(y_t) \\
&= \langle y_t - x_t, v_0 \rangle \leq t\varepsilon \qquad\qquad\qquad \text{by (10)}.
\end{aligned}$$

Hence,

$$\frac{1}{2} d_K(x_0 + hv_0)^2 = f(h) - f(0) = \int_0^h f'(t)\, dt \leq \varepsilon \int_0^h t\, dt = \frac{\varepsilon h^2}{2}.$$

This implies that

$$\liminf_{h \to 0+} \frac{d_K(x_0 + hv_0)}{h} \leq \sqrt{\varepsilon}.$$

Since ε was chosen arbitrarily, we deduce that $\liminf_{h \to 0+} [d_K(x_0 + hv_0)/h] = 0$, i.e., that $v_0 \in D_K(x_0)$. We have proved that $\overline{co}\, D_K(x_0) = D_K(x_0)$. ∎

We mention the following consequence.

THEOREM 2 (Cornet). *Let us assume that X is finite dimensional and that*

$$x \in \bar{K} \mapsto D_K(x) \text{ is lower semicontinuous at } x_0. \tag{11}$$

Then $D_K(x_0)$ is a closed convex cone and

$$\forall v_0 \in D_K(x_0), \qquad \lim_{\substack{x \to x_0 \\ x \in K}} \liminf_{h \to 0+} \frac{d_K(x + hv_0)}{h} = 0. \qquad (12)$$

Proof. We recall that the lower semicontinuity of $x \mapsto D_K(x)$ at x_0 implies that the negative polar cone $D_K(\cdot)^-$ satisfies the following.

For all sequence of elements $x_n \in K$ converging to $x_0 \in K$ and for all sequence of elements $p_n \in N_K(x_0)$ converging to p_0, we have $p_0 \in N_K(x_0)$. When X is finite dimensional, this latter property implies that the set-valued map $x \mapsto \overline{\text{co}}\, D_K(x)$ is lower semicontinuous at $x_0 \in K$ (see Aubin *et al.* [9]). Hence $D_K(x_0)$ is a closed convex cone by Theorem 1. Let v_0 be chosen in $D_K(x_0)$ and let us associate with any $x \in \overline{K}$ an element $v \in D_K(x)$ such that $\|v_0 - v\| = d_{D_K(x)}(v_0)$. Since $x \mapsto D_K(x)$ is lower semicontinuous at x_0, there exists $\eta > 0$ such that $\|v - v_0\| = d_{D_K(x)}(v_0) \le d_{D_K(x_0)}(v_0) + \varepsilon = \varepsilon$ whenever $\|x - x_0\| \le \eta$. Therefore for all $h > 0$,

$$\frac{d_K(x + hv_0) - d_K(x + hv)}{h} \le \|v - v_0\| \le \varepsilon.$$

Since $v \in D_K(x)$, we deduce that for all $x \in x_0 + \eta B$,

$$\liminf_{h \to 0+} \frac{d_K(x + hv_0)}{h} \le \liminf_{h \to 0+} \frac{d_K(x + hv)}{h} + \varepsilon = \varepsilon. \qquad \blacksquare$$

2. Calculus on Contingent Cones

We state and prove several properties of contingent cones.

PROPOSITION 1. *Let $K \subset L \subset X$ be two nonempty subsets. Then,*

$$\forall x \in K, \qquad D_K(x) \subset D_L(x). \qquad (1)$$

Proof. It is left as an exercise.

PROPOSITION 2 (Hess). *Let $K = \bigcup_{i \in I} K_i$ be the union of subsets K_i. If $x \in K$, we set $I(x) = \{i \in I \text{ such that } x \in K_i\}$. Then*

$$\forall x \in K, \qquad \bigcup_{i \in I(x)} D_{K_i}(x) \subset D_K(x). \qquad (2)$$

If I is finite or, more generally, locally finite in the sense that there exists $r > 0$ such that $x + rB$ meets only a finite number of K_i, and if the subsets K_i are closed, then

$$\bigcup_{i \in I(x)} D_{K_i}(x) = D_K(x). \qquad (3)$$

Proof. The first inclusion is obvious. For proving the opposite inclusion, take $v \in D_K(x)$. Then, there exists a sequence of elements $u_n \in X$ and $h_n > 0$ such that $x + h_n u_n \in K$, $\lim_{h \to \infty} h_n = 0$ and $\lim_{h \to \infty} u_n = v$. There exists n_r such that, for all $n \geq n_r$, $x + h_n u_n \in x + rB$. Let $I_r(x) \doteq \{i \in I \,|\, (x + rB) \cap K_i \neq \varnothing\}$, which is finite by assumption. Then, $\forall n \geq n_r, x + h_n u_n \in \bigcup_{i \in I_r(x)} K_i$. Therefore, there exists at least an index $i_0 \in I_r(x)$ and an infinite subsequence such that

$$x + h_{n_p} u_{n_p} \in K_{i_0}.$$

This proves two facts. First, that $v \in D_{K_{i_0}}(x)$ and second, that $x \in \mathrm{cl}(K_{i_0}) = K_{i_0}$ (by assumption). Hence $i_0 \in I(x)$. ∎

PROPOSITION 3. *Let $K \doteq \bigcap_{i \in I} K_i$ be the intersection of subsets K_i. We set $J(x) \doteq \{i \in I$ such that $x \notin \mathrm{Int}\, K_i\}$. Then*

$$\forall x \in K, \qquad D_K(x) \subset \bigcap_{i \in J(x)} D_{K_i}(x). \tag{4}$$

Proof. It is left as an exercise.

PROPOSITION 4. *Let $K = \prod_{i \in I} K_i$ be the product of a family of nonempty subsets K_i of Hilbert spaces X_i. Then*

$$\forall x \in K, \qquad D_K(x) \subset \prod_{i \in I} D_{K_i}(x_i). \tag{5}$$

Proof. It is left as an exercise.

The following proposition gives some information about the contingent cone of the image of a subset by a smooth map.

PROPOSITION 5. *Let X and Y be two Hilbert spaces, $K \subset X$ be a subset of X and A be a continuously differentiable map from an open neighborhood of K to Y. Then*

$$\forall x \in K, \qquad \nabla A(x) D_K(x) \subset D_{A(K)}(Ax). \tag{6}$$

Proof. Let $v \in D_K(x)$, $\varepsilon > 0$, and $\alpha > 0$. Since A is continuously differentiable, there exist $\beta > 0$, $\eta > 0$ such that, $\forall h \leq \beta$ and for all $u \in v + \eta B$, $A(x + hu) = A(x) + h \nabla A(x)u + he$ where $e \in \frac{1}{2}\varepsilon B$. Let $\delta \doteq \min(\varepsilon/2\|\nabla A(x)\|, \eta)$ and $\gamma \doteq \min(\alpha, \beta)$. Since $v \in D_K(x)$, there exist $h < \gamma$ and $u \in v + \delta B$ such that $x + hu \in K$. Hence $A(x) + h(\nabla A(x)u + e) = A(x + hu) \in A(K)$ and $\|\nabla A(x)u + e - \nabla A(x)v\| \leq \varepsilon/2 + \|\nabla A(x)\|\delta \leq \varepsilon$. This states that $\nabla A(x)v \in D_{A(K)}(Ax)$. ∎

In particular, if $A \in \mathscr{L}(X, Y)$, we obtain the formula

$$\forall x \in K, \qquad AD_K(x) \subset D_{A(K)}(A(x)). \tag{7}$$

We study now the contingent cone to the preimage of a set by a smooth map:

PROPOSITION 6. *Let X and Y be two Hilbert spaces, $L \subset X$ and $M \subset Y$ be two subsets, and A be a continuously differentiable map from an open neighborhood of L to Y. We set*

$$K \doteq \{x \in L \mid A(x) \in M\} = L \cap A^{-1}(M). \tag{8}$$

Then,

$$\forall x \in K, \qquad D_K(x) \subset D_L(x) \cap \nabla A(x)^{-1} \cdot D_M(A(x)). \tag{9}$$

Proof. By Proposition 1, $D_K(x) \subset D_L(x)$ because $K \subset L$. By Proposition 5, $\nabla A(x) D_K(x) \subset D_{A(K)}(Ax) \subset D_M(Ax)$ because $A(K) \subset M$. Hence $D_K(x) \subset \nabla A(x)^{-1} D_M(Ax)$ and consequently, formula (9) holds true. ∎

3. CONTINGENT DERIVATIVE OF A SET-VALUED MAP

We adapt to the case of a set-valued map the intuitive definition of a derivative of a function in terms of the tangent to its graph.

Let F be a *proper set-valued map from $K \subset X$ to Y*. [We say that F is *proper* if its images $F(x)$ are nonempty for all $x \in K$.] Let $x_0 \in K$ and $y_0 \in F(x_0)$.

We denote by $DF(x_0, y_0)$ the set-valued map from X to Y whose graph is the contingent cone $D_{\mathrm{graph}(F)}(x_0, y_0)$ to the graph of F at (x_0, y_0).

In other words,

$$v_0 \in DF(x_0, y_0)(u_0) \quad \text{if and only if} \quad (u_0, v_0) \in D_{\mathrm{graph}(F)}(x_0, y_0). \tag{1}$$

DEFINITION 1. We shall say that the set-valued map $DF(x_0, y_0)$ from X to Y is the *contingent derivative* of F at $x_0 \in K$ and $y_0 \in F(x_0)$.

It is a positively homogeneous set-valued map (since its graph is a cone) with closed graph. Also, we note that

$$\mathrm{Dom}\, DF(x_0, y_0) \subset D_K(x_0), \tag{2}$$

i.e., that the domain of $DF(x_0, y_0)$ is contained in the contingent cone to K at x_0.

We first point out that

$$\forall x_0 \in K, \quad \forall y_0 \in F(x_0), \quad DF(x_0, y_0)^{-1} = D(F^{-1})(y_0, x_0). \tag{3}$$

Indeed, to say that $(u_0, v_0) \in D_{\mathrm{graph}(F)}(x_0, y_0)$ amounts to saying that $(v_0, u_0) \in D_{\mathrm{graph}(F^{-1})}(y_0, x_0)$.

Example: *Indicator of a Set and Its Contingent Derivative*

Among the set-valued maps from X to a Hilbert space Y, we single out the "indicator" of K which is the set-valued map Ψ_K from X to Y defined by

$$\Psi_K(x) = \begin{cases} 0 & \text{if } x \in K, \\ \varnothing & \text{if } x \notin K. \end{cases}$$

Note that Ψ_K depends upon the choice of the Hilbert space Y. We recall the following conventions:

$$\text{If } M \subset Y, \quad \text{then} \quad M + \varnothing = \varnothing + M = \varnothing$$

and

$$d(x, \varnothing) = +\infty.$$

Note also that if F is a set-valued map from X to Y, then $F + \Psi_K$ yields the restriction of F to K (since $(F + \Psi_K)(x) = \varnothing$ when $x \notin K$).

PROPOSITION 1. *The contingent derivative of the indicator of a set K is the indicator of the contingent cone to K*:

$$\forall x \in K, \qquad D\Psi_K(x, 0) = \Psi_{D_K}(x). \tag{4}$$

Proof. $\text{graph}[D\Psi_K(x)] = D_{\text{graph}(\Psi_K)}(x, 0) = D_{K \times \{0\}}(x, 0)$. It is easy to check that $D_{K \times \{0\}}(x, 0) = D_K(x) \times \{0\}$. So, $\text{graph}[D\Psi_K(x)] = D_K(x) \times \{0\} :=$ graph $[\Psi_{D_k(x)}]$. ∎

Characterization of the Contingent Derivative

We shall give an analytical characterization of $DF(x_0, y_0)$, which justifies that the above definition is a reasonable candidate for capturing the idea of a derivative as a (suitable) limit of differential quotients. We extend F to X by setting $F(x) = \varnothing$ when $x \notin K$.

THEOREM 1. *Let F be a set-valued map from $K \subset X$ to Y and $(x_0, y_0) \in$ graph(F). Then $v_0 \in DF(x_0, y_0)(u_0)$ if and only if*

$$\liminf_{\substack{h \to 0+ \\ u \to u_0}} d\left(v_0, \frac{F(x_0 + hu) - y_0}{h}\right) = 0 \tag{5}$$

Proof. To say that $v_0 \in DF(x_0, y_0)(u_0)$, i.e., that $(u_0, v_0) \in D_{\text{graph}(F)}(x_0, y_0)$, amounts to saying that for all $\varepsilon_1, \varepsilon_2 > 0$ and $\alpha > 0$, there exists $u \in u_0 + \varepsilon_1 B$ and $v \in v_0 + \varepsilon_2 B$ such that $v \in [F(x_0 + hu) - y_0]/h$.

This is equivalent to saying that $\forall \varepsilon_1 > 0, \varepsilon_2 > 0, \alpha > 0$, we have

$$\inf_{h \leq \alpha} \inf_{\|u - u_0\| \leq \varepsilon_1} d\left(v_0, \frac{F(x_0 + hu) - y_0}{h}\right) \leq \varepsilon_2.$$

The last statement is equivalent to (5). ∎

When F is a single-valued map, we set

$$DF(x_0, y_0) = DF(x_0) \tag{6}$$

since $y_0 = F(x_0)$. The above formula shows that in this case, $v_0 \in DF(x_0)(u_0)$ if and only if

$$\liminf_{\substack{h \to 0+ \\ u \to u_0}} \frac{\|F(x_0 + hu) - F(x_0) - hv_0\|}{h} = 0. \tag{7}$$

If F is a single-valued map which is *regularly Gâteaux differentiable*, in the sense that there exists $\nabla F(x) \in \mathscr{L}(X, Y)$ satisfying:

$$\forall u_0 \in X, \qquad \lim_{\substack{h \to 0+ \\ u \to u_0}} \frac{F(x_0 + hu) - F(x_0)}{h} = \nabla F(x_0)u_0, \tag{8}$$

then the contingent derivative coincides with the Gâteaux erivative:

$$\forall u_0 \in X, \qquad DF(x_0)(u_0) = \nabla F(x_0)u_0. \tag{9}$$

Example: Contingent Derivatives of Locally Lipschitzean Maps

This formula has a simpler form when the set-valued map F is *upper locally Lipschitzean*:

DEFINITION 2. We say that a set-valued map F is *upper locally Lipschitzean* at $x_0 \in \text{Int } K$ if there exist a neighborhood $N(x_0)$ of x_0 and a constant $l > 0$ such that

$$\forall x \in N(x_0), \qquad F(x \subset F(x_0) + l\|x - x_0\|B. \tag{10}$$

Naturally, any locally Lipschitzean (and, a fortiori, Lipschitzean) set-valued map is upper locally Lipschitzean.

PROPOSITION 2. *Let F be an upper locally Lipschitzean set-valued map from $\text{Int } K \subset X$ to Y, $x_0 \in \text{Int } K$ and, $y_0 \in F(x_0)$. Then $v_0 \in DF(x_0, y_0)(u_0)$ if and only if*

$$\liminf_{h \to 0+} d\left(v_0, \frac{F(x_0 + hu_0) - y_0}{h}\right) = 0. \tag{11}$$

Remark. If F is a locally Lipschitzean single-valued map, this formula becomes

$$\liminf_{h \to 0+} \frac{\|F(x_0 + hu_0) - F(x_0) - hv_0\|}{h} = 0. \tag{12}$$

Proof. Let $v_0 \in DF(x_0, y_0)(u_0)$. Then for all $\varepsilon_1 > 0$, $\varepsilon_2 > 0$, $\alpha > 0$, there exists $h < \alpha$ such that

$$v_0 \in \bigcup_{\|u - u_0\| \le \varepsilon_1} \frac{F(x_0 + hu) - y_0}{h} + \varepsilon_2 B.$$

But, F being upper locally Lipschitzean, we know that

$$F(x_0 + hu) \subset F(x_0 + hu_0) + lh\|u - u_0\|B \subset F(x_0 + hu_0) + lh\varepsilon_1 B.$$

Hence, for all $\varepsilon_1, \varepsilon_2 > 0$,

$$v_0 \in \bigcap_{\alpha > 0} \bigcup_{h \le \alpha} \left(\frac{F(x_0 + hu_0) - y_0}{h} + (\varepsilon_1 l + \varepsilon_2)B \right).$$

This implies formulas (11). ∎

4. Calculus on Contingent Derivatives

We shall derive from the properties of the contingent cones a calculus on contingent derivatives of set-valued maps. We shall start naturally with the chain rule:

PROPOSITION 1. *Let F be a set-valued map from $K \subset X$ to Y and A be a continuously differentiable map from an open neighborhood of $F(K) \subset Y$ to Z. Then*

$$\forall u_0 \in X, \quad \nabla A(y_0) \cdot DF(x_0, y_0)(u_0) \subset D(AF)(x_0, Ay_0)(u_0). \tag{1}$$

Proof. Let $(1 \times A)$ be the map: $(x, y) \in X \times F(K) \mapsto (x, A(y)) \in X \times Z$. The graph $\mathscr{G}$ of the set-valued map $x \mapsto AF(x)$ is related to the graph $\mathscr{F}$ of F by the relation $\mathscr{G} = (1 \times A)\mathscr{F}$. By Proposition 2.5, we know that $(1 \times \nabla A(y_0))D_{\mathscr{F}}(x_0, y_0) \subset D_{\mathscr{G}}(x_0, Ay_0)$. This states that for all $v_0 \in DF(x_0, y_0)(u_0)$, $\nabla A(y_0)v_0 \in D(FA)(x_0, Ay_0)(u_0)$. ∎

PROPOSITION 2. *Let F be a set-valued map from $K \subset X$ to Y and A be a continuously differentiable map from an open subset $\Omega \subset Z$ to X. Then*

$$\forall u_0 \in Z, \quad D(FA)(x_0, y_0)(u_0) \subset DF(Ax_0, y_0)(\nabla A(x_0)(u_0)). \tag{2}$$

Proof. Let $(A \times 1): \Omega \times Y \to K \times Y$ be the map defined by $(A \times 1)(z, y) = (Az, y)$. The graph $\mathscr{G}$ of the set-valued map FA from Ω to Y is related to the graph $\mathscr{F}$ of F by the relation $\mathscr{G} = (A \times 1)^{-1}\mathscr{F}$. By Proposition 2.6, we know that $D_{\mathscr{G}}(x_0, y_0) \subset (\nabla A(x_0) \times 1)^{-1}D_{\mathscr{F}}(Ax_0, y_0)$. This states that for all $u_0 \in Z$ and $v_0 \in D(FA)(x_0, y_0)(u_0)$, then $v_0 \in DF(Ax_0, y_0))(\nabla A(x_0)(u_0))$. ∎

We state now the properties of contingent derivatives of unions and intersections of maps.

PROPOSITION 3. *Let us consider a family of set-valued maps* $F_i (i \in I)$ *from* K *to* X *and set* $F(x) = \bigcup_{i \in I} F_i(x)$. *We associate with any* $x_0 \in K$ *and* $y_0 \in F(x_0)$ *the subset* $I(x_0, y_0) = \{i \in I \mid y_0 \in F_i(x_0)\}$. *Then*

$$\forall u_0 \in X, \qquad \bigcup_{i \in I(x_0, y_0)} DF_i(x_0, y_0)(u_0) \subset DF(x_0, y_0)(u_0). \tag{3}$$

Equality holds when the graphs of the maps F_i *are closed and when the family is locally finite.*

Proof. We note that the graph $\mathscr{F}$ of F is the union of the graphs $\mathscr{F}_i$ of F_i. Hence Proposition 3 follows from Proposition 2.2. ∎

PROPOSITION 4. *Let us consider a family of set-valued maps* $F_i (i \in I)$ *from* K *to* X *such that* $F(x) \doteq \bigcap_{i \in I} F_i(x) \neq \varnothing$ *for all* $x \in K$. *Let* $J(x_0, y_0) = \{i \in I$ *such that* $(x_0, y_0) \notin \text{int}(\text{graph}(F_i))\}$. *Then*

$$\forall u_0 \in X, \qquad DF(x_0, y_0)(u_0) \subset \bigcap_{i \in J(x_0, y_0)} DF_i(x_0, y_0)(u_0). \tag{4}$$

Proof. We note that the graph of F is the intersection of the graphs of the maps F_i and we apply Proposition 2.3. ∎

We turn our attention to the study of contingent derivatives of restrictions. First, we begin with this remark.

PROPOSITION 5. *If* $F \subset G$, *in the sense that* graph $F \subset$ graph G, *then* $\forall(x_0, y_0) \in \text{graph}(F)$, *we have*

$$DF(x_0, y_0)(u_0) \subset DG(x_0, y_0)(u_0). \tag{5}$$

Proof. It follows from the fact that $D_{\text{graph}(F)}(x_0, y_0) \subset D_{\text{graph}(G)}(x_0, y_0)$.

In particular, if $f(\cdot)$ is a regularly Gâteaux-differentiable selection of $F(\cdot)$, its Gâteaux-derivative $\nabla f(x)$ is a selection of the set-valued map $DF(x, f(x))$. ∎

We note that the restriction $F|_L \doteq F + \Psi_L$ of a set-valued map F to a subset L is contained in F. In this case, we obtain a more precise result.

PROPOSITION 6. *Let F be a set-valued map from $K \subset X$ to Y, L be a subset of K and $F|_L$ be the restriction of F to L. Then, for any $x_0 \in L$ and $y_0 \in F(x_0)$,*

$$DF|_L(x_0, y_0)(u_0) \subset DF(x_0, y_0)|_{D_L(x_0)}(u_0). \tag{6}$$

In other words, the contingent derivative of the restriction of F to L is contained in the restriction of the contingent derivative to the contingent cone to L.

Proof. Indeed, $v_0 \in DF|_L(x_0, y_0)(u_0)$ if and only if for all $\varepsilon > 0$, for all $\alpha > 0$, there exists $h > 0$ such that

$$v_0 \in \bigcup_{\substack{\|u - u_0\| \leq \varepsilon \\ u \in L}} \frac{F|_L(x_0 + hu) - y_0}{h} \subset \bigcup_{\|u - u_0\| \leq \varepsilon} \frac{F(x_0 + hu) - y_0}{h}.$$

This requires that there exists $u \in u_0 + \varepsilon B$ such that $x + hu \in L$. Hence u_0 belongs necessarily to $D_L(u_0)$ and formula (6) ensues. ∎

We give now a formula on the contingent derivative of the sum of a set-valued map and a smooth single-valued map.

PROPOSITION 7. *Let us assume that the set-valued map F is defined on K by $F(x) = G(x) + H(x)$ where H is a continuously differentiable map from a neighborhood of K to Y and G is any set-valued map from K to Y. Then, for any $x_0 \in K$, $z_0 \in G(x_0)$ and $y_0 = z_0 + H(x_0) \in F(x_0)$, we have*

$$DG(x_0, z_0)(u_0) + \nabla H(x_0) \cdot u_0 \subset DF(x_0, y_0)(u_0). \tag{7}$$

Proof. Let $u_0 \in \operatorname{Dom} DG(x_0, y_0)$ be given and $v_0 \in DG(x_0, y_0)(u_0)$. Since H is continuously differentiable, for any $\varepsilon > 0$, there exist $\delta \in {]0, \varepsilon]}$ and $\alpha > 0$ such that for all $u \in u_0 + \delta B$ and for all $h \leq \alpha$, $H(x_0 + hu) = H(x_0) = h\nabla H(x_0)u + he$, with $\|e\| \leq \varepsilon$. Since $v_0 \in DG(x_0, y_0)(u_0)$, there exists $u \in u_0 + \delta B$ and $h \leq \alpha$ such that

$$v_0 \in \frac{G(x_0 + hu) - z_0}{h} + \varepsilon B.$$

Hence

$$v_0 + \nabla H(x_0)u_0 \in \frac{F(x_0 + hu) - z_0}{h} + 2\varepsilon B + \nabla H(x_0)(u_0 - u) \in \frac{F(x_0 + hu) - z_0}{h}$$

$$+ \varepsilon(2 + \|\nabla H(x_0)\|)B. \quad ∎$$

By taking $G = \Psi_K$, we obtain the following corollary.

PROPOSITION 8. *Let F be a continuously differentiable map on a neighborhood of K and $F|_K$ be its restriction to K. Then*

$$\forall x_0 \in K, \quad DF|_K(x_0)(u_0) = \begin{cases} \{\nabla F(x_0) \cdot u_0\} & \text{when} \quad u_0 \in D_K(x_0), \\ \varnothing & \text{when} \quad u_0 \notin D_K(x_0). \end{cases} \quad (8)$$

Variational Principle

We generalize to the case of set-valued maps the fundamental fact that the derivative of a function at a point where it achieves the minimum vanishes.

Let $P \subset Y$ be a closed convex cone defining a preorder on Y (by calling nonnegative elements those elements of P). Let F be a set-valued map from $K \subset X$ to Y. We say that $x_0 \in K$ achieves the minimum of F on K at $y_0 \in F(x_0)$ if

$$\forall x \in K, \quad F(x) \subset y_0 + P. \quad (9)$$

PROPOSITION 9. *Let us assume that $x_0 \in K$ achieves the minimum of F on K at y_0. Then*

$$\forall u_0 \in X, \quad DF(x_0, y_0)(u_0) \subset P. \quad (10)$$

Proof. Let $v_0 \in DF(x_0, y_0)(u_0)$. For all $\varepsilon > 0$ and $\alpha > 0$, there exists $u \in u_0 + \varepsilon B$ such that

$$v_0 \in \frac{F(x_0 + hu) - y_0}{h} + \varepsilon B \subset P + \varepsilon B$$

by (9). Hence $v_0 \in \text{cl}(P) = P$. ∎

Remark. This inclusion is trivial when $u_0 \notin \text{Dom } DF(x_0, y_0)$, since in this case $DF(x_0, y_0)(u_0) = \varnothing$.

The following property will play an important role for defining upper contingent derivatives of a real-valued function.

Let P be a closed convex cone of Y. For any subset L, we set $L_+ = L + P$; we say that L is "comprehensive" if $L_+ = L$.

PROPOSITION 10 *Let F be a set-valued map from K to Y. Then, for any $(x_0, y_0) \in \text{graph}(F)$, we have*

$$DF(x_0, y_0)_+ \subset DF_+(x_0, y_0). \quad (11)$$

If the images of F are comprehensive, the images of $DF(x_0, y_0)$ are also comprehensive.

Proof. Let $v_0 \in DF(x_0, y_0)(u_0)$ and $z \in P$. Let $\varepsilon > 0$ and $\alpha > 0$. We know that there exist $u \in u_0 + \varepsilon B$ and $h \leq \alpha$ such that

$$v_0 \in \frac{F(x_0 + hu) - y_0}{h} + \varepsilon B.$$

Hence

$$v_0 + z \in \frac{F(x_0 + hu) + hz - y_0}{h} + \varepsilon B \subset \frac{F_+(x_0 + hu) - y_0}{h} + \varepsilon B,$$

and thus, $v_0 + z \in DF_+(x_0, y_0)(u_0)$. ∎

5. Upper Contingent Derivative of a Real-Valued Function

We associate with the function $V: X \to]\infty, +\infty]$ the set-valued map V_+ defined by $V_+(x) = V(x) + \mathbb{R}_+$ when $V(x) < +\infty$ and $V_+(x) = \emptyset$ when $V(x) = +\infty$. Its domain is the domain of V and its graph is the epigraph of V. We consider its contingent derivative $DV_+(x, V(x))$, which has comprehensive values by Proposition 4.10. Therefore, for all $u_0 \in X$, $DV_+(x, V(x))(u_0)$ is either $\mathbb{R}$, or a half line $[v_0, \infty[$, or empty. We set

$$D_+ V(x)(u) \doteq \inf\{v \mid v \in DV_+(x, V(x))(u)\}. \tag{1}$$

It is equal to $-\infty$ if $DV_+(x, V(x)) = \mathbb{R}$, to v_0 if $DV_+(x, V(x))(u) = [v_0, \infty[$ and to $+\infty$ if $DV_+(x, V(x))(u) = \emptyset$.

DEFINITION 1. We shall say that $D_+ V(x)(u)$ is the *upper contingent derivative* of V at x in the direction u.

Remark. We can define as well $V_-(x) = V(x) - \mathbb{R}_+$ and $D_- V(x)(u) \doteq \sup\{v \mid v \in DV_-(x, V(x))(u)\}$. We say that $D_- V(x)(u)$ is the *lower contingent derivative* of V at x in the direction u.

We begin by computing upper contingent derivatives.

THEOREM 1. *If V is a real-valued function, then*

$$D_+ V(x_0)(u_0) = \liminf_{\substack{h \to 0_+ \\ u \to u_0}} \frac{V(x_0 + hu) - V(x_0)}{h}. \tag{2}$$

Proof. Indeed, let $v_0 \in DV_+(x_0, V(x_0))(u_0)$; then, $\forall \varepsilon_1 > 0, \varepsilon_2 > 0, \forall \alpha > 0$, there exist $u \in u_0 + \varepsilon_2 B$ and $h < \alpha$ such that

$$v_0 \in \frac{V_+(x_0 + hu) - V(x_0)}{h} + \varepsilon_1 B.$$

This implies that

$$v_0 \geq \frac{V(x_0 + hu) - V(x_0)}{h} - \varepsilon_1 \geq \inf_{h \leq \alpha} \inf_{\|u - u_0\| \leq \varepsilon_2} \frac{V(x_0 + hu) - V(x_0)}{h} - \varepsilon_1.$$

Therefore

$$v_0 \geq \liminf_{\substack{h \to 0+ \\ u \to u_0}} \frac{V(x_0 + hu) - V(x_0)}{h} - \varepsilon_1.$$

Let us set for the time

$$a \doteq \liminf_{\substack{h \to 0+ \\ u \to u_0}} \frac{V(x_0 + hu) - V(x_0)}{h}.$$

So, we have proved that $a \leq D_+ V(x_0)(u_0)$. On the other hand, we know that for any $M > a$, there exists $\delta > 0$ such that

$$\sup_{\substack{\alpha > 0 \\ \delta > 0}} \inf_{h \leq \alpha} \inf_{\|u_0 - u\| \leq \delta} \frac{V(x_0 + hu) - V(x_0)}{h} \leq M.$$

This shows that $M > a$, $\forall \delta > 0$, there exist $h < \alpha$, and $u \in u_0 + \delta B$ such that

$$\frac{V(x_0 + hu) - V(x_0)}{h} \leq M.$$

Hence $M \in [V_+(x_0 + hu) - V(x_0)]/h$. This proves that $a \in DV_+(x_0, V(x_0))(u_0)$. Since it is smaller than all the other ones, we infer that $a = D_+ V(x_0)(u_0)$. ∎

PROPOSITION 1. *If the function V is locally Lipschitzean, we have*

$$D_+ V(x_0)(v_0) = \liminf_{h \to 0+} \frac{V(x_0 + hu_0) - V(x_0)}{h}. \tag{3}$$

Proof. It is a consequence of Proposition 3.2, since in this case the set-valued map V_+ is upper locally Lipschitzean. One can see it directly for in this case,

$$\liminf_{h \to 0+} \frac{V(x_0 + hu_0) - V(x_0)}{h} = \liminf_{\substack{h \to 0+ \\ u \to u_0}} \frac{V(x_0 + hu) - V(x_0)}{h}. \quad ∎$$

So, in this case, the upper contingent derivative coincides with one of the Dini derivatives.

Remark. We can compute in the same way the lower contingent derivative of V: we obtain

$$D_- V(x_0)(u_0) = \lim_{\substack{h \to 0+ \\ u \to u_0}} \sup \frac{V(x_0 + hu) - V(x_0)}{h}. \tag{4}$$

Therefore, we always have

$$D_+ V(x_0)(u_0) \leq D_- V(x_0)(u_0). \tag{5}$$

We shall say that the interval-valued map:

$$u_0 \mapsto [D_+ V(x_0)(u_0), D_- V(x_0)(u_0)]$$

is the *contingent gap* map.

Let us mention also that

$$D_+ V(x_0)(-u_0) = \lim_{\substack{h \to 0+ \\ u \to u_0}} \sup \frac{V(x_0) - V(x_0 - hu)}{h}. \tag{6}$$

Remark. Let V be a function from X to $]-\infty, +\infty]$. We set

$$K \doteq \{x \in X \text{ such that } V(x) \leq c\}. \tag{7}$$

We can characterize the contingent cone to K at x in the following way:

PROPOSITION 2. *If $V(x) = c$, then*

$$D_K(x) \subset \{v \mid D_+ V(x)(v) \leq 0\}. \tag{8}$$

Proof. If $v_0 \in D_K(x)$, then $\forall \varepsilon > 0$, $\forall \alpha > 0$, there exist $h < \alpha$ and $v \in v_0 + \varepsilon B$ such that $x + hv \in K$, i.e., such that

$$\frac{V(x + hv) - V(x)}{h} \leq 0.$$

This implies that $D_+ V(x)(v_0) \leq 0$. ∎

Remark. The indicator $\phi_K : X \to]-\infty, +\infty]$ of a subset $K \subset X$ is the function defined by $\phi_K(x) = 0$ when $x \in K$ and $\phi_K(x) = +\infty$ when $x \notin K$.

PROPOSITION 3. *The upper contingent derivative of the indicator ϕ_K of $K \subset X$ is the indicator of the contingent cone $D_K(x)$:*

$$D_+ \phi_K(x)(\cdot) = \phi_{D_K(x)}(\cdot). \tag{9}$$

Proof. It follows from Proposition 3.1, for $\phi_{K+} = \psi_K$ is the indicator of K. ∎

6. CALCULUS ON UPPER CONTINGENT DERIVATIVES

The upper contingent derivative inherits the properties of the contingent derivatives of set-valued maps. In the above corollaries, we use the fact that $a \leq b$ if and only if $[b, \infty[\subset [a, \infty[$ and that $\bigcup_i [a_i, \infty[= [\min_i a_i, \infty[$.
We begin by the variational principle.

PROPOSITION 1. *Let V be a function from K to $\mathbb{R}$. If $x_0 \in K$ minimizes V on K, then*

$$\forall u_0 \in X, \qquad 0 \leq D_+ V(x_0)(u_0). \tag{1}$$

Proof. We apply Proposition 4.9 with $F(x) = V_+(x)$, $P = \mathbb{R}_+$, and $y_0 = V(x_0)$. ∎

PROPOSITION 2. *Let V be a function from K to $\mathbb{R}$ and $L \subset K$. Let $V|_L$ be the restriction of V to L. Then*

$$\forall x_0 \in L, \quad \forall v_0 \in D_L(x_0), \qquad D_+ V(x_0)(u_0) \leq D_+ V|_L(x_0)(u_0). \tag{2}$$

Proof. It follows from Proposition 4.6 with $F(x) \doteq V_+(x)$. ∎

We estimate now the upper contingent derivative of the sum of two functions.

PROPOSITION 3. *Let V and W be two functions from K to $\mathbb{R}$ and $U \doteq V + W$. Let $L \subset K$ be a subset of K. Then*

$$\begin{align}
\text{(i)} \quad & D_+ U(x_0)(u_0) \leq D_+ V(x_0)(u_0) + D_- W(x_0)(u_0), \\
\text{(ii)} \quad & \forall u_0 \in D_L(x_0), \quad D_+ V|_L(x_0)(u_0) \leq D_- V(x_0)(u_0).
\end{align} \tag{3}$$

Therefore, when $D_+ V(x_0) = D_- V(x_0)$ (which is the case when V is convex continuous or continuously differentiable), we obtain the formula

$$\forall u_0 \in D_L(x_0), \qquad D_+ V|_L(x_0)(u_0) = D_+ V(x_0)(u_0). \tag{4}$$

Proof. Inequality (3, i) follows from the fact that

$$\liminf_{\substack{h \to 0+ \\ v \to v_0}} (f(h, v) + g(h; v)) \leq \liminf_{\substack{h \to 0+ \\ v \to v_0}} f(h, v) + \limsup_{\substack{h \to 0+ \\ v \to v_0}} g(h, v).$$

where we set

$$f(h, v) = \frac{V(x_0 + hv) - V(x_0)}{h} \quad \text{and} \quad g(h, v) = \frac{W(x_0 + hv) - W(x_0)}{h}.$$

We deduce inequality (3, ii) by taking for function W the indicator $\phi_L(\cdot)$ of L. So, equality (4) follows from (2) and (3) when $D_+V(x_0)(u_0) = D_-V(x_0)(u_0)$. ∎

Remark. We deduce from Proposition 4.7 that when W is continuously differentiable, we have equality

$$D_+U(x_0)(u_0) = D_+V(x_0)(u_0) + \langle \nabla W(x_0), u_0 \rangle.$$

We shall now prove the chain rule formulas.

PROPOSITION 4. *Let $V: K \to \mathbb{R}$ be a function and φ be a continuously differentiable nondecreasing function from an open neighborhood of $V(K)$ to $\mathbb{R}$. Then*

$$D_+(\varphi V)(x_0)(u_0) \le \varphi'(V(x_0))D_+V(x_0)(v_0). \tag{5}$$

Proof. Since φ is nondecreasing, $\varphi(V_+(x)) = (\varphi V)_+(x)$. Hence, we apply Proposition 4.1: with $A = \varphi$, $F(x) = V_+(x)$ and $y = V(x)$. We obtain $\varphi'(V(x_0))DV_+(x_0, V(x_0))(u_0) \subset D(\varphi V)_+(x_0, \varphi V(x_0))(u_0)$, i.e., (5). ∎

PROPOSITION 5. *Let V be a function from K to $\mathbb{R}$ and A a continuously differentiable map from an open subset $\Omega \subset Z$ to K. Then*

$$D_+V(Ax_0)(\nabla A(x_0)u_0) \le D_+(VA)(x_0)(u_0). \tag{6}$$

Proof. We apply Proposition 4.2 with $F(x) = V_+(x)$. ∎

The following formula on the contingent derivative of the pointwise minimum of a finite number of real-valued functions is very useful.

PROPOSITION 6. *Let us consider n functions V_i from K to $\mathbb{R}$ ($i = 1, \ldots, n$). We set $V(x) \doteq \min_{i=1,\ldots,n} V_i(x)$ and $I(x) \doteq \{i \mid V_i(x) = V(x)\}$. Then*

$$D_+V(x_0)(u_0) = \min_{i \in I(x)} D_+V_i(x_0)(u_0). \tag{7}$$

Proof. We note that $V_+(x) = \bigcup_{i \in I} V_{i+}(x)$ and that $I(x, V(x)) = I(x)$. We apply Proposition 4.3. ∎

PROPOSITION 7. *Let us consider n functions V_i from K to $\mathbb{R}$. We set*

(i) $W(x) \doteq \max_i V_i(x)$,

(ii) $J(x_0) \doteq (i = 1, \ldots, n \mid (x_0, W(x_0)) \notin \text{Int } \mathscr{E}p(V_i)\}$.

$$\tag{8}$$

Then

$$D_+W(x_0)(v) \ge \max_{i \in J(x_0)} D_+V_i(x_0)(v). \tag{9}$$

Proof. We note that $W_+(x) = \bigcap_{j=1}^n V_{j+}(x)$ and that $J(x_0) = J(x_0, W(x_0))$. Then Proposition 4.4 implies that

$$DW_+(x_0, W(x_0))(v) \subset \bigcap_{i \in J(x_0, w(x_0))} DV_{i+}(x_0, W(x_0))(v).$$

This inclusion implies inequality (9). ■

Remark. If $x_0 \in \bigcap_{i=1}^n \text{Int Dom } V_i$, we note that

$$J(x_0) = \{i = 1, \ldots, n \,|\, W(x_0) = V_i(x_0)\}. \tag{10}$$

We shall study now the chain rule for the composition of a function V from X to $\mathbb{R}$ and an absolutely continuous function $t \to x(t)$. We reall that almost all t are Lebesgue points, i.e., satisfy $x'(t_0) = \lim_{h \to 0} [(1/h)\int_{t_0}^{t_0+h} x'(\tau)\,d\tau]$.

PROPOSITION 8.   *Let $x(\cdot)$ be an absolutely continuous function from $[t_0 - \eta, t_0 + \eta)$ to $K \subset X$ and assume that*

$$x'(t_0) = \lim_{h \to 0+} \frac{1}{h} \int_{t_0}^{t_0+h} x'(\tau)\,d\tau. \tag{11}$$

[This limit belongs to $D_K(x(t_0))$.]
Set $v(t) \doteq V(x(t))$. Inequality

$$D_+V(x(t_0))(x'(t_0)) \le \limsup_{h \to 0+} \frac{v(t_0 + h) - v(t_0)}{h} \tag{12}$$

always holds true.
    Moreover, if V is the restriction to K of a locally Lipschitzean function $\tilde{V}$ defined on a neighborhood of K, then

$$\liminf_{h \to 0+} \frac{v(t_0 + h) - v(t_0)}{h} \le D_+V(x(t_0))(x'(t_0)). \tag{13}$$

Therefore, if we know that $v'(t_0) = \lim_{h \to 0+} \{[v(t_0 + h) - v(t_0)]/h\}$ exists, we get

$$v'(t_0) = D_+V(x(t_0))(x'(t_0)). \tag{14}$$

Proof.   (a)   We set $v_h = (1/h)\int_0^h x'(\tau)\,d\tau$. So, we can associate to any $\varepsilon > 0$ a positive number $\beta > 0$ such that

$$\forall h \le \beta, \qquad \|v_h - x'(t_0)\| \le \varepsilon. \tag{15}$$

We observe that $x(t_0 + h) = x(t_0) + hv_h \in K$. Hence $x'(t_0) \in D_K(x(t_0))$. We set

$$v'_\#(t_0) \doteq \inf_{\alpha > 0} \sup_{h < \alpha} \frac{v(t_0 + h) - v(t_0)}{h}.$$

We have

$$D_+V(x(t_0))(x'(t_0)) = \liminf_{\substack{h \to 0+ \\ v \to v_0}} \frac{V(x(t_0) + hv) - V(x(t_0))}{h}$$

$$\leq \sup_{\varepsilon > 0} \inf_{\alpha > 0} \sup_{h < \alpha} \inf_{v \in v_0 + \varepsilon B} \frac{V(x(t_0) + hv) - V(x(t_0))}{h}$$

$$\leq \sup_{\varepsilon > 0} \inf_{\alpha > 0} \sup_{h < \alpha} \frac{V(x(t_0) + hv_h) - V(x(t_0))}{h}$$

$$= v'_{\#}(t_0).$$

(b)   Let $v'_b(t_0) \doteq \liminf_{h \to 0+} \{[v(t_0 + h) - v(t_0)]/h\}$. Then, for all $\varepsilon > 0$, there exists $\gamma > 0$ such that,

$$\forall h \leq \gamma, \ v'_b(t_0) \leq \frac{v(t_0 + h) - v(t_0)}{h} + \varepsilon.$$

Let l be the Lipschitz constant of V at $x(t_0)$. Then, for all $v \in x'(t_0) + \varepsilon B$ and $h \leq \alpha \doteq \min(\beta, \gamma)$, we have, thanks to (15),

$$v'_b(t_0) \leq \frac{V(x(t_0) + hv_h) - V(x(t_0))}{h} + \varepsilon \leq \frac{V(x(t_0) + hv) - V(x(t_0))}{h} + 2l\varepsilon.$$

On the other hand, we know that there exists $h \leq \alpha$ and $v \in x'(t_0) + \varepsilon B$ such that

$$\frac{\tilde{V}(x(t_0) + hv) - \tilde{V}(x(t_0))}{h} \leq D_+\tilde{V}(x(t_0))(x'(t_0)) + \varepsilon.$$

Hence

$$v'_b(t_0) \leq D_+\tilde{V}(x(t_0))(x'(t_0)) + (2l + 1)\varepsilon.$$

By letting $\varepsilon \to 0$, we obtain $v'_b(t_0) \leq D_+\tilde{V}(x(t_0)(x'(t_0))$. Since $x'(t_0) \in D_K(x(t_0))$, Proposition 2 implies that $D_+\tilde{V}(x(t_0))(x'(t_0)) \leq D_+V(x(t_0))(x'(t_0))$. Hence (11) holds true.   ■

Remark.   If both V and x are locally Lipschitzean, then v is also locally Lipschitzean and inequality (13) can be written, by setting $D_+v(t_0) \doteq D_+v(t_0)(1)$,

$$D_+v(t_0) \leq D_+V(x(t_0))(x'(t_0)). \tag{16}$$

We can "integrate" inequalities involving contingent derivatives.

PROPOSITION 9. *Let v be a continuous function from $[0, T]$ to $\mathbb{R}$ and w be an upper semicontinuous function from $[0, T[$ to $\mathbb{R}$ which is bounded above. We assume that*

$$\forall t \in [0, T[, \qquad D_+v(t) + w(t) \leq 0. \tag{17}$$

Then, for all $0 \leq a < b < T$, we obtain the inequality

$$v(b) - v(a) + \int_a^b w(\tau)\, d\tau \leq 0. \tag{18}$$

Proof. Let $t \in [a, b]$ and $\varepsilon > 0$ be fixed. Since w is upper semicontinuous, there exists $\eta \in]0, \varepsilon[$ such that, $\forall h \leq \eta$,

$$\frac{1}{h} \int_t^{t+h} w(\tau)\, d\tau < w(t) + \frac{\varepsilon}{2}, \tag{19}$$

and there exists $h_t \leq \eta$ and a_t such that $|a_t - 1| \leq \varepsilon$ that satisfy

$$\frac{v(t + h_t a_t) - v(t)}{h_t} < \liminf_{\substack{h \to 0+ \\ a \to 1}} \frac{v(t + ha) - v(t)}{h} + \frac{\varepsilon}{2}. \tag{20}$$

Hence t belongs to the subset

$$N(t) \doteq \left\{ s \in [a, b] \,\middle|\, v(s + h_t a_t) - v(s) + \int_s^{s + h_t} w(\tau)\, d\tau < \varepsilon h_t \right\}, \tag{21}$$

which is open since v is continuous. Let us set

$$m = \sup_{t \in [a,b]} w(t) < +\infty.$$

Hence the compact interval $[a, b]$ can be covered by n open subsets $N(t_i)$. We set $a_i \doteq a_{t_i}$, $h_i \doteq h_{t_i}$, and $h_0 = \min_{i=1,\ldots,n} h_i > 0$.

We construct by induction the following sequence. We set $\tau_0 \doteq a$; it belongs to some $N(t_{i_1})$ and thus, by taking $s \doteq \tau_0 \, (=a)$ and $\tau_1 = h_{i_1} a_{i_1}$, we obtain

$$v(\tau_1) - v(\tau_0) + \int_{\tau_0}^{\tau_1} w(\tau)\, d\tau \leq \varepsilon h_1 + \int_{h_1 + a}^{h_1 a_1 + a} w(\tau)\, d\tau. \tag{22}$$

Assume that for $j \leq k$, we have constructed a sequence $\tau_j \in [a, b[\; (1 \leq j \leq k)$ such that

$$v(\tau_j) - v(\tau_{j-1}) + \int_{\tau_{j-1}}^{\tau_j} w(\tau)\, d\tau \leq \varepsilon h_j + \int_{h_j + \tau_{j-1}}^{h_j a_j + \tau_{j-1}} w(\tau)\, d\tau. \tag{23}$$

Then τ_k belongs to some $N(t_{i_{k+1}})$; we set

$$s \doteq \tau_k \qquad \text{and} \qquad \tau_{k+1} \doteq \tau_k + h_{i_{k+1}} a_{i_{k+1}}$$

and we deduce that

$$v(\tau_{k+1}) - v(\tau_k) + \int_{\tau_k}^{\tau_{k+1}} w(\tau)\,d\tau \leq \varepsilon h_{i_{k+1}} + \int_{h_{i_{k+1}}+\tau_{k-1}}^{h_{i_{k+1}}a_{i_{k+1}}+\tau_{k-1}} w(\tau)\,d\tau. \quad (24)$$

If $\tau_k < b \leq \tau_{k+1}$, we stop the construction. Otherwise, we continue. Since $\tau_{k+1} - \tau_k = a_{i_{k+1}} h_{i_{k+1}} \geq h_{0/2} > 0$, we are sure that eventually, after a finite number of steps, we shall have an index k such that $\tau_k < b \leq \tau_{k+1}$.

By adding the above inequalities from $j = 1$ to k, we obtain

$$v(\tau_{k+1}) - v(a) + \int_a^{\tau_{k+1}} w(\tau)\,d\tau \leq \varepsilon\left(\sum_{i=1}^{k+1} h_i\right)m' \quad \text{where } m' \text{ is a constant.} \quad (25)$$

When ε converges to 0, τ_k and τ_{k+1} converge to b (for $\tau_{k+1} - \tau_k = h_{i_{k+1}} \leq \eta(1 + \varepsilon) \leq \varepsilon(1 + \varepsilon)$) and thus, we deduce that

$$v(b) - v(a) + \int_a^b w(\tau)\,d\tau \leq 0. \quad \blacksquare$$

In particular, we obtain the following useful consequence.

PROPOSITION 10.   *Let v be a continuous function from $[0, T]$ to $\mathbb{R}$ satisfying*

$$\forall t \in \,]0, T[, \qquad D_+v(t) \leq 0. \quad (26)$$

Then the function v is nonincreasing.

7.   CONTINGENT DERIVATIVES OF MARGINAL FUNCTIONS AND MARGINAL MAPS

Let us consider a family of minimization problems depending upon a parameter y: Minimize the function $x \mapsto U(x, y)$ on a subset $F(y)$. We define the marginal function V by

$$V(y) \doteq \inf_{x \in F(y)} U(x, y) \quad (1)$$

and the marginal map G by

$$G(y) \doteq \{x \in F(y) \mid U(x, y) = V(y)\}. \quad (2)$$

Sensibility analysis deals with the behavior of the marginal functions V and the marginal G when the parameter y varies around a fixed value y_0. This is of upmost relevance in economics, for instance, as well as in other fields. In the convex case, we refer to Rockafellar [22]. In the locally Lipschitzean case, to Aubin–Clarke [6] and to Aubin [2]. We shall study in this section the properties of the contingent derivatives of the marginal function V and the marginal map G.

For simplicity, we assume that

(i) F is a compact-valued map from a subset M of a Hilbert space Y to a Hilbert space X. (3)

(ii) $\forall y \in M, \qquad x \to U(x, y)$ is lower semicontinuous.

Hence the marginal map G is well defined on M.

PROPOSITION 1. *Let* $y_0 \in K$ *and* $x_0 \in G(y_0)$ *achieve the minimum of* $U(\cdot, y_0)$ *on* $F(y_0)$. *Then*

$$\forall v_0 \in \text{Dom } DF(y_0, x_0), \qquad \forall u_0 \in DF(y_0, x_0)(v_0),$$
$$D_+ V(y_0)(v_0) \le D_- U(x_0, y_0)(u_0, v_0). \tag{4}$$

Furthermore,

$$\forall v_0 \in \text{Dom } DG(y_0, x_0), \qquad \forall u_0 \in DG(y_0, x_0)(v_0),$$
$$D_+ U(x_0, y_0)(u_0, v_0) \le D_- V(y_0)(v_0). \tag{5}$$

Proof. (a). Let $u_0 \in DF(y_0, x_0)(v_0)$. Then, for all $\varepsilon > 0, \alpha > 0$, there exist $h < \alpha, v \in v_0 + \varepsilon B, u \in u_0 + \varepsilon B$ such that $x_0 + hu \in F(y_0 + hv)$, and therefore, such that $V(y_0 + hv) \le U(x_0 + hu, y_0 + hv)$. Since $V(y_0) = U(x_0, y_0)$, we deduce that

$$\frac{V(y_0 + hu) - V(y_0)}{h} \le \frac{U(x_0 + hu, y_0 + hv) - U(x_0, y_0)}{h}.$$

This implies inequality (4).

(b). Let $u_0 \in DG(y_0, x_0)(v_0)$. Then, for all $\varepsilon > 0, \alpha > 0$, there exist $h < \alpha, .$ $v \in v_0 + \varepsilon B, u \in u_0 + \varepsilon B$ such that $x_0 + hu \in G(y_0 + hv)$. Hence $V(y_0 + hv) = U(x_0 + hu, y_0 + hv)$. Since $V(y_0) = U(x_0, y_0)$, we deduce that

$$\frac{U(x_0 + hu, y_0 + hv) - U(x_0, y_0)}{h} = \frac{V(y_0 + hu) - V(y_0)}{h}.$$

This implies inequality (5).

8. EKELAND'S VARIATIONAL PRINCIPLE

We shall derive the approximate variational principle of Ekeland in the following form:

THEOREM 1. *Let* $K \subset X$ *be a closed subset of a Hilbert space and* $V : K \to [0, \infty[$ *be a lower semicontinuous function. Then we can associate with any* $\varepsilon > 0$ *and any* $x_\varepsilon \in K$ *satisfying* $V(x_\varepsilon) \le \inf_{x \in K} V(x) + \varepsilon^2$ *an element* $\bar{x}_\varepsilon \in K$

which satisfies

(i) $\|x_\varepsilon - \bar{x}_\varepsilon\| \leq \varepsilon$,

(ii) $\forall u \in X, \qquad 0 \leq D_+ V(\bar{x}_\varepsilon)(u) + \varepsilon\|u\|.$ \hfill (1)

Proof. We derive this result from Ekeland's variational principle (see Ekeland [16] or Aubin [1], p. 174]).

THEOREM (Ekeland). *Let K be a closed subset of a Hilbert space and V be a lower semicontinuous function from K to $[0, \infty[$. Then we can associate with any $\varepsilon > 0$ and any $x_\varepsilon \in K$ satisfying $V(x_\varepsilon) \leq \inf_{x \in K} V(x) + \varepsilon^2$ an element $\bar{x}_\varepsilon \in K$ which satisfies $\|x_\varepsilon - \bar{x}_\varepsilon\| \leq \varepsilon$ and $V(\bar{x}_\varepsilon) = \min_{x \in K}[V(x) + \varepsilon\|x - \bar{x}_\varepsilon\|].$*

Let $u \in \text{Dom } D_+ V(\bar{x}_\varepsilon)$. Then, for any $\eta > 0, \delta > 0, \alpha > 0, \exists h \leq \alpha, \exists v \in u + \delta B$ such that

$$\frac{V(\bar{x}_\varepsilon + hv) - V(\bar{x}_\varepsilon)}{h} \leq D_+ V(\bar{x}_\varepsilon)(u) + \eta.$$

By Ekeland's variational principle, we have

$$-\varepsilon\delta - \varepsilon\|u\| \leq -\varepsilon\|v\| \leq \frac{V(\bar{x}_\varepsilon + hv) - V(\bar{x}_\varepsilon)}{h}.$$

Therefore, we infer that

$$0 \leq D_+ V(\bar{x}_\varepsilon)(u) + \varepsilon\|u\| + \varepsilon\delta + \eta.$$

By letting δ and η converge to 0, we obtain the desired inequality.

9.   SURJECTIVITY THEOREMS

We devote this section to the generalization to the case of upper semicontinuous maps with compact values of the inverse function theorem. We shall begin by proving theorems of existence of stationary points, then deduce surjectivity theorems and we shall end with a theorem ensuring that the image by F of a neighborhood of $\bar{x}$ is a neighborhood of $F(\bar{x})$. These theorems, due to Ekeland, are simple consequences of his variational principle.

It is convenient to start with the following lemma.

LEMMA 1.   *Let G be a set-valued map from $K \subset X$ to $L \subset Y$ and V be a continuous convex function defined on a neighborhood of L. We define the set-valued map $H \doteq V(G)$ from K to $\mathbb{R}$ by*

$$\forall x \in K, \qquad H(x) = \{V(y)\}_{y \in G(x)}. \hfill (1)$$

Assume that

$$\exists x_0 \in K, \quad \exists y_0 \in G(x_0), \quad \text{and} \quad a_0 \in \mathbb{R} \quad \text{such that} \tag{2}$$
$$\forall x \in K, \quad H(x) \subset V(y_0) + a_0\|x - x_0\| + \mathbb{R}_+.$$

Let $DV(y)(\cdot)$ denote the derivative of V at y. Then

$$\forall u_0 \in X, \quad \forall v_0 \in DG(x_0, y_0)(u_0), \qquad a_0\|u_0\| \leq DV(y_0)(v_0). \tag{3}$$

Proof. Let $v_0 \in DG(x_0, y_0)(u_0)$. Hence, for all $\alpha, \beta, \gamma > 0$, there exist $h \leq \alpha$, $u \in u_0 + \gamma B$ such that

$$v_0 \in \frac{G(x_0 + hu) - y_0}{h} + \beta B.$$

so, we can write $y_0 + hv_0 = y_h + \beta hb$ where $y_h \in G(x_0 + hu)$ and $b \in B$.

Since V is convex, we deduce that

$$V(y_h) - V(y_h - hv_0) \leq h\, DV(y_h)(v_0) \tag{4}$$

and since V is continuous (and thus, locally Lipschitzean), there exists $l > 0$ such that

$$V(y_h - hv_0) - V(y_0) \leq l\|y_h - y_0 - hv_0\| \leq l\beta h. \tag{5}$$

We recall that $y \mapsto DV(y)(v_0)$ is upper semicontinuous. Then for all $\varepsilon > 0$, $\exists \eta > 0$ such that

$$DV(y_h)(v_0) \leq DV(y_0)(v_0) + \varepsilon \qquad \text{when} \quad \|y_h - y_0\| \leq \eta. \tag{6}$$

By taking $\alpha \leq \eta/(\beta + \|v_0\|)$ if necessary, inequalities (4), (5), and (6) imply that

$$V(y_h) - V(y_0) \leq h(DV(y_0)(v_0) + \varepsilon + l\beta)$$

or

$$DV(y_0)(v_0) \in \frac{H(x_0 + hu) - V(y_0)}{h} + \mathbb{R}_+ + (\varepsilon + l\beta)B.$$

We use now assumption (2): we obtain

$$DV(y_0)(v_0) \in a_0\|u\| + (\varepsilon + l\beta)B + \mathbb{R}_+.$$

By letting ε, β, and γ go to 0, we infer that

$$DV(y_0)(v_0) \geq a_0\|u_0\|.$$

When L is a subset of X, we set

$$m(L) = \left\{ x \in L \text{ such that } \|x\| = \min_{y \in L} \|y\| \right\}.$$

THEOREM 1. *Let F be an upper semicontinuous map with compact values from a compact subset K of a Hilbert space X to a Hilbert space Y. We assume that*

$$\forall x \in K, \quad \exists y \in m(F(x)), \quad \exists u \in X \quad \text{such that} \quad -y \in DF(x, y)(u). \tag{7}$$

Then there exists a stationary point $\bar{x} \in K$ of F.

Proof. Since the function $x \mapsto \|m(F(x))\|$ is lower semicontinuous (for F is upper semicontinuous with compact values) and since K is compact, there exists $x_0 \in K$ which achieves the minimum of $x \to \|m(F(x))\|$ on K. Let us choose $y_0 \in m(F(x_0))$. We set $V(x) = \|x\|$ and $H(x) = \{V(y)\}_{y \in F(x)}$. It is clear that

$$\forall x \in K, \qquad H(x) \subset V(y_0) + \mathbb{R}_+$$

since, if $c \in H(x)$, then $c \geq \|m(F(x))\|^2 \geq \|m(F(x_0))\| = V(y_0)$. So, we apply Lemma 1 with $a_0 = 0$. We deduce that

$$\forall u_0 \in X, \qquad \forall v_0 \in DF(x_0, y_0)(u_0), \qquad 0 \leq DV(y_0)(v_0).$$

By assumption (7), we can take $v_0 = -y_0$; since

$$DV(y_0)(-y_0) = \lim_{h \to 0+} \frac{\|y_0 - hy_0\| - \|y_0\|}{h} = -\|y_0\|,$$

we deduce that $\|y_0\| \leq 0$. Hence $y_0 = 0 \in F(x_0)$. ∎

COROLLARY 1. *Let F be a Gâteaux differentiable continuous map from a neighborhood of a compact subset K to Y. Assume that*

$$\forall x \in K, \quad \exists u \in D_K(x) \quad \text{such that} \quad \nabla F(x)u = -F(x). \tag{8}$$

Then there exists a stationary point $x_0 \in K$ of F.

By using Ekeland's theorem, we can replace the compactness assumption on K in Theorem 1 by another assumption on the growth of the inverse of the contingent derative.

THEOREM 2 (Ekeland). *Let F be an upper semicontinuous map with compact values from a closed subset K of a Hilbert space X to a Hilbert space Y. We assume that*

$$\exists c > 0 \quad \text{such that} \quad \forall x \in K, \quad \exists y \in m(F(x)),$$
$$\exists u \in X \quad \text{such that} \quad -y \in DF(x, y)(u) \quad \text{and} \quad c\|u\| \leq \|y\|. \tag{9}$$

Then there exists a stationary point $\bar{x} \in K$ of F.

Proof. By Ekeland's theorem, we can associate with any $\varepsilon < c$ an element $x_0 \in K$ such that, for all $x \in K$,

$$\left\| m(F(x_0)) \right\| \leq \left\| m(F(x)) \right\| + \varepsilon \| x - x_0 \|. \tag{10}$$

If $m(F(x_0)) \doteq y_0 = 0$, the theorem is proved. Otherwise we take $V(x) \doteq \|x\|$ and $H(x) = \{V(y)\}_{y \in F(x)}$. Inequality (10) can be written

$$\forall x \in K, \qquad H(x) \subset V(y_0) - \varepsilon \| x - x_0 \| + \mathbb{R}_+. \tag{11}$$

Hence we apply Lemma 1 with $a_0 = -\varepsilon$. We obtain: $\forall u_0 \in X$, $\forall v_0 \in DF(x_0, y_0)(u_0)$, $-\varepsilon \| u_0 \| \leq DV(y_0)(v_0)$. By assumption (9), we can take $v_0 = -y_0$ and $u_0 \in X$ satisfying $c\| u_0 \| \leq \| y_0 \|$. Since $DV(y_0)(-y_0) = -\| y_0 \|$, we obtain the contradiction $-\varepsilon \| u_0 \| \leq -\| y_0 \| \leq -c\| u_0 \|$. So $y_0 = 0$. ∎

COROLLARY 2. *Let F be a Gâteaux-differentiable continuous map from a neighborhood of a closed subset $K \subset X$ to Y. Assume that*

$$\exists c > 0 \quad \text{such that} \quad \forall x \in K,$$
$$\exists u \in D_K(x) \quad \text{such that} \quad \nabla F(x)u = -F(x) \quad \text{and} \quad c\|u\| \leq \|F(x)\|. \tag{12}$$

then there exists a stationary point $x_0 \in K$ of F.

In particular, we can take $K = X$. We obtain

COROLLARY 3. *Let F be a Gâteaux-differentiable continuous map from X to Y. Assume that*

$$\exists c > 0 \quad \text{such that} \quad \forall x \in X,$$
$$\exists u \in X \quad \text{satisfying} \quad \nabla F(x)u = -F(x) \quad \text{and} \quad c\|u\| \leq \|F(x)\|. \tag{13}$$

Then there exists a stationary point x_0 of F.

By replacing the set-valued map F by $G(x) = F(x) - y$, we obtain solutions to the inclusion $y \in F(x)$. Therefore, we obtain the following surjectivity theorems.

THEOREM 3. *Let F be an upper semicontinuous map with compact values from a closed subset K of a Hilbert space X to a Hilbert space Y. We assume that*

$$\exists c > 0 \quad \text{such that} \quad \forall x \in K, \quad \forall y \in F(x), \quad \forall v \in Y,$$
$$\exists u \in X \quad \text{satisfying} \quad v \in DF(x, y)(u) \quad \text{and} \quad c\|u\| \leq \|v\|. \tag{14}$$

Then for all $y \in Y$, there exists a solution $x \in K$ to the inclusion $y \in F(x)$.

In other words, $F(K) = Y$. Theorem 3 says that F is a surjective set-valued map.

For smooth single-valued maps, we obtain the following corollaries.

COROLLARY 4.   *Let F be a Gâteaux-differentiable continuous map from a neighborhood of a closed subset $K \subset X$ to Y. Assume that*

$$\exists c > 0 \quad \text{such that} \quad \forall x \in K, \quad \forall v \in Y,$$
$$\exists u \in D_K(x) \quad \text{satisfying} \quad \nabla F(x)(u) = v \quad \text{and} \quad c\|u\| \le \|v\|. \tag{15}$$

Then $F(K) = Y$.

COLOLLARY 5.   *Let F be a Gâteaux-differentiable continuous map from X to Y. Assume that*

$$\exists c > 0 \quad \text{such that} \quad \forall x \in X, \quad \forall v \in Y,$$
$$\exists u \in X \quad \text{satisfying} \quad \nabla F(x)u = v \quad \text{and} \quad c\|u\| \le \|v\|. \tag{16}$$

Then F is surjective.

We prove now an adaptation of the inverse function theorem.

THEOREM 4   (Ekeland).   *Let F be an upper semicontinuous map with compact values from a neighborhood U of $\bar{x} \in X$ to Y. We assume that*

$$\exists c > 0 \quad \text{such that} \quad \forall x \in u, \quad \forall y \in F(x), \quad \forall v \in Y,$$
$$\exists u \in X \quad \text{satisfying} \quad DF(x, y)(u) = v \quad \text{and} \quad c\|u\| \le \|v\|. \tag{17}$$

Then $F(U)$ is a neighborhood of $F(\bar{x})$.

Proof.   Let U contain a closed ball of center $\bar{x}$ and radius $\eta > 0$ and let $\bar{y} \in F(\bar{x})$. We claim that $F(\bar{x} + \eta B)$ contains the balls $\bar{y} + \varepsilon B$ with $\varepsilon < c\eta$.

Indeed, pick $y_1 \in \bar{y} + \varepsilon B$ and set $G(x) \doteq F(x) - y_1$. We apply Ekeland's theorem in the stronger form to the function $\|m(F(x) - y_1)\|$, taking ε as above. Noting that $\|m(F\bar{x}) - y_1)\| \le \varepsilon$, we get some point x_0 such that

(i) $\|x_0 - \bar{x}\| \le \eta$,

(ii) $\forall x \in \bar{x} + \eta B, \|m(F(x_0) - y_1)\| \le \|m(F(x) - y_1)\| + \varepsilon \eta^{-1}\|x - x_0\|$. $\tag{18}$

Take $y_0 \in m(F(x_0) - y_1)$. Either it is zero [and $y_1 \in F(x_0)$] or $V(y_0) = \|m(F(x_0) - y_1)\| > 0$. In this case we obtain a contradiction. Indeed, inequality (18, ii) implies that

$$\forall x \in K, \quad H(x) \subset V(y_0) - \varepsilon \eta^{-1}\|x - x_0\| + \mathbb{R}_+ \quad \text{where} \quad H(x) \doteq V(G(x)).$$

Hence we apply Lemma 1 with $a_0 = -\varepsilon \eta^{-1}$: We get

$$\forall u_0 \in X, \quad \forall v_0 \in DG(x_0, y_0)(u_0), \quad -\varepsilon \eta^{-1}\|u_0\| \le DV(y_0)(u_0).$$

But $DG(x_0, y_0)(u_0) = DF(x_0, y_0 + y_1)(u_0)$. By assumption (17), we can choose $v_0 = -y_0$ and $u_0 \in X$ such that $c\|u_0\| \le \|y_0\|$. Hence, since $DV(y_0)(-y_0) = -\|y_0\|$, we obtain the contradiction $-\varepsilon \eta^{-1}\|u_0\| \le -\|y_0\| \le -cu_0$. So $y_0 = 0$, i.e., $y_1 \in F(x_0)$. ■

10. THE NEWTON METHOD

We proved in Corollary 9.1 that when F is a continuously differentiable map from a neighborhood of a compact subset $K \subset \mathbb{R}^n$ to $\mathbb{R}^m$ that satisfies the condition

$$\forall x \in K, \quad \exists v \in D_K(x) \quad \text{such that} \quad \nabla F(x)v = -F(x), \tag{1}$$

then there exists a stationary point $x_* \in K$ of F.

These assumptions imply also that there exist trajectories $x(\cdot)$ of the differential inclusion

$$\nabla F(x)x' = -F(x), \qquad x(0) = x_0, \tag{2}$$

that remain in K and that converge to a stationary point of F when $t \to \infty$ (see Haddad [18]).

We can consider such trajectories as the continuous analogs of the classical Newton method, which yields the discrete trajectory defined recursively by

$$\nabla F(x_n)(x_{n+1} - x_n) = -F(x_n); \qquad x_0 \text{ is given.} \tag{3}$$

THEOREM 1. *Let F be a continuously differentiable map from a neighborhood of a closed subset $K \subset \mathbb{R}^n$ to $\mathbb{R}^m$ satisfying*

$$\exists c > 0 \quad \text{such that} \quad \forall x \in K,$$
$$\exists v \in D_K(x) \cap cB \quad \text{satisfying} \quad \nabla F(x)v = -F(x). \tag{4}$$

Then there exists a viable trajectory of the implicit differential equation (2) that satisfies

$$F(x(t)) = e^{-t}F(x(0)).$$

Thus the cluster points of $x(t)$ (if any) are stationary points of F.

Proof. We set $G(x) \doteq -\nabla F(x)^{-1}F(x)$. Trajectories of the differential inclusion $x' \in G(x)$ are the trajectories of the implicit differential equation (2). Assumptions of Haddad's theorem (see [18]) on differential inclusions are satisfied. Hence there exists viable trajectories of the implicit differential equation (2). Consider any such trajectory. Then, since F is continuously differentiable, we have

$$\frac{d}{dt} F(x(t)) = \nabla F(x(t))x'(t) = -F(x(t)).$$

So $y(t) \doteq F(x(t))$ is equal to $e^{-t}y(0)$. Therefore, $F(x(t))$ converges to 0 when $t \to \infty$. Any cluster point $x_* \in K$, limit of a subsequence $x(t_n)$ when $t_n \to \infty$, satisfies $F(x_*) = \lim F(x(t_n)) = 0$, and thus, is a stationary point of F. ∎

Recall that the above sufficient condition for existence of a stationary point of F can be extended to set-valued maps (see Theorem 9.1): We replace the tangential condition (4) by

$$\forall x \in K, \quad \exists y \in F(x), \quad \exists u \in X \quad \text{such that} \quad -y \in DF(x, y)(u), \qquad (5)$$

where $DF(x, y)$ is the contingent derivative of the set-valued map F.

We can generalize the Newton method if we assume, for instance, that there exists a bounded upper semicontinuous proper set-valued map G from graph(F) to the closed convex subsets of $\mathbb{R}^n$ such that

$$\forall x \in K, \quad \forall y \in F(x), \quad \exists u \in G(x, y) \quad \text{such that} \quad -y \in DF(x, y)(u). \qquad (6)$$

THEOREM 2 (Saint-Pierre). *Let F be a proper map from $K \subset \mathbb{R}^n$ to $\mathbb{R}^m$ with closed graph. We posit Assumption (6). Then, for any $x_0 \in K$ and $y_0 \in F(x_0)$, there exists a solution to the differential inclusion*

$$x'(t) \in G(x(t), e^{-t}y_0) \qquad (7)$$

that satisfies

$$\forall t \geq 0, \quad x(t) \in K, \quad \text{and} \quad e^{-t}y_0 \in F(x(t)). \qquad (8)$$

Thus the cluster points of $x(t)$ (if any) are stationary points of F.

Proof. We consider the differential inclusion

$$x' \in G(x, y), \qquad y' = -y \qquad (9)$$

with the initial condition $x(0) = x_0$, $y(0) = y_0$.
Condition (6) implies that

$$\forall (x, y) \in \text{graph}(F), \qquad (G(x, y), -y) \cap D_{\text{graph}(F)}(x, y) \neq \varnothing. \qquad (10)$$

Then Haddad's theorem (see Haddad [18]) implies that there exists a trajectory $(x(t), y(t))$ of this differential inclusion which remains in graph(F). Furthermore, $y(t) = e^{-t}y_0$. Hence $e^{-t}y_0 \in F(x(t))$. The rest of the theorem ensues. ∎

Remark. We note that we can devise a whole family of algorithms that converge to a stationary point of F. Let H be any map from $\mathbb{R}^n$ to itself such that

$$\text{the solution of } y' = H(y), \ y(0) = y_0 \text{ is unique and} \atop \text{converges to 0, when } t \to \infty. \tag{11}$$

We associate with such a map H a bounded continuous map G (single valued for the sake of simplicity) such that

$$\forall (x, y) \in \text{graph}(F), \qquad H(y) \in DF(x, y)(G(x, y)). \tag{12}$$

Then there exist solutions to the differential equation

$$x'(t) \in G(x(t), y(t)), \quad y'(t) = H(y(t)), \quad x(0) = x_0, \quad y(0) = y_0, \tag{13}$$

such that

$$\forall t \geq 0, \qquad y(t) \in F(x(t)). \tag{14}$$

Since $\lim_{t \to \infty} y(t) = 0$ by assumption, the cluster points of $x(t)$ (if any) are stationary points of F.

11. LIAPUNOV FUNCTIONS AND EXISTENCE OF STATIONARY POINTS

We shall use the existence of Liapunov function which we define below for proving the existence of stationary points. In the next sections, we shall prove that the existence of these Liapunov functions implies also that solutions to the associated differential inclusion do converge in some sense to stationary points of F when $t \to 0$.

We begin with a particular case, which is a "stationary point" version of the Aubin–Siegel fixed-point theorem, which is an extension to the case of set-valued maps of the Caristi fixed-point theorem (see Aubin and Siegel [8], Brézis and Browder [10] and Caristi [11]).

THEOREM 1. *Let K be a closed subset of the Hilbert space X, $F:K \to X$ be a set-valued map and $V:K \to [0, \infty[$ be a lower semicontinuous function satisfying*

$$\forall x \in K, \quad \exists v \in F(x) \quad \text{such that} \quad D_+ V(x)(v) + \|v\| \leq 0. \tag{1}$$

Then there exists a stationary point $\bar{x} \in K$ of F.

Proof. Take $\varepsilon < 1$ and $\bar{x}_\varepsilon \in K$ satisfying, thanks to Theorem 7.1,

$$\forall u \in X, \quad 0 \leq D_+ V(\bar{x}_\varepsilon)(u) + \varepsilon \|u\|.$$

By assumption, we can take $u \in F(\bar{x}_\varepsilon)$ satisfying $D_+ V(\bar{x}_\varepsilon)(u) \leq -\|u\|$. Hence $(1 - \varepsilon)\|u\| \leq 0$, i.e., $u = 0 \in F(\bar{x}_\varepsilon)$. ∎

Remark. This theorem can be regarded as the "stationary point" version of the "Caristi's fixed point theorem."

THEOREM 2   (Caristi). *Let K be a closed subset, $g:K \to K$ be a single-valued map and V be a lower semicontinuous map from K to $\mathbb{R}_+$. If*

$$\forall x \in K, \quad V(g(x)) - V(x) + \|g(x) - x\| \le 0,$$

then g has a fixed point.

So, we have proved that if there exists a lower semicontinuous function V such that $D_+ V(x)(v) + \|v\| \le 0$ for all $x \in K$, there exist a stationary point. The question arises whether existence of stationary points implies the existence of a function V satisfying the above condition.

THEOREM 3   (Moreau). *Let K be a closed subset and F be a set-valued map from K to X satisfying $-F$ is monotone*

$$\forall x, y \in K, \quad \forall u \in F(x), \quad \forall v \in F(y), \quad \langle u - v, x - y \rangle \le 0. \tag{2}$$

We assume that the set $F^{-1}(0)$ of stationary points of F has a nonempty interior. Let us associate with any $x_0 \in \operatorname{Int} F^{-1}(0)$ the function V defined by $V(x) = (1/2\rho)\|x - x_0\|^2$ where $\rho = d(x_0, \complement F^{-1}(0)) > 0$. Then

$$\forall x \in K, \quad \forall v \in F(x), \quad DV(x)(v) + \|v\| \le 0. \tag{3}$$

and, consequently, for any $x \in x_0 + \rho \operatorname{Int} B$, $F(x) = \{0\}$.

Proof. Let us take $x \in K$ and $v \in F(x)$. Since $x_0 + \rho B \subset F^{-1}(0)$, then $x_0 - \rho v/\|v\|$ is a stationary point of F. The monotonicity of $-F$ implies that

$$\langle v, x - x_0 \rangle + \rho\|v\| = \langle v - 0, x - x_0 + \rho v/\|v\|\rangle \le 0. \tag{4}$$

Let $x = x_0 + \rho u \in x_0 + \rho \operatorname{Int} B$, and $v \in F(x)$. We infer that $\rho\langle v, u \rangle + \rho\|v\| \le 0$. Hence $\|v\| \le \|v\|\,\|u\| < \|v\|$, which is impossible when $v \ne 0$. ∎

We may generalize Theorem 1 by introducing the concept of Liapunov function V with respect to a set-valued map F and a given function W defined on graph (F).

DEFINITION 1.   Let F be a set-valued map from $K \subset X$ to X and W be a function defined on graph(F). We shall say that the function V defined on K is a Liapunov function for F with respect to W if it satisfies the following *Liapunov property*

$$\forall x \in K, \quad \exists v \in F(x) \quad \text{such that} \quad D_+ V(x)(v) + W(x, v) \le 0. \tag{5}$$

When $W \equiv 0$, we say simply that V is a Liapunov function for F.

THEOREM 4. *Let F be a set-valued map from a closed subset $K \subset X$ to X, W be a nonnegative function from* graph (F) *to $\mathbb{R}$ and V be a lower semicontinuous function from K to $\mathbb{R}_+$. Assume that V is a Liapunov function for F with respect to W. Then*

$$\forall \varepsilon > 0, \quad \exists x_\varepsilon \in K \text{ and } v_\varepsilon \in F(x_\varepsilon) \text{ satisfying } W(x_\varepsilon, v_\varepsilon) \le \varepsilon \|v_\varepsilon\|. \tag{6}$$

If we assume moreover that V is lower semicompact (this means that for all $x \in \mathbb{R}$ the subsets $\{x \in K \,|\, V(x) \le \lambda\}$ are relatively compact), then

$$\exists x_* \in K, \quad \exists v_* \in F(x_*) \quad \text{such that} \quad W(x_*, v_*) = 0. \tag{7}$$

Proof. (a) We apply Ekeland's variational principle: $\forall \varepsilon > 0$, $\exists x_\varepsilon \in K$ such that, $\forall v \in X$, $0 \le D_+ V(x_\varepsilon)(v) + \varepsilon \|v\|^2$. Since V is a Liapunov function with respect to W, there exists $v_\varepsilon \in F(x_\varepsilon)$ such that $D_+ V(x_\varepsilon)(v_\varepsilon) \le -W(x_\varepsilon, v_\varepsilon)$. Hence property (6) holds true.

(b) When V is also lower semicompact, there exists $x_* \in K$ that achieves the minimum of V. So $\forall v \in X$, $0 \le D_+ V(x_*)(v)$ by the variational principle. Since V is a Liapunov function, we choose $v_* \in F(x_*)$ such that $D_+ V(x_*)(v_*) \le -W(x_*, v_*)$. ∎

So, Theorem 1 is the particular case when $W(x, v) \doteq \|v\|$. By taking $W(x, v) \doteq \|v\|^\alpha$, $\alpha > 1$, we obtain the existence of approximate stationary points:

COROLLARY 1. *Let F be a set-valued map from a closed subset $K \subset X$ to X and V be a lower semicontinuous Liapunov function for F with respect to $\|\cdot\|^\alpha$, $\alpha > 1$. Then*

$$\forall \varepsilon > 0, \quad \exists x_\varepsilon \in K \quad \text{such that} \quad F(x_\varepsilon) \cap \varepsilon B \ne \varnothing. \tag{8}$$

Note also that when W satisfies the condition

$$\forall x \in K, \quad \forall v \ne 0, \quad W(x, v) > 0 \tag{9}$$

we obtain the existence of stationary points.

COROLLARY 2. *Let F be a set-valued map from a closed subset $K \subset X$ to X, $W: \text{graph}(F) \to \mathbb{R}_+$ be a nonnegative function satisfying property (9) and V be a lower semicontinuous and lower semicompact Liapunov function for F with respect to W. Then there exists a stationary point $x_* \in K$ of F.*

Remarks. We shall prove that in this case, under some supplementary continuity assumptions, solutions to the differential inclusion $x' \in F(x)$, $x(0) = x_0$, converge to a stationary point of F in some sense when $t \to \infty$.

Note that the tangential condition

$$\forall x \in K, \quad F(x) \cap D_K(x) \neq \varnothing \tag{10}$$

are involved in the Liapunov condition (5), since the domain of the upper contingent derivative $D_+V(x)(\cdot)$ is contained in the contingent cone $D_K(x)$ to the domain K of V. Indeed, if V is a Liapunov function, there exist $v \in F(x)$ such that $D_+V(x)(v) \leq -W(x,v) \leq 0$. Hence $v \in \operatorname{Dom} D_+V(x)(\cdot) \subset D_K(x)$. Therefore the tangential condition (10) holds true.

Actually, if we take $V \doteq \phi_K$ the indicator of K, defined by $\phi_K(x) \doteq 0$ when $x \in K$ and by $\phi_K(x) \doteq \infty$ when $x \notin K$, the Liapunov condition can be written

$$\forall x \in K, \quad \exists v \in F(x) \cap D_K(x) \quad \text{satisfying} \quad W(x,v) = 0. \tag{11}$$

We mention also that when K is convex and compact and when F is upper semicontinuous with closed convex values, the Browder–Ky Fan theorem states that the tangential condition (10) is sufficient for establishing the existence of a stationary point $x_* \in K$ of F (see, e.g., Aubin [3, Chapter 15]).

EXAMPLE. By taking for $V(x)$ the restriction to K of $x \to \frac{1}{2}\|x - x_0\|^2$, we obtain the following corollary, after noticing that

$$D_+V(x)(v) = (x - x_0, v) \qquad \text{when} \quad v \in D_K(x).$$

COROLLARY 3. *Let K be a weakly closed subset of X and W be a function from X to $[0, \infty]$ that is strictly positive when $v \neq 0$. Let a set-valued map F from K to X and $x_0 \in X$ satisfy*

$$\forall x \in K, \quad \exists v \in F(x) \cap D_K(x) \quad \text{such that} \quad \langle x - x_0, v \rangle - W(v) \leq 0. \tag{12}$$

Then the best approximations of x_0 by elements of K are stationary points of F.

As a particular case, we obtain the following result. Let F^{-1} denote the inverse of F and $\sigma(F^{-1}(v), p) \doteq \sup\{\langle p, x \rangle \mid x \in F^{-1}(v)\}$ denote the support function of $F^{-1}(v)$.

COROLLARY 4. *Let K be a weakly closed subset of X and $F: K \to X$ be a set-valued map satisfying*

$$\forall v \in F(K), \quad v \neq 0, \quad \text{then} \quad \sigma(F^{-1}(v), v) < \langle x_0, v \rangle \tag{13}$$

and

$$\forall x \in K, \quad F(x) \cap D_K(x) \neq \varnothing. \tag{14}$$

Then the best approximations of x_0 by elements of K are stationary points of F.

Proof. We apply Corollary 3 when W is defined by

$$W(v) = \langle x_0, v \rangle - \sigma(F^{-1}(v), v) \quad \text{when} \quad v \in F(K), \quad v \neq 0,$$
$$W(0) = 0, \tag{15}$$
$$W(v) = +\infty \quad \text{when} \quad v \notin F(K) \quad \text{and} \quad v \neq 0.$$

Then for any $v \in F(x)$, we deduce that $\langle x, v \rangle \leq \sigma(F^{-1}(v), v)$ since $x \in F^{-1}(v)$. Therefore

$$\langle x - x_0, v \rangle \leq \sigma(F^{-1}(v), v) - \langle x_0, v \rangle \leq -W(v). \quad \blacksquare$$

12. Monotone Trajectories of a Differential Inclusion

Let K and L be subsets of $X = \mathbb{R}^n$. Let $V: K \to \mathbb{R}_+$ and $W: K \times L \to \mathbb{R}$ be two given functions and $F: K \to X$ be a set-valued map.

DEFINITION 1. We say that a trajectory $x(\cdot)$ of the differential inclusion

$$x'(t) \in F(x(t)); \qquad x(0) = x_0 \tag{1}$$

is *monotone* (with respect to V and W) if

$$\forall s > t \geq 0, \qquad V(x(s)) - V(x(t)) + \int_t^s W(x(\tau), x'(\tau)) \, d\tau \leq 0. \tag{2}$$

Note that this condition implicitly implies that

$$\forall t \geq 0, \qquad x(t) \in K \quad \text{(i.e., } x(\cdot) \text{ is "viable")} \tag{3}$$

since V is defined on K.

PROPOSITION 1. *If* $W: K \times L \to \mathbb{R}_+$ *is nonnegative, then*

$$t \to V(x(t)) \text{ decreases and converges to } \alpha \doteq \lim_{t \to \infty} V(x(t)) \tag{4}$$

and

$$\int_0^\infty W(x(\tau), x'(\tau)) \, d\tau \doteq \lim_{t \to \infty} \int_0^t W(x(\tau), x'(\tau)) \, d\tau < +\infty. \tag{5}$$

Remark. Note that (4) implies that when V is lower semicontinuous, all the cluster points x_* of the trajectory when $t \to \infty$ (if any) satisfy $\alpha = V(x_*)$.

Proof. The first statement is obvious. Also, since

$$\int_t^s W(x(\tau), x'(\tau)) \, d\tau \leq V(x(t)) - V(x(0)) \to \alpha - \alpha = 0,$$

when $t, s \to \infty$, the Cauchy criterion implies that when W is nonnegative

$$\int_0^\infty W(x(\tau), x'(\tau)) \, d\tau \doteq \lim_{t \to \infty} \int_0^t W(x(\tau), x'(\tau)) \, d\tau < +\infty$$

(where the integral is a Riemann improper integral). ∎

We shall see in Section 14 that this latter condition implies that $W(x(t), x'(t))$ converges to zero in some sense when $t \to \infty$.

According to the assumptions relating V and W, monotonicity property (2) yields useful information on the behavior of the trajectory.

Example: Trajectories with Finite Length

Let us consider the case where

$$W(x, v) \doteq \|v\|. \tag{6}$$

THEOREM 1. *The trajectories $x(\cdot)$ on $[0, \infty[$ that are monotone with respect to V and $W: (x, v) \to \|v\|$ have finite length $\int_0^\infty \|x'(\tau)\| \, d\tau$ and converge to $x_* \in \bar{K}$ when $t \to \infty$. If K is closed and F is upper semicontinuous with compact convex values, then x_* is a stationary point of F.*

Proof. By (5), $\int_0^\infty \|x'(\tau)\| \, d\tau$, which is the length of the trajectory, is finite. Furthermore, inequality

$$\|x(t) - x(s)\| \leq \int_t^s \|x'(\tau)\| \, d\tau \to 0 \qquad \text{when} \quad t, s \to \infty$$

and the Cauchy criterion imply that $\lim_{t \to \infty} x(t) = x_*$ does exist. The following theorem shows that x_* is a stationary point. ∎

THEOREM 2. *Let F be an upper semicontinuous map from a closed subset $K \subset X$ to X with compact convex values and $x(\cdot)$ be a trajectory of the differential inclusion (1) that converges to some $x_* \in K$. Then x_* is a stationary point of F.*

Proof. Assume that $0 \notin F(x_*)$: There exists $\varepsilon > 0$ such that

$$\varepsilon B \cap (F(x_*) + \varepsilon B) = \varnothing \tag{7}$$

[for $F(x_*)$ is a closed subset]. Since F is upper semicontinuous, there exists $\delta > 0$ such that $F(y) \subset F(x_*) + \varepsilon B$ whenever $\|y - x_*\| \leq \delta$. Hence there exists $T > 0$ such that, $\forall t \geq T$, $\|x(t) - x_*\| \leq \delta$. Consequently,

$$\forall t \geq T, \qquad F(x(t)) \subset F(x_*) + \varepsilon B. \tag{8}$$

Since $x'(t) \in F(x(t))$ for almost all $t \geq 0$, the mean value theorem implies that for all $t \geq T$

$$\frac{x(t) - x(T)}{t - T} = \frac{1}{t - T} \int_T^t x'(\tau) \, d\tau \subset \overline{\text{co}}(F(x_*) + \varepsilon B) = F(x_*) + \varepsilon B, \qquad (9)$$

since $F(x_*) + \varepsilon B$ is closed and convex. Therefore, statements (7) and (9) imply that $\forall t \geq T$, $\| [x(t) - x(T)]/[t - T] \| \geq \varepsilon$, which is a contradiction of the fact that $x_* = \lim_{t \to \infty} x(t)$. ∎

Example

We shall illustrate the importance of monotone trajectories by the following theorem.

THEOREM 3. *Let φ be a continuous bounded function from $[0, \infty[$ to $\mathbb{R}$ and let $W(x, v) \doteq \varphi(V(x))$. Let x be a trajectory satisfying property (2) and $w(\cdot)$ be a solution to the differential equation*

$$w'(t) + \varphi(w(t)) = 0, \qquad \forall t \geq 0, \qquad w(0) = V(x(0)). \qquad (10)$$

Then, we obtain the following estimate:

$$\forall t \geq 0, \quad V(x(t)) \leq w(t). \qquad (11)$$

This statement is an obvious consequence of the following theorem, of which we give a proof due to Antosiewicz.

THEOREM 4. *Let $\Omega \subset \mathbb{R}$ be an open interval and $T \leq +\infty$. We consider a function φ from $[0, T[\times \Omega$ to $\mathbb{R}$ satisfying the following properties.*

(i) $\forall t \in [0, T[, x \mapsto \varphi(t, x)$ is continuous,
(ii) $\forall x \in \Omega, t \mapsto \varphi(t, x)$ is measurable, $\qquad\qquad\qquad$ (12)
(iii) $\exists a \in L^1(0, T)$ such that $\forall (t, x) \in [0, T[\times \Omega, |\varphi(t, x)| \leq a(t)$.

Let $v: [0, T[\to \Omega$ be a continuous function satisfying

$$\forall s > t, \qquad v(s) - v(t) + \int_t^s \varphi(\tau, v(\tau)) \, d\tau \leq 0. \qquad (13)$$

Then, there exists a maximal interval $[0, T_1[$ such that the differential equation $w' + \varphi(w) = 0$, $w(0) = v(0)$ has at least one solution w on $[0, T_1]$ satisfying

$$\forall t \geq 0, \qquad v(t) \leq w(t). \qquad (14)$$

Proof. (a)   We introduce the subsets

$$K = \{(t, x) \in [0, T[\times \Omega \text{ such that } x \geq v(t)\}$$
$$L = \{(t, x) \in [0, T[\times \Omega \text{ such that } x \leq v(t)\}.$$

Since $K \cup L = [0, T[\times \Omega$ and $K \cap L = \{(t, x) | x = v(t)\}$, we can define a function Ψ on $[0, T[\times \Omega$ by

$$\Psi(t, x) = \begin{cases} \varphi(t, x) & \text{if } x \geq v(t), \text{ i.e., if } (t, x) \in K, \\ \varphi(t, v(t)) & \text{if } x \leq v(t), \text{ i.e., if } (t, x) \in L. \end{cases} \tag{15}$$

(b)   The function Ψ inherits the properties of φ. To see this, we associate with any $t \in [0, T[$ the subsets

$$K(t) \doteq \{x \in \Omega | x \geq v(t)\}, \qquad L(t) \doteq \{x \in \Omega | x \leq v(t)\}.$$

They are closed and cover Ω. Thus, when $K(t) \neq \emptyset$ [resp., $L(t) \neq \emptyset$], the restriction of $\Psi(t, \cdot)$ to $K(t)$ [resp., to $L(t)$] is continuous. Hence we conclude that for all $t \in [0, T[, x \mapsto \Psi(t, x)$ is continuous. Similarly, we introduce the subsets:

$$K(x) \doteq \{t \in [0, T[| x \geq v(t)\} \qquad \text{and} \qquad L(x) \doteq \{t \in [0, T[| x \leq v(t)\}.$$

Hence, when $K(x) \neq \emptyset$ [resp., $L(x) \neq \emptyset$], the restriction of $\Psi(\cdot, x)$ to $K(x)$ [resp., $L(x)$] is measurable. Consequently, for all $x \in \Omega$, the function $t \to \Psi(t, x)$ is measurable. Obviously, $|\Psi(t, x)| \leq a(t)$ for all $(t, x) \in [0, T[\times \Omega$.

(c)   We choose now $T_0 < T$ such that the set of points $(t, x) \in [0, T_0] \times \mathbb{R}$ satisfying $|x - x_0| \leq \int_0^t a(\tau) d\tau$ is contained in $[0, T[\times \Omega$. Then, by Carathéodory's theorem, there exists at least a solution $w(\cdot): [0, T_0] \to \Omega$ to the differential equation

$$w'(t) + \Psi(t, w(t)) = 0; \qquad w(0) = v(0). \tag{16}$$

(d)   We assert that for all $t \in [0, T_0]$, $v(t) \leq w(t)$. If not, there would exist $t_2 \in]0, T_0]$ such that $v(t_2) > w(t_2)$. Let $t_1 \doteq \inf\{s \in [0, T_0] | v(t) > w(t)$ for all $t \in [s, t_2]\}$. By continuity, we have $w(t_1) = v(t_1)$ and $v(t) > w(t)$ for all $t \in]t_1, t_2]$. Since w is a solution to (16), then

$$w(t_2) = w(t_1) - \int_{t_1}^{t_2} \Psi(s, w(s)) \, ds.$$

Since $v(s) > w(s)$, we deduce that $\psi(s, w(s)) = \varphi(s, v(s))$ for all $s \in]t_1, t_2]$. Hence

$$w(t_2) = v(t_1) - \int_{t_1}^{t_2} \varphi(s, v(s)) \, ds.$$

By assumption (13), we deduce that $w(t_2) \geq v(t_2)$ (because $t_1 \leq t_2$). This contradicts inequality $v(t_2) > w(t_2)$. Hence $v(t) \leq w(t)$ on $[0, T_0]$. ∎

13. ALMOST CONVERGENCE OF MONOTONE TRAJECTORIES TO STATIONARY POINTS

We recall that when a Liapunov function V for F (with respect to W) satisfies

$$V \text{ is lower semicontinuous and lower semicompact,} \qquad (1)$$

there exists x_* and v_* satisfying

$$x_* \in K, \qquad v_* \in F(x_*), \qquad \text{and} \qquad W(x_*, v_*) = 0. \qquad (2)$$

So, when the function W satisfies the property

$$\forall x \in K, \qquad \forall v \neq 0, \qquad W(x, v) > 0, \qquad (3)$$

the existence of a Liapunov function V for F with respect to W and assumptions (1) and (3) imply the existence of stationary points.

We wish to prove that cluster points x_* and v_* of $x(t)$ and $x'(t)$ when $t \to \infty$ do satisfy the property (2). [In this case, property (3) guarantees that such cluster points x_* of $x(t)$ are stationary points of F.]

We already noted that when $W(x, v) = \|v\|$, the trajectory $x(t)$ converges to a limit x_*, which is a stationary point by Theorem 11.2. The proof of this theorem does not yield the fact that cluster points of $x(t)$ are stationary points.

A difficulty arises right now: The derivative $x'(t)$ is only defined almost everywhere. So we shall use an adaptation proposed by A. Cellina of the concepts of limit and cluster points for measurable functions to prove that measurable functions taking their values in a compact subset do have such "almost" cluster points and that "almost cluster points" x_* and v_* of $x(t)$ and $x'(t)$ satisfy $W(x_*, v_*)$ when $x(\cdot)$ is a monotone trajectory with respect to V and W.

Let $\mu(A)$ denote the Lebesgue measure of a subset $A \subset \mathbb{R}$.

DEFINITION 1. Let $x: [0, \infty[\to X$ be a measurable function and $x_* \in X$. We say that x_* is *the almost limit of* $x(\cdot)$ *when* $t \to \infty$ (and we write $x_* = a \lim_{t \to \infty} x(t)$) if

$$\forall \varepsilon > 0, \quad \exists T > 0 \quad \text{such that} \quad \mu\{t \in [T, \infty[\,|\, \|x(t) - x_*\| \geq \varepsilon\} = 0. \qquad (4)$$

We say that x_* is an *almost cluster point* of $x(t)$ when $t \to \infty$ if

$$\forall \varepsilon > 0, \qquad \mu\{t \in [0, \infty[\,|\, \|x(t) - x_*\| \leq \varepsilon\} = \infty. \qquad (5)$$

These concepts are justified by the following theorem:

THEOREM 1. *Let F be an upper semicontinuous map from $K \subset \mathbb{R}^n$ to the compact subsets of $\mathbb{R}^n$, W be a nonnegative lower semicontinuous function* *defined on graph (F) and V be a nonnegative lower semicontinuous and lower semicompact function defined on K. For any monotone trajectory x(t) of F with respect to V and W, the functions x(t) and x'(t) have almost cluster points x_* and v_* which satisfy*

$$x_* \in K, \quad v_* \in F(x_*) \quad and \quad W(x_*, v_*) = 0. \tag{6}$$

If W satisfies the condition

$$\forall x \in K, \quad \forall v \neq 0, \quad W(x, v) > 0 \tag{3}$$

then such an almost cluster point x_* is a stationary point.

The proof of this theorem will be obtained by tying up the following properties of almost convergence. For simplicity, we restrict our study to the case of functions of a real variable. Adaptation for functions defined on a measured space is quite easy.

We begin by showing that the usual concepts of limit and cluster point are particular cases of almost limit and almost cluster point.

PROPOSITION 1. *Any limit of x_* of $x(\cdot):[0, \infty[\to X$ is an almost limit point. If $x(\cdot)$ is uniformly continuous, any cluster point x_* of $x(\cdot)$ is an almost cluster point*

Proof. (a) To say that $x_* = \lim_{t \to \infty} x(t)$ amounts to saying that $\forall \varepsilon > 0$, $\exists T > 0$ such that $[T, \infty[\cap \{t \mid \|x(t) - x_*\| \geq \varepsilon\} = \emptyset$. Hence the measure of this set is equal to 0.

(b) Let x_* be a cluster point of $x(\cdot)$: Since $x(\cdot)$ is uniformly continuous, there exists η such that $|s - t| \leq \eta$ implies $\|x(t) - x(s)\| \leq \varepsilon/2$. Also, there exists a sequence $t_n \to \infty$ (which satisfies $t_{n+1} - t_n \geq 2\eta$) such that $\|x_* - x(t_n)\| \leq \varepsilon/2$ when $n \geq N_\varepsilon$. So, for any $n \geq N_\varepsilon$, the disjoint intervals $[t_n - \eta, t_n + \eta]$ are contained in $\{t \in [0, \infty[\mid \|x(t) - x_*\| \leq \varepsilon\}$. Hence

$$\infty = \mu\left(\bigcup_{n \geq N_\varepsilon} [t_n - \eta, t_n + \eta]\right) \leq \mu(\{t \in [\cdot, \infty[\mid \|x(t) - x_*\| \leq \varepsilon\}. \quad \blacksquare$$

The following example justifies the introduction of the concept of almost cluster point.

PROPOSITION 2. *If W is a nonnegative function belonging to $L^1(0, \infty)$, then 0 is an almost cluster point of W when $t \to \infty$.*

Proof. If not, there exists $\varepsilon > 0$ such that the measure of

$$A_\varepsilon = \{t \in [0, \infty[\mid W(t) \leq \varepsilon\}$$

is finite. Hence the measure of $\complement A_\varepsilon = \{t \in [0, \infty[\,|\,W(t) > \varepsilon\}$ is infinite. Therefore:

$$\int_0^\infty w(\tau)\,d\tau \geq \int_{A_\varepsilon} w(\tau)\,d\tau + \int_{\complement A_\varepsilon} w(\tau)\,d\tau \geq \varepsilon \mu(\complement A_\varepsilon) = \infty$$

which is a contradiction. ∎

PROPOSITION 3. *An almost limit x_* of a measurable function $x(\cdot):[0, \infty[\to X$ is the unique almost cluster point.*

Proof. Let y_* be an almost cluster point difference from x_*. We choose $\varepsilon < \|x_* - y_*\|/2$ and T such that the subset $K \doteq \{t \in [T, \infty[\,|\,\|x(t) - x_*\| \geq \varepsilon\}$ has a measure equal to 0.

The subset $L = \{t \in [T, \infty[\,|\,\|x(t) - y_*\| \leq \varepsilon\}$ is obviously contained in K and has an infinite measure for y_* is an almost cluster point. Hence $\infty = \mu(L) \leq \mu(K) = 0$, which is impossible. So, $x_* = y_*$. ∎

PROPOSITION 4. *Let f be a continuous (single-valued) map from X to Y and $x(\cdot):[0, \infty[\to X$ be a measurable function. If x_* is the almost limit (resp., an almost cluster point) of $x(\cdot)$, then $f(x_*)$ is the almost limit [resp., an almost cluster point) of $f(x)\cdot)$].*

Proof. It is left as an exercise.

THEOREM 2 (Cellina). *Let K be a compact subspace of X and $x(\cdot):[0, \infty[\to K$ be a measurable function. There exists an almost cluster point $x_* \in K$ of $x(\cdot)$ when $t \to \infty$.*

Proof. We define inductively decreasing sequences of measurable sets $\Delta_n \subset [0, \infty[$ and of closed subsets $E_n \subset K$ such that

$$\Delta_n = x^{-1}(E_n), \qquad \mu(\Delta_n) = \infty, \qquad \text{diam}(E_n) \leq 1/n. \tag{7}$$

For $n = 1$, we cover the compact set K with a finite number of sets B_j^1 of diameter at most 1. Thus the subsets $x^{-1}(B_j^1)$ cover $[0, \infty[$ and, consequently, one of them, denoted Δ_1, has an infinite measure. We call E_1 the corresponding set B_j^1.

Having defined the subsets Δ_k and E_k up to n, we cover the compact set E_n by a finite number of closed subsets B_j^{n+1} of diameter at most $1/(n + 1)$.

Their preimages $x^{-1}(B_j^n)$ form a finite covering of Δ_n. Since $\mu(\Delta_n) = \infty$, at least one of these sets, denoted Δ_{n+1}, has an infinite measure. Call E_{n+1} the corresponding B_j^{n+1}. Hence $\bigcap_{n \geq 0} E_n = \{x_*\}$. It remains to show that x_* is an almost cluster point. Fix $\varepsilon > 0$ and $T > 0$. Then, a neighborhood $N_\varepsilon(x_*)$

contains the subsets E_n for $n \geq n(\varepsilon)$. Consequently, $x^{-1}(N_\varepsilon(x_*)) \supset x^{-1}(E_n) = \Delta_n$ for all $n \geq n_0(\varepsilon)$. Hence,

$$\mu\{t \in [0, \infty[\,|\,x(t) \in N_\varepsilon(x_*)\} \geq \mu_1(\Delta_n) = \infty. \quad\blacksquare$$

PROPOSITION 5.   *Let W be a nonnegative lower semicontinuous map from $L \subset X$ to $\mathbb{R}$. If $x(\cdot)$ is a measurable function from $[0, \infty[$ to L such that*

$$\int_0^\infty W(x(\tau))\,d\tau < +\infty, \tag{8}$$

then any almost cluster point x_* of $x(t)$ when $t \to \infty$ satisfies the equation $W(x_*) = 0$.

Proof.   Let x_* be an almost cluster point of $x(\cdot)$ when $t \to \infty$ and assume that $W(x_*) > 0$. We take $\varepsilon = W(x_*)/2 > 0$. Since W is lower semicontinuous, there exists η such that $W(x_*)/2 \leq W(y)$ when $\|y - x_*\| \leq \eta$. So the subset $A_\eta = \{t \in [0, \infty[\,|\,\|x(t) - x_*\| \leq \eta\}$, whose measure is infinite, is contained in the set $B_\varepsilon = \{t \in [0, \infty[\,|\,W(x_*)/2 \leq W(x(t))\}$. Hence

$$\int_0^\infty W(x(\tau))\,d\tau \geq \int_{B_\varepsilon} W(x(\tau))\,d\tau \geq \frac{W(x_*)}{2}\,\mu(B_\varepsilon) = \infty.$$

This is a contradiction. $\blacksquare$

We are ready to prove Theorem 1.

Proof of Theorem 1.   Since V is lower semicontinuous and lower semicompact, then $x(t)$ remains in the compact subset $Q \doteq \{x \in K\,|\,V(x) \leq V(x_0)\}$. Because F is upper semicontinuous with compact values, the set

$$\mathscr{F}_Q = \{(x, v) \in \mathbb{R}^n \times \mathbb{R}^n\,|\,x \in Q, v \in F(x)\},$$

which is the graph of the restriction $F|_Q$ of F to Q, is compact. Hence the function $t \to (x(t), x'(t))$ is a measurable function taking its values in the compact set $\mathscr{F}_Q$. By Theorem 2, there exists an almost cluster point $(x_*, v_*) \in \mathscr{F}_Q$. Since $x(\cdot)$ is a monotone trajectory with respect to V and W, we know that $\int_0^\infty W(x(\tau), x'(\tau))\,d\tau < +\infty$. Hence Proposition 5 implies that $W(x_*, v_*) = 0$. $\blacksquare$

14.   NECESSARY CONDITIONS FOR THE EXISTENCE OF MONOTONE TRAJECTORIES

We shall prove that the existence of monotone trajectories with respect to V and W of the differential inclusion $x' \in F(x)$, $x(0) = x_0$ for all initial value $x_0 \in K$ implies that V is a Liapunov function for F with respect to W.

THEOREM 1. *Let F be an upper semicontinuous map from $K \subset \mathbb{R}^n$ to $\mathbb{R}^n$ with compact convex values. Let W be a lower semicontinuous nonnegative function on $K \times \operatorname{co} F(K)$ which is convex with respect to v. We assume that for all $x_0 \in K$, there exist $T > 0$ and a trajectory $x(\cdot)$ on $[0, T[$ of the differential inclusion $x' \in F(x)$, $x(0) = x_0$ satisfying*

$$\forall s > t, \quad V(x(s)) - V(x(t)) + \int_t^s W(x(\tau), x'(\tau)) \, d\tau \leq 0. \tag{1}$$

Then, V is a Liapunov function for F with respect to W:

$$\forall x \in K, \quad \exists v \in F(x) \quad \text{such that} \quad D_+ V(x)(v) + W(x, v) \leq 0. \tag{2}$$

Proof. It is analogous to the proof of Haddad's proposition. Since F is upper semicontinuous, we associate to $\varepsilon > 0$ an $\eta > 0$ such that, for all $\tau \leq \eta$, $F(x(\tau)) \subset F(x_0) + \varepsilon B$, which is compact and convex. Since $[x(h) - x_0]/h = (1/h) \int_0^h x'(\tau) \, d\tau$ and since, for almost all τ, $x'(\tau)$ belongs to $F(x_0) + \varepsilon B$, the mean-value theorem implies that $[x(h) - x_0]/h$ belongs to this compact subset.

Hence there exists a subsequence $h_n \to 0$ such that $v_n \doteq [x(h_n) - x_0]/h_n$ converges to some v_0 in $F(x_0) + \varepsilon B$.

Since this inclusion holds true for all $\varepsilon > 0$, we deduce that $v_0 \in F(x_0)$. We also observe that $x_0 + h_n v_n = x(h_n)$ belongs to K, the domain of V.

Hence property (1) implies that

$$\frac{V(x_0 + h_n v_n) - V(x_0)}{h_n} + \frac{1}{h_n} \int_0^{h_n} W(x(\tau), x'(\tau)) \, d\tau \leq 0. \tag{3}$$

Let us assume for a while that the following properties hold true:

PROPOSITION 1. *Let W be a lower semicontinuous function on $K \times L$ which is convex with respect to $v \in L$. Let $x \in \mathscr{C}(0, T; K)$ and $v \in L^\infty(0, T, M)$ be given, where M is compact. Let*

$$x_0 \doteq \lim_{t \to 0} x(t) \quad \text{and} \quad v_0 \doteq \lim_{h_n \to 0} \frac{1}{h_n} \int_0^{h_n} v(t) \, dt. \tag{4}$$

Then

$$W(x_0, v_0) \leq \liminf_{h_n \to 0} \frac{1}{h_n} \int_0^{h_n} W(x(\tau), v(\tau)) \, d\tau. \tag{5}$$

End of the proof of Theorem 1. By Proposition 1,

$$W(x_0, v_0) \leq \liminf_{h_n \to 0} \frac{1}{h_n} \int_0^{h_n} W(x(\tau), x'(\tau)) \, d\tau.$$

Hence

$$D_+V(x_0)(v_0) + W(x_0, v_0)$$

$$\le \liminf_{\substack{h_n \to 0+ \\ v_n \to v_0}} \frac{V(x_0 + h_n v_n) - V(x_0)}{h_n} + \liminf_{h_n \to 0+} \frac{1}{h_n} \int_0^{h_n} W(x(\tau), x'(\tau)) \, d\tau$$

$$\le \liminf_{\substack{h_n \to 0+ \\ v_n \to v_0}} \left[\frac{V(x_0 + h_n v_n) - V(x_0)}{h_n} + \frac{1}{h_n} \int_0^{h_n} W(x(\tau), x'(\tau)) \, d\tau \right] \le 0. \quad \blacksquare$$

Before proving Proposition 1, we need the following result of Ekeland–Temam [17].

When M is a subset of X, we denote by $W_M(x, \cdot)$ the restriction to M of the function $v \mapsto W(x, v)$ and by $\mathscr{E}pW_M(x, \cdot)$ its epigraph.

PROPOSITION 2. *Let K and L be two nonempty subsets of X and $W : K \times L \to \mathbb{R}$ satisfy*

(i)   *W is lower semicontinuous,*

(ii)   *$\forall x \in K$, $v \mapsto W(x, v)$ is convex.*   (6)

Then, for any compact convex subset $M \subset L$, the restriction $W_M(x, \cdot)$ of $v \to W(x, v)$ to M satisfies the property

$x \to \mathscr{E}pW_M(x, \cdot)$ *is upper semicontinuous with closed convex values.* $\blacksquare$ (7)

Proof of Proposition 2. To say that $x \to \mathscr{E}pW_M(x, \cdot)$ is upper semicontinuous at x_0 means that for all $\varepsilon > 0$, there exists $\eta > 0$ such that, $\forall x \in K \cap (x_0 + \eta B)$, $\forall v \in M$, $\exists v_0 \in M \cap (v + \varepsilon B)$ satisfying $W(x_0, v_0) \le W(x, v) + \varepsilon$. Since W is lower semicontinuous, we can associate to each $v \in M$ an $\eta_v \in [0, \varepsilon[$ such that

$$\forall x \in x_0 + \eta_v B, \quad \forall w \in v + \eta_v B, \quad W(x_0, v) \le W(x, w) + \varepsilon. \quad (8)$$

Since M is compact, it is covered by a finite number p of balls $v_i + \eta_{v_i} B$. Let $\eta \doteq \min_{i=1,\dots,p} \eta_{v_i}$. Hence for all $x \in x_0 + \eta B$, for all $v \in M$, there exists v_i such that $\|v - v_i\| \le \varepsilon$ satisfying $W(x_0, v_i) \le W(x, v) + \varepsilon$. Since $v \to W_M(x, v)$ is convex and continuous, then $\mathscr{E}pW(x, \cdot)$ is closed and convex. $\blacksquare$

We are now ready to prove Proposition 1.

Proof of Proposition 1. By Proposition 2, $x \to \mathscr{E}pW_M(x, \cdot)$ is upper semicontinuous. Then there exists h_ε such that, for all $t \in [0, h_\varepsilon]$, $\mathscr{E}pW_M(x(t), \cdot) \subset \mathscr{E}pW_M(x_0, \cdot) + \varepsilon(B \times B)$. Therefore, for all $\tau \in [0, h_\varepsilon]$, $(v(\tau), W(x(\tau), v(\tau))) \in \mathscr{E}pW_M(x(\tau), \cdot) \subset \mathscr{E}pW_M(x_0, \cdot) + \varepsilon(B \times B)$. Hence, by the mean-value theorem, we deduce that, for all $h_n \leq h_\varepsilon$,

$$\left(\frac{1}{h_n} \int_0^{h_n} v(\tau)\, d\tau, \frac{1}{h_n} \int_0^{h_n} W(x(\tau), v(\tau))\, d\tau \right) \in \overline{\mathrm{co}}(\mathscr{E}pW_M(x_0, \cdot)) + \varepsilon(B \times B)). \quad (9)$$

Therefore, by taking the limit, we obtain

$$\left(v_0, \liminf_{h_n \to 0} \frac{1}{h_n} \int_0^{h_n} W(x(\tau), v(\tau))\, d\tau \right) \in \overline{\mathrm{co}}(\mathscr{E}pW_M(x_0, \cdot)) + \varepsilon(B \times B)). \quad (10)$$

Since this is true for all $\varepsilon > 0$ and since $\mathscr{E}pW_M(x_0, \cdot)$ is closed and convex, we get:

$$W(x_0, v_0) \leq \liminf_{h_n \to 0} \frac{1}{h_n} \int_0^{h_n} W(x(\tau), v(\tau))\, d\tau. \quad \blacksquare \quad (11)$$

15. Sufficient Conditions for the Existence of Monotone Trajectories

We shall prove that, conversely, if V is a Liapunov function for F with respect to W there exist monotone trajectories of the differential inclusion $x' \in F(x)$ that are monotone with respect to V and W.

THEOREM 1. *Let K be a closed subset of $\mathbb{R}^n$ and F be an upper set-valued map from K to the nonempty compact convex subsets of $\mathbb{R}^n$. Let W be a function defined on $K \times \mathrm{co}\, F(K)$ satisfying*

(i) *W is nonnegative and continuous,*
(ii) *$\forall x \in K,\ v \to W(x, v)$ is convex.* $\qquad (1)$

Let $V : K \to \mathbb{R}_+$ be a Liapunov function with respect to F and W:

$$\forall x \in K, \quad \exists v \in F(x) \quad \text{such that} \quad D_+ V(x)(v) + W(x, v) \leq 0. \quad (2)$$

We also assume that

$$V \text{ is continuous (for the topology induced on } K). \quad (3)$$

Then, for every $x_0 \in K$, there exist $T > 0$ and a monotone trajectory $x(\cdot)$ on $[0, T[$ of the differential inclusion $x' \in F(x)$ and $x(0) = x_0$.
If $F(K)$ is bounded, then we can take $T = \infty$.

Proof. It is analogous to the proof of Haddad's theorem (see [18]). Since K is locally compact, there exists $r > 0$ such that $K_0 = K \cap (x_0 + rB)$

is compact. We set $T = r/(\|F(K_0)\| + 1)$. If $F(K)$ is bounded, we take T arbitrary and we set $K_0 = K \cap \mathrm{cl}(x_0 + TF(K) + B)$, which is compact.

Let us take $y \in K$ and $v_y \in F(y)$ satisfying

$$D_+ V(y)(v_y) + W(y, v_y) \le 0. \tag{4}$$

This is possible thanks to assumption (2).

By the very definition of $D_+ V(y)(v_y)$, we know that there exist $h_y \le 1/k$ and $u_y \in v_y + 1/kB$ such that

$$\frac{V(y + h_y u_y) - V(y)}{h_y} \le D_+ V(y)(v_y) + \frac{1}{k}. \tag{5}$$

Hence

$$\frac{V(y + h_y u_y) - V(y)}{h_y} + W(y, v_y) < \frac{1}{k}. \tag{6}$$

We set

$$N(y) = \left\{ x \in K \,\middle|\, \frac{V(x + h_y u_y) - V(x)}{h_y} + W(y, v_y) < \frac{1}{k} \right\}. \tag{7}$$

Since the function V is continuous, the sets $N(y)$ are open and $y \in N(y)$. So we can find a ball of radius $\alpha_y \in \,]0, 1/k]$ such that $(y + \alpha_y B) \cap K \subset N(y)$.

Since K_0 is compact, it can be covered by q such balls $y_j + \alpha_{y_j} B$. We set $\alpha_j \doteq \alpha_{y_j}$, $h_j \doteq h_{y_j}$, $u_j \doteq u_{y_j}$, and $v_j \doteq v_{y_j}$. Let us take now any $x \in K_0$. It belongs to a ball $y_j + \alpha_j B$ for some $j = 1, \ldots, q$. Then there exists u_j such that

(i) $\dfrac{V(x + h_j u_j) - V(x)}{h_j} + W(y_j, v_j) < \dfrac{1}{k}$,

(ii) $\|x - y_j\| \le \alpha_j \le 1/k, \qquad \|u_j - v_j\| \le 1/k.$ $\qquad\qquad$ (8)

Let $h_0(k) = \min_{j=1,\ldots,q} h_j > 0$. So, canceling the index j, we have proved that for all $x \in K_0$, there exist $h \in [h_0(k), 1/k]$ and $u \in \mathbb{R}^n$ satisfying the two following properties

(i) $\exists y \in K \quad$ and $\quad v \in F(y)$ such that

$$\|x - y\| \le 1/k, \quad \|u - v\| \le 1/k,$$

(ii) $\dfrac{V(x + hu) - V(x)}{h} + W(y, v) \le \dfrac{1}{k}.$ $\qquad\qquad$ (9)

Therefore, we can construct inductively a sequence of elements $x_p \in K_0$, $v_p \in F(y_p)$, $h_p \in [h_0(k), 1/k]$ and $u_p \in \mathbb{R}^n$ satisfying

(i) $x_{p+1} = x_p + h_p u_p \in K_0$,

(ii) $(x_p, u_p) \in (y_p, v_p) + (1/k)(B \times B) \subset \mathrm{graph}(F) + (1/k)(B \times B)$, (10)

(iii) $\dfrac{V(x_{p+1}) - V(x_p)}{h_p} + W(y_p, v_p) < \dfrac{1}{k}$.

We are sure that there exists an integer m such that

$$h_0 + h_1 + \cdots + h_m \leq T < h_1 + \cdots + h_m + h_{m+1}$$

for $h_p \in [h_0(k), 1/k]$ for all p.

Let us set $\tau_k^q = h_0 + \cdots + h_q$. We interpolate this sequence by the piecewise linear function $x_k(\cdot)$ defined on each interval $]\tau_k^{p-1}, \tau_k^p[$ by $x_k(t) \doteq x_{p-1} + (t - \tau_k^{p-1})u_{p-1}$. We denote by $y_k(\cdot)$ and $v_k(\cdot)$ the step functions defined on this interval by $y_k(t) = y_p$ and $v_k(t) = v_p$.

When t is fixed in $]\tau_k^{p-1}, \tau_k^p[$, we have $|t - \tau_k^p| \leq 1/k$ and there exists $(y_p, v_p) \in \mathrm{graph}(F)$ such that $\|x_k'(t) - v_p\| = \|u_p - v_p\| \leq 1/k$ and

$$
\begin{aligned}
\|x_k(t) - y_p\| &\leq \|x_k(t) - x_p\| + \|x_p - y_p\| \\
&\leq |t - \tau_k^p|(\|u_p - v_p\| + \|v_p\|) + \|x_p - y_p\| \\
&\leq 1/k(\|F(K_0)\| + 2).
\end{aligned}
$$

By setting $\tilde{F}(t, x) \doteq F(x)$, we have proved that $\forall t \geq 0$,

$$
\begin{aligned}
(t, x_k(t), x_k'(t)) &\in (t, y_k(t), v_k(t)) + \varepsilon(k)(B \times B \times B) \\
&\subset \mathrm{graph}(\tilde{F}) + \varepsilon(k)(B \times B \times B).
\end{aligned}
$$ (11)

where $\varepsilon(k) \to 0$ when $k \to \infty$. We also know that $\|x_k'(t)\| \leq \|F(K_0)\| + 1$ and $x_k(t) \in \overline{\mathrm{co}}(K_0)$, which is compact. Hence the assumptions of the convergence theorem (see Aubin et al. [9]) are satisfied: A subsequence of $x_k(\cdot)$ converges uniformly over compact intervals to a solution $x(\cdot)$ of the differential inclusion $x' \in F(x)$. Moreover, the sequence of derivatives $x_k'(\cdot)$ converges to $x'(\cdot)$ in $L^\infty(0, T; \mathbb{R}^n)$ supplied with the weak topology $\sigma(L^\infty, L^1)$.

On each point τ_k^p of the grid, the following inequality holds:

$$V(x_k(\tau_k^{p+1})) - V(x_k(\tau_k^p)) + h^p W(x_k(\tau_k^p), v_k(\tau_k^p)) \leq h^p/k.$$ (12)

By summing these inequalities from $p = q$ to $p = r - 1$, we obtain,

$$V(x_k(\tau_k^r)) - V(x_k(\tau_k^q)) + \sum_{p=q}^{r-1} h^q W(x_k(\tau_k^p), v_k(\tau_k^p)) \leq \frac{(\tau_k^r - \tau_k^q)}{k}.$$

We remark that $v_k(\tau) = x_k'(\tau_k^p)$ on the interval $[\tau_k^p, \tau_k^{p+1}[$. So, we can write the above inequality in the form:

$$V(y_k(\tau_k^r)) - V(y_k(\tau_k^q)) + \int_{\tau_k^q}^{\tau_k^n} W(y_k(\tau), v_k(\tau)) \, d\tau \leq \frac{(\tau_k^r - \tau_k^q)}{h}. \tag{13}$$

We recall that $x_k(\cdot)$ converges to $x(\cdot)$ uniformly on compact intervals; so does $y_k(\cdot)$. We also know that $x_k'(\cdot)$ converges weakly to $x'(\cdot)$ in $L^\infty(0, T; X)$; so does $v_k(\cdot)$.

We assume for a while that the following proposition holds true (see Ekeland and Temam [17]).

PROPOSITION 1.   *Assume that the function $W: K \times L \to \mathbb{R}_+$ is nonnegative, lower semicontinuous, and convex with respect to v. Then, for all compact convex subset $M \subset L$, the functional $\mathbf{W}$ defined by*

$$\mathbf{W}(x, v) \doteq \int_0^\infty W(x(\tau), v(\tau)) \, d\tau \in [0, \infty] \tag{14}$$

is lower semicontinuous *on $\mathscr{C}(0, \infty; K) \times L^\infty(0, \infty; M)$ when $\mathscr{C}(0, \infty; X)$ is supplied with the topology of uniform convergence on compact intervals and $L^\infty(0, \infty; M)$ with the topology induced by the weak topology $\sigma(L^\infty, L^1)$ on $L^\infty(0, \infty; X)$.*

End of the proof of Theorem 1.   By Proposition 1, we deduce that for all $s > t$,

$$\int_t^s W(x(\tau), x'(\tau)) \, d\tau \leq \liminf_{k \to \infty} \int_t^s W(y_k(\tau), v_k(\tau)) \, d\tau. \tag{15}$$

Since we can approximate s by τ_k^r and t by τ_k^q with $\tau_k^r \geq \tau_k^q$ for k large enough, we deduce from the continuity of V that $V(x_k(\tau_k^r))$ converges to $V(x(s))$ and $V(x_k(\tau_k^q))$ converges to $V(x(t))$. Also, since W is continuous, it is bounded on $K_0 \times F(K_0)$ and thus, $\int_{\tau_k^q}^{\tau_k^r} W(x(\tau), x'(\tau)) \, d\tau$ converges to $\int_t^s W(x(\tau), x'(\tau)) \, d\tau$. Hence, we can take the limit when $k \to \infty$ in inequalities (13): we find that $x(\cdot)$ satisfies the monotonicity condition

$$\forall s > t', V(x(s)) - V(x(t)) + \int_t^s W(x(\tau), x'(\tau)) \, d\tau \leq 0.$$

When $F(K)$ is bounded, we have chosen T independent of x_0. So we can extend the trajectory $x(\cdot)$ on $[0, T]$ on a trajectory on $[0, 2T]$, $[0, 3T]$, etc. So, there exists a trajectory defined on $[0, \infty[$. ∎

Proof of Proposition 1.   Let $x_k(\cdot)$ converging to $x(\cdot)$ uniformly on compact intervals and $v_k(\cdot)$ converging weakly to $v(\cdot)$ in $L^\infty(0, \infty; X)$. Hence

(i) $\forall t \in [0, \infty[, x_k(t)$ converges to $x(t)$,

(ii) $\forall T > 0, v_k(\cdot)$ converges weakly to $v(\cdot)$ in $L^1(0, T; X)$. $\qquad$ (16)

Actually, it is sufficient to suppose the latter properties (16) hold true. If $\liminf_{k \to \infty} \mathbf{W}(x_k, v_k) = +\infty$, the theorem is true; if not, let

$$c \doteq \liminf_{k \to \infty} \mathbf{W}(x_k, v_k). \tag{17}$$

There exist subsequences (again denoted x_k and v_k) of x_k and v_k such that

$$\forall k \in \mathbb{N}, \qquad \mathbf{W}(x_k, v_k) \le c + 1/k. \tag{18}$$

By Mazur's theorem, there exists a sequence of elements

$$w_h(\cdot) = \sum_{k=h}^{\infty} a_h^k v_k(\cdot), \, a_h^k \ge 0, \qquad \sum_{k=h}^{\infty} a_h^k = 1 \tag{19}$$

(where $a_h^k = 0$ but for a finite number of indexes k) that converges strongly to $v(\cdot)$ in $L^1(0, T; X)$:

$$\forall h \in \mathbb{N}, \qquad \left\| w_h - v \right\|_{L^1(0,T;X)} \le 1/h.$$

Hence, a subsequence (again denoted) w_h converges almost everywhere to $v(\cdot)$. Let $t \ge 0$ be any point where

$$v(t) = \lim_{h \to \infty} w_h(t). \tag{20}$$

By Proposition 14.2, there exists η such that $\mathscr{E}pW(x, \cdot) \subset \mathscr{E}pW(x(t), \cdot) + \varepsilon(B \times B)$ when $\left\| x - x(t) \right\| \le \eta$. Let k_0 such that $\left\| x_k(t) - x(t) \right\| \le \eta$ whenever $k \ge k_0$. Since $\mathscr{E}pW(x_k(t), \cdot)$ is convex, we deduce that

$$\left(w_h(t), \sum_{k=h}^{\infty} \alpha_h^k W(x_k(t), v_k(t)) \right) \in \mathscr{E}pW(x_k(t), \cdot) \subset \mathscr{E}pW(x(t), \cdot) + \varepsilon(B \times B).$$

Hence, by letting $h \to \infty$, we obtain

$$\left(v(t), \liminf_{h \to \infty} \sum_{k=h}^{\infty} \alpha_h^k W(x_k(t), v_k(t)) \right) \in \text{cl}(\mathscr{E}pW(x(t), \cdot) + \varepsilon(B \times B).$$

Since this holds true for all $\varepsilon > 0$, and since $\mathscr{E}pW(x(t), \cdot)$ is closed, it follows that

$$\left(v(t), \liminf_{h \to \infty} \sum_{k=h}^{\infty} \alpha_h^k W(x_k(t), v_k(t)) \right) \in \mathscr{E}pW(x(t), \cdot), \tag{21}$$

i.e., that

$$\liminf_{h \to \infty} \sum_{k=h}^{\infty} \alpha_h^k W(x_k(t), v_k(t)) \ge W(x(t), v(t)). \tag{22}$$

We integrate this inequality and we apply Fatou's lemma, which is possible for W is nonnegative. We obtain

$$\mathbf{W}(x,v) \leq \int_0^\infty \left(\liminf_{h \to \infty} \sum_{k=h}^\infty \alpha_h^k W(x_k(t), v_k(t)) \right) dt$$

$$\leq \liminf_{h \to \infty} \sum_{k=h}^\infty \alpha_h^k \mathbf{W}(x_k, v_k) \leq \lim_{h \to \infty} \left(c + \frac{1}{h} \right) = c$$

by (18). Hence $\mathbf{W}(x,v) \leq c \doteq \liminf_{k \to \infty} \mathbf{W}(x_k, v_k)$. ∎

16.   STABILITY AND ASYMPTOTIC STABILITY

We consider the case when $W \equiv 0$; in this particular case, we say that V is a *Liapunov function with respect to F* if

$$\forall x \in K, \quad \exists v \in F(x) \quad \text{such that} \quad D_+ V(x)(v) \leq 0 \tag{1}$$

and that a trajectory $x(\cdot)$ of $x' \in F(x)$ is *monotone* with respect to V if

$$\text{the function } t \mapsto V(x(t)) \text{ is nonincreasing.} \tag{2}$$

Hence monotone trajectories remain in the "level sets"

$$\{x \in K \,|\, V(x) \leq V(x_0)\}. \tag{3}$$

So, we obtain the following stability property.

THEOREM 1.   *Let K be a closed subset of $\mathbb{R}^n$ and F be a bounded upper semicontinuous map from K to the nonempty compact convex subsets of $\mathbb{R}^n$. Let $V:K \to \mathbb{R}_+$ be a continuous lower semicompact Liapunov function. Let $P_* \doteq \{x \in K \,|\, V(x_*) = \min_{x \in K} V(x)\}$. Then the following stability property holds:*

For any open neighborhood M of P_*, there exists a neighborhood $N \subset M$ of P_* such that, for all $x_0 \in N$, there exists a trajectory starting at x_0 and remaining in M.

Proof.   We set $Q \doteq \{x \in K \,|\, V(x) \leq \min_{y \in K} V(y) + 1\}$, which is compact. Hence, since M is an open neighborhood of P_*, $Q \cap M$ is compact and $c = \min_{x \in Q \cap M} V(x)$ is finite. Thus the subset $N \doteq \{x \in Q \,|\, V(x) < c\}$ is contained in M and is a neighborhood of P_* (for V is continuous). Now, if $x_0 \in N$, there exists a trajectory $x(\cdot)$ which is monotone (by Theorem 15.1) and thus, which remains in $N \subset M$ because $V(x(t)) \leq V(x_0) < c$. ∎

We shall give now conditions implying asymptotic stability when $W \equiv 0$.

THEOREM 2. *Let F be an upper semicontinuous map from $K \subset \mathbb{R}^n$ to the compact convex subsets of $\mathbb{R}^n$ and V be the restriction to K of a locally Lipschitzean function $\tilde{V}$ which is lower semicompact. We assume that V is a Liapunov function with respect to F satisfying*

$$\forall x \in K, \quad \text{if there exists } v \in F(x) \text{ such that } D_+ V(x)(v) \geq 0, \text{ then}$$
$$V(x) = \min_{y \in K} V(y). \tag{4}$$

We also assume that

$$\text{the function } (x, v) \in \text{graph}(F) \to D_+ \tilde{V}(x)(v) \text{ is upper semicontinuous.} \tag{5}$$

(*This is the case when, for instance, $\tilde{V}$ is continuously differentiable or convex continuous.*) *Then any monotone trajectory satisfies*

$$\lim_{t \to \infty} V(x(t)) = \min_{y \in K} V(y). \tag{6}$$

Proof. We set $v(t) \doteq V(x(t))$. Let us assume that $\alpha \doteq \lim_{t \to \infty} v(t) > \min_{x \in K} V(x) \geq 0$. Let $Q \doteq \{x \in K \,|\, V(x) \leq V(x_0)\}$, which is compact because V is lower semicompact and lower semicontinuous. Therefore, the graph $\mathscr{F}_Q$ of the restriction to Q of F is compact, for F is upper semicontinuous from K to the compact subsets of $\mathbb{R}^n$.

Assumption (4) implies that $D_+ \tilde{V}(x)(v) < 0$ for all $(x, v) \in \mathscr{F}_Q$; so, we deduce from assumption (5) that there exists $\mu > 0$ such that

$$\forall x \in Q, \quad \sup_{v \in F(x)} D_+ \tilde{V}(x)(v) \leq -\mu. \tag{7}$$

Let $x(\cdot) \in \mathscr{C}(0, \infty; \mathbb{R}^n)$ be a trajectory such that $x'(\cdot) \in L^\infty(0, \infty; \mathbb{R}^n)$. Since $x(t) \in Q$ for all $t \geq 0$, we deduce that $D_+ \tilde{V}(x(t))(x'(t)) \leq -\mu$. Also, because V is the restriction of a locally Lipschitzean function $\tilde{V}$, Proposition 6.8 implies that

$$\liminf_{h \to 0+} \frac{v(t + h) - v(t)}{h} \leq D_+ \tilde{V}(x(t))(x'(t)) \leq -\mu. \tag{8}$$

Therefore, we deduce from Proposition 6.9 that for all $T = v(0)/\mu$, we have

$$v(T) - v(0) \leq \int_0^T -\mu \, dt = -v(0). \tag{9}$$

Thus $v(T) \leq 0 \leq \min_{x \in K} V(x)$ and $v(T) \geq \alpha > \min_{x \in K} V(x)$.
The theorem follows from this contradiction. ∎

17. LIAPUNOV FUNCTIONS FOR U-MONOTONE MAPS

Let $U: K \times K \to \mathbb{R}_+$ be a nonnegative continuous function such that $U(y, y) = 0$ for all $y \in K$, which plays the role of a semidistance (without having to obey the triangle inequality). We shall associate with U the class of "U-monotone" maps F which enjoy the following property. If we know in advance that $x_* \in K$ is a stationary point of F, then the "distance function" to $x_*: x \to U(x, x_*)$ is a Liapunov function. When $U(x, y) \doteq \frac{1}{2}\sum_{i=1}^{r} |x_i - y_i|^2$, the class of U-monotone maps coincides with the class of monotone maps in the usual sense.

We list some examples of functions U.

$$U_p(x, y) \doteq \sum_{i=1}^{n} |x_i - y_i|^p, \qquad 1 \leq p < +\infty, \tag{1}$$

and, in particular, for $p = 2$:

$$U_2(x, y) \doteq \frac{1}{2} \sum_{i=1}^{p} |x_i - y_i|^2. \tag{2}$$

We mention also

$$U_\infty(x, y) \doteq \max_{i=1,\ldots,n} (x_i - y_i) \tag{3}$$

and, if $K = \mathring{\mathbb{R}}_+^n$,

$$U_0(x, y) = \max_{i=1,\ldots,n} \frac{x_i}{y_i} - \min_{i=1,\ldots,n} \frac{x_i}{y_i}. \tag{4}$$

We associate with U the function U' defined on $K \times K \times X$ by

$$U'(x, y)(v) \doteq D_+(x \mapsto U(x, y))(x)(v). \tag{5}$$

Now, we introduce the class of U-monotone set-valued maps.

DEFINITION 1. Let $U: K \times K \to \mathbb{R}_+$ be a continuous function such that $U(y, y) = 0$ for all $y \in K$. We say that a set-valued map from K to $\mathbb{R}^n$ is U-monotone if

$$\forall x, y \in K, \quad u \in F(x), \quad v \in F(y), \quad U'(x, y)(v - u) \leq 0. \tag{6}$$

We say that F is *strictly U-monotone* if

$$\forall x, y \in K \quad \text{such that} \quad U(x, y) > 0, \qquad\qquad\qquad (7)$$
$$\forall u \in F(x), \quad v \in F(y), \quad U'(x, y)(v - u) < 0$$

and strongly U-monotone if there exists $c > 0$ such that

$$\forall x, y \in K, \quad u \in F(x), \quad v \in F(y), \quad U'(x, y)(v - u) + c(U(x, y)) \leq 0 \tag{8}$$

Finally, if φ is a bounded continuous map from $\mathbb{R}_+$ to $\mathbb{R}_+$, we say that F is (φ, U)-monotone if

$$\forall x, y \in K, \quad u \in F(x), \quad v \in F(y), \quad U'(x, y)(v - u) + \varphi(U(x, y)) \leq 0. \tag{9}$$

EXAMPLES OF U-MONOTONE OPERATORS. When we take $K = \mathbb{R}^n_+$ and

$$U_2(x, y) = \frac{1}{2} \sum_{i=1}^{n} |x_i - y_i|^2 \tag{10}$$

we see that F is (φ, U_2)-monotone if

$$\forall x, y \in K, \quad u \in F(x), \quad v \in F(y), \quad \langle u - v, x - y \rangle \geq \varphi(U_2(x, y)). \tag{11}$$

So, usual monotone maps are the U_2-monotone maps.

When $1 < p < \infty$, we set

$$U_p(x, y) \doteq \frac{1}{p} \sum_{i=1}^{n} |x_i - y_i|^p \tag{12}$$

and

$$I_0(x, y) \doteq \{i \,|\, x_i = y_i\}. \tag{13}$$

Therefore, F is (φ, U_p)-monotone if and only if, for all $x, y \in \mathbb{R}^n$, $u \in F(x)$, $v \in F(y)$, we have

$$\sum_{i \notin I_0(x,y)} |x_i - y_i|^{p-2}(x_i - y_i)(u_i - v_i) \geq \varphi(U_p(x, y)) \tag{14}$$

when $p = \infty$, we set

$$U_\infty(x, y) \doteq \max_{i=1,\ldots,n} (x_i - y_i) \tag{15}$$

and

$$J(x, y) \doteq \{i \,|\, x_i - y_i = \max_{i=1,\ldots,n} (x_i - y_i)\}. \tag{16}$$

Then F is (φ, U_∞)-monotone if and only if for all $x, y \in \mathbb{R}^n$, $u \in F(x)$, $v \in F(y)$, we have

$$\min_{j \in J(x,y)} (u_i - v_i) \geq \varphi(U_\infty(x, y)). \tag{17}$$

Finally, if x and $y \in \mathring{\mathbb{R}}^n_+$, we set

$$U_0(x, y) = \max_{i=1,\ldots,n} \frac{x_i}{y_i} - \min_{i=1,\ldots,n} \frac{x_i}{y_i} \tag{18}$$

and

$$K_+(x, y) = \left\{ i \,\Big|\, \frac{x_i}{y_i} = \max_i \frac{x_i}{y_i} \right\}, \qquad K_-(x, y) = \left\{ i \,\Big|\, \frac{x_i}{y_i} = \min_i \frac{x_i}{y_i} \right\}. \tag{19}$$

[Note that $U_0(x, y) = 0$ if and only if $\exists \lambda \geq 0$ such that $x = \lambda y$.]

Then F is (φ, U_0)-monotone if and only if for all $x, y \in \mathring{\mathbb{R}}^n_+$, $u \in F(x)$, $v \in F(y)$,

$$\min_{i \in K_+(x,y)} \frac{u_i - v_i}{y_i} - \max_{j \in K_-(x,y)} \frac{u_j - v_j}{y_j} \geq \varphi(U_0(x, y)). \tag{20}$$

Proof. The above characterizations follow obviously from the computation of the contingent derivatives of $x \to U(x, y)$. Since these functions are convex and continuous, the contingent derivatives coincide with the directional derivatives from the right.

When $1 < p < +\infty$, we set $I_+(x, y) = \{i \mid x_i > y_i\}$ and $I_-(x, y) = \{i \mid y_i < x_i\}$. Then it is clear that

$$U'_p(x, y)(v) = \sum_{i \in I_+(x,y)} (x_i - y_i)^{p-1} v_i - \sum_{i \in I_-(x,y)} (y_i - x_i)^{p-1} v_i$$

$$= \sum_{i \notin I_0(x,y)} |x_i - y_i|^{p-2}(x_i - y_i)v_i.$$

If $p = 1$, $U'_1(x, y)(v) = \sum_{i=1}^n v_i$ and, if $p = \infty$, we have $W'(x, y)(v) = \max_{j \in J(x,y)} v_j$. Finally, for $p = 0$, we have

$$U'_0(x, y)(v) = \max_{i \in K_+(x,y)} \frac{v_i}{y_i} - \min_{i \in K_-(x,y)} \frac{v_i}{y_i}. \quad \blacksquare$$

The U-monotone maps enjoy the following fundamental properties.

THEOREM 1. *Let U be a continuous nonnegative function from $K \times K \to \mathbb{R}^n$ such that $U(y, y) = 0$ for all $y \in K$. Let F be a proper bounded upper semicontinuous map from K to the compact convex subsets of $\mathbb{R}^n$. We also posit the following assumptions*

(i) *there exists a stationary point $x_* \in K$ of F,*

(ii) *$-F$ is (φ, U) monotone.* $\qquad(21)$

Let $w \in \mathscr{C}(0, \infty; \mathbb{R})$ be a solution to the differential equation

$$w' + \varphi(w) = 0; \quad w(0) = U(x_0, x_*), \quad x_0 \text{ is given in } K. \tag{22}$$

Then there exists a solution $x(\cdot)$ to the differential inclusion $x' \in F(x)$, $x(0) = x_0$ such that $t \to U(x(t), x_*)$ is nonincreasing, such that

$$\int_0^\infty \varphi(U(x(t), x_*)) \, dt < +\infty$$

and such that

$$\forall t \geq 0, \quad U(x(t), x_*) \leq w(t). \tag{23}$$

Proof. We set $V(x) \doteq U(x, x_*)$ and $W(x, v) \doteq \varphi(U(x, x_*))$. Since $0 \in F(x_*)$, then, for all $v \in F(x)$, we get $D_+V(x)(v) + W(x, v) = U'(x, x_*)(v - 0) + \varphi(U(x, x_*))$. The right-hand side of this inequality is nonpositive since $-F$ is (φ, U)-monotone. Hence V is a Liapunov function for F with respect to W. So, we apply successively Theorems 15.1 and 12.3. ∎

We mention now a result on asymptotic stability. For simplicity, we prove only a special case.

THEOREM 2.   *Let $L \subset \mathbb{R}^n$ be an open convex subset of $\mathbb{R}^n$ and $U: L \times L \to \mathbb{R}_+$ be a nonnegative continuous function such that*

(i) $\forall y \in L, U(y, y) = 0$;

(ii) $\forall y \in L, x \mapsto U(x, y)$ is convex.
$\qquad\qquad\qquad\qquad\qquad\qquad\qquad\qquad\qquad\qquad\qquad$ (24)

Let K be a closed subset of L and F be a bounded upper semicontinuous map from K to the compact convex subsets of $\mathbb{R}^n$. We assume also that

(i)   there exists a stationary point $x_* \in K$ of F;

(ii) $\forall x \in K, D_K(x) \cap F(x) \neq \varnothing$;

(iii) $-F$ is strictly U-monotone.
$\qquad\qquad\qquad\qquad\qquad\qquad\qquad\qquad\qquad\qquad\qquad$ (25)

Then, for any $x_0 \in K$, there exists a solution to the differential inclusion $x' \in F(x), x(0) = x_0$ satisfying

$$\lim_{t \to \infty} U(x(t), x_*) = 0. \qquad\qquad (26)$$

Proof.   We take for Liapunov function the restriction $V|_K$ of the function V defined on L by $V(x) \doteq U(x, x_*)$, where x_* is a stationary point of F. Since V is convex and continuous, then $D_+V(x)(v) = D_-V(x)(v)$ is upper semicontinuous with respect to (x, v). It is a Liapunov function since there exists $v \in D_K(x) \cap F(x)$ by assumption (26, ii): Hence $D_+V|_K(x)(v) \leq D_-V(x)(v) = D_+V(x)(v) = U'(x, x_*)(v - 0) \leq 0$ because $v \in F(x)$, $0 \in F(x_*)$, and $-F$ is U-monotone. Also, let us assume that there exists $v \in F(x)$ such that $D_+V(x)(v) \geq 0$. Then $U'(x, x_*)(v - 0) \geq 0$ and, since $v \in F(x)$, $0 \in F(x_*)$, and $-F$ is strictly monotone, we deduce from (7) that $V(x) \doteq U(x, x_*) = 0 = \min_{y \in K} V(y)$. Therefore assumptions of Theorem 16.2 are satisfied and thus, any monotone trajectory statisfies property (27). They do exist by Theorem 15.1. ∎

Remark.   We recall that the Browder–Ky Fan theorem states that when K is convex and compact, the tangential condition (26, ii) implies the existence of a stationary point $x_* \in K$.

18. CONSTRUCTION OF LIAPUNOV FUNCTIONS

If the dynamical system described by the set-valued map $F: K \subset \mathbb{R}^n \to \mathbb{R}^n$ and the function $W: K \times F(K) \to \mathbb{R}_+$ are given, the problem arises whether there exist a Liapunov function V. By Theorems 14.1 and 15.1, we have to find functions V satisfying the property

$$\forall x \in K, \quad \exists v \in F(x) \quad \text{such that} \quad D_+V(x)(v) + W(x, v) \le 0. \tag{1}$$

For this purpose, we denote by $\mathcal{T}_\infty(x_0)$ the set of viable trajectories of the differential inclusion

$$x' \in F(x), \quad x(0) = x_0 \quad \text{given in} \quad K. \tag{2}$$

We define the function V_F from K to $[0, +\infty]$ by

$$\forall x_0 \in K, \quad V_F(x_0) = \inf_{x(\cdot) \in \mathcal{T}_\infty(x_0)} \int_0^\infty W(x(\tau), x'(\tau)) \, d\tau. \tag{3}$$

We begin by pointing out the following remark.

PROPOSITION 1. *Let $V: K \to \mathbb{R}_+$ and $W: K \times F(K) \to \mathbb{R}_+$ be nonnegative functions. If there exists a monotone trajectory $x(\cdot) \in \mathcal{T}_\infty(x_0)$ (with respect to V and W) then*

$$0 \le V_F(x_0) \le V(x_0). \tag{4}$$

Proof. It follows from Proposition 12.1 and from inequality

$$\forall t \ge 0, \quad \int_0^t W(x(\tau), x'(\tau)) \, d\tau \le V(x_0) - V(x(t)) \le V(x_0). \quad \blacksquare$$

We now prove that V_F does satisfy the Liapunov condition for F with respect to W.

PROPOSITION 2. *Let $K \subset \mathbb{R}^n$ be closed, $F: K \to \mathbb{R}^n$ be a proper upper semi-continuous map with compact convex images and $W: K \times \mathrm{co}(F(K)) \to \mathbb{R}_+$ be a nonnegative lower semicontinuous function that is convex with respect to v. If the minimum in $V_F(x_0)$ is achieved for $x_0 \in K$, V_F satisfies the Liapunov condition*

$$\exists v_0 \in F(x_0) \quad \text{such that} \quad D_+V_F(x_0)(v_0) + W(x_0, v_0) \le 0. \tag{5}$$

Proof. Let us assume that there exists $x(\cdot) \in \mathcal{T}_\infty(x_0)$ such that $V_F(x_0) = \int_0^\infty W(x(\tau), x'(\tau)) \, d\tau$. Since F is upper semicontinuous, we can associate with any $\varepsilon > 0$ and $\eta > 0$ such that, for all $x \in x_0 + \varepsilon B$, $F(x) \subset F(x_0) + \varepsilon B$. So, for h small enough, $x'(\tau) \in F(x(\tau)) \subset F(x_0) + \varepsilon B$ for all $\tau \in [0, h]$. On the other hand, $(x(h) - x_0)/h = (1/h) \int_0^h x'(\tau) \, d\tau$ belongs to $F(x_0) + \varepsilon B$ by the mean-value theorem, because the latter set is compact and convex. So, a subsequence $v_n = [x(h_n) - x_0]/h_n$ converges to some $v_0 \in F(x_0)$. On the other hand,

$$
\begin{aligned}
V_F(x(h_n)) &\leq \int_{h_n}^{\infty} W(x(\tau), x'(\tau)) \, d\tau \\
&= \int_0^{\infty} W(x(\tau), x'(\tau)) \, d\tau - \int_0^{h_n} W(x(\tau), x'(\tau)) \, d\tau \leq V_F(x_0) \quad (6) \\
&\quad - \int_0^{h_n} W(x(\tau), x'(\tau)) \, d\tau.
\end{aligned}
$$

Therefore,

$$
\frac{V_F(x_0 + h_n v_n) - V_F(x_0)}{h_n} + \frac{1}{h_n} \int_0^{h_n} W(x(\tau), x'(\tau)) \, d\tau \leq 0.
$$

By the very definition of the upper contingent derivative, we have

$$
D_+ V_F(x_0, v_0) = \liminf_{\substack{h \to 0 \\ v \to v_0}} \frac{V_F(x_0 + hv) - V_F(x_0)}{h}
$$

and, by Proposition 14.1,

$$
W(x_0, v_0) \leq \liminf_{h_n \to 0} \frac{1}{h_n} \int_0^{h_n} W(x(\tau), x'(\tau)) \, d\tau.
$$

So, by taking the limit in inequalities (6), we obtain the Liapunov condition (5). ∎

Therefore, by tying up some of the preceding results, we can prove that $V_F(\cdot)$ is the smallest Liapunov function with respect to F and W.

We begin by making more precise Theorem 14.1 on necessary conditions.

THEOREM 1. *Let F be a bounded upper semicontinuous map from $K \subset \mathbb{R}^n$ to the compact convex subsets of $\mathbb{R}^n$. Let W be a nonnegative lower semicontinuous function from $K \times \mathrm{co}\, F(K)$ which is convex with respect to v.*

Let V be a nonnegative function on K. If for all $x_0 \in K$, there exists a monotone trajectory $x(\cdot) \in \mathcal{T}_\infty(x_0)$ with respect to V and W, then not only V is a Liapunov function with respect to F and W, but V_F is also a nonnegative lower semicontinuous Liapunov function smaller than or equal to V.

Proof.   Since there exist monotone trajectories $x(\cdot) \in \mathcal{T}_\infty(x_0)$ with respect to V and W for all $x_0 \in K$, we deduce from Theorem 14.1 that V is a Liapunov function for F with respect to W, from Proposition 1 that $V_F(x_0) \leq V(x_0)$ for all $x_0 \in K$ and from Proposition 2 that V_F is a Liapunov function. The set-valued map $x_0 \to \mathcal{T}_\infty(x_0)$ from K to the space of functions $x(\cdot) \in \mathscr{C}(0, \infty; \mathbb{R}^n)$ whoser derivatives $x'(\cdot) \in L^\infty(0, \infty; \mathbb{R}^n)$ is upper semicontinuous with compact values when $\mathscr{C}(0, \infty; \mathbb{R}^n)$ is supplied with the topology of uniform convergence on compact intervals and when $L^\infty(0, \infty; \mathbb{R}^n)$ is supplied with the weak topology $\sigma(L^\infty, L^1)$. Proposition 15.1 states that the functional $x \mapsto \int_0^\infty W(x(\tau), x'(\tau)) d\tau$ is lower semicontinuous on that space. Hence the maximum theorem implies that the function $V_F(\cdot)$ is lower semicontinuous. ■

Another combination of the same arguments yield the following statement:

THEOREM 2. *Let F be a bounded upper semicontinuous map from a closed subset* $K \subset \mathbb{R}^n$ *to the compact convex subsets of* $\mathbb{R}^n$, *satisfying the tangential condition*

$$\forall x \in K, \qquad F(x) \cap D_K(x) \neq \varnothing. \tag{7}$$

Let $W: K \times \mathrm{co}\, F(K) \to \mathbb{R}_+$ *be a nonnegative lower semicontinuous function which is convex with respect to v. If for* $x_0 \in K$, $V_F(x_0)$ *is finite, then* $V_F(x_0)$ *is lower semicontinuous at* x_0 *and satisfies*

$$\exists v_0 \in F(x_0) \quad \text{such that} \quad D_+ V_F(x_0) + W(x_0, v_0) \leq 0. \tag{8}$$

Consequently, if $V_F(\cdot)$ *is finite on K, it is a lower semicontinuous (l.s.c.) Liapunov function for F with respect to W.*

The fact that V_F is a Liapunov function yields the following characterization of trajectories achieving the minimum of V_F.

PROPOSITION 3. *Let us assume that the function* V_F *defined by (3) is finite on K and is a Liapunov function for F with respect to W. Then the trajectories of the differential inclusion* $x' \in F(x)$, $x(0) = x_0$, *which are monotone with respect to* V_F *and W achieve the minimum in* $V_F(x_0)$.

Proof. If $x(\cdot)$ is a monotone trajectory with respect to V_F and W, we obtain the inequality

$$\int_0^\infty W(x(\tau), x'(\tau)) d\tau \leq V_F(x_0) = \inf_{y(\cdot) \in \mathcal{T}_\infty(x_0)} \int_0^\infty W(y(\tau), y'(\tau)) d\tau. \quad \blacksquare \tag{9}$$

This simple statement has a very important consequence in control theory. We associate with V_F the new map G defined by

$$G(x) \doteq \{v \in F(x) \,|\, D_+ V_F(x)(v) + W(x, v) = \min_{w \in F(x)} (D_+ V_F(x)(w) + W(x, w))\}.$$

Note that $G(x)$ is single valued when $v \to D_+ V_F(x)(v)$ is convex and $v \mapsto W(x, v)$ is strictly convex for all $x \in K$. One can devise sufficient conditions implying that $(x, v) \to D_+ V_F(x)(v)$ is upper semicontinuous. This and the continuity of W guarantees that G is an upper semicontinuous map.

In any case, by Theorem 16.1, solutions $x(\cdot)$ of the differential inclusion $x' \in G(x)$, $x(0) = x_0$ yield trajectories of $x' \in F(x)$, $x(0) = x_0$ which achieve the minimum in $V_F(x_0)$. ∎

19. Construction of Dynamical Systems Having Monotone Trajectories

The question arises whether V and W being given, there exists a continuous single-valued map f such that the differential equation $x' = f(x)$ has monotone trajectory with respect to V and W.

In this section, we shall assume that

(i) K is compact and convex,
(ii) V is the restriction to K of a convex continuous function $\tilde{V}$,   (1)
(iii) W is continuous and convex with respect to v.

We recall that a necessary and sufficient condition for f to have monotone trajectories with respect to V and W is that

$$\forall x \in K, \quad D_+ V(x)(f(x)) + W(x, f(x)) \leq 0. \tag{2}$$

Since $D_+ V(x)(v)$ is the restriction to the tangent cone $D_K(x)$ of $D_+ \tilde{V}(x)(v)$, we set

$$S(x) \doteq \{ v \in X \mid D_+ \tilde{V}(x)(v) + W(x, v) \leq 0 \}. \tag{3}$$

So, the necessary and sufficient conditions can be written

$$\forall x \in K, \quad f(x) \in S(x) \cap D_K(x). \tag{4}$$

In order to exclude the obvious solution $f \equiv 0$, we introduce the cones

$$\mathring{S}(x) = \{ v \in X \mid D_+ \tilde{V}(x)(v) + W(x, v) < 0 \}. \tag{5}$$

which may be empty. We also set

$$K_0 \doteq \{ x \in K \mid \mathring{S}(x) \cap D_K(x) \neq \varnothing \}; \quad K_1 \doteq K \cap \complement K_0. \tag{6}$$

Theorem 1   (Cornet).   *Let K be a compact convex subset, V be the restriction to K of a continuous convex function $\tilde{V}$ and W be a nonnegative continuous function on $K \times \mathbb{R}^n$ which is convex with respect to v. We assume that the subset K_0 defined by (6) is nonempty.*

Then there exists a continuous function f whose set of stationary points is K_1 such that the differential equation $x' = f(x)$, $x(0) = x_0$ has a monotone trajectory with respect to V and W.

Proof. Since $(x, v) \mapsto D_{+}\mathring{V}(x)(v)$ is upper semicontinuous, the graph of the map $\mathring{S}$, which is equal to

$$\text{graph } \mathring{S} = \{(x, v) \in K_0 \times \mathbb{R}^n | D_{+}V(x)(v) + W(x, v) < 0\}$$

is open. Since the set-valued map $x \rightarrow D_K(x)$ is l.s.c. and has convex values $x \rightarrow \mathring{S}(x) \cap D_K(x)$ is locally selectionable from K_0 to $\mathbb{R}^n$ and its images are convex cones (see Cornet [15]). Hence, by results of Cornet [15], there exists a continuous function f from K to $\mathbb{R}^n$ satisfying

$$\forall x \in K_0, \quad f(x) \in D_K(x) \cap \mathring{S}(x) \quad \text{and} \quad \forall x \in K_1, \quad f(x) = 0. \tag{7}$$

So, such a function f satisfies the assumptions of Theorem 15.1. Hence there exists monotone trajectories of the differential equation $x' = f(x)$. ∎

20. Feedback Controls Yielding Monotone Trajectories

Let U be a set of controls u and $f : K \times U \rightarrow X$ the map that assigns to each state x and to each control u the velocity $f(x, u)$ of the state.

A feedback control is a map $\mathbf{u} : x \in K \rightarrow \mathbf{u}(x) \in U$ associating with each state of the system a control according to a fixed rule for achieving a given purpose.

The example of such a purpose is the requirement that the trajectories of the differential equation

(i) $\quad x'(t) = f(x(t), \mathbf{u}(x(t)))$,

(ii) $\quad x(0) = 0$

$$\tag{1}$$

exist and satisfy the monotonicity condition:

$$\forall s > t, \quad V(x(s)) - V(x(t)) + \int_t^s W(x(\tau), x'(\tau)) \, d\tau \leq 0. \tag{2}$$

We assume that V is the restriction to K of a convex continuous function $\tilde{V}$.

We introduce the following set-valued map $\mathring{S}$ defined by

$$\mathring{S}(x) = \{v \in X | D_{+}\tilde{V}(x)(v) + W(x, v) < 0\} \tag{3}$$

and we set

$$K_0 = \{x \in K | \mathring{S}(x) \cap D_K(x) \neq \varnothing\}, \, K_1 = K \cap \complement K_0. \tag{4}$$

Theorem 1. *Let us assume that $K \subset X$ and U are both convex compact subsets, that U contains 0, and that $F : K \times U \rightarrow X$ is a continuous map that is linear with respect to the controls. We assume that $K_0 \neq \varnothing$ and that there* *exists $\gamma > 0$ such that*

$$\forall x \in K_0, \quad \forall y \in X, \quad \|y\| \leq \gamma,$$
$$\exists u \in U \quad \text{such that} \quad f(x,u) - y \in D_K(x) \cap \mathring{S}(x). \tag{5}$$

Then there exists a feedback control $\mathbf{u} \in \mathscr{C}(K, U)$, vanishing on K_1 that provides monotone trajectories with respect to V and W.

Proof. By *Theorem* 15.1, we have to prove the existence of a feedback control $\mathbf{u} \in \mathscr{C}(K, U)$ such that

(i) $\forall x \in K_0, f(x, \mathbf{u}(x)) \in D_K(x) \cap \mathring{S}(x),$
(ii) $\forall x \in K_1, \mathbf{u}(x) = 0.$ $\qquad\qquad$ (6)

Let us set, if $x \in K_0$,

$$G(x) \doteq \{u \in U \mid f(x,u) \in D_K(x) \cap \mathring{S}(x)\}. \tag{7}$$

We already mentioned that $x \mapsto D_K(x) \cap \mathring{S}(x)$ is locally selectionable and, thus, lower semicontinuous. Assumption (5) implies that G is lower semicontinuous on K_0 (see Aubin *et al.* [9]). Michael's theorem states that there exists a continuous selection $\mathbf{v} \in \mathscr{C}(K_0, U)$ of the set-valued map C. We denote by $d_{K_1}(x)$ the distance from x to K_1 and we set:

$$\mathbf{u}(x) = \begin{cases} d_{K_1}(x)\mathbf{v}(x) & \text{if } x \in K_0, \\ 0 & \text{if } x \in K_1. \end{cases} \tag{8}$$

This function $\mathbf{u}$ is continuous on K. It is obviously true when $x \in K_0$. Let us check that it is continuous at $x \in K_1$. Let $\varepsilon > 0$ and $y \in x + \varepsilon B$. Then $\mathbf{u}(x) = 0$ and either $\mathbf{u}(y) = 0$ (when $y \in K_1$) or $\mathbf{u}(y) \leq d_{K_1}(y)M \leq \varepsilon$ where $M = \|U\|$. Then $\|\mathbf{u}(x) - \mathbf{u}(y)\| = \|\mathbf{u}(y)\| \leq \varepsilon M$. Hence $\mathbf{u}$ is continuous. Since $\mathbf{v}(x) \in G(x)$ and since f is linear with respect to the controls, we deduce that

$$f(x, \mathbf{u}(x)) = d_{K_1}(x)f(x, \mathbf{v}(x)) \in D_K(x) \cap \mathring{S}(x)$$

when $x \in K_0$ and that $f(x, \mathbf{u}(x)) = 0$ when $x \in K_1$. ∎

21. THE TIME DEPENDENT CASE

We shall adapt to the time dependent case the results we proved for the time independent case. We only have to use the classical transformation which amounts to observing that the solutions to the differential inclusion

$$x'(t) \in F(t, x); \qquad x(t_0) = x_0 \tag{1}$$

and the solutions $\tau \mapsto (t(\tau), x(\tau)) \doteq \hat{x}(\tau)$ of the differential inclusion

$$\hat{x}' \in \hat{F}(\hat{x}), \qquad x(0) = (t_0, x_0) \tag{2}$$

where we set

$$\forall (t, x) \doteq \hat{x} \in \mathrm{Dom}(F), \qquad \hat{F}(\hat{x}) \doteq (1, F(t, x)). \tag{3}$$

We shall denote by $\hat{K} \doteq \mathrm{Dom}\, F$ the domain of F, which is the domain of $\hat{K}$. We introduce

   (i)   a nonnegative function V from $\hat{K}$ to $\mathbb{R}$
   (ii) a nonnegative function W from $\hat{K} \times \mathrm{co}\, F(\hat{K})$ to $\mathbb{R}$. $\qquad$ (4)

Now, the symbol $D_+ V(t, x)(1, v)$ denotes the upper contingent derivative of V at (t, x) in the direction $(1, v)$. We recall that when V is differentiable, we have

$$D_+ V(t, x)(1, v) = \frac{\partial}{\partial t} V(t, x) + \sum_{i=1}^{n} \frac{\partial}{\partial x_i} V(t, x) v_i. \tag{5}$$

PROPOSITION 1. *$\hat{x}(\cdot)$ is a monotone trajectory of the differential inclusion (2) with respect to V and W if and only if $x(\cdot)$ is a trajectory of the differential inclusion (1) which is monotone with respect to V and W in the sense that*

$$\forall s \geq t, \quad V(s, x(s)) - V(t, x(t)) + \int_0^\infty W(\tau, x(\tau), x'(\tau))\, d\tau \leq 0. \tag{6}$$

Proof. Indeed, $\hat{x}(\cdot)$ is a solution to (2) if and only if $\hat{x}(t) = (t, x(t))$ where $x(\cdot)$ is a solution to (1). Note that $\hat{x}'(t) = (1, x'(t))$. So, condition (6) is equivalent to $\forall s \geq t$, $V(\hat{x}(s)) - V(\hat{x}(t)) + \int_0^\infty W(\hat{x}(\tau), \hat{x}'(\tau))\, d\tau \leq 0.$ ■

We introduce now the concept of Liapunov function.

DEFINITION 1. Let F be a set-valued map from $\hat{K} \subset \mathbb{R}_+ \times \mathbb{R}^n$ to $\mathbb{R}^n$, V be a nonnegative function from $\hat{K}$ to $\mathbb{R}$ and W be a nonnegative function from $\hat{K} \times \mathrm{co}(F(K))$ to $\mathbb{R}$. We say that V is a *Liapunov function* for F with respect to W if

$$\forall (t, x) \in \hat{K}, \quad \exists v \in F(t, x) \quad \text{such that} \quad D_+ V(t, x)(1, v) + W(t, x, v) \leq 0. \tag{7}$$

We can consider $\hat{K} \subset \mathbb{R} \times \mathbb{R}^n$ as the graph of a set-valued map $t \to K(t)$. Then monotonicity condition (6) implies in particular that

$$\forall t, \quad x(t) \in K(t) \tag{8}$$

and the Liapunov condition (8) implies in particular that

$$\forall t \geq 0, \quad \forall x \in K(t), \quad v \in DK(t, x)(1), \tag{9}$$

since the latter condition is equivalent to the tangential condition $(1, v) \in D_K(t, x)$. When V is the restriction to $\hat{K}$ of a continuous convex function $\tilde{V}$, the Liapunov condition can be written

$$\forall t \geq 0, \quad \forall x \in K(t), \quad \exists v \in F(t, x) \cap DK(t, x)(1)$$
$$\text{such that} \quad D_+ \tilde{V}(t, x)(1, v) + W(t, x, v) \leq 0. \tag{10}$$

When V is the restriction to $\hat{K}$ of a continuously differentiable function $\tilde{V}$, the Liapunov condition can be written

$$\forall t \geq 0, \quad \forall x \in K(t), \quad \exists v \in F(t, x) \cap DK(t, x)(1) \tag{11}$$
$$\text{such that} \quad \frac{\partial}{\partial t} V(t, x) + \sum_{i=1}^n \frac{\partial}{\partial x_i} V(t, x) v_i + W(t, x, v) \leq 0.$$

We deduce from Theorem 14.1 and 15.1 the following characterization of existence of monotone trajectories.

THEOREM 1 *Let F be a bounded upper semicontinuous map from $\hat{K} \subset \mathbb{R}_+ \times \mathbb{R}^n$ to the compact convex subsets of $\mathbb{R}^N$, V be a nonnegative continuous function from $\hat{K}$ to $\mathbb{R}$ and W be a nonnegative continuous function from $K \times \mathrm{co}\, F(K)$ to $\mathbb{R}$ which is convex with respect to the last argument. Then the differential inclusion*

$$x'(t) \in F(t, x(t)); \qquad x(t_0) = x_0 \tag{12}$$

has a monotone trajectory $x(\cdot) \in \mathscr{C}(t_0, \infty; \mathbb{R}^n)$ for all $t_0 \geq 0$ and $x_0 \in K(t_0)$ if and only if V is a Liapunov for F with respect to W.

We mention also the following adaptation of Theorem 18.2.

We denote by $\mathscr{T}_\infty(t_0, x_0)$ the set of solutions $x(\cdot) \in \mathscr{C}(t_0, \infty; x)$ of the differential inclusion (12). We introduce the function

$$V_F(t_0, x_0) = \inf_{x(\cdot) \in \mathscr{T}_\infty(t_0, x_0)} \int_0^\infty W(\tau, x(\tau), x'(\tau)) \, d\tau. \tag{13}$$

THEOREM 2 *Let F be a bounded upper semicontinuous map from a closed graph of a set-valued map $K(\cdot) : \mathbb{R}_+ \to \mathbb{R}^n$ to the compact convex subsets of $\mathbb{R}^n$, satisfying*

$$\forall t \geq 0, \quad \forall x \in K(t), \quad F(t, x) \cap DK(t, x)(1) \neq \varnothing. \tag{14}$$

Let $W : \hat{K} \times \mathrm{co}(F(\hat{K})) \to \mathbb{R}_+$ be a nonnegative lower semicontinuous function which is convex with respect to the last argument. If for all $(t_0, x_0) \in \hat{K}$, the function $V_F(t_0, x_0)$ is finite, it is the smallest nonnegative lower semicontinuous Liapunov function for F with respect to W.

If $V(t_0, x_0)$ is a Liapunov function for F with respect to W, then

$$\inf_{v \in F(t, x)} [D_+ V(t, x)(1, v) + W(t, x, v)] \leq 0. \tag{15}$$

JEAN PIERRE AUBIN

When V is the restriction to $\hat{K}$ of a differentiable function $\tilde{V}$, this inequality can be written

$$\frac{\partial V}{\partial t}(t,x) + \inf_{v \in F(t,x) \cap DK(t,x)(1)} \left[\sum_{i=1}^{n} \frac{\partial V}{\partial x_i}(t,x)v_i + W(t,x,v) \right] \leq 0. \quad (16)$$

This is the Carathéodory–Hamilton–Jacobi–Bellman equation of control theory.

ACKNOWLEDGMENT

I would like to thank Georges Haddad for his hidden but important contribution as well as Arrigo Cellina, Bernard Cornet, and Ivar Ekeland.

REFERENCES

1. J. P. AUBIN, "Applied Abstract Analysis," Wiley (Interscience), New York, 1977.
2. J. P. AUBIN, Further properties of Lagrange multipliers in nonsmooth optimization, *Appl. Math. Opt.*, **6** (1980), 79–90.
3. J. P. AUBIN, "Applied Functional Analysis," Wiley (Interscience), New York, 1979.
4. J. P. AUBIN AND F. H. CLARKE, Monotone invariant solutions to differential inclusions, *J. London Math. Soc.* **16** (1977), 357–366.
5. J. P. AUBIN AND F. H. CLARKE, Points critiques et méthode de Newton sur des compacts non convexes, to appear.
6. J. P. AUBIN AND F. H. CLARKE, Multiplicateurs de Lagrange en optimisation non convexe et applications, *C. R. Acad. Sci. Sér.* **285** (1977), 951–959.
7. J. P. AUBIN AND I. EKELAND, "Applied Nonlinear Analysis," Wiley (Interscience), New York, to appear.
8. J. P. AUBIN AND J. SIEGEL, Fixed points and stationary points of dissipative multivalued maps, *Bull. Am. Math. Soc.*, **78** (1980), 391–398.
9. J. P. AUBIN, A. GELLINA, AND J. NOHEL, Monotone trajectories of multivalued dynamical system, *Ann. Mat. Pura Appl.* **115** (1977), 99–117.
10. H. BRÉZIS AND F. BROWDER, A general principle on ordered sets in nonlinear functional analysis, *Adv. in Math.* **21** (1976), 355–369.
11. J. CARISTI, Fixed point theorems for mappings satisfying inwardness conditions, *Trans. Amer. Math. Soc.* **78** (1972), 186–197.
12. F. H. CLARKE, Generalized gradients and applications, *Trans. Amer. Math. Soc.* **205** (1975), 247–262.
13. F. H. CLARKE, Generalized gradients of Lipschitz functionals, *Adv. in Math.*, to appear.
14. F. H. CLARKE, A new approach to Lagrange multipliers, *Math. Oper. Res.* **1** (1976), 97–102.
15. B. CORNET, Thesis, Université de Paris–Dauphiné, 1981.
16. I. EKELAND, On the variational principle, *J. Math. Anal. Appl.* **47** (1974), 324–353.
17. I. EKELAND AND R. TEMAM, "Analyse convexe et problèmes variationals," Dunod, Paris, 1974.
18. G. HADDAD, Tangential conditions for existence theorems in differential inclusions and functional differential inclusions with memory, *Israel J. Math.*, to appear.

19. A. D. IOFFÉ, Differentielles généralisées d'applications localement Lipschitziennes d'un epace de Banach dans un autre, *C. F. Acad. Sci Sér.* **289** (1979), 637–640.
20. J. J. MOREAU, Un cas de convergence des iterees d'ure contraction d'un espace Hilbertien, *C. . Acad. Sci. Sér.* **286** (1978), 143–144.
21. M. NAGUMO, Über die laga der Integralkurven gewöhnlicher Differentialgleichungen, *Proc. Phys. Math. Soc. Japan* **24** (1942), 551–559.
22. R. T. ROCKAFELLAR, "Convex Analysis," Princeton Univ. Press, Princeton, New Jersey, 1970.
23. R. T. ROCKAFELLAR, Monotone processes of convex and concave type, *Mem. Amer. Math. Soc.* **77** (1967).
24. R. T. ROCKAFELLAR, Generalized directional derivatives and subgradients of nonconvex functions, *Can. J. Math.* **32** (1980), 257–280.
25. R. T. ROCKAFELLAR, Clarke's tangent cones and the boundaries of closed sets in $\mathbb{R}$, *Nonlinear Anal.* **3** (1979), 145–159.
26. R. T. ROCKAFELLAR, "La théorie des sous-gradients et ses applications a l'optimisation," Presses Univ. Montréal, Montréal, 1978.
27. S. Saks, "Theory of the Integral," Hafner, New York, 1937.
28. L. Schwartz, "Théorie des distributions," Hermann, Paris, 1966.

AMS (MOS) Subject Classification: 47H10, 47H15, 47H99, 49B30, 49B99

Sur des équations intégrales d'évolution

M. S. BAOUENDI

Department of Mathematics
Purdue University
West Lafayette, Indiana

ET

C. GOULAOUIC[†]

Centre de Mathématiques
École Polytechnique
Palaiseau, France

EN HOMMAGE À LAURENT SCHWARTZ

Résumé

Nous prouvons des résultats d'existence et d'unicité locale des solutions (du type des théorèmes de Cauchy–Kovalewsky et de Holmgren), pour des équations d'évolution fonctionnelles, telles que, par exemple, des équations différentielles avec retard.

Abstract

We prove existence and uniqueness results (similar to the theorems of Cauchy–Kovalewsky and Holmgren) for solutions of functional evolution equations, for example, differential equations with delay.

Nous traitons ici des problèmes de Cauchy pour des équations intégrales avec une régularité minimale en la variable d'évolution et analyticité en les variables d'espace; les résultats obtenus sont des généralisations des théorèmes de Cauchy–Kovalewsky et Hölmgren dans un cadre général d'équations fonctionnelles, incluant des équations "différentielles avec retard" par exemple.

La méthode de démonstration comporte une formulation abstraite d'un théorème de Cauchy dans une chaîne d'espaces de Banach qui apparaît comme une généralisation d'un théorème de Nishida [3]. Nous passons d'abord d'hypothèses très faibles (intégrabilité en temps) à des hypothèses plus fortes (caractère borné) par un changement de variables; ensuite on utilise une technique de démonstration très voisine de [2].

[†] Laboratoire associé au C.N.R.S. n° 169.

Dans la démonstration d'un théorème d'unicité pour des équations avec retard, nous avons eu recours à une méthode de troncature déjà utilisée dans [1] pour des problèmes fuchsiens.

Les hypothèses de régularité en la variable d'évolution et leur manipulation sont inspirées par la remarque suivante bien connue sur le théorème classique de Cauchy–Lipschitz:

Remarque. Soient X un espace de Banach où $B(R)$ désigne la boule $\{u \in X; \|u\| < R\}$, f une fonction mesurable de $[0, T] \times B(R)$ dans X vérifiant:

$$\|f(t, u) - f(t, v)\| \le C(t)\|u - v\|$$

pour u, v dans $B(R)$ et où C est intégrable sur $[0, T]$ ($C \in L^1(0, T)$).

Alors il existe au plus une fonction u absolument continue solution de

$$\begin{aligned} u'(t) &= f(t, u(t)) \quad \text{sur } [0, T], \\ u(0) &= 0. \end{aligned} \tag{0}$$

Si de plus $\|f(., 0)\|$ est intégrable sur $[0, T]$, il existe $a \in \,]0, T]$ et une solution de l'équation (0) sur $[0, a[$.

L'unicité est immédiate; si u_1 et u_2 sont deux solutions de (0), en notant $v_1 = u_1'$ et $v_2 = u_2'$ on a

$$\|(v_1 - v_2)(t)\| \le C(t) \int_0^t \|(v_1 - v_2)(\sigma)\| \, d\sigma,$$

$$\int_0^{t_0} \|(v_1 - v_2)(t)\| \, dt \le \int_0^{t_1} C(t) \, dt \int_0^{t_1} \|(v_1 - v_2)(\sigma)\| \, d\sigma$$

et donc $v_1 = v_2$ sur $[0, t_1]$ tel que $\int_0^{t_1} C(t) \, dt < 1$.

L'existence d'une solution se montre en effectuant un changement de variable d'évolution $(y = \int_0^t (C(\sigma) + \|f(\sigma, 0)\| + 1) \, d\sigma)$ qui permet de se ramener à la situation classique du théorème de Cauchy–Lipschitz.

Déjà ici on a préféré remplacer l'étude de l'équation différentielle (0) par celle de l'équation intégrale

$$v(t) = f\left(t, \int_0^t v(\sigma) \, d\sigma\right).$$

Nous allons considérer ci-après des équations intégrales plus générales de la forme:

$$v(t) = f\left(t, \int_0^t K(t, \sigma)v(\sigma) \, d\sigma\right)$$

avec des hypothèses convenables.

I. Théorème de Cauchy pour une équation integrate abstraite

Soient $(X_s)_{0 < s \le 1}$ une chaîne décroissante d'espaces de Banach (pour $s' < s$, X_s s'injecte dans $X_{s'}$, l'injection étant de norme ≤ 1), R, T, des constantes positives.

Notons $B(s, R) = \{u \in X_s; \|u\|_s < R\}$, en désignant par $\|\ \|_s$ la norme dans X_s.

Soit F une fonction de $[0, T] \times \bigcup_{1 \ge s > 0} B(s, R)$ dans $\bigcup_{1 \ge s > 0} X_s$ telle que, pour tous $0 < s' < s \le 1$, F envoie $[0, T] \times B(s, R)$ dans $X_{s'}$ et vérifie:

$$\|F(t, u) - F(t, v)\|_{s'} \le \frac{C(t)}{s - s'} \|u - v\|_s \qquad \text{pour } t \in [0, T] \text{ et } u, v \text{ dans } B(s, R)$$

$$\text{avec } C \in L^1([0, T]). \tag{1}$$

$$\|F(t, 0)\|_s \le \frac{M(t)}{1 - s} \qquad \text{pour } 0 < s < 1 \quad \text{avec } M \in L^1([0, T]). \tag{2}$$

Pour toute fonction mesurable u de $[0, T]$ dans $B(s, R)$, la fonction $t \mapsto F(t, u(t))$ est mesurable. $\tag{3}$

Nous considérons aussi une fonction K mesurable bornée de $]0, T[\times]0, T[$ dans l'espace $\mathscr{L}(X_s, X_s)$ des opérateurs linéaires continus de X_s dans lui-même pour tout $s \in]0, 1]$ et vérifiant

$$\sup_{\substack{s \in]0, 1] \\ (t, \sigma) \in]0, T[\times]0, T[}} \|K(t, \sigma)\|_{\mathscr{L}(X_s, X_s)} = m < \infty. \tag{4}$$

Nous notons $\varphi(t) = \int_0^t (1 + C(\sigma) + M(\sigma)) \, d\sigma$ et ψ la fonction inverse de φ; nous avons le résultat:

Théorème 1. *Sous les hypothèses* (1) (2) (3) (4) *ci-dessus, il existe* $a \in]0, \varphi(T)[$ *et une unique fonction v qui pour tout $s \in]0, 1[$ est intégrable sur* $]0, \psi(a(1 - s))[$ *à valeurs dans X_s et vérifie*

$$\sup_{0 \le \varphi(t) < a(1 - s)} \left\| \int_0^t K(t, \sigma) v(\sigma) \, d\sigma \right\|_s < R, \tag{5}$$

$$v(t) = F\left(t, \int_0^t K(t, \sigma) v(\sigma) \, d\sigma\right) \qquad pour \quad \varphi(t) \in [0, a[. \tag{6}$$

Démonstration. Nous nous ramenons d'abord au cas où C et M sont bornées par le changement de variables $y = \varphi(t)$; nous notons

$$w(y) = v(\psi(y))\psi'(y),$$
$$\tilde{K}(y, \tau) = K(\psi(y), \psi(\tau)),$$
$$\tilde{F}(y, .) = \psi'(y)F(\psi(y), .)$$

et nous vérifions que (5) (6) équivaut à

$$\sup_{0 \leq y < a(1-s)} \left\| \int_0^y \tilde{K}(y, \tau)w(\tau)\,d\tau \right\|_s < R,$$

$$w(y) = \tilde{F}\left(y, \int_0^y \tilde{K}(y, \tau)w(\tau)\,d\tau \right) \quad \text{pour} \quad 0 \leq y < a$$

et que $\tilde{F}$ vérifie (1), (2) avec $C = M = 1$. Pour démontrer le théorème 1 il suffit donc de montrer que si F et K vérifient (3), (4) et

Il existe $C > 0$ telle que

$$\left\| F(t, u) - F(t, v) \right\|_{s'} \leq \frac{c}{s - s'} \left\| u - v \right\|_s$$

$$\text{pour } t \in [0, T] \text{ et } u, v \text{ dans } B(s, R), \tag{1'}$$

Il existe $M > 0$ telle que

$$\left\| F(t, 0) \right\|_s < \frac{M}{1 - s} \quad \text{pour} \quad 0 < s < 1, \tag{2'}$$

alors il existe $a \in \,]0, T]$ et une unique fonction v qui, pour tout $s \in \,]0, 1[$ est localement bornée sur $[0, a(1 - s)[$ à valeurs dan X_s et vérifie

$$\sup_{0 < t < a(1-s)} \left\| \int_0^t K(t, \sigma)v(\sigma)\,d\sigma \right\|_s < R, \tag{5'}$$

$$v(t) = F\left(t, \int_0^t K(t, \sigma)v(\sigma)\,d\sigma \right) \quad \text{pour} \quad t \in [0, a[. \tag{6'}$$

Pour $a \in \,]0, T]$, nous désignons alors par E_a l'espace des fonctions u de $[0, a[$ dans $\bigcup_{s > 0} X_s$ qui, pour tout $s \in \,]0, 1[$, sont localement bornées de $[0, a(1 - s)[$ dans X_s et vérifient:

$$\|\|u\|\|_a = \sup_{\substack{0 \leq t < a(1-s) \\ 0 < s < 1}} \|u(t)\|_s (1 - s) \sqrt{1 - \frac{t}{a(1 - s)}} < \infty. \tag{7}$$

Muni de la norme $\|\| \ \|\|_a$, l'espace E_a est un espace de Banach; pour $a \leq b \leq T$ nous avons $E_b \subset E_a$, l'inclusion étant de norme ≤ 1.

Notons $Gv(t) = F(t, \int_0^t K(t, \sigma)v(\sigma)\,d\sigma)$; nous allons montrer que G est une contraction dans un espace métrique complet, à savoir un sous ensemble fermé de E_a pour a assez petit; pour cela nous utilisons les résultats:

LEMME 1.   *Soient* $a \in \,]0, T\,]$, $v \in E_a$, $s \in \,]0, 1[$, $0 \le t < a(1 - s)$; *alors*

$$\left\| \int_0^t K(t, \sigma)v(\sigma)\,d\sigma \right\|_s \le 2ma \||v\||_a.$$

Démonstration

$$\left\| \int_0^t K(t, \sigma)v(\sigma)\,d\sigma \right\|_s \le m\||v\||_a \int_0^t \frac{\sqrt{a(1 - s)}}{(1 - s)\sqrt{a(1 - s) - \sigma}}\,d\sigma$$

$$\le 2ma\||v\||_a.$$

LEMME 2.   *Soient* $a \in \,]0, T[$, $v \in E_a$, $s \in \,]0, 1[$, $0 \le t < a(1 - s)$; *alors, avec* $s(\sigma) = \frac{1}{2}(1 + s - \sigma/a)$, *nous avons*

$$\int_0^t \frac{\|v(\sigma)\|_{s(\sigma)}}{s(\sigma) - s}\,d\sigma \le 8a\||v\||_a \frac{1}{1 - s}\sqrt{\frac{a(1 - s)}{a(1 - s) - t}}.$$

Démonstration.   (Voir Figure 1.) Nous avons:

$$\int_0^t \frac{\|v(\sigma)\|_{s(\sigma)}}{s(\sigma) - s}\,d\sigma \le \||v\||_a \int_0^t \frac{1}{(1 - s(\sigma))(s(\sigma) - s)}\sqrt{\frac{a(1 - s(\sigma))}{a(1 - s(\sigma)) - \sigma}}\,d\sigma$$

$$\le \||v\||_a \int_0^t 4a^2(a(1 - s) + \sigma)^{-1/2}(a(1 - s) - \sigma)^{-3/2}\,d\sigma$$

$$\le \||v\||_a \frac{4a}{(1 - s)} \int_0^{t/a(1 - s)} (1 + \tau)^{-1/2}(1 - \tau)^{-3/2}\,d\tau$$

$$\le 8a\||v\||_a \frac{1}{1 - s}\sqrt{\frac{a(1 - s)}{a(1 - s) - t}}$$

LEMME 3.   *Soient* $a \in \,]0, T/2[$, $s \in \,]0, 1[$, $0 < t < a(1 - s)$, $u \in E_a$ *vérifiant* $\||u\||_a < R/4ma$, $v \in E_{2a}$ *vérifiant* $\||v\||_{2a} < R/8ma$; *alors:*

$$\left\| F\left(t, \int_0^t K(t, \sigma)u(\sigma)\,d\sigma\right) - F\left(t, \int_0^t K(t, \sigma)v(\sigma)\,d\sigma\right) \right\|_s$$

$$\le Cm \int_0^t \frac{\|u(\sigma) - v(\sigma)\|_{s(\sigma)}\,d\sigma}{s(\sigma) - s}$$

avec $s(\sigma) = \frac{1}{2}(1 + s - \sigma/a)$.

Démonstration   Notons $t_j = jt/n$ pour $0 \le j \le n$; nous avons

$$F\left(t, \int_0^t K(t, \sigma)u(\sigma)\,d\sigma\right) - F\left(t, \int_0^t K(t, \sigma)v(\sigma)\,d\sigma\right)$$

$$= \sum_{j=1}^n \left\{F\left(t, \int_0^{t_j} K(t, \sigma)u(\sigma)\,d\sigma + \int_{t_j}^t K(t, \sigma)v(\sigma)\,d\sigma\right)\right.$$

$$\left. - F\left(t, \int_0^{t_{j-1}} K(t, \sigma)u(\sigma)\,d\sigma + \int_{t_{j-1}}^t K(t, \sigma)v(\sigma)\,d\sigma\right)\right\}.$$

Le lemme 1 implique

$$\left\|\int_0^{t_j} K(t, \sigma)u(\sigma)\,d\sigma + \int_{t_j}^t K(t, \sigma)v(\sigma)\,d\sigma\right\|_{s_j} < R \qquad \text{pour} \quad j = 1, \ldots, n \quad (9)$$

avec $s_j = \inf_{t_{j-1} \le \sigma \le t_j} s(\sigma)$.
De (8) et (1′) nous déduisons:

$$\left\|F\left(t, \int_0^t K(t, \sigma)u(\sigma)\,d\sigma\right) - F\left(t, \int_0^t K(t, \sigma)v(\sigma)\,d\sigma\right)\right\|_s$$

$$\le \sum_{j=1}^n \frac{C}{s_j - s} \left\|\int_{t_{j-1}}^{t_j} K(t, \sigma)(u(\sigma) - v(\sigma))\,d\sigma\right\|_{s_j}$$

$$\le Cm \sum_{j=1}^n \int_{t_{j-1}}^{t_j} \frac{\|u(\sigma) - v(\sigma)\|_{s_j}}{s_j - s}\,d\sigma$$

$$\le Cm \int_0^t \frac{\|u(\sigma) - v(\sigma)\|_{\tilde{s}_n(\sigma)}}{s_n(\sigma) - s}\,d\sigma$$

où $\tilde{s}_n$ est définie par $\tilde{s}_n(\sigma) = s_j$ pour $\sigma \in [t_{j-1}, t_j[$ et $1 \le j \le n$; enfin un passage à la limite quand $n \to \infty$ donne l'inégalité annoncée.

Preuve de l'existence d'une solution: Pour $b \in \,]0, T[$, $u \in E_b$ avec $\|\|u\|\|_b < R/4mb$, $s \in \,]0, 1[$, et $t \in \,]0, b(1 - s)[$, nous avons:

$$\|Gu(t)\|_s \le \|Gu(t) - F(t, 0)\|_s + \|F(t, 0)\|_s,$$

$$\|Gu(t)\|_s \le 8b\,Cm\|\|u\|\|_b \frac{1}{1 - s} \sqrt{\frac{b(1 - s)}{b(1 - s) - t}} + \frac{M}{1 - s}, \qquad (10)$$

$$\|\|Gu\|\|_b \le 8b\,Cm\|\|u\|\|_b + M.$$

Il résulte de (10) que pour $a \in \,]0, T/2[$, $u \in E_a$, et $v \in E_{2a}$ vérifiant $\|\|u\|\|_a < R/4ma$ et $\|\|v\|\|_{2a} < R/8ma$, Gu et Gv sont respectivement dans E_a et E_{2a} et

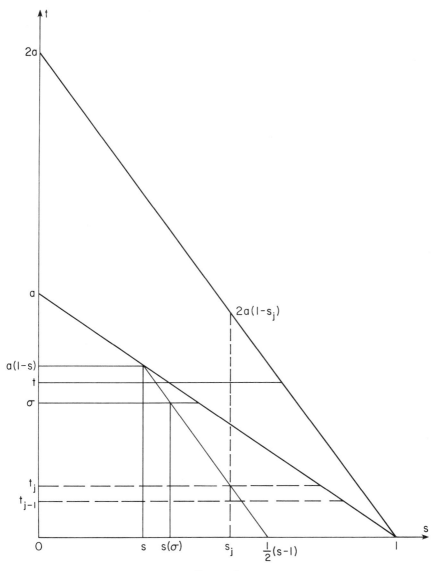

FIGURE 1

vérifient

$$\||Gu\||_a < 2CR + M,$$
$$\||Gv\||_{2a} < 2CR + M,$$

(11)

$$\||Gu - Gv\||_a \leq 8a\,Cm\||u - v\||_a.$$

(12)

Notons S_a la fermeture dans E_a de la boule $\{u \in E_{2a};\ \||u\||_{2a} < R/8ma\}$ et choisissons

$$a < \inf\left(\frac{T}{2}, \frac{R}{8m(M + 2CR)}\right).$$

(13)

Alors (11) (12) impliquent que G est une contraction de S_a dans lui-même de rapport $< \frac{1}{2}$, et a donc un unique point fixe $\mathbf{u} \in S_a$; il est immédiat que $\mathbf{u}$ vérifie (5') (6').

Montrons maintenant l'unicité de la solution de (5') (6'). Nous remarquons d'abord que S_b contient S_a lorsque $b \leq a$; supposons alors (5') (6') vérifiées par v pour un certain $a \in\ ]0, T[$; nous en déduisons:

$$\|v(t)\|_s \leq \left\|F\left(t, \int_0^t K(t,\sigma)v(\sigma)\,d\sigma\right) - F(t,0)\right\|_s + \|F(t,0)\|_s,$$

$$\|v(t)\|_s \leq \frac{C}{s(t) - s}\left\|\int_0^t K(t,\sigma)v(\sigma)\,d\sigma\right\|_{s(t)} + \frac{M}{1 - s},$$

$$\|v(t)\|_s \leq \frac{M}{s(t) - s} + \frac{M}{1 - s} \quad \text{avec} \quad s < s(t) < 1 - \frac{t}{a},$$

d'où

$$\|v(t)\|_s \leq \frac{acR}{a(1 - s) - t} + \frac{M}{1 - s}.$$

Pour $b < a/2$, nous obtenons

$$\||v\||_b \leq \sup_{t < 2b(1-s)} (1 - s)\sqrt{1 - \frac{t}{2b(1 - s)}}\left(\frac{aCR}{(a - 2b)(1 - s)} + \frac{M}{1 - s}\right),$$

$$\||v\||_b \leq aCr/(a - 2b) + M$$

et donc $v \in S_b$ pourvu que

$$aCR/(a - 2b) + M < \frac{R}{8mb}$$

c'est-à-dire par example pour $b = a/4$ et

$$a < R/[2M_1(M + 2CR)].$$

(14)

Nous remarquons que (13) implique (14); en conclusion, pour a vérifiant (13), les relations (5′) (6′) impliquent que $v \in S_b$ avec $b = a/4$ et donc que v est l'unique solution de l'équation intégrale dans tout S_c pour $c \leq a$.

Le théorème est complètement démontré.

Remarque 1.   Dans le théorème 1 avec les hypothèses (1′)(2′), il est possible de remplacer l'hypothèse (4) sur K par:

Il existe une fonction k mesurable et ≥ 0 sur $]0, T[\times]0, T[$ vérifiant:

(i) $\left\| \int_\alpha^\beta K(t, \sigma)v(\sigma)\, d\sigma \right\|_s \leq \int_\alpha^\beta k(t, \sigma) \|v(\sigma)\|_s\, d\sigma,$

pour tous $s \in]0, 1[, v \in X_s, 0 \leq \alpha \leq \beta \leq t < T$

(ii) $\displaystyle \sup_{t/2 < \sigma < t < T} k(t, \sigma) < \infty,$ (14′)

(iii) $\displaystyle \sup_{t < T} \frac{1}{t} \int_0^{t/2} k(t, \sigma)\, d\sigma < \infty.$

Ceci permet de traiter en particulier les problèmes caractéristiques du type de Fuchs et de généraliser quelque peu les résultats de [2].

Remarquons que cette hypothèse (4′) ne suffit pas lorsque sont faites seulement les hypothèses (1) (2).

Remarque 2.   Le théorème 1 se généralise au cas où (6) est remplacée par:

$$v(t) = F\left(t, \int_0^t K_0(t, \sigma)v(\sigma)\, d\sigma, \ldots, \int_0^t K_n(t, \sigma)v(\sigma)\, d\sigma \right),$$

où chaque K_j vérifie (4) et F vérifie:

(i) $\left\| F(t, u_0, \ldots, u_n) - F(t, v_0, \ldots, v_n) \right\|_{s'} \leq [C(t)/(s - s')] \sum_{j=0}^n \|u_j - v_j\|_s$
pour tous $0 < s' < s \leq 1, 0 \leq t < T, u_0, \ldots, v_n$ dans $B(s, R)$, avec $C \in L^1([0, T])$.
(ii) $\left\| F(t, 0, \ldots, 0) \right\|_s \leq M(t)/(1 - s)$ pour $0 < s < 1$ avec $M \in L^1([0, T])$.
(iii)   Pour toutes fonctions $u_0, \ldots, u_n$ mesurables de $[0, T]$ dans $B(s, R)$, la fonction $t \mapsto F(t, u_0(t), \ldots, u_n(t))$ est mesurable.

II.   Quelques applications (equations avec retard)

(a)   Considérons le problème de Cauchy:

$$\begin{aligned} u'(t) &= F(t, u(\rho(t))), \\ u(0) &= 0, \end{aligned}$$ (15)

où F est une fonction vérifiant les hypothèses (1) (2) (3) dans une chaîne $(X_s)_{0 < s \leq 1}$ d'espaces de Banach et où ρ est une fonction mesurable vérifiant

$$0 \leq \rho(t) \leq t \qquad \text{p.p. sur } [0, T[. \qquad (16)$$

Le théorème 1 implique le résultat:

COROLLAIRE 1. *Le problème* (15) *a une solution unique au sens suivant: Il existe* $a \in {]}0, T[$ *et une unique fonction* u *qui, pour tout* $s \in {]}0, 1[$, *est absolument continue de* $[0, \psi(a(1 - s))[$ *dans* X_s *et vérifie*:

$$\sup_{0 \leq \varphi(t) < a(1-s)} \left\| u(\rho(t)) \right\|_s < R$$

$$u'(t) = F(t, u(\rho(t))) \qquad pour \quad 0 \leq \varphi(t) < a(1 - s).$$

Démonstration: En posant $u' = v$, le problème (15) équivaut à l'équation intégrale

$$v(t) = F\left(t, \int_0^{\rho(t)} v(\sigma) \, d\sigma \right)$$

qui entre dans le cadre du théorème 1 avec

$$K(t, \sigma) = \begin{cases} I & \text{(l'identité de } \mathcal{L}(X_s, X_s)) & \text{si} \quad 0 < \sigma < \rho(t), \\ 0 & & \text{si} \quad \rho(t) < \sigma < t. \end{cases}$$

Evidemment ce noyau K vérifie (4) avec $m = 1$.

Un résultat analogue vaut lorsque F dépend de u par différents retards $\rho_0, \ldots, \rho_N$ qui vérifient (16), conformément à la remarque 2.

(b) En particulier, nous déduisons du corollaire 1 un théorème de Cauchy–Kovalewsky avec retard, que nous énonçons pour une équation (ou un système d'équations) du premier ordre:

COROLLAIRE 2. *Soient* $\rho_0, \ldots, \rho_n$ *des fonctions measurables vérifiant* (16), u_0 *une fonction analytique au voisinage de* 0 *dans* $\mathbb{R}^n$, $F(t, x, u, u_1, \ldots, u_n)$ *une fonction intégrable en* t *sur* $[0, T[$ *à valeurs dans les fonctions analytiques au voisinage de* $(0, u_0(0), (\partial u_0/\partial x_1)/(0), \ldots, (\partial u_0/\partial x_n)/(0))$.
Alors il existe $a \in {]}0, T[$ *et une unique fonction* $u(t, x)$ *absolument continue en* t *sur* $[0, a[$ *à valeurs dans les fonctions analytiques au voisinage de* 0 *et telle que*

$$\frac{\partial u}{\partial t}(t, x) = F\left(t, x, u(\rho_0(t), x), \ldots, \frac{\partial u}{\partial x_n}(\rho_n(t), x) \right),$$

$$u(0, x) = u_0(x).$$

Pour démontrer ce résultat il suffit d'utiliser le corollaire 2 étendu au cas de plusieurs retards et de choisir une chaîne d'espaces de Banach $(X_s)_{0 < s}$, où X_s est l'espace des fonctions holomorphes bornées sur la boule de centre 0 et de rayon s.

Notons que, lorsque $\rho_0, \ldots, \rho_n$ et F sont aussi holomorphes en t, la solution u est holomorphe en t (il suffit de remarquer que la démonstration du théorème peut s'adapter pour $t \in \mathbb{C}$).

(c) *Unicité dans les équations linéaires avec retard*

Il s'agit ici d'une extension du théorème de Holmgren aux équations avec retard; le cas général d'un système d'équations d'ordre quelconque pouvant toujours être réduit à un système d'équations du premier ordre, nous nous plaçons immédiatement dans cette situation en supposant de plus une très faible régularité (intégrabilité en la variable d'évolution; dans le cas d'un seul retard (pour simplifier les notations) nous obtenons donc:

THÉORÈME 2. *Soit ρ une fonction measurable vérifiant* (16). *Soient $A_0(t, x), \ldots, A_n(t, x)$ des matrices $N \times N$ à coefficients intégrables en t sur $[0, T[$ à valeurs dan les fonctions analytiques en x au voisinage de 0 dans $\mathbb{R}^n$. Si u est une fonction absolument continue sur $[0, T[$ à valeurs dans les distributions $\mathscr{D}'(\Omega))^N$ où Ω est un voisinage de 0 dans $\mathbb{R}^n$, vérifiant:*

$$D_t u(t, x) + \sum_{j=1}^{n} A_j(t, x) D_j u(\rho(t), x) + A_0(\rho(t), x) u = 0 \quad sur\ [0, T[\times \Omega,$$

$$u(0, \cdot) = 0,$$

(17)

Alors $u = 0$ au voisinage de 0 dans $[0, \infty[\times \mathbb{R}^n$.

Démonstration. Comme nous ne pouvons pas nous servir ici d'un changement de variables pour nous ramener au cas de fonctions à valeurs dans les distributions à support compact, nous utilisons une troncature comme dans [1] en utilisant une fonction $\varphi \in \mathscr{D}(\mathbb{R}^n)$ qui vaut 1 sur un voisinage ω_1 de 0 et à support dans une boule Ω_1 assez petite (dans l'intersection de Ω avec les domaines complexes d'analyticité des données).

Nous notons $\omega = \Omega_1 - \bar{\omega}_1$ et $v(t, x) = \varphi(x) u(t, x)$. Si $\mathcal{O}$ est un ouvert borné de $\mathbb{R}^n$, désignons par $\mathcal{O}_s$ la réunion des polydisques de $\mathbb{C}^n$ de rayon s et centrés en des points de $\mathcal{O}$ et par $F_s(\mathcal{O})$ l'adhérence des fonctions holomorphes sur $\mathbb{C}^n$ dans l'espace des fonctions holomorphes bornées sur $\mathcal{O}_s$; $(F_s(\mathcal{O}))_{s > 0}$ est une chaîne décroissante d'espaces de Banach et les espaces duaux constituent une chaîne croissante $(F'_s(\mathcal{O}))_{s > 0}$ d'espaces de Banach; notons que les distributions à support compact dans $\mathcal{O}$ s'injectent dans $F'_s(\mathcal{O})$. De (17) nous déduisons que v est solution de:

$$D_t v(t,x) + \sum_{j=1}^{n} A_j(t,x) D_j v(\rho(t),x) + A_0(t,x) v(\rho(t),x) = f(t,x),$$
$$v(0,\cdot) = 0,$$

(18)

où f est une fonction intégrable à valeurs dans la chaine $(F_s'(\omega))^N$.

Il résulte du corollaire 1 que (18) a une solution unique absolument continue, pour t assez petit, à valeurs dans la chaîne $(F_s'(\omega))^N$; d'autre part on sait que pour t fixé, $v(t,\cdot) \in (\mathscr{E}'(\mathbb{R}^n))^N$, où $\mathscr{E}'(\mathbb{R}^n)$ désigne l'espace des distributions à support compact sur $\mathbb{R}^n$; nous déduisons alors que $v(t,\cdot) = 0$ au voisinage de l'origine pour t assez petit (c'est-à-dire la conclusion du théorème), du lemme suivant:

Lemme 4 [1]. *Soient ω un ouvert borné de $\mathbb{R}^n$ et $\mathcal{O}$ un ouvert relativement compact de $\mathbb{R}^n - \bar\omega$; il existe $r > 0$ tel que, si $V \in \mathscr{E}'(\mathbb{R}^n)$ et vérifie la condition: il existe $L > 0$ tel que l'on ait*

$$\left| \langle V, \theta \rangle \right| \le L \sup_{z \in \omega_r} |\theta(z)| \qquad \text{pour tout } \theta \text{ holomorphe sur } \mathbb{C}^n,$$

alors $V = 0$ sur $\mathcal{O}$.

Déonstration du lemme. Soit la suite de fonctions entières θ_k définies pour $k \ge 1$ et $z \in \mathbb{C}^n$ par

$$\theta_k(z) = k^n e^{-k^2 z^2} \qquad \text{avec} \quad z^2 = \sum_{j=1}^{n} z_j^2.$$

Choisissons $r > 0$ tel que

$$\inf_{\substack{z = x + iy \in \omega_r \\ a \in \mathcal{O}}} ((a - x)^2 - y^2) > 0.$$

Cela est toujours possible; alors

$$\sup_{a \in \mathcal{O}} \left| (v * \theta_k)(a) \right| = \sup_{a \in \mathcal{O}} \left| \langle V(x), \theta_k(a - x) \rangle \right|$$

$$\le L \sup_{a \in \mathcal{O}} \sup_{z \in \omega_r} k^n e^{-k^2} ((x-a)^2 - y^2)$$

et donc $V * \theta_k$ converge vers 0 uniformément sur $\mathcal{O}$, donc $V = 0$ sur $\mathcal{O}$.

Remarque 3. Lorsque les coefficients de (17) et le retard ρ sont de classe C^∞ en la variable d'évolution, nous déduisons du théorème 2 un résultat d'unicité pour des distributions, en montrant d'abord la régularité de u en la variable t.

RÉFÉRENCES

1. M. S. BAOUENDI ET C. GOULAOUIC, Cauchy problems with characteristic initial hypersurface, *Comm. Pure Appl. Math.* **26** (1973), 455–475.
2. M. S. BAOUENDI ET C. GOULAOUIC, Remarks on the abstract form of nonlinear Cauchy–Kavalewsky theorems, *Comm. Partial Differential Equations* **2**, (1977), 1151–1162.
3. T. NISHIDA, A note on the Nirenberg's theorem as an abstract form of the nonlinear Cauchy–Kovalewsky theorem in a scale of Banach spaces, *J. Differential Geom.* **12** (1977), 629–633.

MATHEMATICAL ANALYSIS AND APPLICATIONS, PART A
ADVANCES IN MATHEMATICS SUPPLEMENTARY STUDIES, VOL. 7A

A Local Constancy Principle for the Solutions of Certain Overdetermined Systems of First-Order Linear Partial Differential Equations[†]

M. S. Baouendi

Department of Mathematics
Purdue University
West Lafayette, Indiana

AND

F. Treves

Department of Mathematics
Rutgers University
New Brunswick, New Jersey

Dedicated to Laurent Schwartz on His 65th Birthday

Introduction

In the present article we look at systems of *real analytic* complex vector fields in an open subset Ω of $\mathbb{R}^{m+1}$. We deal with m such vector fields, $L_1, \ldots, L_m$. We suppose that they are linearly independent at each point, and that they satisfy the Frobenius integrability condition, in the complex sense, by which we mean that the commutation brackets $[L_j, L_k] = L_j L_k - L_k L_j$ are linear combinations of $L_1, \ldots, L_m$ at every point. Our study is purely local, concerned with the local distribution solutions of the system of homogeneous equations

$$L_j h = 0, \qquad j = 1, \ldots, m. \tag{1}$$

We are therefore willing to replace the system $L = (L_1, \ldots, L_m)$ by nonsingular linear substitutions, with analytic coefficients, of the originally given one. This, plus suitable contractions of the open set in which the study is carried out, allows us to suppose that

$$L_j = \partial/\partial t_j + \lambda_j(t, x)\partial/\partial x, \qquad j = 1, \ldots, m, \tag{2}$$

† This work was partly supported by NSF grants, #7804006 for M. S. Baouendi and #7903545 for F. Treves.

245

where $t_1, \ldots, t_m$, x are the coordinates in $\mathbb{R}^{m+1}$, and the λ_j are analytic (which always mean real analytic—we use "holomorphic" for complex analytic) and complex valued in Ω. In the form (2) the Frobenius integrability condition simply states that the L_j pairwise commute, i.e., $[L_j, L_k] = 0$ for all j, $k = 1, \ldots, m$. It allows us to integrate in complex space, after extending holomorphically the coefficients, and thus construct a solution $z(t, x)$ of (1) whose differential is nowhere zero. Two such solutions are related to one another by a local biholomorphic transformation, and therefore the sets of points (t, x) on which they are constant are the same. They can be called *the fibers of* the system $(L_1, \ldots, L_m)$ (in the neighborhood under consideration). The local constancy principle, to which the title of this article refers, states that all continuous functions h that are solutions of (1) (in the same neighborhood) are also constant on those fibers. This is essentially the content of our main theorem (Theorem 2.1); Theorem 2.2 states that the principle, in a sense, extends also to distribution solutions, though in this case, of course, local constancy must be given a "generalized" meaning. From these two basic results a number of consequences are drawn: necessary and sufficient conditions in order that the system L be analytic hypoelliptic (they are necessary in order that L be C^∞ hypoelliptic; see Theorems 2.5, 2.6); local approximation of solutions of (1) by polynomials with respect to the solution $z(t, x)$ above (Theorems 2.7, 2.8); characterization of the *admissible Cauchy data* for L, that is to say, of the traces on the line $t = 0$ of the distribution solutions of (1) (Theorem 2.9). In the last section, Section 5, we show by some examples what are the "natural limits" on this kind of results.

Since the writing of the present article the authors have been able to extend the fundamental theorem (Theorem 2.1) to all locally integrable systems of C^∞ vector fields [1]. In the general case, however, the consequences of that result are not as many nor as precise as in the present particular case. To a large extent this is due to the main effect of the fundamental theorem: that of transforming the given system of vector fields into the induced Cauchy–Riemann equations on some generally complicated subset of $\mathbb{C}^n$ (at least for the purpose of studying the solutions of the homogeneous equations). In the present situation $n = 1$, and thus the induced Cauchy–Riemann equations are the standard ones in the interior of a subset of the complex plane. Besides, the description of the latter is fairly straightforward. Much of the arguments developed in the present article are based on such a description. Not so in the general case, where they are based on the approximation of arbitrary solutions by polynomials in the "first integrals" (cf. Theorems 2.7 and 2.8 here). For these and some other reasons we think it worthwhile to present to the public the results gathered here.

1. BASIC NOTATION AND INGREDIENTS

We shall deal with a *Frobenius Lie algebra* $\mathbb{L}$ *of complex vector fields* in an open subset Ω of $\mathbb{R}^{m+1}$. This means that, in the neighborhood of each point of Ω, $\mathbb{L}$ is generated by a fixed number (the *dimension* of $\mathbb{L}$) of vector fields, and that it is closed under commutation bracket. In the present work we reason under two fundamental hypotheses: first, that all vector fields we deal with are (real) analytic; second, that the dimension of $\mathbb{L}$ is exactly equal to m (i.e., $\mathbb{L}$ has *codimension* one).

Our results will be local and thus we may assume that Ω is an open neighborhood of the origin. To simplify the statements we assume it to be connected. Possibly after some contracting of Ω, we may and shall suppose that the coordinates in $\mathbb{R}^{m+1}$, which we denote by $t_1, \ldots, t_m$, x, have been so chosen that our Lie algebra $\mathbb{L}$ is generated, over the ring of analytic functions in Ω, by the vector fields

$$L_j = \partial/\partial t_j + \lambda_j(t,x)\partial/\partial x, \qquad j = 1, \ldots, m, \tag{1.1}$$

with the λ_j analytic in Ω, and complex-valued. The fact that $\mathbb{L}$ is a Lie algebra for the commutation bracket is expressed by the commutation relations

$$[L_j, L_k] = 0, \qquad j, k = 1, \ldots, m, \tag{1.2}$$

which express that $\mathbb{L}$ is integrable in complex space (after we have extended the coefficients λ_j as holomorphic functions). By integrating along the complex characteristics of $\mathbb{L}$ we can construct the (unique) holomorphic solution of the Cauchy problem

$$L_j z = 0, \qquad j = 1, \ldots, m, \qquad z|_{t=0} = x, \tag{1.3}$$

in an open neighborhood $\Omega^{\mathbb{C}}$ of Ω (suitably contracted) in $\mathbb{C}^{m+1}$. The initial condition shows that $(\partial/\partial x)\mathrm{Re}\, z \neq 0$ and this allows us to introduce $\mathrm{Re}\, z(t,x)$ as a new coordinate in the neighborhood of the origin, in $\mathbb{R}^{m+1}$. In other words we may assume that, in Ω (possibly further contracted),

$$z = x + i\Phi(t,x), \tag{1.4}$$

$$\Phi \text{ real valued}, \quad \Phi(0,x) \equiv 0. \tag{1.5}$$

From (1.1) and (1.4) we derive at once

$$\lambda_j = -i(1 + i\Phi_x)^{-1}(\partial\Phi/\partial t_j), \qquad j = 1, \ldots, m. \tag{1.6}$$

Let us now introduce the following (complex) vector field:

$$L_0 = z_x^{-1}(\partial/\partial x). \tag{1.7}$$

We contend that *the commutation relations* (1.2) *continue to hold if* k *is allowed to take the value zero, i.e.,*

$$[L_j, z_x^{-1}(\partial/\partial x)] = 0, \qquad j = 1, \ldots, m. \tag{1.8}$$

Indeed, (1.8) is equivalent to

$$L_j(z_x^{-1}) - z_x^{-1}(\partial/\partial x)\lambda_j = 0,$$

that is to say, to

$$L_j z_x + [(\partial/\partial x)\lambda_j]z_x = 0,$$

which follows from (1.3) by differentiation with respect to x. Note that

$$L_0 z = 1. \tag{1.9}$$

The map from Ω to $\mathbb{R}^2$, $(t, x) \mapsto z = z(t, x)$, has rank either equal to *one* or to *two*. Its critical set is

$$\Sigma = \{(t, x) \in \Omega; d_t\Phi(t, x) = 0\}. \tag{1.10}$$

It is either the whole domain Ω, or else a proper analytic subset of it. We shall exclude the case where $\Sigma = \Omega$, for then the system $L = (L_1, \ldots, L_m)$ is simply the exterior derivative with respect to t, whose theory is fairly well known. We shall therefore make the assumption that the complement of Σ in Ω, which we shall denote by $\mathcal{O}$, is open and dense in Ω. The following statements are easy to ascertain:

Over $\mathcal{O}$, the characteristic set of $\mathbb{L}$, that is to say, the set of points
(t, x, τ, ξ), with $(t, x) \in \mathcal{O}$ and

$$\tag{1.11}$$

$$(\tau, \xi) \in \mathbb{R}^{m+1}\backslash\{0\}; \qquad \tau_j + \lambda_j(t, x)\xi = 0, \qquad j = 1, \ldots, m, \tag{1.12}$$

is empty.

In other words, $\mathbb{L}$ is *elliptic in $\mathcal{O}$*. And

$$z(\mathcal{O}) \text{ is a dense open subset of } z(\Omega). \tag{1.13}$$

Indeed, $\mathcal{O}$ is the subset of Ω in which the mapping z has maximum rank.

We select a number $r > 0$ and an open interval J in the real line, containing the origin, such that the closure of the product set

$$\mathscr{B}_r \times J, \qquad \mathscr{B}_r = \{t \in \mathbb{R}^m; |t| < r\}, \tag{1.14}$$

is contained in Ω. The critical set

$$\mathscr{N}_0 = \{x \in \bar{J}; \forall t \in \mathscr{B}_r, \Phi(t, x) = 0\} \tag{1.15}$$

consists of finitely many points $x_1, \ldots, x_\mu$. It is easy to visualize the image via the mapping z of the closure $\bar{\mathscr{B}}_r \times \bar{J}$-image which we denote by W

throughout the sequel. Indeed, we have

$$z(\bar{\mathscr{B}}_r \times \{x_j\}) = \{x_j\}, \qquad j = 1, \ldots, \mu,$$

and if $x_0 \notin \mathscr{N}_0$,

$$z(\bar{\mathscr{B}}_r \times \{x_0\}) = \{x_0\} \times I(x_0),$$

where $I(x_0)$ is a compact interval containing more than one point. Note that zero might belong to the boundary of $I(x_0)$.

By the *fibers* of z in a subset A of Ω we shall mean the preimages of points via $z|_A$, that is to say, the sets $\{(t, x) \in A; z(t, x) = z_0\}$. Because of (1.4)–(1.5) these sets are of the form

$$\{(t, x_0) \in A; \Phi(t, x_0) = y_0\}. \tag{1.16}$$

A last remark, which might illuminate a little the situation: the push forward of the vector field L_0 via the mapping z is the vector field $\partial/\partial z + z_x^{-1}\bar{z}_x \partial/\partial\bar{z}$. That of L_j is

$$(L_j z)\frac{\partial}{\partial z} + (L_j \bar{z})\frac{\partial}{\partial\bar{z}} = (L_j \bar{z})\frac{\partial}{\partial\bar{z}} = [(L_j - \bar{L}_j)\bar{z}]\frac{\partial}{\partial\bar{z}} = 2\sqrt{-1}(\operatorname{Im}\lambda_j)\bar{z}_x\frac{\partial}{\partial\bar{z}},$$

and therefore, by (1.6),

$$-2\sqrt{-1}(1 + \sqrt{-1}\Phi_x)^{-1}\frac{\partial\Phi}{\partial t_j}\frac{\partial}{\partial\bar{z}} = 2\lambda_j\frac{\partial}{\partial\bar{z}}. \tag{1.17}$$

Thus, generically speaking, the solutions of $\mathbb{L}u = 0$ will be transformed (via the mapping z) into holomorphic functions of z, and among the latter, the polynomials in z will correspond to distributions annihilated by high enough powers of L_0.

2.   STATEMENT OF THE RESULTS

As we have said, let W denote the image via z of the closure $\bar{\mathscr{B}}_r \times \bar{J}$. We call $\mathring{W}$ the interior of W.

THEOREM 2.1.   *The following two properties of a continuous function u in $\bar{\mathscr{B}}_r \times \bar{J}$ are equivalent*:

$$\mathbb{L}u = 0 \qquad \text{in} \quad \mathscr{B}_r \times J; \tag{2.1}$$

there is a continuous function $\tilde{u}$ in W, holomorphic in $\mathring{W}$, whose pullback to $\bar{\mathscr{B}}_r \times \bar{J}$, $\tilde{u} \circ z$, is equal to u. $\tag{2.2}$

By (2.1) we mean that $Mu = 0$ in $\mathcal{B}_r \times J$ whatever the vector field M belonging to $\mathbb{L}$; this is equivalent to saying that $L_j u = 0$ in $\mathcal{B}_r \times J$ for all $j = 1, \ldots, m$.

A restatement of Theorem 2.1 could go as follows:

THEOREM 2.1′. *The pullback $\tilde{u} \to \tilde{u} \circ z$ is an isometry of the Banach space $B(W)$ of the continuous functions in W holomorphic in $\mathring{W}$, onto the closed subspace of $C^0(\bar{\mathcal{B}}_r \times \bar{J})$ consisting of the functions satisfying (2.1).*

Let L_0 be the vector field in (1.7).

THEOREM 2.2. *Let u be a distribution in Ω satisfying $\mathbb{L}u = 0$ in Ω. There exist, then, an integer $N \geq 0$ and a continuous function v in $\mathcal{B}_r \times J$ such that, there,*

$$u = L_0^N v, \qquad \mathbb{L}v = 0. \tag{2.3}$$

Theorems 2.1 and 2.2 are the main results of the present work, from which all others follow (as shown in Section 4).

The next two statements embody what we call the "local constancy principle." They are immediate consequences of Theorems 2.1 and 2.2:

THEOREM 2.3. *Let $u \in C^0(\bar{\mathcal{B}}_r \times \bar{J})$ satisfying (2.1). Then u is constant on each fiber of z in $\bar{\mathcal{B}}_r \times \bar{J}$.*

THEOREM 2.4. *Let u be a distribution like the one in Theorem 2.2. The restriction of u to $z^{-1}(\mathring{W}) \cap (\mathcal{B}_r \times J)$ is an analytic function, constant on the fibers of z, in that open set.*

Theorems 2.3 and 2.4 show that the fibers of z, in a suitably small open neighborhood of the origin, are essentially independent of z: in particular they are equal, in a subneighborhood, to the fibers of another analytic solution w of $\mathbb{L}w = 0$ such that $dw \neq 0$. It is therefore more appropriate to call them *the fibers of the Lie algebra $\mathbb{L}$ near the origin.*

We look now at the Lie algebra $\mathbb{L}$ from the viewpoint of pointwise analytic hypoellipticity:

DEFINITION 2.1. We say that the Lie algebra $\mathbb{L}$ is *analytic hypoelliptic* at a point ω_0 of Ω if, given any distribution u in an open neighborhood $U \subset \Omega$ of ω_0 such that, given any vector field $M \in \mathbb{L}$, Mu is an analytic function in some open neighborhood $V \subset U$ of ω_0 (possibly depending on M), then u itself is an analytic function in some open neighborhood $U_0 \subset U$ of ω_0.

We shall say that $\mathbb{L}$ is analytic hypoelliptic in a subset A of Ω if this is so at every point of A. We shall also use, below, the obvious C^∞ analogs of these definitions. Next we introduce a new concept:

DEFINITION 2.2. We say that the Lie algebra $\mathbb{L}$ is *open* at a point ω_0 of Ω if there is an analytic function w in an open neighborhood $U \subset \Omega$ of ω_0 having the following two properties:

$$\mathbb{L}w = 0, \qquad dw \neq 0 \quad \text{at every point of } U; \tag{2.4}$$

whatever the neighborhood $V \subset U$ of ω_0, $w(V)$ is a neighborhood of $v(\omega_0)$ in $\mathbb{C}$. $\hspace{4cm}$ (2.5)

We say that $\mathbb{L}$ is open in a subset A of Ω if it is open at every point of A. Introducing Definition 2.2 is motivated by the following.

THEOREM 2.5. *In order that the Lie algebra $\mathbb{L}$ be analytic hypoelliptic at a point ω_0 (resp., in an open subset Ω') of Ω, it is necessary and sufficient that it be open at ω_0 (resp., in Ω').*

Theorem 2.5 is incorporated in a more comprehensive statement in the next theorem. We say that a C^∞ function is *flat* on a set if all its derivatives vanish on that set.

THEOREM 2.6. *The following properties are equivalent:*

(a) $\mathbb{L}$ is not analytic hypoelliptic at the origin.
(b) There is a neighborhood $U \subset \Omega$ of the origin such that $z(U)$ is not a neighborhood of the origin in $\mathbb{C}$.
(c) There is a number $r', 0 < r' < r$, such that zero is not an interior point of the image of $\mathscr{B}_{r'}$ under the map $t \to \Phi(t, 0)$.
(d) There is a C^∞ function u in an open neighborhood $V \subset \Omega$ of 0 having the following properties (in V):

$$\mathbb{L}u = 0; \tag{2.6}$$

$$u \text{ is flat on the fiber of } 0, F_0; \tag{2.7}$$

$$u \text{ is analytic and nowhere zero in } V \backslash F_0. \tag{2.8}$$

(e) There is an open neighborhood $V \subset \Omega$ of 0 such that, given any integer $k \geq 0$, there is $u \in C^k(V)$ satisfying (2.6) and not of class C^{k+1} in any sub-neighborhood of the origin.

Among other things, Theorem 2.6 shows that if $\mathbb{L}$ is C^∞ hypoelliptic at the origin, it is analytic hypoelliptic at the same point. The converse is not true: in Section 5 we shall give an example of a single vector field (i.e., a

Lie algebra $\mathbb{L}$ of dimension $m = 1$) which is analytic hypoelliptic but not C^∞ hypoelliptic at the origin. If instead of dealing with pointwise hypoellipticity, we consider hypoellipticity in open sets, such examples would not be possible, according to the results of [3]. We do not know, however, whether, when $m > 1$, analytic hypoellipticity in an open set implies C^∞ hypoellipticity in the same open set. In [1] we show that if there is *more than one* space variable x (in which case $\mathbb{L}$ is a Lie algebra of real-analytic complex vector fields on an open subset of $\mathbb{R}^{m+n}$), such examples can be constructed (the example in [1] corresponds to $m = n = 2$).

From Theorems 2.1 and 2.2 we also derive some *approximation theorems*:

Let us denote by $\mathbb{C}[z(t,x)]$ the vector space of all polynomials, with complex coefficients, with respect to $z(t,x)$. It is obvious that all these functions satisfy (2.6) in Ω.

THEOREM 2.7.   *The restrictions to* $\bar{\mathcal{B}}_r \times \bar{J}$ *of the functions belonging to* $\mathbb{C}[z(t,x)]$ *make up a dense subspace of the* (closed) *subspace of* $C^0(\bar{\mathcal{B}}_r \times \bar{J})$ *consisting of the functions that satisfy* (2.1).

THEOREM 2.8.   *The restrictions to* $\mathcal{B}_r \times J$ *of the functions belonging to* $\mathbb{C}[z(t,x)]$ *make up a dense subspace of the space of distributions in the same set that satisfy* (2.1).

Finally we discuss the existence of solutions to the Cauchy problem.

LEMMA 2.1.   *If a distribution u in* $\mathcal{B}_r \times J$ *satisfies* (2.1) *it is a* C^∞ *function of t in* $\mathcal{B}_r$ *valued in the space* $\mathscr{D}'(J)$ *of distributions of x.*

The proof of Lemma 2.1 is standard: use Eq. (2.1) to trade derivatives in x for derivatives in t. We leave the details to the reader. Because of Lemma 2.1 the following definition makes sense.

DEFINITION 2.3.   We say that a distribution u_0 in an open neighborhood of zero in $\mathbb{R}^1$, J_0, is an *admissible Cauchy datum* for $\mathbb{L}$ at the origin if there is a distribution u in an open neighborhood of the origin in Ω, such that $\mathbb{L}u = 0$ and

$$u\big|_{t=0} = u_0 \tag{2.9}$$

in an open neighborhood of zero [in which both sides of (2.9) are defined].

Although in the next statement r and J are allowed to vary, we continue to write $W = z(\bar{\mathcal{B}}_r \times \bar{J})$, $\mathring{W}$ = interior of W. Notice that $\bar{J} = W \cap \mathbb{R}$.

THEOREM 2.9.   *In order that* $u_0 \in \mathscr{D}'(J_0)$ *be an admissible Cauchy datum for* $\mathbb{L}$ *at the origin, it is necessary and sufficient that there be a number* $r > 0$, *an open interval* $J \subset J_0$ *containing zero, an integer* $N \geq 0$, *and a function* $v \in C^0(W)$ *holomorphic in* $\mathring{W}$ *such that*

$$u_0 = (d/dx)^N v(0, x) \quad in \ J. \tag{2.10}$$

The reader should not think that in order that a distribution, or even a continuous function u_0 in J_0 be an admissible Cauchy datum for $\mathbb{L}$ at the origin, it suffices that u_0 extend as a holomorphic function to $\mathring{W}$. A counter-example to this will be given in Section 5 (Example 5.3).

3. Proof of the Main Theorems

3.1. *Proof That* (2.2) *Implies* (2.1)

Suppose that (2.2) holds. Then, for any $(t, x) \in \mathcal{U} = (\mathcal{B}_r \times J) \cap z^{-1}(\mathring{W})$, $u(t, x) = \tilde{u}(z(t, x))$, and therefore obviously $\mathbb{L}u = 0$ in $\mathcal{U}$. But the complement of $\mathcal{U}$ in $\mathcal{B}_r \times J$ is contained in a proper analytic subset. Since u is a continuous function we must have (2.1).

3.2. *Proof That* (2.1) *Implies* (2.2)

Actually we shall prove Theorem 2.3, that is to say the *constancy of* u *on the fibers of* z *in* $\bar{\mathcal{B}}_r \times \bar{J}$. For then we may conclude that $u = \tilde{u} \circ z$ with $\tilde{u}$ a function on W. An elementary lemma in point set topology implies then that $\tilde{u}$ is continuous (for any convergent sequence $\{z_j\}$ in W contains a subsequence which is the image, via z, of a convergent sequence in $\bar{\mathcal{B}}_r \times \bar{J}$; on the subsequence $\tilde{u}$ necessarily converges to its value at the limit of the z_j). We shall also have to prove that $\tilde{u}$ is holomorphic in $\mathring{W}$.

We shall subdivide the proof into two parts, according to whether $m = 1$ or $m > 1$.

Observe that, regardless of the value of m, each fiber of the mapping z is contained in a hyperplane $x = x_0$. If $x_0 \in \mathcal{N}_0$, the set (1.15), there is only one fiber, $\mathcal{B}_r \times \{x_0\}$. According to (1.1) and (1.6) on this fiber we have $L_j = \partial/\partial t_j$, $j = 1, \ldots, m$, and thus (2.1) implies that $d_t u(t, x_0) = 0$ for all t in $\mathcal{B}_r$, hence u is constant on the fiber in question. Because of this, in the remainder of the proof we restrict our attention to the case $x_0 \notin \mathcal{N}_0$. For the sake of simplicity we shall assume that $x_0 = 0$.

Case $m = 1$.   We select a smooth closed curve γ in the complex t-plane, winding once around the real interval $[-r, r]$, contained and contractible in the section $\Omega_0^{\mathbb{C}} = \{t \in \mathbb{C}; (t, 0) \in \Omega^{\mathbb{C}}\}$ ($\Omega^{\mathbb{C}}$ is the open neighborhood of Ω in $\mathbb{C}^2$ in which the function $z(t, x)$ is defined and holomorphic), such further-more that $\Phi_t(t, 0)$ (Φ_t: first partial derivative with respect to t) is nowhere zero on γ. If then $\varepsilon > 0$ is small enough we shall have $\Phi_t(t, x) \neq 0$ for all t in γ and $|x| \leq \varepsilon$. The elementary symmetric functions of the zeros of $\Phi_t(\cdot, x)$ inside γ are then well defined by

$$S^k(x) = (2i\pi)^{-1} \oint_\gamma t^k \Phi_t(t, x)^{-1} \Phi_{tt}(t, x)\, dt.$$

The S^k are analytic functions of x in $[-\varepsilon, \varepsilon]$ and so, therefore, are the coefficients of the unique monic polynomial in t, $p(t, x)$, having those zeros, counted according to multiplicity, as its own roots. We have, in the rectangle $[-r, r] \times [-\varepsilon, \varepsilon]$:

$$\Phi_t(t, x) = E(t, x)p(t, x), \qquad (3.1)$$

with E and p both real valued, and E nowhere zero. (We have just recalled the "semiglobal" version of the Weierstrass preparation theorem, and one of its standard proofs.)

Possibly after decreasing $\varepsilon > 0$ we may assume that

the polynomial in t, $p(t, x)$, has exactly N_+ (resp., N_-) real roots if $0 < x \leq \varepsilon$ (resp., $-\varepsilon \leq x < 0$). $\qquad (3.2)$

We order the real roots of *odd* multiplicity of $p(t, x)$ in increasing order:

$$\rho_j(x) < \rho_{j+1}(x), \qquad j = 1, \ldots, N'_+ - 1, \qquad 0 < x \leq \varepsilon,$$

and similarly for $-\varepsilon \leq x < 0$, with N'_- in the place of N'_+ (the primes indicate that we disregard the roots with even multiplicity). With the agreement that $\rho_0(x) \equiv -r$ and $\rho_{N'_+ + 1}(x) \equiv +r$, set

$$A_j = \{(t, x): 0 < x \leq \varepsilon, \rho_j(x) \leq t \leq \rho_{j+1}(x)\}.$$

If $0 < x_0 \leq \varepsilon$, $t \mapsto \Phi(t, x_0)$ is a strictly monotone map of the interval $[\rho_j(x_0), \rho_{j+1}(x_0)]$ into $\mathbb{R}$, and z is a homeomorphism of A_j onto

$$B_j = \{x + iy \in \mathbb{C}; 0 < x \leq \varepsilon, \Phi(\rho_j(x), x) \leq y \leq \Phi(\rho_{j+1}(x), x)\},$$

if we assume, for instance, that Φ is *increasing* in the above interval. Of course, in the open subset of A_j where $\Phi_t \neq 0$, z is a diffeomorphism. The open subset in question is the complement of a finite number of analytic curves $t = \varphi(x)$. Let us denote by $L = \partial/\partial t + \lambda(t, x)(\partial/\partial x)$ the single generator of the Lie algebra $\mathbb{L}$ in the present case. Its push forward, via the mapping z, is equal, according to (1.17), to

$$2\lambda(\partial/\partial \bar{z}). \qquad (3.3)$$

Set then

$$\tilde{u}_j = u \circ z^{-1} \quad \text{in } B_j.$$

By virtue of (1.6) and (3.3) we derive that $\tilde{u}_j$ is holomorphic in the complement, in B_j, of a finite number of analytic curves $y = \Phi(\varphi(x), x)$, hence in the whole interior, $\overset{\circ}{B}_j$, of B_j (since $\tilde{u}_j$ is continuous).

The next remark is that, since Φ_t changes sign when (t, x) crosses the curve $t = \rho_{j+1}(x)$, the curve which we denote momentarily by Γ_j is a piece of the boundary of both B_j and B_{j+1}, and also of $B_j \cap B_{j+1}$. Not only is the interior of $B_j \cap B_{j+1}$ not empty but the intersection of the interior of $B_j \cap B_{j+1}$ with each vertical line $x = x_0$, $0 < x_0 < \varepsilon$, is a nonempty open interval. Since $\tilde{u}_j = \tilde{u}_{j+1}$ on Γ_j we have $\tilde{u}_j = \tilde{u}_{j+1}$ in the whole interior of $B_j \cap B_{j+1}$, therefore in the whole set $B_j \cap B_{j+1}$. But this means that there is a continuous function $\tilde{u}_{j, j+1}$ in $B_j \cup B_{j+1}$ whose restriction to B_j (resp., to B_{j+1}) is equal to $\tilde{u}_j$ (resp., to $\tilde{u}_{j+1}$). Such a function $\tilde{u}_{j, j+1}$ is holomorphic in the interior of $B_j \cup B_{j+1}$. By repeating this kind of reasoning we reach the conclusion that there is a unique continuous function $\tilde{u}$ in the set

$$\bigcup_{j=0}^{N'_+} (B_j \cup B_{j+1}) = z([-r, r] \times \,]0, \varepsilon]), \tag{3.4}$$

whose restriction to $B_j \cup B_{j+1}$ is equal to $\tilde{u}_{j, j+1}$—for each j. By the argument used often—that a continuous function that satisfies the Cauchy–Riemann equations in the complement of a proper analytic subset (here, in fact, a union of finitely many analytic curves) satisfies those equations everywhere—we conclude that $\tilde{u}$ is holomorphic in the interior of (3.4). We may repeat the same argument when $0 > x \geq -\varepsilon$.

We have thus obtained a continuous function $\tilde{u}$ in

$$z([-r, r] \times [-\varepsilon, \varepsilon]) \backslash z([-r, r] \times \{0\}), \tag{3.5}$$

holomorphic in the interior of (3.5) and such that

$$u(t, x) = \tilde{u}(z(t, x)), \qquad |t| \leq r, \quad |x| \leq \varepsilon, \quad x \neq 0. \tag{3.6}$$

Finally consider the set of points t where $\Phi_t(t, 0) \neq 0$: z induces a diffeomorphism of an open neighborhood of $(t, 0)$ onto an open neighborhood of $\sqrt{-1}y$, with $y = \Phi(t, 0)$. The push forward of u is a holomorphic function of z in such a neighborhood, which is of course equal to $\tilde{u}$ when $\mathrm{Re}\, z \neq 0$. This shows that $\tilde{u}$ extends as a holomorphic function in the complement of a finite number of points of the form $\sqrt{-1}y_j$ $(j = 1, \ldots, v)$. But we also know that $\tilde{u}$ is bounded where it is defined [because of (3.6)], and therefore extends now as a holomorphic function in the whole interior of $z([-r, r] \times [-\varepsilon, \varepsilon])$.

The last step is to prove that $u(t, 0)$ is constant on the fibers of z in $[-r, r] \times \{0\}$. We know that this is true if the value of z on the fiber lies in the interior of the interval $z([-r,r] \times \{0\})$. Let therefore $\sqrt{-1}y_0$ be one of the two end-points of that interval, and $t_1 < t_2$ two points in $[-r,r]$ such that $\Phi(t_j,0) = y_0, j = 1, 2$. Assume also, as one may, that there is no other point, in the segment $[t_1,t_2]$, where $\Phi(t,0)$ takes the value y_0. This implies that we can find $t'_j, j = 1, 2$, with

$$t_1 < t'_1, \qquad t'_2 < t_2,$$

such that $t \mapsto \Phi(t,0)$ is a homeomorphism of $[t_1,t'_1]$, and also of $[t'_2,t_2]$, onto one and the same segment $[y_0, y_*]$, such furthermore that $\Phi_t(t,0) \neq 0$ except possibly at the points $t_j, j = 1, 2$. But then, since $u = \tilde{u} \circ z$ in the set

$$(]t_1,t'_1] \cap [t'_2,t_2[) \times \{0\}, \tag{3.7}$$

we see that u is constant on each fiber of z in (3.7). This implies at once, by continuity, that $u(t_1,0) = u(t_2,0)$. ∎

Case m > 1. Let $z_0 = x_0 + iy_0 \in W$ and $t^{(j)} \in \bar{\mathscr{B}}_r$ be two points such that $z(t^{(j)},x_0) = z_0, j = 1, 2$. Let $\gamma: s \mapsto t(s)$ be an analytic curve, from $[-1,1]$ to $\bar{\mathscr{B}}_r$, such that $t(-1) = t^{(1)}, t(+1) = t^{(2)}, t(0) = 0$. We suppose that $s \mapsto t(s)$ is an immersion, that is to say, that $\dot{t}(s) \neq 0$ at every $s \in [-1,1]$. We denote by $u^\#$ the pullback to $[-1,1] \times J$ of the restriction of u on the immersed two-dimensional manifold $\mathfrak{M} = \gamma \times J$, via the map $(s,x) \mapsto (t(s),x)$. Note that the vector field $\sum_{j=1}^m (dt^j/ds)(s)L_j$ is tangent to $\mathfrak{M}$, and that we can pull it back to $[-1,1] \times J$; we denote by $L^\#$ its pullback. The reader will easily check [by using (1.6)] that

$$L^\# = \partial/\partial s - \sqrt{-1}[1 + \sqrt{-1}\Phi_x^\#(s,x)]^{-1} \Phi_s^\#(s,x)(\partial/\partial x), \tag{3.8}$$

where we have used the notation

$$\Phi^\#(s,x) = \Phi(t(s),x).$$

Of course we have

$$L^\# u^\# = 0 \quad \text{in }]-1,1[\times J. \tag{3.9}$$

Furthermore, $z^\# = x + \sqrt{-1}\,\Phi^\#(s,x)$ is the unique analytic solution of

$$L^\# z^\# = 0, \qquad z^\#|_{s=0} = x. \tag{3.10}$$

This reduces the study of $u^\#$ to the case where $m = 1$. Inspection of the proof in the case $m = 1$ shows that it did not require the number r to be small (nor does the statement of Theorem 2.1: the only requirement is that Φ be defined and analytic in an open neighborhood of $\bar{\mathscr{B}}_r \times \bar{J}$). We immediately conclude that $u(t^{(1)},x_0) = u(t^{(2)},x_0)$. As indicated at the beginning, this implies the existence and the continuity of $\tilde{u}$. We must now prove its holomorphy in $\mathring{W}$.

Let then $z_0 = x_0 + iy_0$ belong to $\mathring{W}$; we necessarily have $x_0 \notin \mathcal{N}_0$. The map $t \mapsto \Phi(t,x_0)$ maps $\bar{\mathscr{B}}_r$ onto $I(x_0)$ whose interior contains y_0. Let γ be the analytic curve used above. We make now the additional hypothesis that there is $t_0 \in \gamma$ such that $\Phi(t_0, x_0) = y_0$, and that $\Phi_t(t, x_0) \neq 0$ for some (therefore for all except finitely many) t in γ. Then z_0 is an interior point of $z^\#([-1, 1] \times \bar{J})$ (see the preceding). By restricting u to $\mathfrak{M}$ as we have done above we conclude that $\tilde{u}$ is holomorphic in an open neighborhood of z_0.

The proof of Theorem 2.1 is complete.

3.3. Proof of Theorem 2.2

Let u be a distribution in Ω satisfying $\mathbb{L}u = 0$ there. Let us apply Lemma 2.1 with $r' > r$ substituted for r and the open interval $J' \supset \bar{J}$ substituted for J, with the proviso that $\mathscr{B}_{r'} \times J' \subset \Omega$. Then u can be regarded as a continuous function of t in $\mathscr{B}_{r'}$ valued in $\mathscr{D}'(J')$. Since $\bar{\mathscr{B}}_r$ is compact in $\mathscr{B}_{r'}$, its image via u is a compact subset of $\mathscr{D}'(J')$, whose image under the restriction mapping to a relatively compact open subset of J' containing $\bar{J}$ is contained in a space of distributions of finite order. This shows (by means of a finite number of integrations with respect to x) that there is an integer $N \geq 0$ and a continuous function v_1 in $\bar{\mathscr{B}}_r \times \bar{J}$, such that

$$u = (\partial/\partial x)^N v_1 \quad \text{in} \quad \mathscr{B}_r \times J. \tag{3.11}_N$$

This implies that there is $v_2 \in C^0(\bar{\mathscr{B}}_r \times \bar{J})$ such that

$$u = L_0^N v_2 \quad \text{in} \quad \mathscr{B}_r \times J. \tag{3.12}_N$$

We prove this implication by induction on N, taking into account that it is trivial for $N = 0$. Suppose we have proved the implication up to N inclusively, and that $(3.11)_{N+1}$ is valid. The transposed Leibniz formula yields:

$$z_x(\partial/\partial x)^{N+1} v_1 = \sum_{j=0}^{N+1} (-1)^j \binom{N+1}{j} (\partial/\partial x)^{N+1-j} [v_1 (\partial/\partial x)^{j+1} z]$$

$$= (\partial/\partial x)^{N+1} v_1^\#, \tag{3.13}$$

with

$$v_1^\# = \sum_{j=0}^{N+1} (-1)^j \binom{N+1}{j} (\partial/\partial x)^{-j} [v_1 (\partial/\partial x)^{j+1} z],$$

where $(\partial/\partial x)^{-1} f$ is the antiderivative of f which vanishes at the origin. Thus $v_1^\#$ belongs to $C^0(\bar{\mathscr{B}}_r \times \bar{J})$. By the induction on N the right-hand side in (3.13) can be written $(\partial/\partial x) L_0^N v_2$, whence $(3.12)_{N+1}$ after division of both sides in (3.13) by z_x.

Since L_0 and the L_j $(j \geq 1)$ commute [(1.8)] we derive from (3.12) that

$$L_0^N(L_j v_2) = 0, \quad j = 1, \ldots, m, \tag{3.14}$$

in $\mathscr{B}_r \times J$, and therefore

$$L_j v_2 = \sum_{v=0}^{N-1} c_{j,v}(t) z(t,x)^v, \qquad c_{j,v} \in \mathscr{D}'(\mathscr{B}_r). \tag{3.15}$$

By using the fact that $[L_j, L_k] = 0$ and $L_j z = 0$ for all j, k, we get

$$(\partial/\partial t_k) c_{j,v} = (\partial/\partial t_j) c_{k,v}, \qquad j, k = 1, \ldots, m. \tag{3.16}$$

This means that there are distributions $C_v (v = 0, \ldots, N-1)$ in $\mathscr{B}_r$ such that, for each j and v,

$$c_{v,j} = (\partial/\partial t_j) C_v. \tag{3.17}$$

If we combine this with (3.15) we get that

$$L_j \left(v_2 - \sum_{v=0}^{N-1} C_v z^v \right) = 0, \qquad j = 1, \ldots, m. \tag{3.18}$$

We apply once again Lemma 2.1:

$$v = v_2 - \sum_{v=0}^{N-1} C_v z^v \tag{3.19}$$

is a C^∞-function of t in $\mathscr{B}_r$ valued in $\mathscr{D}'(J)$. Since $L_0 z = 1$ we derive from (3.19) that

$$C_k = L_0^k (v_2 - v) - \sum_{v=k+1}^{N-1} \frac{v!}{(v-k)!} C_v z^{v-k} \tag{3.20}$$

for $k = N - 1$ [in which case there is no sum at the right in (3.20)], $N - 2, \ldots, 0$. By induction we derive that C_k is a continuous function of t in $\mathscr{B}_r$—with values in $\mathscr{D}'(J)$, but C_k being independent of x, we conclude that $C_k \in C^0(J)$. Thus v, given by (3.19), belongs to $C^0(\mathscr{B}_r \times J)$. By (3.18) it verifies $\mathbb{L} v = 0$ in $\mathscr{B}_r \times J$, and of course $L_0^N v = L_0^N v_2 = u$.

The proof of Theorem 2.2 is complete.

4. PROOF OF THEOREMS 2.5–2.9

Theorem 2.5 follows at once from the equivalence of (a) and (b) in Theorem 2.6. We begin therefore by the proof of the latter:

4.1. *Proof of Theorem 2.6*

The implications (d) → (a) and (e) → (a) are obvious. Let us prove right away that (c) → (d) & (e). The image of $\mathscr{B}_{r'}$ via $\Phi(\cdot, 0)$ is an interval containing zero. By (c) zero must be one of its endpoints, from which it follows that, if J is an open interval as before (such that $\bar{\mathscr{B}}_r \times \bar{J} \subset \Omega$), the image

$z(\mathscr{B}_{r'} \times J)$ does not intersect one, at least, of the two open half imaginary axes $\operatorname{Re} z = 0$, $\operatorname{Im} z > 0$, or $\operatorname{Re} z = 0$, $\operatorname{Im} z < 0$, and therefore, given any real number $\alpha \geq 0$, a branch of the α-root, z^{α}, defines, by restriction, a continuous function on that image. If we take $V = \mathscr{B}_{r'} \times J$ and $u(t, x) = z(t, x)^{\alpha}$, condition (e) will be satisfied for a suitable choice of α, say $\alpha = k + \frac{1}{2}$. If, instead, we take $u(t, x) = \exp[-z(t, x)^{-1/4}]$, where now $z^{-1/4}$ is the branch that is positive on the positive real numbers, (d) is satisfied.

Suppose now that (c) does not hold. Given any r', $0 < r' < r$, there are two points $t^{(1)}$, $t^{(2)}$ in $\mathscr{B}_{r'}$ such that

$$\Phi(t^{(1)}, x) < 0 < \Phi(t^{(2)}, x) \tag{4.1}$$

when $x = 0$, therefore also for $x \in [-\varepsilon, \varepsilon]$ if $\varepsilon > 0$ is small enough. Since the image of $\mathscr{B}_r$ under $\Phi(\cdot, x)$ is an interval, by (4.1) zero must be an interior point of that interval whatever x, $|x| \leq \varepsilon$. This means that (b) cannot be true. Next we show that (a) cannot be true.

Let u be a distribution in an open neighborhood U of the origin such that $L_j u = f_j$ is an analytic function in U for each $j = 1, \ldots, m$. Possibly after contracting U we can define, for (t, x) in U,

$$v(t, x) = \int_0^t \sum_{j=1}^m f_j(t', w(t, x; t')) \, dt', \tag{4.2}$$

where the integration is performed, say, on the straight line segment joining 0 to t, and where $w(t, x; t')$ is the unique analytic solution of the Cauchy problem:

$$L_j w = 0, \qquad j = 1, \ldots, m, \qquad w|_{t = t'} = x. \tag{4.3}$$

We have $L_j v = f_j$ for every j, and therefore $L(u - v) = 0$ in U. Since v is analytic, it will be enough to show that $u - v$ also is analytic. In other words we may restrict our attention to those distributions u such that $Lu = 0$ in a neighborhood of the origin. But then we have the right to use the representation (2.3) for such a distribution u, and since in (2.3) v is continuous, and satisfies $Lv = 0$, we may apply Theorem 2.1, for a suitable choice of r and J. Since (b) is not true the interior $\mathring{W}$ of W contains the origin; its preimage is an open neighborhood V of the origin in (t, x)-space, in which $v = \tilde{v} \circ z$ is analytic (since $\tilde{v}$ is holomorphic in $\mathring{W}$), and so must u be.

The proof of Theorem 2.6 is complete.

4.2. Proof of Theorems 2.7 and 2.8

Theorem 2.7 follows at once from Theorem 2.1' and from the results of Mergelyan [2]. Indeed, observe that the intersection of W with each vertical line $x = x_0$ is an interval, and therefore W is a Runge compact subset of $\mathbb{C}$.

In order to prove Theorem 2.8 we combine Theorems 2.7 and 2.2. We apply the latter with $\mathscr{B}_r \times J$ substituted for Ω, $\mathscr{B}_{r'} \times J'$ for $\mathscr{B}_r \times J$, with $0 < r' < r$, $J' \subset\subset J$. Suppose that we have the representation (2.3) in $\mathscr{B}_{r'} \times J'$. Let r'' be a number such that $0 < r'' < r'$, and let J'' be an interval whose closure is contained in J'. Then v is the uniform limit over $\mathscr{B}_{r''} \times J''$, of a sequence of polynomials with respect to $z(t, x)$—according to Theorem 2.7. By the diagonal process we derive that u is the limit in $\mathscr{D}'(\mathscr{B}_r \times J)$ of such a sequence.

Theorem 2.9 follows at once from Theorems 2.1 and 2.2, if we observe that, when $t = 0$, $L_0 = \partial/\partial x$.

5. Some Examples

We begin by discussing an example of a vector field L which is analytic hypoelliptic at the origin, but is not analytic hypoelliptic in any neighborhood of the origin:

Example 5.1.   Consider the function in $\mathbb{R}^2$,

$$z(t, x) = x + \sqrt{-1}\,\Phi(t, x), \qquad \Phi(t, x) = t(t - x)^2. \tag{5.1}$$

Here $\mathbb{L}$ will be the Lie algebra of real analytic, complex vector fields in $\mathbb{R}^2$, M, such that $Mz = 0$. Thus $m = 1$, and $\mathbb{L}$ is generated by the single vector field

$$L = \partial/\partial t + \lambda(t, x)(\partial/\partial x), \tag{5.2}$$

$$\lambda = -\sqrt{-1}\,[(t - x)(3t - x)]/[1 - 2\sqrt{-1}\,t(t - x)]. \tag{5.3}$$

Observe that $\Phi(t, 0) = t^3$. It follows from Theorem 2.6 that L is analytic hypoelliptic at the origin. But, on the other hand, given any number $\theta > 0$, we have $\Phi(t, \theta) \geq 0$ for all $t \geq 0$, and $\Phi(0, \theta) = 0$. By applying once again Theorem 2.6 we conclude that L is not analytic hypoelliptic at the point (θ, θ). Same conclusion if $\theta < 0$.

In Section 2 we have promised an example of a vector field which is analytic hypoelliptic, but not C^∞ hypoelliptic at some point. This is again provided by the vector field (5.2):

Example 5.2.   Let L denote the vector field (5.2) in $\mathbb{R}^2$. Select arbitrarily a sequence of numbers $\theta_j > 0$ converging to zero, and an integer $N \geq 0$. Take $\alpha = N + \frac{1}{2}$ and denote by z^α the branch that is positive on positive real numbers. For each $j = 1, 2, \ldots$, select a function $g_j \in C_c^\infty(]0, +\infty[\times]0, +\infty[)$

equal to one in a neighborhood of (θ_j, θ_j) and whose support does not contain any point $(\theta_{j'}, \theta_{j'})$ for $j' \neq j$. The continuous function

$$u_j(t, x) = g_j(t, x)[z(t, x) - \theta_j]^\alpha \tag{5.4}$$

satisfies $Lu_j \in C_c^\infty(\mathbb{R}^2)$. We can find a sequence of numbers $c_j > 0$ such that $\sum_j c_j u_j$ converges uniformly, therefore to a compactly supported continuous function u, furthermore such that $\sum_j c_j Lu_j$ converges in $C_c^\infty(\mathbb{R}^2)$. But the singular support of u is equal to the convergent sequence of points (θ_j, θ_j) to which one must adjoin its limit point $(0, 0)$.

Last we take up the question of the admissible Cauchy data (see the end of Section 2):

EXAMPLE 5.3. Again we consider a single vector field in $\mathbb{R}^2$, now the following one:

$$L = \partial/\partial t - \sqrt{-1}x/(1 + \sqrt{-1}t)\partial/\partial x. \tag{5.5}$$

Here the solution of (1.3) is given by

$$z = x(1 + \sqrt{-1}t), \tag{5.6}$$

and W will be the closed sector in $\mathbb{C}$ defined by

$$\operatorname{Re} z \in \bar{J}, \qquad |\operatorname{Im} z| \leq r|\operatorname{Re} z|. \tag{5.7}$$

Let us then take

$$u_0(x) = xe^{i/x}. \tag{5.8}$$

It is a continuous function on $\mathbb{R}$, and extends as a holomorphic function to $\mathbb{C}\backslash\{0\}$, in particular to $\mathring{W}$.

We can define the function

$$u(t, x) = x(1 + it)e^{i/(x + ixt)} \tag{5.9}$$

for $x \neq 0$; it is a solution, there, of $Lu = 0$. If we fix $t > 0$ arbitrarily and let x tend to $+0$ (5.9) "blows up" exponentially, and therefore does not extend as a distribution in $\mathcal{B}_r \times J$. But any distribution h there that would solve $Lh = 0$ and be such that $h(0, x) = u_0(x)$ for all x in J would be such an extension (by Holmgren's uniqueness theorem, or by Theorems 2.1 and 2.2). Consequently such distributions h cannot exist, and u_0 cannot be an admissible Cauchy datum for L at the origin.

Remark 5.1. The Dirac measure at the origin in $\mathbb{R}^1$, $\delta(x)$, is an admissible Cauchy datum for the vector field (5.5). Indeed, if we set

$$h(t, x) = (1 + it)^{-1} \otimes \delta(x), \tag{5.10}$$

we have $Lh = 0$ in $\mathbb{R}^2$, and $h(0, x) = \delta(x)$. In fact, set

$$\tilde{v}(z) = z \quad for \quad \mathrm{Re}\, z \geq 0, \qquad \tilde{v}(z) = 0 \quad for \quad \mathrm{Re}\, z < 0. \qquad (5.11)$$

The restriction of $\tilde{v}$ to W, which, we recall, is given by (5.7), is continuous, and $\tilde{v}$ is holomorphic in $\mathring{W}$. We have $\delta(x) = (\partial/\partial x)^2 \tilde{v}(x)$, and

$$h = L_0^2 v, \qquad (5.12)$$

where L_0 is defined in (1.7) and $v(t, x) = \tilde{v}(z(t, x))$. This is an illustration of Theorems 2.2 and 2.9.

REFERENCES

1. M. S. BAOUENDI AND F. TREVES, A property of the functions and distributions anihilated by a locally integrable system of complex vector fields, *Ann. of Math.*, to appear.
2. S. N. MERGELYAN, Uniform approximation of functions of a complex variable, *Usp. Mat. Nauk* 7, No. 2 (48) (1952), 31–122; also *Amer. Math. Soc. Transl.* **101**.
3. F. TREVES, Analytic-hypoelliptic PDE's of principal type, *Comm. Pure Appl. Math.* **24** (1971), 537–570.

MATHEMATICAL ANALYSIS AND APPLICATIONS, PART A
ADVANCES IN MATHEMATICS SUPPLEMENTARY STUDIES, VOL. 7A

A Note on Isolated Singularities for Linear Elliptic Equations

HAIM BRÉZIS AND PIERRE-LOUIS LIONS

Département de Mathématiques
Université de Paris VI
Paris, France

DEDICATED TO LAURENT SCHWARTZ

Set $B_R = \{x \in \mathbb{R}^N; |x| < R\}$ with $N \geq 2$. Our main result is the following:

THEOREM 1.   *Let $u \in L^1_{\text{loc}}(B_R \backslash \{0\})$ be such that*

$$\Delta u \in L^1_{\text{loc}}(B_R \backslash \{0\}) \quad \text{in the sense of distributions on } B_R \backslash \{0\}, \tag{1}$$

$$u \geq 0 \qquad \qquad \text{a.e. in } B_R, \tag{2}$$

$$\Delta u(x) \leq au(x) + f(x) \quad \text{a.e. in } B_R, \tag{3}$$

where a is a positive constant and $f \in L^1_{\text{loc}}(B_R)$. Then $u \in L^1_{\text{loc}}(B_R)$ and there exists $\varphi \in L^1_{\text{loc}}(B_R)$ and $\alpha \geq 0$ such that

$$-\Delta u = \varphi + \alpha \delta_0 \quad \text{in } \mathcal{D}'(B_R), \tag{4}$$

where δ_0 is the Dirac mass at the origin. In particular, it follows that $u \in M^p_{\text{loc}}(B_R)^\dagger$ where $p = N/(N-2)$ when $N > 2$ and $p < \infty$ is arbitrary when $N = 2$.

Remarks. (1)   Theorem 1 turns out to be useful in the study of isolated singularities in semilinear elliptic equations. For example, it implies the following fact—which extends a result of [1]. Assume f is a continuous function from $\mathbb{R}_+$ into $\mathbb{R}$ such that $\liminf_{t \to +\infty} f(t)/t > -\infty$; assume $u \in L^1_{\text{loc}}(B_R \backslash \{0\})$, $f(u) \in L^1_{\text{loc}}(B_R \backslash \{0\})$, $u \geq 0$ a.e. and $-\Delta u = f(u)$ in $\mathcal{D}'(B_R \backslash \{0\})$. Then $f(u) \in L^1_{\text{loc}}(B_R)$, and $-\Delta u = f(u) + \alpha \delta_0$ in $\mathcal{D}'(B_R)$ for some $\alpha \geq 0$.
(2)   Suppose $N \geq 3$; if instead of (2), we assume that

$$u \geq g \quad \text{a.e. in } B_R \qquad \text{with} \quad g \in M^{N/N-2}_{\text{loc}}(B_R). \tag{2'}$$

---

† M^p denotes the weak L^p (or Marcinkiewicz) space for $1 < p < \infty$.

Then (4) holds with some $\alpha \in \mathbb{R}$. If in addition $g \in L^{N/(N-2)}_{\text{loc}}(B_R)$, then $\alpha \geq 0$. The proof is an easy modification of the proof of Theorem 1, and we shall omit it.

The proof of Theorem 1 is divided into three steps.

Step 1.   Let $\bar{u}(r)$ be the average of u over the sphere $|x| = r$, i.e.,

$$u(r) = \frac{1}{s_N r^{N-1}} \int_{|x|=r} u(x)\, ds,$$

where s_N denotes the measure of the unit sphere in $\mathbb{R}^N$. It follows from (1), (2), and (3) that

$$\frac{1}{r^{N-1}} (r^{N-1}\bar{u}_r)_r \in L^1_{\text{loc}}(0, R), \qquad \bar{u} \geq 0,$$

and

$$\frac{1}{r^{N-1}} (r^{N-1}\bar{u}_r)_r \leq a\bar{u} + \bar{f} \quad \text{on } (0, R). \tag{5}$$

In particular $\bar{u} \in C^1(0, R)$; let $R' < R$ be fixed. Integrating (5) over (r, R') we find

$$-r^{N-1}\bar{u}_r(r) \leq a\psi(r) + C \qquad \text{for} \quad r \in (0, R'), \tag{6}$$

where $\psi(r) = \int_r^R s^{N-1}\bar{u}(s)\, ds$ is a nonincreasing function and C is a constant.

Dividing (6) through by r^{N-1} and integrating we obtain

$$\bar{u}(r) \leq a \int_r^{R'} \frac{\psi(s)}{s^{N-1}}\, ds + \frac{C}{r^{N-2}} + C \tag{7}$$

(when $N = 2$, C/r^{N-2} must be replaced by $C|\log r|$). In particular we have

$$-r^{N-1}\bar{u}(r) \leq a \int_r^{R'} \psi(s)\, ds \leq aR'\psi(r),$$

and integrating again we see that

$$\psi(r) \leq a(R')^2\psi(r) + C \qquad \text{for} \quad 0 < r < R'.$$

Choosing R' small enough we obtain that $\psi(r)$ remains bounded as $r \to 0$ and consequently $u \in L^1_{\text{loc}}(B_R)$. Going back to (7) we also find that

$$\bar{u}(r) \leq \frac{C}{r^{N-2}} + C \qquad \text{if} \quad N \geq 3,$$

$$\bar{u}(r) \leq C|\log r| + C \qquad \text{if} \quad N = 2. \tag{8}$$

Step 2.   We define a measurable function φ by

$$\varphi(x) = -\Delta u(x) \quad \text{a.e. in } B_R.$$

By (1) we know that $\varphi \in L^1_{\text{loc}}(B_R\backslash\{0\})$; we shall now prove that in fact $\varphi \in L^1_{\text{loc}}(B_R)$. Consider functions $\zeta_\varepsilon(x)$ satisfying the following properties:

$$\zeta_\varepsilon \in C^\infty(B_R), \qquad 0 \leq \zeta_\varepsilon \leq 1 \quad \text{on } B_R, \tag{9}$$

$$\zeta_\varepsilon(x) \to 1 \quad \text{as } \varepsilon \to 0 \quad \text{for every} \quad x \neq 0$$

and

$$\nabla\zeta_\varepsilon \to 0 \quad \text{as } \varepsilon \to 0, \quad \text{uniformly on compact subsets of } B_R\backslash\{0\}, \tag{10}$$

$$\zeta_\varepsilon(x) = 0 \qquad \text{for} \quad |x| < \varepsilon, \tag{11}$$

$$\Delta\zeta_\varepsilon \geq 0 \quad \text{on } B_R. \tag{12}$$

To see that such functions exist it suffices to choose ζ_ε of the form

$$\zeta_\varepsilon(x) = \begin{cases} \Phi((\varepsilon/|x|)^{N-2}) & \text{when} \quad N \geq 3, \\ \Phi(\log|x|/\log\varepsilon) & \text{when} \quad N = 2, \end{cases}$$

where Φ is a C^∞ convex function on $[0, \infty)$ such that $\Phi(0) = 1$, $\Phi(t) = 0$ for $t \geq 1$. Let $\eta \in \mathscr{D}(B_R)$, $0 \leq \eta \leq 1$, $\eta(x) = 1$ near $x = 0$. We have

$$\int_{B_R} \varphi\zeta_\varepsilon\eta = -\int_{B_R} u\Delta(\zeta_\varepsilon\eta)$$

$$= -\int_{B_R} u(\Delta\zeta_\varepsilon)\eta - 2\int_{B_R} u\nabla\zeta_\varepsilon\nabla\eta - \int_{B_R} u\zeta_\varepsilon\Delta\eta$$

$$\leq -2\int_{B_R} u\nabla\zeta_\varepsilon\nabla\eta - \int_{B_R} u\zeta_\varepsilon\Delta\eta \quad \text{by (2) and (12).}$$

From Fatou's lemma—which is valid since $\varphi \geq -au - f \in L^1_{\text{loc}}(B_R)$—we deduce as $\varepsilon \to 0$ that $\varphi \in L^1_{\text{loc}}(B_R)$ and in addition

$$\int_{B_R} \varphi\eta \leq -\int_{B_R} u\Delta\eta. \tag{13}$$

Step 3.   Consider now the distribution

$$T = -\Delta u - \varphi \in \mathscr{D}'(B_R).$$

The support of T is contained in $\{0\}$; thus by a classical result about distributions (see [2, Theorem XXXV]) we know that

$$T = \sum_{|p| \leq m} c_p D^p \delta_0. \tag{14}$$

We claim that $c_p = 0$ when $|p| \geq 1$. Indeed let $\zeta \in \mathscr{D}(B_R)$ be any fixed function such that

$$(-1)^{|p|}(D^p\zeta)(0) = c_p \qquad \text{for every} \quad |p| \leq m.$$

Multiplying (14) by $\zeta_\varepsilon(x) = \zeta(x/\varepsilon)$ we obtain

$$-\int_{B_R} u\,\Delta\zeta_\varepsilon = \int_{B_R} \varphi\zeta_\varepsilon + \sum_{|p|\le m} \frac{c_p^2}{\varepsilon^{|p|}}. \tag{15}$$

On the other hand we have

$$\int_{B_R} u\,\Delta\zeta_\varepsilon = \frac{1}{\varepsilon^2} \int_{|x|<R\varepsilon} u(x)(\Delta\zeta)\left(\frac{x}{\varepsilon}\right)$$

and therefore

$$\left|\int_{B_R} u\,\Delta\zeta_\varepsilon\right| \le \frac{C}{\varepsilon^2} \int_{|x|<R\varepsilon} u = \frac{C}{\varepsilon^2} \int_0^{R\varepsilon} \bar{u}(r)r^{N-1}\,dr.$$

We deduce from (8) that

$$\left|\int_{B_R} u\,\Delta\zeta_\varepsilon\right| \le C \qquad \text{when} \quad N \ge 3,$$

$$\left|\int_{B_R} u\,\Delta\zeta_\varepsilon\right| \le C|\log\varepsilon| + C \qquad \text{when} \quad N = 2. \tag{16}$$

Comparing (15) and (16) we conclude that $c_p = 0$ when $|p| \ge 1$. In conclusion we have proved that

$$-\Delta u = \varphi + c_0\delta_0 \quad \text{in } \mathscr{D}'(B_R). \tag{17}$$

Finally we choose $\eta \in \mathscr{D}(B_R)$, $0 \le \eta \le 1$, and $\eta(x) = 1$ near $x = 0$. By (13) we have

$$\int_{B_R} \varphi\eta \le -\int_{B_R} u\,\Delta\eta$$

while by (17) we have

$$\int_{B_R} \varphi\eta = -\int_{B_R} u\,\Delta\eta - c_0.$$

It follows that $c_0 \ge 0$.

REFERENCES

1. P. L. LIONS, Isolated singularities in semilinear problems, *J. Differential Equations* (1981), to appear.
2. L. SCHWARTZ, "Theorie des distributions," Hermann, Paris, 1966.

Une étude des covariances mesurables

Pierre Cartier

I.H.E.S.
Bures sur Yvette, France

À Laurent Schwartz,
À Qui Je Dois L'espoir Têtu
D'un Monde Meilleur, En
Fidèle Amitié

A stochastic process (X_t) admits of a *Karhunen representation* $X_t = \sum_{n=1}^{\infty} f_n(t) \cdot Z_n$ with measurable functions f_n and an orthogonal sequence of random variables Z_n if and only if it possesses a measurable version. Its covariance function is defined by $C(t, t') = E[X_t X_{t'}]$ and can be expanded into a series $C(t, t') = \sum_{n=1}^{\infty} f_n(t) \cdot f_n(t')$. In this chapter, we study in depth the class of functions of two variables which occur as such covariance functions. We characterize them by various inequalities, both pointwise and integral inequalities. We give a generalization of Mercer's theorem to integral equations with measurable kernels. We study the Hilbert spaces with measurable self-reproducing kernels. Special attention is to be paid to the occurrence of nonseparable Hilbert spaces. Null sets have to be controlled in a very precise way, the main technical difficulty lying in the fact that the diagonal of a square is a null set.

Contents

Introduction

On ne saurait surestimer l'importance des matrices hermitiennes, et en particulier de celles dont les valeurs propres sont positives. Dans l'étude des équations intégrales, les noyaux symétriques jouent un rôle prédominant,

267

comme l'a montré Schmidt [14] vers 1900. Soit donc K une fonction continue de deux variables réelles, définie dans le produit $I \times I$ où I est un intervalle compact de $\mathbf{R}$, et vérifiant la propriété de symétrie hermitienne $K(y, x) = \overline{K(x, y)}$. Le problème aux valeurs propres correspondant se formule par l'équation

$$\int_I K(x, y) f(y) \, dy = \lambda f(x), \qquad x \in I. \tag{1}$$

Il est assez remarquable que le fait que les valeurs propres soient positives se traduise par une condition ponctuelle: la matrice hermitienne d'éléments $K(x_i, x_j)$ a ses valeurs propres positives quels que soient les nombres distincts $x_1, \ldots, x_n$ dans l'intervalle I.

L'étude de l'équation intégrale (1) permet aussi d'introduire des développements de la forme

$$K(x, y) = \sum_n \lambda_n f_n(x) \cdot \overline{f_n(y)}. \tag{2}$$

Un des résultats culminants de la théorie est le *théorème de Mercer*. Il affirme que lorsque l'équation intégrale (1) n'admet que des valeurs propres positives, on a un développement *uniformément convergent* de la forme (2), où les λ_n sont les valeurs propres, et les f_n les fonctions propres (elles sont continues) normalisées par $\int_I |f_n(t)|^2 \, dt = 1$.

Considérons le cas particulier où I est l'intervalle $[0, 1]$ et K est de la forme $K(x, y) = k(x - y)$, la fonction k étant continue de période 1. Les fonctions propres sont les exponentielles $e^{2\pi i n x}$ (n entier de signe quelconque), et le développement (2) n'est autre que le développement en série de Fourier

$$k(t) = \sum_n \lambda_n e^{2\pi i n t}. \tag{3}$$

Par suite, les coefficients de Fourier λ_n sont positifs si et seulement si la matrice d'éléments $k(x_i - x_j)$ est hermitienne à valeurs propres positives quels que soient les nombres réels $x_1, \ldots, x_n$. La fonction k appartient donc à la classe des *fonctions de type positif*, introduite par Bochner vers 1930.

Dans un domaine tout différent, considérons un processus aléatoire (X_t), où la variable t est réelle. Pour simplifier, supposons le processus réel et centré, c'est-à-dire qu'on a $E[X_t] = 0$ pour tout t; supposons aussi que les variables X_t aient des moments du second ordre finis. La covariance du processus est alors définie par $C(t, t') = E[X_t X_{t'}]$. Les propriétés du processus qu'on appelle "du second ordre" sont celles qui ne dépendent que de la covariance, et qui ne font qu'exprimer la géométrie de la "courbe" formée par les points X_t de l'espace de Hilbert des variables aléatoires avec un moment du second ordre fini. Ce point de vue a été exploré vers 1945 par Loéve [11].

La covariance C est une fonction symétrique à valeurs réelles, et, pour chaque système fini $(t_1, \ldots, t_n)$ de valeurs du paramètre t, la matrice symétrique d'éléments $C(t_i, t_j)$ est à valeurs propres positives. Supposons que le processus soit continu en moyenne quadratique, c'est-à-dire que $E[(X_t - X_{t_0})^2]$ tende vers 0 lorsque t tend vers t_0. La covariance C est alors continue, et par application du théorème de Mercer, se développe sous la forme

$$C(t, t') = \sum_n \lambda_n f_n(t) \cdot f_n(t').$$ (4)

Le processus lui-même se représente sous la forme

$$X_t = \sum_n \lambda_n^{1/2} f_n(t) \cdot Z_n,$$ (5)

où les variables aléatoires Z_n satisfont aux conditions

$$E[Z_n] = 0, \qquad E[Z_n^2] = 1, \qquad E[Z_m Z_n] = 0 \qquad \text{si} \quad m \neq n$$ (6)

("développement de Karhunen"). Lorsque le processus est stationnaire, et admet la période 1, on obtient un développement en série de Fourier aléatoire

$$X_t = \lambda_0^{1/2} Z_0 + 2 \sum_{n=1}^{\infty} \lambda_n^{1/2} [U_n \cdot \cos(2\pi nt) + V_n \cdot \sin(2\pi nt)].$$ (7)

Ces résultats sont le point de départ de l'*analyse harmonique des processus aléatoires*, un sujet très vaste dont le pionnier fut N. Wiener.

Les trois points de vue précédents ont été très développés et généralisés; en particulier, les fonctions de type positif sont un outil fondamental dans l'étude des groupes localement compacts et de leurs représentations (voir, e.g., le livre de Weil [18]). Ce qu'ils ont en commun, c'est la considération d'une certaine classe de fonctions de deux variables, que l'on appelle en général "noyaux de type positif". Comme la terminologie n'est pas tout à fait fixée, et que nous avons en vue les applications probabilistes, nous avons préféré le terme plus bref de "covariance" pour désigner ces fonctions.

Jusqu'à présent, on s'est essentiellement limité aux covariances *continues*. J'y vois au moins deux raisons: en théorie des groupes, les fonctions de type positif considérées sont automatiquement continues, et l'étude des processus aléatoires très discontinus est relativement récente. Cependant, l'introduction de processus aléatoires mesurables conduit à des covariances mesurables. Leur théorie est pleine de pièges; en particulier, la diagonale d'un produit est de mesure nulle dans les cas usuels, et l'on doit donc manier des expressions telles que $C(t, t)$ avec beaucoup de prudence. J'ai fait une chasse attentive aux erreurs de ce genre, mais je ne puis garantir absolument que je les ai toutes dépistées.

Nous utiliserons quatre outils fondamentaux:

(a)   *la représentation d'une covariance sous la forme* $C(t, t') = \langle \varphi(t) | \varphi(t') \rangle$, où φ est une application à valeurs dans un espace de Hilbert;

(b)   *l'espace de Hilbert $\mathscr{E}_C$ de fonctions associé a une covariance*, dont elle est le "noyau reproduisant";

(c)   *les opérateurs de Hilbert–Schmidt*;

(d)   *les covariances séparables*.

La représentation signalée en (a) est bien naturelle du point de vue des processus. La notion de "noyau reproduisant" n'est pas neuve; elle a été introduite par Aronszajn [1] et appliquée par Bergmann aux fonctions holomorphes de plusieurs variables complexes. Son utilisation systématique dans l'étude des covariances et de leurs développements en série me semble plus originale. Il est inutile de souligner l'importance des opérateurs de Hilbert–Schmidt; dans notre exposé, ils interviennent essentiellement à cause du théorème suivant (démontré au n° 20):

Si $\mathscr{H}$ est un espace de Hilbert, et u une application linéaire continue de $\mathscr{H}$ dans un espace $L^2(T, \mu)$, alors u est de Hilbert-Schmidt si et seulement s'il existe une fonction positive h dans $L^2(T, \mu)$ telle que l'on ait $|u(a)| \leq h$ pour tout vecteur a de norme ≤ 1 dans $\mathscr{H}$.

En particulier, on déduit facilement de là une version mesurable du théorème de Mercer.

Une des plaies de la théorie est l'apparition d'espaces de Hilbert non séparables. La notion de covariance séparable est destinée à les éliminer, et il est heureux que la covariance d'un processus mesurable soit automatiquement séparable. Le contre-exemple universel est fourni par les processus monstrueux (X_t) tels que $E[X_t X_{t'}] = 0$ dès que t et t' sont distincts; de tels processus ne peuvent avoir aucune version mesurable.

Nous renvoyons au sommaire le lecteur qui veut se faire une idée rapide du contenu de ce travail, et nous nous contenterons donc d'une brève description. Les parties les plus importantes et les plus nouvelles sont les parties IV à VII. Les propriétés de base des covariances mesurables et séparables sont établies dans IV, en particulier le théorème qui implique que la covariance d'un processus mesurable est mesurable et séparable. La partie V contient les résultats les plus profonds, en particulier une version mesurable du théorème de Mercer, où les ensembles de mesure nulle ont été soigneusement circonscrits, et le théorème selon lequel toute covariance mesurable est égale presque partout à une covariance mesurable et séparable. Dans les parties VI et VII, le thème est la comparaison des inégalités ponctuelles et des inégalités intégrales, avec là aussi le désir de contrôler les ensembles de mesure nulle; une variante du théorème de Dunford-Pettis joue un rôle important.

Les parties I et II sont des préliminaires faciles, où l'on met au point l'outil, et la partie III est un résumé, presque sans démonstration, des propriétés des opérateurs dans un espace de Hilbert que nous utiliserons dans la suite. Enfin, la partie VIII est une série de commentaires et de remarques sur les applications des covariances. C'est ici qu'apparaissent le mieux les limitations de la théorie, en particulier par le fait que nous ne donnons aucun exemple explicite d'une covariance non continue; l'étude des covariances est le chapitre de l'*Analyse Hilbertienne* qui s'occupe des espaces de Hilbert de fonctions où la convergence en norme implique la convergence en tout point. Pour aller plus loin, il faut introduire des espaces de fonctions beaucoup plus discontinues, tels que les espaces de Sobolev.

Dans la théorie générale, un résultat important a été omis. C'est le théorème de Grothendieck [8] selon lequel toute covariance continue C définie sur un espace compact se développe sous la forme

$$C(t, t') = \sum_n \lambda_n f_n(t) \cdot \overline{f_n(t')} \tag{8}$$

avec $|f_n(t)| \leq 1$ et $\sum_n |\lambda_n| \leq sh(\pi/2) \sup_{t,t'} |C(t, t')|$. Par différence avec les résultats précédents, on ne suppose pas les coefficients λ_n positifs. Je n'aurais guère pu que recopier un article ancien de moi-même [3], et pour les progrès récents dans cette direction, je préfère renvoyer aux travaux de Krivine [10].

CONVENTIONS

On note **R** le corps des nombres réels. Nous convenons que 0 est un nombre positif, en réservant le vocable "strictement positif" aux nombres positifs non nuls.

On note **C** le corps des nombres complexes. Si z est un nombre complexe, on note respectivement Re z, Im z et $\bar{z}$ sa partie réelle, sa partie imaginaire et son conjugué.

Dans un espace de Hilbert, on note $\langle u|v \rangle$ le produit scalaire, et $\|u\| = \langle u|u \rangle^{1/2}$ la norme. On convient que $\langle u|v \rangle$ est fonction linéaire de la *deuxième* variable v.

On note $A \backslash B$ la différence des ensembles. Si μ est une mesure, on dit qu'un ensemble A est μ-négligeable si l'on a $\mu(A) = 0$; une fonction f est dite μ-négligeable si l'ensemble $\{f \neq 0\}$ est μ-négligeable. La notation $L^p = L^p(T, \mathscr{A}, \mu)$ est la notation usuelle de la théorie de Lebesgue; en particulier, la norme dans L^p est définie par

$$\|f\|_p = \left\{ \int_T |f(t)|^p \, d\mu(t) \right\}^{1/p}$$

pour $p \geq 1$ fini. Dans l'espace de Hilbert L^2, le produit scalaire est noté $\langle f | g \rangle$ ou parfois $\langle f | g \rangle_2$; il est défini par la formule

$$\langle f | g \rangle = \int_T \overline{f(t)} g(t) \, d\mu(t).$$

Soient T_1 et T_2 deux ensembles, munis respectivement de tribus $\mathscr{A}_1$ et $\mathscr{A}_2$ de parties. On note $\mathscr{A}_1 \otimes \mathscr{A}_2$ la tribu de parties de $T_1 \times T_2$ engendrée par les ensembles de la forme $B_1 \times B_2$ avec $B_1 \in \mathscr{A}_1$ et $B_2 \in \mathscr{A}_2$. De même, si μ_i est une mesure positive sur la tribu $\mathscr{A}_i$, on note $\mu_1 \otimes \mu_2$ la mesure produit sur la tribu $\mathscr{A}_1 \otimes \mathscr{A}_2$; elle est caractérisée par la relation

$$(\mu_1 \otimes \mu_2)(B_1 \times B_2) = \mu_1(B_1) \cdot \mu_2(B_2).$$

I. PROPRIÉTÉS GÉNÉRALES DES COVARIANCES

1.   Soit $A = (a_{ij})$ une matrice à éléments complexes d'ordre n ; supposons-la *hermitienne*, c'est-à-dire qu'on ait $\overline{a}_{ij} = a_{ji}$, et associons-lui la forme hermitienne

$$q(\xi) = \sum_{i,j=1}^{n} a_{ij} \overline{\xi}_i \xi_j, \tag{1}$$

où $\xi = (\xi_1, \ldots, \xi_n)$ parcourt l'espace $\mathbf{C}^n$ des vecteurs à n coordonnées complexes. On dit classiquement que la matrice A est *hermitienne positive* si l'on a $q(\xi) \geq 0$ pour tout vecteur ξ dans $\mathbf{C}^n$.

Soient $\lambda_1, \ldots, \lambda_n$ les valeurs propres de A, éventuellement répétées selon leur multiplicité. Elles sont réelles, et il existe une matrice unitaire $U = (u_{ij})$ telle que l'on ait $A = U^{-1} \Lambda U$, où Λ est la matrice diagonale d'éléments $\lambda_1, \ldots, \lambda_n$. On a

$$a_{ij} = \sum_{k=1}^{n} \lambda_k \overline{u}_{ki} u_{kj}, \tag{2}$$

et donc

$$q(\xi) = \sum_{k=1}^{n} \lambda_k \left| \sum_{i=1}^{n} u_{ki} \xi_i \right|^2. \tag{3}$$

Comme la matrice U est inversible, la matrice A est hermitienne positive si et seulement si l'on a $\sum_{k=1}^{n} \lambda_k |\eta_k|^2 \geq 0$ quels que soient les nombres complexes $\eta_1, \ldots, \eta_n$. Autrement dit, *la matrice hermitienne A est positive si et seulement si ses valeurs propres $\lambda_1, \ldots, \lambda_n$ sont positives*; cette propriété sert parfois de définition.

2. Soient T un ensemble et C une fonction à valeurs complexes sur $T \times T$. On dira que C est une *covariance* sur T si, quels que soient les points $t_1, \ldots, t_n$ distincts de T, la matrice d'éléments $C(t_i, t_j)$ est hermitienne positive. Autrement dit, on impose les deux conditions suivantes:

(a) *on a $C(t', t) = \overline{C(t, t')}$ quels que soient t, t' dans T*;

(b) *quels que soient les points distincts $t_1, \ldots, t_n$ de T, et les nombres complexes $\xi_1, \ldots, \xi_n$, on a $\sum_{i,j=1}^{n} C(t_i, t_j) \bar{\xi}_i \xi_j \geq 0$.*

En fait, l'hypothèse (a) est conséquence de (b). En effet, le cas particulier $n = 1$, $t_1 = t$ et $\xi_1 = 1$ de l'hypothèse (b) s'écrit

$$C(t, t) \geq 0 \qquad \text{pour tout } t \text{ dans } T. \qquad (4)$$

Considérons maintenant le cas $n = 2$ et faisons successivement (ξ_1, ξ_2) égal à $(1, 1)$ et à $(1, i)$; on en déduit que les nombres

$$C(t_1, t_1) + C(t_2, t_2) + C(t_1, t_2) + C(t_2, t_1)$$
$$C(t_1, t_1) + C(t_2, t_2) + i(C(t_1, t_2) - C(t_2, t_1))$$

sont réels, et comme les nombres $C(t_1, t_1)$ et $C(t_2, t_2)$ sont aussi réels, on conclut que $C(t_2, t_1)$ est le conjugué de $C(t_1, t_2)$.

Utilisons le critère classique qui caractérise les matrices hermitiennes positives par la positivité des mineurs principaux (voir Gantmacher [6]); on voit que les covariances sont les fonctions C satisfaisant à la condition (a) ci-dessus et à la suivante:

(b') *quels que soient l'entier $n \geq 1$ et les points $t_1, \ldots, t_n$ de T, le déterminant*

$$D_n(t_1, \ldots, t_n) = \det(C(t_i, t_j))_{1 \leq i, j \leq n} \qquad (5)$$

est positif.

Le cas $n = 1$ redonne l'inégalité (4); le cas $n = 2$ est particulièrement important:

$$|C(t, t')|^2 \leq C(t, t) \cdot C(t', t'). \qquad (6)$$

3. Soit E un espace vectoriel complexe. On appelle d'habitude *forme sesquilinéaire* sur E toute application Γ de $E \times E$ dans $\mathbf{C}$ telle que $\Gamma(f, g)$ et $\overline{\Gamma(g, f)}$ soient des fonctions linéaires de g pour tout f fixé. On dit que Γ est *hermitienne* si l'on a

$$\Gamma(g, f) = \overline{\Gamma(f, g)} \qquad \text{pour } f, g \text{ dans } E. \qquad (7)$$

On a la *formule de polarisation*

$$3\Gamma(f, g) = \sum_j j^{-1} \Gamma(f + jg, f + jg), \qquad (8)$$

où j parcourt les solutions de l'équation $j^3 = 1$; par suite, Γ est hermitienne si et seulement si $\Gamma(f, f)$ est réel pour tout f dans E. On dit que Γ est *positive* si $\Gamma(f, f)$ est positif pour tout f dans E; une forme sesquilinéaire positive est donc hermitienne.

Soit Γ une forme sesquilinéaire positive. On a l'inégalité de Cauchy–Schwarz

$$|\Gamma(f, g)|^2 \le \Gamma(f, f) . \Gamma(g, g) \qquad \text{pour } f, g \text{ dans } E. \tag{9}$$

Par suite, un élément f de E satisfait à la relation $\Gamma(f, f) = 0$ si et seulement si l'on a $\Gamma(f, g) = 0$ pour tout g dans E. L'ensemble N des éléments f de E tels que $\Gamma(f, f) = 0$ est donc un sous-espace vectoriel de E, et il existe une forme sesquilinéaire positive Γ_0 sur l'espace quotient $E_0 = E/N$ caractérisée par la relation

$$\Gamma(f, g) = \Gamma_0(f + N, g + N) \qquad \text{pour } f, g \text{ dans } E. \tag{10}$$

Par construction, on a $\Gamma_0(u, u) > 0$ pour tout u non nul dans E_0. Le procédé de complétion fournit un espace de Hilbert $\mathcal{H}$ contenant E_0 comme sous-espace vectoriel dense, tel que $\Gamma_0(u, v) = \langle u | v \rangle$ pour u, v dans E_0; notons π l'application linéaire de E dans $\mathcal{H}$ qui associe à f la classe $f + N$. En résumé, on a établi le résultat suivant:

LEMME. *Soient E un espace vectoriel complexe et Γ une forme sesquilinéaire positive sur E. Il existe un espace de Hilbert $\mathcal{H}$ et une application linéaire π de E dans $\mathcal{H}$ satisfaisant aux conditions suivantes:*

(a) *l'image $\pi(E)$ de π est dense dans $\mathcal{H}$;*
(b) *on a $\Gamma(f, g) = \langle \pi(f) | \pi(g) \rangle$ pour f, g dans E.*

4. Nous pouvons faire le lien entre les covariances et les espaces de Hilbert.

LEMME. *Soit C une fonction à valeurs complexes sur $T \times T$. Pour que C soit une covariance, il faut et il suffit qu'il existe un espace de Hilbert $\mathcal{H}$ et une application φ de T dans $\mathcal{H}$ tels que l'on ait*

$$C(t, t') = \langle \varphi(t) | \varphi(t') \rangle \qquad \text{pour } t, t' \text{ dans } T. \tag{11}$$

Si C satisfait à la relation précédente, on a

$$\sum_{i, j = 1}^{n} C(t_i, t_j) \overline{\xi}_i \xi_j = \left\| \sum_{i = 1}^{n} \xi_i \varphi(t_i) \right\|^2 \tag{12}$$

pour $t_1, \ldots, t_n$ dans T et $\xi_1, \ldots, \xi_n$ dans $\mathbf{C}$; donc C est une covariance.

Réciproquement, supposons que C soit une covariance. Soit E l'espace vectoriel des fonctions $f: T \to \mathbf{C}$ telles que l'ensemble $\{f \neq 0\}$ soit fini. On définit une forme sesquilinéaire Γ sur E par la formule

$$\Gamma(f, g) = \sum_{t, t' \in T} C(t, t') \overline{f(t)} g(t') \tag{13}$$

et par définition d'une covariance, on a $\Gamma(f, f) \geq 0$ pour tout f dans E. Appliquons le lemme du n° 3. Pour tout point t de T, soit ε_t la fonction égale à 1 en t et nulle ailleurs; si l'on pose $\varphi(t) = \pi(\varepsilon_t)$, on a

$$\langle \varphi(t) | \varphi(t') \rangle = \Gamma(\varepsilon_t, \varepsilon_{t'}) = C(t, t') \qquad \text{pour } t, t' \text{ dans } T.$$

5. Soit C une covariance sur T. Choisissons $\mathscr{H}$ et φ de manière à satisfaire à la relation (11), ce que nous exprimerons en disant que la covariance C est *associée à* $\mathscr{H}$ *et* φ. Soit $\mathscr{F}$ le plus petit sous-espace vectoriel fermé de $\mathscr{H}$ contenant l'image de φ. A tout élément u de $\mathscr{F}$, on associe la fonction Hu sur T par la formule

$$Hu(t) = \langle \varphi(t) | u \rangle. \tag{14}$$

Si Hu est nulle, le vecteur u de $\mathscr{F}$ est orthogonal à l'image de φ, donc à $\mathscr{F}$ tout entier, et l'on a $u = 0$. Par suite, l'application $u \mapsto Hu$ est un isomorphisme de $\mathscr{F}$ sur un espace vectoriel $\mathscr{E}_C$ de fonctions à valeurs complexes sur T. Au moyen de cet isomorphisme, on transporte le produit scalaire de $\mathscr{F}$ en un produit scalaire sur $\mathscr{E}_C$, noté $\langle f | g \rangle_C$; on pose $\|f\|_C = \langle f | f \rangle_C^{1/2}$. Par définition, on a donc

$$\langle Hu | Hv \rangle_C = \langle u | v \rangle \qquad \text{pour } u, v \text{ dans } \mathscr{F}. \tag{15}$$

Le lemme suivant montre que l'espace de Hilbert $\mathscr{E}_C$ ne dépend que de la covariance C, mais non des données auxiliaires $\mathscr{H}$ et φ; on dit parfois que $\mathscr{E}_C$ est *l'espace de noyau reproduisant* C.

LEMME. *Soient C une covariance sur T, f une application de T dans $\mathbf{C}$ et M une constante positive. Pour que f appartienne à $\mathscr{E}_C$ et satisfasse à $\|f\|_C \leq M$, il faut et il suffit que la fonction D_M définie par*

$$D_M(t, t') = M^2 . C(t, t') - f(t) . \overline{f(t')} \tag{16}$$

soit une covariance sur T.

Par construction, pour que f appartienne à $\mathscr{E}_C$ et satisfasse à $\|f\|_C \leq M$, il faut et il suffit qu'il existe un élément u de norme $\leq M$ dans $\mathscr{F}$ tel que $\langle u | \varphi(t) \rangle = \overline{f(t)}$ pour tout $t \in T$. Comme un espace de Hilbert est son propre dual, ceci signifie qu'il existe sur $\mathscr{F}$ une forme linéaire de norme $\leq M$ qui prenne la valeur $\overline{f(t)}$ en l'élément $\varphi(t)$ pour tout t dans T. D'après un critère élémentaire et classique, cela signifie que l'on a

$$\left|\sum_{i=1}^{n} \xi_i \overline{f(t_i)}\right| \leq M \cdot \left\|\sum_{i=1}^{n} \xi_i \varphi(t_i)\right\| \tag{17}$$

quels que soient $t_1, \ldots, t_n$ dans T, et $\xi_1, \ldots, \xi_n$ dans $\mathbf{C}$. Le lemme résulte alors de la relation évidente

$$\sum_{i,j=1}^{n} D_M(t_i, t_j)\bar{\xi}_i\xi_j = M^2 \cdot \left\|\sum_{i=1}^{n} \xi_i \varphi(t_i)\right\|^2 - \left|\sum_{i=1}^{n} \xi_i \overline{f(t_i)}\right|^2. \tag{18}$$

6. Le problème se pose maintenant de caractériser les espaces de Hilbert de fonctions admettant un noyau reproduisant.

Soit d'abord C une covariance. Je dis que la fonction $C_t : t' \mapsto C(t', t)$ appartient à $\mathscr{E}_C$, et qu'elle satisfait aux relations

$$f(t) = \langle C_t | f \rangle_C \qquad \text{pour} \quad f \in \mathscr{E}_C \quad \text{et} \quad t \in T, \tag{19}$$

$$C(t, t') = \langle C_t | C_{t'} \rangle_C \qquad \text{pour} \quad t, t' \text{ dans } T. \tag{20}$$

En effet, les relations (11) et (14) montrent que l'on a $C_t = Hu$ avec $u = \varphi(t)$, d'où $C_t \in \mathscr{E}_C$; par ailleurs, si $f \in \mathscr{E}_C$ est de la forme Hu avec $u \in \mathscr{F}$, on a

$$f(t) = \langle \varphi(t) | u \rangle = \langle H\varphi(t) | Hu \rangle_C = \langle C_t | f \rangle_C$$

d'après la définition (14) de Hu et la définition (15) du produit scalaire dans $\mathscr{E}_C$; enfin, d'après la définition même de C_t, la formule (20) est le cas particulier $f = C_{t'}$ de la formule (19).

On remarquera que la formule (19) caractérise l'élément C_t de $\mathscr{E}_C$; de plus, si un élément de $\mathscr{E}_C$ est orthogonal à tous les C_t, il est nul d'après (19), et par conséquent, l'ensemble des combinaisons linéaires de la forme $\xi_1 . C_{t_1} + \cdots + \xi_n . C_{t_n}$ est dense dans $\mathscr{E}_C$.

LEMME. *Soit $\mathscr{E}$ un espace vectoriel de fonctions complexes sur T, muni d'un produit scalaire $\langle f | g \rangle$ qui en fasse un espace de Hilbert. Les conditions suivantes sont équivalentes:*

(a) *il existe une covariance C sur T telle que $\mathscr{E} = \mathscr{E}_C$ et $\langle f | g \rangle = \langle f | g \rangle_C$ pour f et g dans $\mathscr{E}$;*

(b) *il existe une fonction h sur T à valeurs positives finies, telle que l'on ait*

$$|f(t)| \leq h(t) \tag{21}$$

pour tout point t de T et tout élément f de norme ≤ 1 dans $\mathscr{E}$.

De plus, s'il existe une covariance C comme dans (a), elle est unique.

Les formules (19) et (20) fournissent une caractérisation de la covariance C en seuls termes de l'espace $\mathscr{E}_C$ et de son produit scalaire $\langle f | g \rangle_C$, d'où l'unicité. Par ailleurs, ces mêmes formules et l'inégalité de Cauchy–Schwarz entraînent

$$|f(t)| \leq \|f\|_C . C(t, t)^{1/2} \qquad \text{pour} \quad f \in \mathscr{E}_C \quad \text{et} \quad t \in T, \tag{22}$$

avec égalité seulement si f est proportionnelle à C_t. Par suite, (a) entraîne (b).

Réciproquement, sous l'hypothèse (b), l'application $f \mapsto f(t)$ est une forme linéaire continue sur l'espace de Hilbert $\mathscr{E}$ pour tout $t \in T$. Il existe donc un élément $\varphi(t)$ de $\mathscr{E}$ tel que l'on ait $f(t) = \langle \varphi(t) | f \rangle$ pour $f \in \mathscr{E}$ et $t \in T$; soit C la covariance associée à l'application $t \mapsto \varphi(t)$ de T dans $\mathscr{E}$. Reprenons les notations du n° 5 en y faisant $\mathscr{H} = \mathscr{E}$; si un élément f de $\mathscr{E}$ est orthogonal à l'image de φ, on a $f = 0$ d'après la définition des $\varphi(t)$; on a donc $\mathscr{F} = \mathscr{E}$. D'après la formule (14), on a $Hu = u$ pour tout $u \in \mathscr{E}$, d'où $\mathscr{E} = \mathscr{E}_C$ et $\langle f | g \rangle = \langle f | g \rangle_C$ pour f, g dans $\mathscr{E}$. Par suite, (b) entraîne (a).

7. Considérons toujours l'ensemble T et introduisons un espace de Hilbert $\mathscr{H}$ muni d'une base orthonormale $(e_\alpha)_{\alpha \in A}$. Se donner une application φ de T dans $\mathscr{H}$ revient à se donner une famille $(f_\alpha)_{\alpha \in A}$ d'applications de T dans $\mathbf{C}$ telle que l'on ait

$$\sum_{\alpha \in A} |f_\alpha(t)|^2 < +\infty \qquad \text{pour tout } t \text{ dans } T. \tag{23}$$

La correspondance entre φ et les f_α s'exprime par les formules

$$f_\alpha(t) = \langle \varphi(t) | e_\alpha \rangle, \tag{24}$$

$$\varphi(t) = \sum_{\alpha \in A} \overline{f_\alpha(t)} . e_\alpha. \tag{25}$$

S'il en est ainsi, la covariance sur T associée à $\mathscr{H}$ et φ est définie par la formule

$$C(t, t') = \sum_{\alpha \in A} f_\alpha(t) . \overline{f_\alpha(t')}. \tag{26}$$

De plus, pour tout α dans A, la fonction D_1 définie par

$$D_1(t, t') = C(t, t') - f_\alpha(t) . \overline{f_\alpha(t')}$$

est égale à $\sum_{\beta \neq \alpha} f_\beta(t) . \overline{f_\beta(t')}$, donc est encore une covariance. Compte tenu du lemme du n° 5, on a donc établi le résultat suivant:

LEMME. *Si la famille de fonctions $(f_\alpha)_{\alpha \in A}$ sur T satisfait à la condition (23), la somme dans (26) converge absolument dans $T \times T$, et définit une covariance*

C sur T. Pour tout $\alpha \in A$, *la fonction* f_α *est un élément de norme* ≤ 1 *de l'espace de Hilbert* $\mathscr{E}_C$.

Particularisons au cas où $\mathscr{H} = \mathscr{E}$ est un espace de fonctions sur T satisfaisant à la condition (b) du lemme du n° 6. On a vu qu'il existait une covariance C sur T telle que $\mathscr{E}$ soit l'espace de noyau reproduisant C. Alors, pour toute base orthonormale $(f_\alpha)_{\alpha \in A}$ de $\mathscr{E}$, on a $\sum_{\alpha \in A} |f_\alpha(t)|^2 < +\infty$ pour tout point t de T, et la covariance C est donnée explicitement par la formule (26).

8. Illustrons ce qui précède par l'exemple classique des polynômes orthogonaux. Soit μ une mesure positive sur la droite réelle **R** telle que l'intégrale $\int_{\mathbf{R}} |x|^n \, d\mu(x)$ soit finie pour tout entier $n \geq 0$. Soit $(p_n)_{n \geq 0}$ une famille de polynômes orthogonaux correspondant à μ; on a donc

$$\int_{\mathbf{R}} p_m(x) \cdot p_n(x) \, d\mu(x) = 0 \qquad \text{si} \quad m \neq n, \tag{27}$$

et p_n est un polynôme à coefficients réels de degré égal à n.

Fixons un entier $n \geq 0$, et supposons que la mesure μ ne soit pas localisée sur un ensemble fini ayant au plus n points. Soit $\mathscr{E}$ l'espace vectoriel formé des polynômes à coefficients complexes, de degré au plus égal à n. On définit un produit scalaire sur $\mathscr{E}$ par la formule

$$\langle p | q \rangle = \int_{\mathbf{R}} \overline{p(x)} \cdot q(x) \, d\mu(x). \tag{28}$$

Alors $\mathscr{E}$ est un espace de Hilbert de dimension finie $n + 1$, admettant une base orthogonale $(p_0, p_1, \dots, p_n)$.

Avec les notations classiques, on pose

$$h_v = \int_{\mathbf{R}} p_v(x)^2 \, d\mu(x) \qquad \text{pour} \quad 0 \leq v \leq n, \tag{29}$$

$$K_n(x, y) = \sum_{v=0}^{n} h_v^{-1} p_v(x) \cdot \overline{p_v(y)} \qquad \text{pour} \quad x, y \text{ dans } \mathbf{C}. \tag{30}$$

Rappelons aussi que, si k_n est le coefficient de x^n dans $p_n(x)$ et k_{n+1} celui de x^{n+1} dans $p_{n+1}(x)$, on a la formule de Christoffel–Darboux:

$$K_n(x, y) = (k_n/k_{n+1}h_n) \cdot (p_{n+1}(x)\overline{p_n(y)} - p_n(x)\overline{p_{n+1}(y)})/(x - y) \tag{31}$$

où x et y sont deux nombres complexes distincts. Enfin, on a

$$p(x) = \int_{\mathbf{R}} K_n(x, y) \cdot p(y) \, d\mu(y) \tag{32}$$

pour tout nombre complexe x et tout polynôme p de degré $\leq n$.

Interprétons les polynômes comme des applications de **C** dans **C**. Ce qui précède montre que K_n est une covariance sur **C**, et que $\mathscr{E}$ est l'espace de

Hilbert associé à cette covariance. Soit x_0 un nombre complexe. D'après la formule (22), on a

$$|p(x_0)| \leq K_n(x_0, x_0)^{1/2} \tag{33}$$

pour tout polynôme p de degré $\leq n$ tel que $\int_{\mathbf{R}} |p(x)|^2 \, d\mu(x) \leq 1$, avec égalité si et seulement si l'on a

$$p(x) = u.[K_n(x_0, x_0)]^{-1/2} K_n(x, x_0) \tag{34}$$

avec une constante u de module 1 convenable (voir Szegö [17] pour tout ce n°).

II. Covariances continues

Dans toute cette partie, on note T un espace topologique.

9. Soient φ une application de T dans un espace de Hilbert $\mathscr{H}$, et C la covariance associée. On a donc par définition

$$C(t, t') = \langle \varphi(t) | \varphi(t') \rangle \qquad \text{pour } t, t' \text{ dans } T. \tag{1}$$

On en déduit

$$\|\varphi(t) - \varphi(t')\| = d_C(t, t') \tag{2}$$

où le nombre réel positif $d_C(t, t')$ ne dépend que de la covariance C, puisque l'on a

$$d_C(t, t')^2 = C(t, t) + C(t', t') - 2\operatorname{Re} C(t, t'). \tag{3}$$

LEMME. *L'application φ de T dans $\mathscr{H}$ est continue si et seulement si la fonction C est continue sur $T \times T$.*

Le produit scalaire est une application continue de $\mathscr{H} \times \mathscr{H}$ dans $\mathbf{C}$, donc la continuité de φ entraîne celle de C. Réciproquement, supposons que C soit continue; d'après la formule (3), d_C est une fonction continue sur $T \times T$. Donnons-nous un point t_0 de T et un nombre réel $\varepsilon > 0$. Vu la continuité de d_C, il existe un voisinage U de t_0 dans T tel que l'on ait $d_C(t, t_0) < \varepsilon$ pour tout $t \in U$. On a donc $\|\varphi(t) - \varphi(t_0)\| < \varepsilon$ pour tout t dans U, d'où la continuité de φ.

10. Considérons maintenant une covariance C sur T, et appliquons ce qui précède à l'espace de Hilbert $\mathscr{H} = \mathscr{E}_C$ et à $\varphi(t) = C_t$. On a donc

$$d_C(t, t') = \|C_t - C_{t'}\|_C, \tag{4}$$

d'où les relations

$$d_C(t, t) = 0, \tag{5}$$

$$d_C(t, t') = d_C(t', t), \tag{6}$$

$$d_C(t, t'') \le d_C(t, t') + d_C(t', t'') \tag{7}$$

(pour t, t' et t'' dans T). Il s'en faut de peu que d_C ne soit une distance sur T; simplement, on peut avoir $d_C(t, t') = 0$ même lorsque t et t' sont distincts. Pour tout élément f de $\mathscr{E}_C$, on a $f(t) - f(t') = \langle C_t - C_{t'} | f \rangle_C$ et l'inégalité

$$\left| f(t) - f(t') \right| \le \|f\|_C \cdot d_C(t, t') \tag{8}$$

s'obtient par application de l'inégalité de Cauchy–Schwarz.

11. LEMME. *Pour que la covariance C soit continue sur $T \times T$, il faut et il suffit que $\mathscr{E}_C$ se compose de fonctions continues, et que toute partie bornée de $\mathscr{E}_C$ soit équicontinue.*

Supposons d'abord que C soit continue sur $T \times T$. Alors la fonction d_C est continue sur $T \times T$, et l'on a $d_C(t, t) = 0$ et $d_C(t, t') \ge 0$. L'inégalité (8) fournit un module de continuité pour les fonctions de $\mathscr{E}_C$, uniforme sur toute partie de $\mathscr{E}_C$ bornée en norme.

Réciproquement, supposons que l'ensemble B des fonctions f de $\mathscr{E}_C$ telles que $\|f\|_C \le 1$ soit équicontinu, et montrons que l'application $t \mapsto C_t$ de T dans $\mathscr{E}_C$ est continue. Soient $t_0 \in T$ et $\varepsilon > 0$; par hypothèse, il existe un voisinage U de t_0 dans T tel que l'on ait $\left| f(t) - f(t_0) \right| \le \varepsilon$ pour $t \in U$ et $f \in B$. On a alors

$$\left\| C_t - C_{t_0} \right\| = \sup_{f \in B} \left| \langle C_t - C_{t_0} | f \rangle \right| = \sup_{f \in B} \left| f(t) - f(t_0) \right| \le \varepsilon$$

pour tout t dans U, d'où notre assertion. Comme on a $C(t, t') = \langle C_t | C_{t'} \rangle_C$, la continuité de l'application $t \mapsto C_t$ entraîne celle de C d'après le lemme du n° 9.

12. Gardons les notations précédentes et supposons C continue. D'après le lemme du n° 5, une fonction f sur T appartient à B si et seulement l'on a les inégalités

$$\left| \sum_{i=1}^{n} \xi_i \overline{f(t_i)} \right|^2 \le \sum_{i,j=1}^{n} C(t_i, t_j) \overline{\xi}_i \xi_j \tag{9}$$

où $t_1, \ldots, t_n$ sont des points de T et $\xi_1, \ldots, \xi_n$ des nombres complexes. Par suite, B est fermé pour la topologie de la convergence simple dans T. De plus, on a $\left| f(t) \right| \le C(t, t)^{1/2}$ pour $f \in B$ et $t \in T$, et la fonction $t \mapsto C(t, t)^{1/2}$ est continue, donc bornée sur toute partie compacte de T. Enfin, le lemme précédent montre que B est un ensemble équicontinu de fonctions. Par application du théorème d'Ascoli, on obtient donc le résultat suivant:

LEMME. *Supposons la covariance C continue sur $T \times T$. Alors l'ensemble B des fonctions f de $\mathscr{E}_C$ telles que $\|f\|_C \leq 1$ est compact pour la topologie de la convergence compacte sur T. De plus, la convergence en norme dans $\mathscr{E}_C$ est plus forte que la convergence compacte sur T.*

13. Venons-en aux propriétés de développement en série des covariances continues.

THÉORÈME. *Supposons que l'espace T soit compact et que C soit une covariance continue sur T. Soit $(f_\alpha)_{\alpha \in A}$ une famille de fonctions sur T, telle que l'on ait*

$$C(t, t') = \sum_{\alpha \in A} f_\alpha(t) . \overline{f_\alpha(t')}, \tag{10}$$

la série convergeant absolument sur $T \times T$. Alors:

(a) *Les fonctions f_α sont continues sur T.*
(b) *L'ensemble des indices $\alpha \in A$ tels que $f_\alpha \neq 0$ est dénombrable.*
(c) *La série (10) converge uniformément dans $T \times T$.*

D'après le lemme du n° 7, chaque fonction f_α appartient à $\mathscr{E}_C$, donc est continue d'après le lemme du n° 11. Ceci prouve (a).

Comme la fonction d_C est continue, et que T est compact, il existe pour tout entier $n \geq 1$ une partie finie D_n de T telle que, pour tout $t \in T$, il existe $s \in D_n$ avec $d_C(t, s) \leq 2^{-n}$. Compte tenu de l'inégalité (8), toute fonction appartenant à $\mathscr{E}_C$ qui est nulle en tout point de l'ensemble dénombrable $D = \bigcup_{n=1}^\infty D_n$ est nulle. Pour chaque point s de D, on a $\sum_{\alpha \in A} |f_\alpha(s)|^2 < +\infty$, et par conséquent, il existe une partie dénombrable $A(s)$ de A telle que l'on ait $f_\alpha(s) = 0$ pour α dans $A \backslash A(s)$. L'ensemble $B = \bigcup_{s \in D} A(s)$ est une partie dénombrable de A, et l'on a $f_\alpha(s) = 0$ pour $s \in D$ et $\alpha \in A \backslash B$. On a finalement $f_\alpha = 0$ pour tout indice α dans $A \backslash B$, d'où (b).

Pour prouver (c), on peut donc supposer que A est l'ensemble des entiers $\alpha \geq 1$. Chacune des fonctions f_α est continue d'après (a), et il en est de même de $\sum_{\alpha \in A} |f_\alpha|^2$ puisqu'on a $C(t, t) = \sum_{\alpha \in A} |f_\alpha(t)|^2$ et que la fonction C est continue. D'après le lemme de Dini, la série $\sum_{\alpha \in A} |f_\alpha|^2$ converge uniformément sur T. Soit $\varepsilon > 0$; il existe donc un entier $n \geq 1$ tel que l'on ait $\sum_{\alpha \geq n} |f_\alpha(t)|^2 \leq \varepsilon$ pour tout $t \in T$. D'après l'inégalité de Cauchy–Schwarz, on a donc

$$\left| \sum_{\alpha \geq n} f_\alpha(t) . \overline{f_\alpha(t')} \right|^2 \leq \sum_{\alpha \geq n} |f_\alpha(t)|^2 . \sum_{\alpha \geq n} |f_\alpha(t')|^2 \leq \varepsilon^2$$

quels que soient t et t' dans T. Ceci prouve (c).

14.   Nous allons montrer par un exemple comment la continuité de la covariance et l'uniformité de la convergence de la série (10) sont étroitement liées. Nous allons en effet construire une covariance C sur l'intervalle $[0,1]$ de $\mathbf{R}$ qui a les deux propriétés suivantes:

(a)   *la fonction C est continue en tout point du carré $[0,1] \times [0,1]$, à l'exception du point $(0,0)$;*
(b)   *toute fonction appartenant à $\mathscr{E}_C$ est continue sur l'intervalle $[0,1]$.*

Il n'existe alors aucun développement de la forme $C(t,t') = \sum_{n=1}^{\infty} f_n(t)\overline{f_n(t')}$ qui converge uniformément sur $[0,1] \times [0,1]$. En effet, s'il en était ainsi, chaque fonction f_n appartiendrait à $\mathscr{E}_C$(n° 7), donc serait continue [propriété (b)] et la fonction C serait continue, contrairement à la propriété (a).

Soit $\mathscr{H}$ un espace de Hilbert, muni d'une base orthonormale $(e_n)_{n \geq 0}$. On considère l'application h de $[0,1]$ dans $\mathscr{H}$ admettant les valeurs suivantes:

$$h(0) = 0, \qquad h(2^{-2n}) = 0, \qquad h(2^{-2n-1}) = e_n \qquad \text{pour tout} \quad n \geq 0 \quad (11)$$

et qui est linéaire dans chacun des intervalles $[2^{-m-1}, 2^{-m}]$ (pour $m \geq 0$). La covariance C est la covariance associée à h.

Il est immédiat que l'application h est continue en tout point non nul de $[0,1]$, donc C est continue en tout point (t,t') de $[0,1] \times [0,1]$ pour lequel t et t' sont non nuls. Soit t un point non nul de l'intervalle $[0,1]$, et soit n un entier positif tel que $2^{-2n} < t$; alors la fonction C est identiquement nulle dans le voisinage $[2^{-2n}, 1] \times [0, 2^{-2n}]$ du point $(t,0)$, donc la fonction C est continue en ce point. Pour une raison analogue, la fonction C est continue en tout point $(0,t)$ avec $t \neq 0$. Enfin, on a $C(0,0) = 0$ et $C(2^{-2n-1}, 2^{-2n-1}) = 1$ pour tout entier $n \geq 0$, donc la fonction C n'est pas continue au point $(0,0)$. Ceci établit la propriété (a).

Il est clair que $\mathscr{H}$ est le seul sous-espace vectoriel fermé de lui-même contenant l'image de h. La formule $Hu(t) = \langle h(t)|u \rangle$ définit donc un isomorphisme $u \mapsto Hu$ de $\mathscr{H}$ sur $\mathscr{E}_C$. Il résulte de la formule (11) que l'espace $\mathscr{E}_C$ se compose des fonctions f sur $[0,1]$ telles que $f(0) = f(2^{-2n}) = 0$ pour tout $n \geq 0$, qui sont linéaires dans chacun des intervalles $[2^{-m-1}, 2^{-m}]$ (pour $m \geq 0$), et telles que $\sum_{n=0}^{\infty}|f(2^{-2n-1})|^2$ soit fini. Il est clair qu'une telle fonction est continue en dehors de 0; comme on a $|f(t)| \leq \sup_{n \geq m}|f(2^{-2n-1})|$ pour tout t dans l'intervalle $[0, 2^{-2m}]$, la fonction f est aussi continue en 0. Ceci établit la propriété (b).

15.   Dans ce n°, on suppose que l'espace T est *localement compact* (donc séparé); on note $\mathscr{C}_c(T)$ l'espace vectoriel des fonctions continues (complexes) à support compact sur T. Soient $\mathscr{A}$ la tribu borélienne de T et μ une mesure positive définie sur $\mathscr{A}$; on suppose que le nombre $\mu(K)$ est fini pour toute

partie compacte K de T. Alors les éléments de $\mathscr{C}_c(T)$ sont des fonctions μ-intégrables.

THÉORÈME.   *Soit C une fonction continue sur $T \times T$.*

(a)   *Si C est une covariance, l'intégrale*

$$C[u] = \int_T \int_T C(t, t')\overline{u(t)}u(t') \, d\mu(t) \, d\mu(t') \tag{12}$$

est positive pour toute fonction u dans $\mathscr{C}_c(T)$.

(b)   *Réciproquement, supposons que $\mu(U)$ soit non nul pour tout ouvert U non vide de T, et que l'on ait $C[u] \geq 0$ pour toute fonction $u \in \mathscr{C}_c(T)$. Alors C est une covariance sur T.*

Supposons que C soit une covariance. Soit u une fonction continue sur T, nulle hors d'une partie compacte K de T. Comme la restriction de C à $K \times K$ est une covariance, le théorème du n° 13 montre que C se représente sous forme d'une série

$$C(t, t') = \sum_{n=1}^{\infty} f_n(t) \cdot \overline{f_n(t')} \tag{13}$$

qui converge uniformément sur $K \times K$, les fonctions f_n étant continues sur K. On peut donc intégrer terme à terme, d'où

$$\begin{aligned}
C[u] &= \int_K \int_K C(t, t')\overline{u(t)}u(t') \, d\mu(t) \, d\mu(t') \\
&= \sum_{n=1}^{\infty} \int_K \int_K f_n(t)\overline{f_n(t')}\,\overline{u(t)}u(t') \, d\mu(t) d\mu(t') \\
&= \sum_{n=1}^{\infty} \left| \int_K \overline{u(t)}f_n(t) \, d\mu(t) \right|^2 ;
\end{aligned}$$

on a donc $C[u] \geq 0$.

Réciproquement, faisons les hypothèses de (b). Soient $t_1, \ldots, t_n$ des points distincts de T et $\xi_1, \ldots, \xi_n$ des nombres complexes. Soit $\varepsilon > 0$; comme C est continue, on peut trouver des voisinages ouverts U_1 pour $t_1, \ldots, U_n$ pour t_n, tels que C oscille d'au plus ε sur chacun des ensembles $U_i \times U_j$. Quitte à rétrécir les ensembles U_i, on peut les supposer deux à deux disjoints. Il existe alors, pour chaque i, une fonction continue u_i sur T, positive, non identiquement nulle, à support compact contenu dans U_i. L'ensemble des points t de T tels que $u_i(t) > 0$ est ouvert et non vide, donc n'est pas μ-négligeable; quitte à multiplier u_i par une constante $c_i > 0$, on peut supposer que l'on a

$$\int_T u_i(t) \, d\mu(t) = 1 \qquad \text{pour} \quad 1 \leq i \leq n. \tag{14}$$

Posons $u = \xi_1 u_1 + \cdots + \xi_n u_n$. Alors $\overline{u(t)}u(t')$ est nul, sauf si (t, t') appartient à l'un des ensembles $U_i \times U_j$, auquel cas on a

$$\overline{u(t)}u(t') = \overline{\xi_i}\xi_j\overline{u_i(t)}u_j(t'), \tag{15}$$

$$|C(t, t') - C(t_i, t_j)| \le \varepsilon. \tag{16}$$

On a alors

$$C[u] = \sum_{i,\, j=1}^{n} \overline{\xi_i}\xi_j \int_{U_i} \int_{U_j} C(t, t')\overline{u_i(t)}u_j(t')\, d\mu(t)\, d\mu(t'). \tag{17}$$

Compte tenu des formules (14), (16), et (17), on a finalement

$$\left| C[u] - \sum_{i,\, j=1}^{n} \overline{\xi_i}\xi_j C(t_i, t_j) \right| \le \varepsilon \cdot \left(\sum_{i=1}^{n} |\xi_i| \right)^2. \tag{18}$$

Comme $C[u]$ est positif et que ε est arbitrairement petit, on a

$$\sum_{i,\, j=1}^{n} \overline{\xi_i}\xi_j C(t_i, t_j) \ge 0, \tag{19}$$

donc C est une covariance.

III. Opérateurs dans les espaces de Hilbert

Les notions dont nous avons besoin sont tout à fait classiques, et il existe d'excellents exposés (voir, e.g., Bourbaki [2]). Nous nous contenterons donc d'un bref résumé.

16. Soient $\mathscr{E}$ et $\mathscr{F}$ deux espaces de Hilbert. On appelle *opérateur* de $\mathscr{E}$ dans $\mathscr{F}$ toute application linéaire continue de $\mathscr{E}$ dans $\mathscr{F}$. Si v est un tel opérateur, il existe un opérateur v^* de $\mathscr{F}$ dans $\mathscr{E}$, appelé l'*adjoint* de v, et caractérisé par la relation

$$\langle v(x) | y \rangle = \langle x | v^*(y) \rangle \qquad \text{pour} \quad x \in \mathscr{E}, \quad y \in \mathscr{F}. \tag{1}$$

On dit que l'opérateur u dans $\mathscr{E}$ est *hermitien* si l'on a $u = u^*$; si l'on a de plus $\langle x | u(x) \rangle \ge 0$ pour tout x dans $\mathscr{E}$, on dit que u est *hermitien positif*.

17. Soit u un opérateur hermitien positif dans $\mathscr{E}$. Le nombre

$$\text{Tr}(u) = \sum_{\alpha \in A} \langle e_\alpha | u(e_\alpha) \rangle \tag{2}$$

(positif ou égal à $+\infty$) est indépendant de la base orthonormale $(e_\alpha)_{\alpha \in A}$ de $\mathscr{E}$. On l'appelle la *trace* de u.

18.   *Le théorème de décomposition spectrale* pour les opérateurs de trace finie (D.Hilbert, E.Schmidt) s'énonce comme suit:

Soient $\mathscr{E}$ un espace de Hilbert, u un opérateur hermitien positif de trace finie dans $\mathscr{E}$ et $\mathscr{E}_0$ le noyau de u (ensemble des vecteurs x de $\mathscr{E}$ tels que $u(x) = 0$). Soit $\mathscr{E}_+$ le sous-espace de $\mathscr{E}$ orthogonal à $\mathscr{E}_0$. Il existe alors une base orthonormale $(e_n)_{n \in I}$ de $\mathscr{E}_+$ et une suite décroissante $(\lambda_n)_{n \in I}$ de nombres réels strictement positifs qui satisfont aux conditions suivantes:

(a)   on a $u(e_n) = \lambda_n e_n$ pour tout $n \in I$;
(b)   ou bien I est un intervalle fini $[0, p]$ de l'ensemble des nombres entiers, ou bien I se compose de tous les entiers positifs, et la suite $(\lambda_n)_{n \geq 0}$ tend vers 0.

Sous les hypothèses précédentes, on a

$$u(x) = \sum_{n \in I} \lambda_n \langle e_n | x \rangle . e_n \qquad \text{pour } x \text{ dans } \mathscr{E}, \tag{3}$$

$$\mathrm{Tr}(u) = \sum_{n \in I} \lambda_n. \tag{4}$$

19.   Soient $\mathscr{E}$ et $\mathscr{F}$ deux espaces de Hilbert, et v un opérateur de $\mathscr{E}$ dans $\mathscr{F}$. Soient $(e_\alpha)_{\alpha \in A}$ une base orthonormale de $\mathscr{E}$, et $(f_\beta)_{\beta \in B}$ une base orthonormale de $\mathscr{F}$. On a les égalités

$$\mathrm{Tr}(v^*v) = \sum_{\alpha \in A} \|v(e_\alpha)\|^2 = \sum_{\alpha \in A} \sum_{\beta \in B} |\langle v(e_\alpha) | f_\beta \rangle|^2 = \sum_{\beta \in B} \|v^*(f_\beta)\|^2 = \mathrm{Tr}(vv^*).$$

Si tous ces nombres sont finis, on dit que v est un *opérateur de Hilbert–Schmidt*.

La structure d'un opérateur de Hilbert–Schmidt $v \colon \mathscr{E} \to \mathscr{F}$ se décrit ainsi: Notons $\mathscr{E}_0$ le noyau de v et $\mathscr{F}_0$ celui de v^*; notons aussi $\mathscr{E}_+$ l'orthogonal de $\mathscr{E}_0$ dans $\mathscr{E}$ et $\mathscr{F}_+$ celui de $\mathscr{F}_0$ dans $\mathscr{F}$. Alors $\mathscr{E}_0$ est aussi le noyau de v^*v, et $\mathscr{F}_0$ celui de vv^*. Conformément au n° 18, choisissons une base orthonormale $(e_n)_{n \in I}$ de $\mathscr{E}_+$ et des nombres réels $\lambda_n > 0$ tels que $v^*v(e_n) = \lambda_n e_n$ pour tout $n \in I$. Posons $\mu_n = \lambda_n^{1/2}$ et $f_n = \mu_n^{-1} v(e_n)$ pour tout $n \in I$. Alors $(f_n)_{n \in I}$ est une base orthonormale de $\mathscr{F}_+$ et l'on a

$$v(x) = \sum_{n \in I} \mu_n \langle e_n | x \rangle . f_n \qquad \text{pour } x \text{ dans } \mathscr{E}, \tag{5}$$

$$v^*(y) = \sum_{n \in I} \mu_n \langle f_n | y \rangle . e_n \qquad \text{pour } y \text{ dans } \mathscr{F}. \tag{6}$$

De plus, on a

$$\mathrm{Tr}(vv^*) = \mathrm{Tr}(v^*v) = \sum_{n \in I} \mu_n^2. \tag{7}$$

20.   L'exemple suivant d'opérateur de Hilbert–Schmidt est crucial pour la suite de ce travail.

THÉORÈME. *Soient* $(T, \mathscr{A}, \mu)$ *un espace mesuré et $\mathscr{H}$ un espace de Hilbert.*
Pour qu'un opérateur v de $\mathscr{H}$ dans $L^2 = L^2(T, \mathscr{A}, \mu)$ soit de Hilbert–Schmidt,
il faut et il suffit qu'il existe une fonction positive $h \in L^2$ telle que l'on ait
$|v(x)| \le h$ *μ-presque partout sur T, pour tous les vecteurs x de norme ≤ 1 dans*
$\mathscr{H}$. S'il en est ainsi, on a

$$\operatorname{Tr}(v^*v) \le \|h\|_2^2. \tag{8}$$

Supposons d'abord que v soit un opérateur de Hilbert–Schmidt de $\mathscr{H}$
dans L^2, que l'on représente comme dans (5). Pour tout $n \in I$, la fonction
f_n est de norme 1 dans L^2, d'où

$$\sum_{n \in I} \int_T \mu_n^2 |f_n(t)|^2 \, d\mu(t) = \sum_{n \in I} \mu_n^2 < +\infty. \tag{9}$$

Quitte à modifier chacune des fonctions f_n sur un ensemble μ-négligeable,
on peut supposer que la fonction h définie par

$$h(t) = \left\{ \sum_{n \in I} \mu_n^2 |f_n(t)|^2 \right\}^{1/2} \tag{10}$$

est finie en tout point de T. Soit alors x un vecteur de norme ≤ 1 dans $\mathscr{H}$.
On a donc $\sum_{n \in I} |\langle e_n | x \rangle|^2 \le 1$; d'après l'inégalité de Cauchy–Schwarz, on
a donc

$$\left| \sum_{n \in I} \mu_n \langle e_n | x \rangle \cdot f_n(t) \right| \le h(t), \tag{11}$$

la série du premier membre convergeant absolument pour tout t dans T.
Comme la série $\sum_{n \in I} \mu_n \langle e_n | x \rangle \cdot f_n$ converge en norme dans L^2 vers $v(x)$,
un représentant de la classe de fonctions $v(x)$ est fourni par la fonction
$t \mapsto \sum_{n \in I} \mu_n \langle e_n | x \rangle \cdot f_n(t)$. D'après (11), on a donc $|v(x)| \le h$ μ-presque partout
sur T, quel que soit le représentant choisi pour $v(x)$.

Réciproquement, supposons qu'il existe une fonction positive h dans L^2
telle que l'on ait $|v(x)| \le h$ μ-presque partout sur T, pour tout vecteur x de
norme ≤ 1 dans $\mathscr{H}$. Soit $(e_\alpha)_{\alpha \in A}$ une base orthonormale dans $\mathscr{H}$, et soit
f_α un représentant de la classe de fonctions $v(e_\alpha)$. Soit J une partie finie de
A. Si $(\xi_\alpha)_{\alpha \in J}$ est une famille de nombres complexes telle que $\sum_{\alpha \in J} |\xi_\alpha|^2 \le 1$,
on a

$$\left| \sum_{\alpha \in J} \xi_\alpha f_\alpha(t) \right| \le h(t) \tag{12}$$

μ-presque partout sur T. L'ensemble μ-négligeable exceptionnel dépend de
la famille $(\xi_\alpha)_{\alpha \in J}$, mais il existe un ensemble μ-négligeable N tel que l'inégalité

(12) ait lieu chaque fois que t appartient à $T\backslash N$ et que les parties rèelles et imaginaires des ξ_α sont rationnelles. Par continuité, l'inégalité (12) reste valable pour tout t dans $T\backslash N$ quels que soient les nombres complexes ξ_α tels que $\sum_{\alpha \in J} |\xi_\alpha|^2 \le 1$. On en déduit

$$\sum_{\alpha \in J} |f_\alpha(t)|^2 \le h(t)^2 \qquad \text{pour } t \text{ dans } T\backslash N, \tag{13}$$

d'où par intégration

$$\sum_{\alpha \in J} \int_T |f_\alpha(t)|^2 \, d\mu(t) \le \int_T h(t)^2 \, d\mu(t). \tag{14}$$

En passant à la borne supérieure sur les parties finies J de A, on obtient l'inégalité $\sum_{\alpha \in A} \|f_\alpha\|_2^2 \le \|h\|_2^2 < +\infty$, donc v est un opérateur de Hilbert–Schmidt, et l'on a $\mathrm{Tr}(v^*v) \le \|h\|_2^2$ comme annoncé.

IV. COVARIANCES MESURABLES ET SÉPARABLES

Dans toute cette partie, on note T un ensemble muni d'une tribu $\mathscr{A}$ de parties.

21.   Commençons par élucider les relations entre mesurabilité forte et faible. Soient $\mathscr{H}$ un espace de Hilbert séparable, $(e_n)_{n \in I}$ une base orthonormale de $\mathscr{H}$ et φ une application de T dans $\mathscr{H}$. On pose

$$f_n(t) = \langle e_n | \varphi(t) \rangle, \tag{1}$$

$$C(t, t') = \langle \varphi(t) | \varphi(t') \rangle \tag{2}$$

pour t, t' dans T et n dans I.

LEMME.   *Les conditions suivantes sont équivalentes:*

(a)   *L'application φ de T dans $\mathscr{H}$ est $\mathscr{A}$-mesurable si l'on munit $\mathscr{H}$ de sa tribu borélienne.*

(b)   *Pour chaque élément a de $\mathscr{H}$, la fonction scalaire $t \mapsto \langle a | \varphi(t) \rangle$ est $\mathscr{A}$-mesurable.*

(c)   *Pour chaque n dans I, la fonction scalaire f_n est $\mathscr{A}$-mesurable.*

(d)   *La fonction C est mesurable sur $T \times T$ par rapport à la tribu $\mathscr{A} \otimes \mathscr{A}$.*

Pour tout a dans $\mathscr{H}$, l'application $x \mapsto \langle a | x \rangle$ de $\mathscr{H}$ dans $\mathbf{C}$ est continue, donc borélienne; par suite, (a) entraîne (b). Il est clair que (c) est un cas particulier de (b). Enfin, on a

$$C(t, t') = \sum_{n \in I} \overline{f_n(t)} f_n(t') \qquad \text{pour } t, t' \text{ dans } T; \tag{3}$$

comme l'espace de Hilbert $\mathscr{H}$ est séparable, l'ensemble I est dénombrable, donc la mesurabilité des fonctions f_n entraîne celle de C; par suite (c) entraîne (d).

Montrons que (c) *entraîne* (a). Pour tout vecteur $a = \sum_{n \in I} \alpha_n e_n$ de $\mathscr{H}$ et tout nombre réel $r > 0$, notons $B(a, r)$ la boule fermée de centre a et de rayon r dans $\mathscr{H}$; on sait que de telles boules engendrent la tribu borélienne sur l'espace séparable $\mathscr{H}$. Supposons que les fonctions f_n soient $\mathscr{A}$-mesurables. Pour toute partie fine J de I, la fonction $u_J = \sum_{n \in J} |f_n - \alpha_n|^2$ est $\mathscr{A}$-mesurable; de plus, l'ensemble des parties finies de I est dénombrable. Par suite, l'ensemble

$$\varphi^{-1}(B(a, r)) = \bigcap_J u_J^{-1}([0, r^2]) \tag{4}$$

est $\mathscr{A}$-mesurable. Autrement dit, la fonction φ est $\mathscr{A}$-mesurable.

Enfin, *montrons que* (d) *entraîne* (b).† Supposons C mesurable et notons $\mathscr{H}_1$ l'ensemble des vecteurs a de $\mathscr{H}$ tels que la fonction $t \mapsto \langle a | \varphi(t) \rangle$ soit $\mathscr{A}$-mesurable sur T. Il est clair que $\mathscr{H}_1$ est un sous-espace vectoriel fermé de $\mathscr{H}$. Tout vecteur de $\mathscr{H}$ orthogonal à l'image de φ appartient à $\mathscr{H}_1$; de plus, on a $\varphi(T) \subset \mathscr{H}_1$ car les fonctions partielles $t \mapsto C(t_0, t)$ sont $\mathscr{A}$-mesurables. On a donc $\mathscr{H}_1 = \mathscr{H}$, d'où notre assertion.

22.   Soit C une covariance sur T. On dit que C est *séparable* si l'espace de Hilbert $\mathscr{E}_C$ est séparable; il revient au même de supposer que C admet un développement de la forme

$$C(t, t') = \sum_{n=1}^{\infty} f_n(t) \overline{f_n(t')}. \tag{5}$$

On dit que C est une *covariance mesurable* sur $(T, \mathscr{A})$ si la fonction C est mesurable sur $T \times T$ par rapport à la tribu $\mathscr{A} \otimes \mathscr{A}$. S'il en est ainsi, toute fonction de $\mathscr{E}_C$ est $\mathscr{A}$-mesurable puisque c'est la limite en norme dans $\mathscr{E}_C$, donc en chaque point de T, d'une suite de fonctions dont chacune est de la forme $t \mapsto \sum_{i=1}^n \xi_i C(t, t_i)$.

Le lemme du n° 21 se reformule comme suit:

LEMME.   *Soit C une covariance séparable sur T, avec un développement de la forme* (5). *Les conditions suivantes sont équivalentes*:

(a)   *C est une covariance mesurable sur $(T, \mathscr{A})$;*
(b)   *$\mathscr{E}_C$ se compose de fonctions $\mathscr{A}$-mesurables sur T;*
(c)   *chacune des fonctions f_n est $\mathscr{A}$-mesurable sur T.*

23.   Donnons maintenant une application du théorème de Lusin.

---

† Cette partie de la démonstration reste valable pour des espaces de Hilbert non séparables.

THÉORÈME. *Supposons que T soit un espace polonais*[†] *et μ une mesure positive σ-finie sur la tribu borélienne de T. Soit C une covariance séparable sur T, qui est une fonction borélienne sur T × T. Choisissons un développement de C comme en* (5). *Alors il existe une suite croissante* $(K_p)_{p \geq 1}$ *de parties compactes de T avec les propriétés suivantes*:

(a) *le complémentaire dans T de la réunion des* K_p *est μ-négligeable*;

(b) *pour tout entier* $p \geq 1$, *la restriction de C à* $K_p \times K_p$ *est continue, et la série* (5) *définissant C converge absolument et uniformément sur* $K_p \times K_p$.

Rappelons que la mesure $μ$ satisfait à la propriété de régularité intérieure: la mesure $μ(A)$ d'une partie borélienne A de T est la borne supérieure des mesures $μ(K)$ des parties compactes K de T contenues dans A. De plus, comme on a supposé que la mesure $μ$ est $σ$-finie, il existe une mesure finie ayant les mêmes ensembles négligeables que $μ$, et l'on ne restreint pas la généralité en supposant que $μ(T)$ est fini. Rappelons enfin le théorème de Lusin: si K est une partie compacte de T, f une fonction borélienne sur K et $ε$ un nombre réel strictement positif, il existe une partie compacte L de K telle que $μ(K\backslash L) < ε$ et que la restriction de f à L soit une application continue de L dans $\mathbf{C}$.

Construisons par récurrence une suite de parties compactes L_p de T, deux à deux disjointes. Supposons construites les parties $L_1, \ldots, L_p$ et soit A le complémentaire dans T de leur réunion. Choisissons une partie compacte K de A telle que $μ(A\backslash K) \leq 2^{-p-2}$ et posons $f_0 = \sum_{n=1}^{\infty} |f_n|^2$. Par application du théorème de Lusin, choisissons pour tout entier $n \geq 0$ une partie compacte M_n de K telle que la restriction de f_n à M_n soit continue, et que l'on ait $μ(K\backslash M_n) < 2^{-p-n-3}$. Posons $L_{p+1} = \bigcap_{n=1}^{\infty} M_n$; par construction, L_{p+1} est disjoint de $L_1, \ldots, L_p$, c'est une partie compacte de T, la restriction de chacune des fonctions f_n à L_{p+1} est continue et l'on a

$$\mu(T\backslash(L_1 \cup \cdots \cup L_{p+1})) = \mu(A\backslash L_{p+1}) \leq \mu(A\backslash K) + \sum_{n=0}^{\infty} \mu(K\backslash M_n)$$

$$\leq 2^{-p-2} + \sum_{n=0}^{\infty} 2^{-p-n-3} = 2^{-p-1}.$$

Posons alors $K_p = L_1 \cup \cdots \cup L_p$; la suite des parties compactes K_p de T est croissante, et comme on a $μ(T\backslash K_p) \leq 2^{-p}$, la réunion des K_p a un complémentaire $μ$-négligeable. De plus, la restriction à K_p de chacune des fonctions f_n (pour $n \geq 1$) et $f_0 = \sum_{n=1}^{\infty} |f_n|^2$ est continue.

Appliquons alors le lemme de Dini: la série $\sum_{n=1}^{\infty} |f_n|^2$ converge *uniformément* sur l'espace compact K_p. Soit alors $ε > 0$ et soit $m \geq 1$ un entier

---

[†] Autrement dit, T est un espace métrique séparable et complet.

tel que l'on ait $\sum_{n>m} |f_n(t)|^2 < \varepsilon$ pour tout t dans K_p. L'inégalité de Cauchy–Schwarz montre alors que l'on a $\sum_{n>m} |f_n(t)| \cdot |f_n(t')| < \varepsilon$ pour t, t' dans K_p. Par suite, la série $\sum_{n=1}^{\infty} f_n(t) \cdot \overline{f_n(t')}$ converge absolument et uniformément sur $K_p \times K_p$, et la restriction de C à $K_p \times K_p$ est une fonction continue.

24.   Considérons maintenant un espace mesuré σ-fini $(\Omega, \mathcal{F}, \lambda)$ et une fonction complexe X sur $T \times \Omega$. On fait les deux hypothèses suivantes:

(a) X *est mesurable par rapport à la tribu* $\mathcal{A} \otimes \mathcal{F}$ *sur* $T \times \Omega$;
(b)   *pour tout point t de T, l'intégrale* $\int_{\Omega} |X(t,\omega)|^2 \, d\lambda(\omega)$ *est finie.*

Pour chaque point t de T, la fonction $\omega \mapsto X(t,\omega)$ sur Ω est mesurable par rapport à $\mathcal{F}$ et de carré intégrable par rapport à λ; elle définit donc un élément X_t de l'espace de Hilbert $L^2 = L^2(\Omega, \mathcal{F}, \lambda)$. La covariance C associée à l'application $t \mapsto X_t$ de T dans L^2 est donnée par la formule

$$C(t,t') = \int_{\Omega} \overline{X(t,\omega)} \cdot X(t',\omega) \, d\lambda(\omega). \tag{6}$$

LEMME.   *La covariance C est séparable et mesurable sur* $(T, \mathcal{A})$.

La formule (6) montre à l'évidence que C est mesurable sur $T \times T$ par rapport à la tribu $\mathcal{A} \otimes \mathcal{A}$ (la mesure λ est supposée σ-finie). Pour montrer que C est séparable, il suffit de montrer que les X_t appartiennent à un sous-espace de Hilbert séparable de L^2. Or, par définition, la tribu $\mathcal{A} \otimes \mathcal{F}$ est engendrée par les ensembles de la forme $A \times B$ avec $A \in \mathcal{A}$ et $B \in \mathcal{F}$, et chaque élément de $\mathcal{A} \otimes \mathcal{F}$ appartient à la sous-tribu engendrée par une suite de tels ensembles $A_n \times B_n$. Appliquons ceci à chacun des ensembles $X^{-1}(I)$ où I est un intervalle à extrémités rationnelles de $\mathbf{R}$. Il existe donc une sous-tribu $\mathcal{A}_0$ de $\mathcal{A}$, et une sous-tribu $\mathcal{F}_0$ de $\mathcal{F}$, toutes deux dénombrablement engendrées, et telles que X soit mesurable par rapport à la tribu $\mathcal{A}_0 \otimes \mathcal{F}_0$. Mais alors $L^2(\Omega, \mathcal{F}_0, \lambda_0)$ (où λ_0 est la restriction de λ à $\mathcal{F}_0$) est un sous-espace séparable de l'espace de Hilbert $L^2 = L^2(\Omega, \mathcal{F}, \lambda)$, contenant tous les X_t. Ceci prouve le lemme.

25.   Concluons cette partie en donnant un exemple typique de *covariance mesurable et non séparable*. Prenons pour T l'intervalle $[0,1]$ de $\mathbf{R}$ et pour $\mathcal{A}$ la tribu borélienne sur T.

Définissons une fonction borélienne D sur $T \times T$ par la formule

$$D(t,t') = \begin{cases} 1 & \text{si} \quad t = t', \\ 0 & \text{si} \quad t \neq t'. \end{cases} \tag{7}$$

Autrement dit, D est la fonction caractéristique (ou indicatrice) de la diagonale du carré $[0,1] \times [0,1]$. Si $t_1, \ldots, t_n$ sont des points distincts de

T, et $\xi_1, \ldots, \xi_n$ des nombres complexes, on a

$$\sum_{i,\,j=1}^{n} D(t_i, t_j)\bar{\xi}_i\xi_j = \sum_{i=1}^{n} |\xi_i|^2, \qquad (8)$$

donc D est une covariance sur $[0, 1]$.

Posons $D_t(t') = D(t', t)$. Vu la définition de D, la famille non dénombrable $(D_t)_{t \in T}$ est une base orthonormale de l'espace de Hilbert $\mathscr{E}_D$, donc la covariance D n'est pas séparable.

L'espace $\mathscr{E}_D$ se compose des fonctions f sur T telles que $\sum_{t \in T} |f(t)|^2$ soit fini; une telle fonction est nulle en dehors d'une partie dénombrable de T. Enfin, le produit scalaire dans l'espace $\mathscr{E}_D$ est donné par la formule

$$\langle f | g \rangle_D = \sum_{t \in T} \overline{f(t)} . g(t). \qquad (9)$$

V. Une généralisation du théorème de Mercer

Dans toute cette partie, on note $(T, \mathscr{A}, \mu)$ *un espace mesuré σ-fini.*

26. Voici d'abord le lemme-clé:

Lemme. *Soit C une covariance mesurable sur $(T, \mathscr{A})$. On suppose que l'intégrale*

$$\gamma = \int_T C(t, t)\,d\mu(t) \qquad (1)$$

est finie (ce qui a lieu par exemple si $\mu(T)$ est fini et la fonction C bornée). Alors toute fonction appartenant à $\mathscr{E}_C$ est de carré μ-intégrable. De plus, l'application qui à toute fonction dans $\mathscr{E}_C$ fait correspondre sa classe d'équivalence μ-presque partout est un opérateur de Hilbert–Schmidt j de $\mathscr{E}_C$ dans $L^2 = L^2(T, \mathscr{A}, \mu)$. On a $\mathrm{Tr}(j^*j) \leq \gamma$, avec égalité si C est séparable.

Posons $h(t) = C(t, t)^{1/2}$ pour t dans T; par hypothèse, la fonction h sur T est de carré μ-intégrable et l'on a $\|h\|_2^2 = \gamma$. D'après la formule (22) du n° 6, on a $|f(t)| \leq \|f\|_C . h(t)$ pour $t \in T$ et $f \in \mathscr{E}_C$. Il en résulte que toute fonction $f \in \mathscr{E}_C$ est de carré μ-intégrable, et que j est un opérateur de $\mathscr{E}_C$ dans L^2. Le théorème du n° 20 montre alors que j est un opérateur de Hilbert–Schmidt, et que l'on a $\mathrm{Tr}(j^*j) \leq \|h\|_2^2 = \gamma$.

Supposons maintenant C séparable. Si $(f_n)_{n \in I}$ est une base orthonormale de l'espace de Hilbert $\mathscr{E}_C$, on a $\sum_{n \in I} |f_n(t)|^2 = C(t, t)$ pour tout t dans T, et l'ensemble I est dénombrable. Par intégration, on en déduit

$$\mathrm{Tr}(j^*j) = \sum_{n \in I} \|j(f_n)\|_2^2 = \sum_{n \in I} \int_T |f_n(t)|^2\,d\mu(t) = \gamma.$$

27.   On conserve les notations et les hypothèses du lemme précédent. Par construction, le noyau $\mathcal{N}$ de l'opérateur $j:\mathcal{E}_C \to L^2$ se compose des fonctions μ-négligeables appartenant à $\mathcal{E}_C$; de plus, $\mathcal{E}_C$ est somme directe de $\mathcal{N}$ et du sous-espace $\mathcal{S}$ orthogonal à $\mathcal{N}$. Pour tout $t \in T$, décomposons l'élément $C_t:t' \mapsto C(t',t)$ de $\mathcal{E}_C$ en $C_t = N_t + S_t$ avec $N_t \in \mathcal{N}$ et $S_t \in \mathcal{S}$. Pour toute fonction f dans $\mathcal{E}_C$, on a

$$f(t) = \langle C_t | f \rangle_C = \langle N_t | f \rangle_C + \langle S_t | f \rangle_C. \tag{2}$$

En particulier, on a $g(t) = \langle N_t | g \rangle$ pour g dans $\mathcal{N}$; si l'on pose $N(t,t') = N_{t'}(t)$, la fonction N est une covariance sur T et l'on a $\mathcal{N} = \mathcal{E}_N$. On introduit de même la covariance S par $S(t,t') = S_{t'}(t)$ et l'on a $\mathcal{S} = \mathcal{E}_S$.

Comme j est un opérateur de Hilbert–Schmidt de $\mathcal{E}_C$ dans L^2, il résulte des résultats rappelés au n° 19 que l'orthogonal du noyau de j est un espace de Hilbert séparable, donc la covariance S est séparable. Comme la covariance C est mesurable sur $(T, \mathcal{A})$, l'espace $\mathcal{E}_C$ se compose de fonctions $\mathcal{A}$-mesurables, et il en donc de même du sous-espace $\mathcal{E}_S$ de $\mathcal{E}_C$; comme la covariance S est séparable, elle est donc aussi mesurable sur $(T, \mathcal{A})$. Comme on a $N = C - S$, la covariance N est mesurable sur $(T, \mathcal{A})$. De plus, par construction, aucune fonction non nulle de $\mathcal{E}_S$ n'est μ-négligeable, et N_t est μ-négligeable pour tout t dans T.

On dira dans la suite que $C = N + S$ exprime la μ-*décomposition* de la covariance C mesurable sur $(T, \mathcal{A})$. Montrons que les propriétés énoncées précédemment caractérisent cette décomposition de C.

LEMME.   *Soit C une covariance mesurable sur $(T, \mathcal{A})$ telle que l'intégrale $\int_T C(t,t)\, d\mu(t)$ soit finie. Soient S' et N' deux covariances mesurables sur $(T, \mathcal{A})$, de somme C. On suppose que 0 est la seule fonction μ-négligeable appartenant a $\mathcal{E}_{S'}$, et que la fonction $N'_t:t' \mapsto N'(t',t)$ est μ-négligeable pour tout t dans T. Alors, on a $N' = N$ et $S' = S$.*

Posons $\mathcal{N}' = \mathcal{E}_{N'}$ et $\mathcal{S}' = \mathcal{E}_{S'}$. Toute fonction appartenant à $\mathcal{N}'$ est limite en norme dans $\mathcal{N}'$, donc en tout point de T, d'une suite de combinaisons linéaires finies de fonctions N'_t. Donc $\mathcal{N}'$ se compose de fonctions μ-négligeables, et par hypothése, on a $\mathcal{N}' \cap \mathcal{S}' = 0$. Sur l'espace somme $\mathcal{E}' = \mathcal{N}' + \mathcal{S}'$, on définit un produit scalaire par la formule

$$\langle g + h | g' + h' \rangle = \langle g | g' \rangle_{N'} + \langle h | h' \rangle_{S'} \tag{3}$$

pour g, g' dans $\mathcal{N}'$ et h, h' dans $\mathcal{S}'$. Alors $\mathcal{E}'$ est un espace de Hilbert, somme directe des sous-espaces orthogonaux $\mathcal{N}'$ et $\mathcal{S}'$. Pour tout point t de T, introduisons l'élément $S'_t:t' \mapsto S'(t',t)$ de $\mathcal{S}'$. De la relation $C = N' + S'$, on déduit que $C_t = N'_t + S'_t$ appartient à $\mathcal{E}'$; la définition (3) du produit scalaire dans $\mathcal{E}'$ montre aussitôt que l'on a $f(t) = \langle C_t | f \rangle$ pour $t \in T$ et $f \in \mathcal{E}'$.

On a donc $\mathscr{E}' = \mathscr{E}_C$, et comme $\mathscr{N}'$ se compose visiblement des fonctions μ-négligeables de $\mathscr{E}'$, on a $\mathscr{N}' = \mathscr{N}$ avec égalité des produits scalaires associés respectivement à N' et N. On a donc $N' = N$, d'où $S' = C - N' = C - N = S$.

28. THÉORÈME. *Toute covariance mesurable sur $(T, \mathscr{A})$ est la somme d'une covariance séparable et mesurable sur $(T, \mathscr{A})$ et d'une covariance qui est une fonction sur $T \times T$ négligeable pour la mesure $\mu \otimes \mu$.*

Par hypothèse, la mesure μ est σ-finie sur $(T, \mathscr{A})$. Il existe donc une mesure v sur $(T, \mathscr{A})$ ayant les mêmes ensembles négligeables que μ, et telle que $v(T)$ soit fini. D'après le théorème de Lebesgue–Fubini, les mesures $\mu \otimes \mu$ et $v \otimes v$ ont les mêmes ensembles négligeables.

Soit D une covariance mesurable sur $(T, \mathscr{A})$. L'ensemble T_1 des points de T tels que $D(t, t) \neq 0$ appartient à $\mathscr{A}$, et la fonction u définie sur T par

$$u(t) = \begin{cases} D(t,t)^{1/2} & \text{si } t \in T_1, \\ 1 & \text{si } t \in T \backslash T_1 \end{cases} \tag{4}$$

est $\mathscr{A}$-mesurable et positive. Si l'on pose

$$C(t, t') = u(t)^{-1} u(t')^{-1} D(t, t'), \tag{5}$$

on définit une covariance C mesurable sur $(T, \mathscr{A})$. On a $0 \leq C(t, t) \leq 1$ pour tout t dans T, et comme $v(T)$ est fini, l'intégrale $\int_T C(t, t)\, dv(t)$ est finie. Introduisons la v-décomposition $C = N + S$, et posons

$$D_1(t, t') = u(t) u(t') S(t, t'), \tag{6}$$

$$D_2(t, t') = u(t) u(t') N(t, t') \tag{7}$$

pour t, t' dans T. Alors D_1 et D_2 sont deux covariances mesurables sur $(T, \mathscr{A})$, de somme D. Comme S est séparable, il en est de même de D_1. Par hypothèse, l'application $t' \mapsto N(t', t)$ est v-négligeable pour tout t dans T; d'après le théorème de Lebesgue–Fubini, la fonction N est négligeable pour $v \otimes v$, donc aussi pour $\mu \otimes \mu$. Enfin, la formule (7) montre que D_2 est elle aussi négligeable pour $\mu \otimes \mu$.

29. Dans la discussion qui suit, nous devrons soigneusement distinguer une fonction f mesurable sur $(T, \mathscr{A})$ de sa classe d'équivalence pour l'égalité presque partout, qu'on notera $\tilde{f}$.

Soit de nouveau C une covariance mesurable sur $(T, \mathscr{A})$ telle que l'intégrale $\int_T C(t, t)\, d\mu(t)$ soit finie. Pour tout point t de T, la fonction $C_t : t' \mapsto \overline{C(t, t')}$ appartient à $\mathscr{E}_C$, donc elle est de carré μ-intégrable. L'intégrale

$$KF(t) = \int_T C(t, t') . f(t')\, d\mu(t') \tag{8}$$

est donc définie si f est une fonction de carré μ-intégrable sur T, quel que soit le point t de T.

LEMME. *L'opérateur j^* de L^2 dans $\mathscr{E}_c$ adjoint de j est défini par la relation $j^*(\tilde{f}) = Kf$ pour toute fonction f de carré μ-intégrable.*

Soit t un point de T. On a

$$j^*(\tilde{f})(t) = \langle C_t | j^*(\tilde{f}) \rangle_C = \langle j(C_t) | \tilde{f} \rangle_2 = \langle \tilde{C}_t | \tilde{f} \rangle_2$$
$$= \int_T \overline{C_t(t')} f(t') \, d\mu(t') = \int_T C(t, t') f(t') \, d\mu(t').$$

Ce calcul établit le lemme.

Le lemme précédent montre que, si f est une fonction de carré μ-intégrable sur T, la fonction Kf définie par la formule (8) est aussi de carré μ-intégrable. Par passage aux classes, on définit donc un opérateur $\tilde{K}$ dans L^2, tel que $\tilde{K}\tilde{f} = \widetilde{Kf}$; cet opérateur n'est autre que jj^*.

30. Nous aurons encore besoin d'une remarque sur les vecteurs propres de l'opérateur $\tilde{K}$ dans L^2.

LEMME. *Soient u un élément de L^2 et λ un nombre réel non nul tels que $\tilde{K}u = \lambda u$. Il existe un unique représentant f de u tel que $Kf = \lambda f$.*

La formule de définition de Kf montre aussitôt que la fonction Kf est la même pour tous les représentants f de u; si l'on note Ku cette fonction, on a $\tilde{K}u = \widetilde{Ku}$. L'égalité $\tilde{K}u = \lambda u$ signifie donc que u est la classe de la fonction $f = \lambda^{-1} Ku$; on a $Ku = \lambda f$, et comme f est un représentant de u, on a $Kf = \lambda f$.

Soit f_1 un représentant de u tel que $Kf_1 = \lambda f_1$. Comme f et f_1 diffèrent par une fonction μ-négligeable, on a $Kf_1 = Kf$, d'où $\lambda f_1 = \lambda f$; en simplifiant par la constante $\lambda \neq 0$, on conclut à l'égalité de f_1 et f.

31. Voici maintenant la généralisation annoncée du théorème de Mercer.

THÉORÈME. *Soient $(T, \mathscr{A}, \mu)$ un espace mesuré σ-fini, et C une covariance mesurable sur $(T, \mathscr{A})$ telle que l'intégrale $\gamma = \int_T C(t, t) \, d\mu(t)$ soit finie.*

(a) *Pour toute fonction f de carré μ-intégrable sur T, la fonction Kf définie par*

$$Kf(t) = \int_T C(t, t') \cdot f(t') \, d\mu(t') \tag{9}$$

(pour $t \in T$) est de carré μ-intégrable sur T.

(b) *Par passage aux classes, K définit un opérateur $\tilde{K}$ dans $L^2 = L^2(T, \mathscr{A}, \mu)$ qui est hermitien, positif et de trace finie.*

(c) Soit $\tilde{\mathcal{N}}$ le noyau de $\tilde{K}$. Il existe une suite $(f_n)_{n \in I}$ de fonctions de carré μ-intégrable sur T, et une suite décroissante $(\lambda_n)_{n \in I}$ de nombres réels strictement positifs tels que l'on ait

$$Kf_n = \lambda_n f_n \qquad \text{pour tout} \quad n \in I, \tag{10}$$

et que la famille des classes $(\tilde{f}_n)_{n \in I}$ soit une base orthonormale du sous-espace de L^2 orthogonal à $\tilde{\mathcal{N}}$. L'ensemble I est fini, ou bien la suite $(\lambda_n)_{n \in I}$ tend vers 0.

(d) Soit $C = N + S$ la μ-décomposition de la covariance C. Alors S est donnée par le développement absolument convergent

$$S(t, t') = \sum_{n \in I} \lambda_n f_n(t) \overline{f_n(t')}, \tag{11}$$

et les fonctions S et C sont égales presque partout sur $T \times T$ par rapport à la mesure $\mu \otimes \mu$.

(e) La trace de $\tilde{K}$ est égale à $\sum_{n \in I} \lambda_n$. On a aussi la relation

$$\text{Tr}(\tilde{K}) = \int_T S(t, t)\, d\mu(t) \leq \int_T C(t, t)\, d\mu(t), \tag{12}$$

avec égalité si la covariance C est séparable.

D'après le lemme du n° 26, la formule $j(f) = \tilde{f}$ définit un opérateur de Hilbert–Schmidt de $\mathscr{E}_C$ dans L^2. On a vu au n° 29 que $\tilde{K}$ est égal à jj^*, d'où les assertions (a) et (b). L'assertion (c) résulte du théorème de décomposition spectrale rappelé au n° 18, le choix précisé des fonctions propres se faisant au moyen du lemme du n° 30.

Le noyau de j est l'espace $\mathscr{N}$ introduit au n° 27. Comme j est un opérateur de Hilbert–Schmidt, il en est de même de j^*. En appliquant le théorème de structure des opérateurs de Hilbert–Schmidt rappelé au n° 19, on voit que les fonctions $\lambda_n^{-1/2} j^*(\tilde{f}_n)$ forment une base orthonormale du sous-espace $\mathscr{S}$ de $\mathscr{E}_C$ orthogonal à $\mathscr{N}$. Or on a

$$\lambda_n^{-1/2} j^*(\tilde{f}_n) = \lambda_n^{-1/2} Kf_n = \lambda_n^{1/2} f_n, \tag{13}$$

et $\mathscr{S}$ est l'espace de Hilbert $\mathscr{E}_S$ associé à la covariance S. La formule (11) résulte aussitôt de là. On sait que, pour tout point t de T, l'application $t' \mapsto N(t', t)$ est μ-négligeable; d'après le théorème de Lebesgue–Fubini, la fonction $N = C - S$ est donc négligeable pour $\mu \otimes \mu$. Ceci établit l'assertion (d).

La formule $\text{Tr}(\tilde{K}) = \sum_{n \in I} \lambda_n$ résulte des rappels du n° 18. De plus, d'après la formule (11), on a $S(t, t) = \sum_{n \in I} \lambda_n |f_n(t)|^2$ pour tout $t \in T$; par intégration de cette série de fonctions positives, on obtient

$$\int_T S(t, t)\, d\mu(t) = \sum_{n \in I} \lambda_n \int_T |f_n(t)|^2\, d\mu(t) = \sum_{n \in I} \lambda_n. \tag{14}$$

Enfin, on a $\tilde{K} = jj^*$, d'où par application du lemme du n° 26 la relation

$$\text{Tr}(\tilde{K}) = \text{Tr}(jj^*) = \text{Tr}(j^*j) \le \gamma, \tag{15}$$

avec égalité si la covariance C est séparable.

32.   Il n'y a pas toujours égalité dans la formule (12). Revenons en effet à l'exemple du n° 25. Nous supposons donc que T est l'intervalle $[0, 1]$ de **R**, que $\mathscr{A}$ est la tribu borélienne sur T, et que μ est la mesure de Lebesgue. La covariance D est l'indicatrice de la diagonale dans $T \times T$. Pour tout $t \in T$, la fonction D_t est nulle en tout point de T différent de t, donc est μ-négligeable. Par suite, la μ-décomposition de D est donnée par $N = D$ et $S = 0$. L'opérateur intégral associé à D par la formule

$$Kf(t) = \int_T D(t, t')f(t') \, dt' \tag{16}$$

est nul. On a donc $\tilde{K} = 0$, d'où $\text{Tr}(\tilde{K}) = 0$. Cependant, on a $\int_T D(t, t) \, dt = 1$.

33.   Pour retrouver le théorème de Mercer usuel, supposons désormais que T soit un espace topologique, $\mathscr{A}$ la tribu borélienne sur T, et que la mesure μ satisfasse à $\mu(U) > 0$ pour toute partie ouverte non vide U de T. On suppose aussi que la covariance C est continue, et que l'intégrale $\int_T C(t, t) \, d\mu(t)$ est finie.

Comme la covariance C est continue, l'espace $\mathscr{E}_C$ se compose de fonctions continues (voir le n° 11). Soit $\mathscr{N}$ l'ensemble des fonctions μ-négligeables appartenant à $\mathscr{E}_C$. Si f appartient à $\mathscr{N}$, l'ensemble des points où f est non nulle est un ouvert U de T tel que $\mu(U) = 0$, d'ou $U = \varnothing$ et $f = 0$. L'orthogonal $\mathscr{S}$ de $\mathscr{N}$ dans $\mathscr{E}_C$ est donc égal à $\mathscr{E}_C$, et dans la μ-décomposition $C = N + S$, on a $N = 0$ et $S = C$; en particulier, la covariance C est séparable.

Dans le théorème du n° 31, on peut donc remplacer les assertions (d) et (e) par les suivantes:

(d')   *Les fonctions f_n sont continues et C admet un développement convergeant absolument et uniformément sur toute partie compacte de T:*

$$C(t, t') = \sum_{n \in I} \lambda_n f_n(t) \cdot \overline{f_n(t')}. \tag{17}$$

(e')   *On a les égalités*

$$\text{Tr}(\tilde{K}) = \sum_{n \in I} \lambda_n = \int_T C(t, t) \, d\mu(t). \tag{18}$$

Voici quelques propriétés supplémentaires:

(f)   La famille $(\lambda_n^{1/2} f_n)_{n \in I}$ est une base orthonormale de l'espace de Hilbert $\mathscr{E}_C$. D'aprés les n°s 11 et 12, toute fonction f de $\mathscr{E}_C$ est continue et elle admet un développement orthogonal $f = \sum_{n \in I} c_n f_n$ avec $\sum_{n \in I} |c_n|^2/\lambda_n$ fini, qui converge en norme dans les espaces de Hilbert $L^2(T, \mathcal{A}, \mu)$ et $\mathcal{E}_C$, et uniformément sur toute partie compacte de T. Les coefficients c_n sont donnés par la formule usuelle

$$c_n = \int_T \overline{f_n(t)} f(t) \, d\mu(t). \tag{19}$$

(g) D'après le lemme du n° 29, la fonction $f = Kg$ appartient à $\mathcal{E}_C$ pour toute fonction g de carré μ-intégrable; on peut donc lui appliquer les résultats de (f).

Pour les résultats classiques sur le théorème de Mercer, on pourra consulter Riesz et Nagy [13].

VI. Covariances mesurables et bornées

Dans toute cette partie, on note $(T, \mathcal{A}, \mu)$ un espace mesuré σ-fini.

34.   Commençons par un résultat facile sur les covariances bornées.

Lemme.   *Soient C une covariance sur T, et M une constante positive. Les conditions suivantes sont équivalentes:*

(a)   *on a $C(t, t) \le M$ pour tout $t \in T$;*
(b)   *on a $|C(t, t')| \le M$ pour t, t' dans T;*
(c)   *on a $|f(t)| \le M^{1/2}$ pour toute fonction f dans $\mathcal{E}_C$ telle que $\|f\|_C \le 1$.*

L'équivalence de (a) et (b) résulte de l'inégalité

$$|C(t, t')|^2 \le C(t, t) . C(t', t'). \tag{1}$$

Par ailleurs, on a $f(t) = \langle C_t | f \rangle_C$, et la condition (c) équivaut à $\|C_t\|_C \le M^{1/2}$; l'équivalence de (a) et (c) provient alors de la formule $C(t, t) = \|C_t\|_C^2$.

35.   On suppose maintenant que C est une covariance mesurable sur $(T, \mathcal{A})$, qui est bornée comme fonction sur $T \times T$. Si u est une fonction μ-intégrable sur T, la fonction Ku sur T définie par

$$Ku(t) = \int_T C(t, t') u(t') \, d\mu(t') \tag{2}$$

est $\mathcal{A}$-mesurable et bornée.

Lemme.   *La fonction Ku appartient à $\mathcal{E}_C$ et l'on a*

$$\langle Ku | f \rangle_C = \int_T \overline{u(t)} f(t) \, d\mu(t) \qquad \text{pour toute fonction} \quad f \in \mathcal{E}_C \tag{3}$$

$$\|Ku\|_C^2 = \int_T \int_T C(t, t') \overline{u(t)} u(t') \, d\mu(t) \, d\mu(t'). \tag{4}$$

Soit M une constante positive telle que l'on ait $|C(t, t')| \le M$ pour t, t' dans

T. D'après le lemme précédent, on a $|f(t)| \le M^{1/2} \cdot \|f\|_C$ pour $t \in T$ et $f \in \mathscr{E}_C$; par intégration on en déduit l'inégalité

$$\left| \int_T \overline{u(t)} f(t) \, d\mu(t) \right| \le M^{1/2} \cdot \|u\|_1 \|f\|_C \tag{5}$$

pour tout $f \in \mathscr{E}_C$. Comme l'espace de Hilbert $\mathscr{E}_C$ est son propre dual, il existe un élément g de $\mathscr{E}_C$ tel que l'on ait

$$\int_T \overline{u(t)} f(t) \, d\mu(t) = \langle g \, | \, f \rangle_C \qquad \text{pour tout} \quad f \in \mathscr{E}_C. \tag{6}$$

Prenons en particulier $f = C_{t'}$, d'où $f(t) = \overline{C(t',t)}$ et $\langle g \, | \, f \rangle_C = \overline{g(t')}$; on en déduit l'égalité de g et de Ku, d'où la formule (3). Pour obtenir la formule (4), il suffit de faire $f = Ku$ dans la formule (3) et d'appliquer le théorème de Lebesgue–Fubini.

36.   *Théorème. Soient C une covariance séparable et mesurable sur $(T, \mathscr{A})$ et M une constante positive telles que l'on ait $|C(t,t')| \le M$ presque partout sur $T \times T$ par rapport à la mesure $\mu \otimes \mu$. Il existe alors un ensemble μ-négligeable N tel que l'on ait $|C(t,t')| \le M$ lorsque t et t' appartiennent à $T \backslash N$. En particulier, on a $C(t,t) \le M$ μ-presque partout sur T.*

Faisons d'abord la démonstration dans le cas ou C est bornée. Vu l'hypothèse faite sur C, la formule (4) entraîne $\|Ku\|_C^2 \le M \cdot \|u\|_1^2$; d'après la formule (3) et l'inégalité de Cauchy–Schwarz, on a donc

$$\left| \int_T \overline{u(t)} f(t) \, d\mu(t) \right| \le M^{1/2} \cdot \|f\|_C \|u\|_1 \tag{7}$$

quelles que soient la fonction $f \in \mathscr{E}_C$ et la fonction μ-intégrable u. Comme il est bien connu, on déduit de là l'existence d'un ensemble μ-négligeable N_f (dépendant de f) tel que l'on ait $|f(t)| \le M^{1/2} \|f\|_C$ pour tout point t de $T \backslash N_f$. Comme la covariance C est séparable, il existe une partie dénombrable dense D dans l'espace de Hilbert $\mathscr{E}_C$. Posons $N = \bigcup_{f \in D} N_f$; c'est un ensemble μ-négligeable. D'après ce qui précède, on a donc

$$|\langle C_t \, | \, f \rangle_C| \le M^{1/2} \cdot \|f\|_C \qquad \text{pour} \quad f \in D \quad \text{et} \quad t \in T \backslash N. \tag{8}$$

Comme D est dense dans $\mathscr{E}_C$, on a donc $\|C_t\| \le M^{1/2}$ pour t dans $T \backslash N$. En utilisant l'inégalité de Cauchy–Schwarz et la formule $C(t,t') = \langle C_t \, | \, C_{t'} \rangle_C$, on en déduit l'inégalité $|C(t,t')| \le M$ pour t, t' dans $T \backslash N$.

Passons au cas général. Pour tout entier $p \ge 1$, soit T_p l'ensemble des points t de T tels que $C(t,t) \le p$. On a $|C(t,t')| \le p$ pour t, t' dans T_p, et la restriction de C à $T_p \times T_p$ est donc une covariance bornée. D'après la première partie de la' démonstration, il existe une partie μ-négligeable N_p de T_p telle que l'on ait $|C(t,t')| \le M$ pour t, t' dans $T_p \backslash N_p$. L'ensemble $N = \bigcup_{p=1}^{\infty} N_p$ satisfait aux exigences du théorème.

37.   On ne peut supprimer l'hypothèse de séparabilité dans le théorème précédent. Reprenons notre exemple familier, où T est l'intervalle $[0, 1]$ de **R**, $\mathscr{A}$ est la tribu borélienne de T et μ la mesure de Lebesgue. Soit D l'indicatrice de la diagonale dans $T \times T$. Alors D est nulle presque partout sur $T \times T$, et l'on a $D(t, t) = 1$ pour tout point t de T.

38.   Le théorème suivant renforce le théorème classique de Dunford-Pettis [4] en ce qu'il évite toute hypothèse de séparabilité, mais il ne s'applique qu'aux espaces de Hilbert. Il est essentiel que la mesure μ soit σ-finie, comme le montre l'inclusion de L^1 dans L^2 lorsque T est non dénombrable et que tout point de T est mesurable de mesure 1.

THÉORÈME.   *Soient $\mathscr{H}$ un espace de Hilbert, M une constante positive et F une application linéaire continue de $L^1 = L^1(T, \mathscr{A}, \mu)$ dans $\mathscr{H}$ telle que l'on ait $\|F(u)\| \leq M . \|u\|_1$ pour tout $u \in L^1$. Il existe alors un sous-espace de Hilbert séparable $\mathscr{H}_0$ de $\mathscr{H}$ et une application mesurable φ de $(T, \mathscr{A})$ dans $\mathscr{H}_0$ muni de sa tribu borélienne avec les propriétés suivantes:*

(a)   *on a $\|\varphi(t)\| \leq M$ pour tout point t de T;*
(b)   *l'application F prend ses valeurs dans $\mathscr{H}_0$;*
(c)   *on a*

$$\langle F(u) | a \rangle = \int_T \overline{u(t)} \langle \varphi(t) | a \rangle \, d\mu(t) \qquad pour \quad u \in L^1 \quad et \quad a \in \mathscr{H}. \qquad (9)$$

Soit L^∞ l'espace des classes de fonctions mesurables et bornées sur $(T, \mathscr{A}, \mu)$; il s'identifie au dual de L^1 de la manière bien connue. Par suite, il existe une application linéaire Φ de $\mathscr{H}$ dans L^∞ qui satisfait aux relations suivantes:

$$|\Phi a(t)| \leq M . \|a\| \qquad \mu\text{-presque partout sur } T, \qquad (10)$$

$$\langle F(u) | a \rangle = \int_T \overline{u(t)} \Phi a(t) \, d\mu(t), \qquad (11)$$

pour tout $a \in \mathscr{H}$ et tout $u \in L^1$.

Comme la mesure μ est σ-finie, il existe une fonction mesurable h sur $(T, \mathscr{A})$ telle que $h(t) > 0$ pour tout $t \in T$ et que $\int_T h(t)^2 \, d\mu(t)$ soit fini. Définissons une application linéaire j de L^2 dans $\mathscr{H}$ par $j(f) = F(\tilde{h}f)$. Il est clair qu'on a $j^*(a) = \tilde{h} . \Phi a$ pour tout $a \in \mathscr{H}$. D'après l'inégalité (10), on a $|j^*(a)| \leq M . \tilde{h}$ μ-presque partout pour chaque vecteur a de norme ≤ 1 dans $\mathscr{H}$. D'après le théorème du n° 20, l'opérateur j^* est donc de Hilbert-Schmidt, et il en est par suite de même de j. Par ailleurs, l'ensemble des classes de fonctions $\tilde{h}f$, où f parcourt L^2, est dense dans L^1, et l'image d'un opérateur de Hilbert–Schmidt est séparable d'après les résultats rappelés au n° 19. Autrement dit, il existe un sous-espace de Hilbert séparable $\mathscr{H}_0$ de $\mathscr{H}$ contenant l'image de F.

Le reste de la démonstration est classique. Choisissons une base ortho-normale $(e_n)_{n \in I}$ de $\mathcal{H}_0$ et des représentants f_n pour les classes Φe_n. Soient J une partie finie de I et $\xi = (\xi_j)_{j \in J}$ une famille de nombres complexes. On a $\Phi(\sum_{j \in J} \xi_j e_j) = \sum_{j \in J} \xi_j \tilde{f}_j$, et compte tenu de (10), il existe un ensemble μ-négligeable N_ξ tel que l'on ait

$$\left| \sum_{j \in J} \xi_j f_j(t) \right|^2 \leq M^2 \sum_{j \in J} |\xi_j|^2 \qquad (12)$$

pour tout t dans $T \backslash N_\xi$. Soit N_J l'ensemble μ-négligeable, réunion des ensembles N_ξ pour toutes les familles ξ où chaque ξ_j ait une partie réelle et une partie imaginaire rationnelles. Par continuité, l'inégalité (12) reste satisfaite pour t dans $T \backslash N_J$ et toute famille ξ; on a donc $\sum_{j \in J} |f_j(t)|^2 \leq M^2$ pour tout t dans $T \backslash N_J$. L'ensemble $N = \bigcup_J N_J$ est μ-négligeable, et l'on a $\sum_{n \in I} |f_n(t)|^2 \leq M^2$ pour tout t dans $T \backslash N$. Soit φ l'application de T dans $\mathcal{H}$ qui est nulle en tout point de N et telle que $\varphi(t) = \sum_{n \in I} \overline{f_n(t)} . e_n$ pour tout t dans $T \backslash N$. On a $\|\varphi(t)\| \leq M$ pour tout point t de T, et l'application φ est mesurable de $(T, \mathcal{A})$ dans $\mathcal{H}_0$ muni de sa tribu borélienne (voir par exemple le n° 21).

Il reste à vérifier la formule (9). Elle est vraie par construction lorsque a est l'un des vecteurs e_n, et elle est trivialement satisfaite lorsque a est ortho-gonal à $\mathcal{H}_0$. Pour u fixée dans L^1, les deux membres de (9) sont des formes linéaires continues du vecteur a de $\mathcal{H}$; le cas général se déduit aussitô de là.

39. D'après un théorème connu de Grothendieck [7], toute forme sequilinéaire continue sur L^1 admet la représentation

$$\Gamma(u, v) = \int_T \int_T C(t, t') \overline{u(t)} v(t') \, d\mu(t) \, d\mu(t'), \qquad (13)$$

où C est une fonction mesurable et bornée sur $(T \times T, \mathcal{A} \otimes \mathcal{A})$. Si M est une constante positive, les relations

$$|\Gamma(u, v)| \leq M . \|u\|_1 \|v\|_1 \qquad \text{pour } u, v \text{ dans } L^1 \qquad (14)$$

et

$$|C(t, t')| \leq M \qquad \text{presque partout sur } T \times T \text{ pour } \mu \otimes \mu \qquad (15)$$

sont équivalentes. Si la forme sesquilinéaire Γ est positive, elle satisfait à l'inégalité de Cauchy–Schwarz $|\Gamma(u, v)|^2 \leq \Gamma(u, u) . \Gamma(v, v)$ et la relation (14) équivaut encore à la suivante

$$\Gamma(u, u) \leq M . \|u\|_1^2 \qquad \text{pour } u \text{ dans } L^1. \qquad (16)$$

THÉORÈME. *Soit Γ une forme sesquilinéaire positive et continue sur L^1. Il existe une covariance C séparable, mesurable et bornée sur $(T, \mathscr{A})$ satisfaisant à la relation (13) ci-dessus.*

D'après le lemme du n° 3, il existe un espace de Hilbert $\mathscr{H}$ et une application linéaire continue F de L^1 dans $\mathscr{H}$ tels que l'on ait

$$\Gamma(u, v) = \langle F(u) | F(v) \rangle \qquad \text{pour } u, v \text{ dans } L^1. \qquad (17)$$

D'après le théorème du n° 38, on peut supposer que l'espace $\mathscr{H}$ est séparable et qu'il existe une application mesurable et bornée φ de $(T, \mathscr{A})$ dans $\mathscr{H}$ satisfaisant à la relation

$$\langle F(u) | a \rangle = \int_T \overline{u(t)} \langle \varphi(t) | a \rangle \, d\mu(t) \qquad (18)$$

pour $u \in L^1$ et $a \in \mathscr{H}$.

Notons C la covariance associée à $\mathscr{H}$ et φ; elle est séparable, bornée et mesurable sur $(T \times T, \mathscr{A} \otimes \mathscr{A})$. Soit $\mathscr{F}$ le sous-espace vectoriel fermé de $\mathscr{H}$ engendré par l'image de φ. Comme au n° 5, définissons un isomorphisme H de $\mathscr{F}$ sur $\mathscr{E}_C$ par la relation $Ha(t) = \langle \varphi(t) | a \rangle$ pour $a \in \mathscr{F}$ et $t \in T$. La formule (18) équivaut donc à la suivante

$$\langle HF(u) | f \rangle_C = \int_T \overline{u(t)} f(t) \, d\mu(t) \qquad \text{pour} \quad u \in L^1 \quad \text{et} \quad f \in \mathscr{E}_C. \quad (19)$$

Si l'on prend en particulier pour f une fonction de la forme $C_{t'} : t \mapsto C(t, t')$, on voit que $HF(u)$ n'est autre que la fonction notée Ku au n° 35. On a donc

$$\Gamma(u, v) = \langle F(u) | F(v) \rangle = \langle HF(u) | HF(v) \rangle_C = \langle Ku | Kv \rangle_C \qquad (20)$$

pour u, v dans L^1. Enfin si l'on fait $f = Kv$ dans la formule (3) du n° 35, on obtient

$$\langle Ku | Kv \rangle_C = \int_T \int_T C(t, t') \overline{u(t)} v(t') \, d\mu(t) \, d\mu(t') \qquad (21)$$

d'après le théorème de Lebesgue–Fubini. Le théorème résulte des formules (20) et (21).

40. Terminons cette partie par un corollaire important du théorème précédent. Soit C une fonction mesurable et bornée sur $(T \times T, \mathscr{A} \otimes \mathscr{A})$. On définit la forme sesquilinéaire Γ sur L^1 par la formule (13). Si A et B sont deux éléments de $\mathscr{A}$, on pose

$$\Gamma[A, B] = \int_A \int_B C(t, t') \, d\mu(t) \, d\mu(t'). \qquad (22)$$

COROLLAIRE. *Les conditions suivantes sont équivalentes:*

(a) *La fonction C est égale presque partout (pour $\mu \otimes \mu$) à une covariance mesurable et bornée.*

(b) *La fonction C est égale presque partout à une covariance séparable, mesurable et bornée.*

(c) *On a* $\Gamma(u, u) \geq 0$ *pour toute fonction μ-intégrable u sur T.*

(d) *La matrice d'éléments $\Gamma[A_i, A_j]$ est hermitienne positive pour toute suite $(A_i)_{1 \leq i \leq n}$ d'éléments de $\mathscr{A}$.*

Si C est une covariance, on a $\Gamma(u, u) = \|Ku\|_C^2$ avec les notations du n° 35, donc (a) implique (c). Le théorème du n° 39 exprime que (c) implique (b), et il est clair que (b) implique (a).

Avec les notations de (d), on a

$$\sum_{i, j = 1}^{n} \Gamma[A_i, A_j]\overline{\xi_i}\xi_j = \Gamma(u, u) \tag{23}$$

où la fonction u prend la valeur constante ξ_j sur A_j, et est nulle hors de $A_1 \cup \cdots \cup A_n$. Comme les classes de telles fonctions sont denses dans L^1, les conditions (c) et (d) sont équivalentes.

D'après le corollaire précédent, toute covariance mesurable et bornée est égale presque partout à une covariance séparable, mesurable et bornée. On a donc retrouvé un cas particulier du théorème du n° 28.

VII. Une inégalité intégrale

Dans toute cette partie, on note $(T, \mathscr{A}, \mu)$ un espace mesuré σ-fini.

41.   Soit C une covariance mesurable sur $(T, \mathscr{A})$. On pose

$$C[u] = \int_T \int_T C(t, t')\overline{u(t)}u(t')\, d\mu(t)\, d\mu(t') \tag{1}$$

pour toute fonction mesurable u sur $(T, \mathscr{A})$ pour laquelle l'intégrale précédente converge absolument. Nous nous proposons de montrer que $C[u]$ *est toujours positif.*

Nous connaissons déjà plusieurs cas particuliers:

(a) Supposons que la mesure μ soit portée par un ensemble fini $A = \{t_1, \ldots, t_n\}$ au sens que $\mu(T \backslash A)$ est nul, et que toute partie de A appartient à $\mathscr{A}$. Posons $\xi_i = u(t_i)\mu(\{t_i\})$ pour $1 \leq i \leq n$. On a alors $C[u] = \sum_{i, j = 1}^{n} C(t_i, t_j)\overline{\xi_i}\xi_j$ et l'inégalité $C[u] \geq 0$ exprime la définition d'une covariance.

(b) Au n° 15, nous avons considéré le cas où T est un espace localement compact, muni de sa tribu borélienne $\mathscr{A}$, où la mesure μ donne une masse finie à toute partie compacte de T, où C est continue et u continue à support compact.

(c)   Supposons que les intégrales $\int_T C(t,t)\,d\mu(t)$ et $\int_T |u(t)|^2\,d\mu(t)$ soient finies. D'après l'inégalité de Cauchy–Schwarz, l'intégrale $\int_T C(t,t)^{1/2}|u(t)|\,d\mu(t)$ est finie et l'inégalité $|C(t,t')| \le C(t,t)^{1/2}C(t',t')^{1/2}$ montre que l'intégrale définissant $C[u]$ converge absolument. Utilisons les notations du n° 31; d'après le théorème de Lebesgue–Fubini, on a $C[u] = \langle \tilde{u} | \tilde{K}\tilde{u} \rangle$, et comme l'opérateur $\tilde{K}$ est hermitien positif dans $L^2(T, \mathscr{A}, \mu)$, on a bien $C[u] \ge 0$.

(d)   D'après le corollaire du n° 40, on a encore $C[u] \ge 0$ lorsque la covariance C est bornée et que la fonction u est μ-intégrable.

42.   Pour démontrer l'inégalité $C[u] \ge 0$, nous ferons une réduction préliminaire.

Comme la mesure μ est σ-finie, il existe une fonction $\mathscr{A}$-mesurable v sur T, à valeurs strictement positives et finies, et une mesure v sur $\mathscr{A}$ telles que $v(T)$ soit fini et que l'on ait $\mu(A) = \int_A v(t)\,dv(t)$ pour tout A dans $\mathscr{A}$. Posons

$$D(t,t') = C(t,t')\overline{u(t)v(t)}u(t')v(t') \qquad (2)$$

pour t, t' dans T. Il est immédiat que D est une covariance mesurable sur $(T, \mathscr{A})$ et que l'on a

$$C[u] = \int_T \int_T D(t,t')\,dv(t)\,dv(t'). \qquad (3)$$

Par ailleurs, soit T_p l'ensemble des points t de T tels que $D(t,t) \le p$. Alors T est la réunion de la suite croissante $(T_p)_{p \ge 1}$, et l'on a donc

$$C[u] = \lim_{p \to \infty} \int_{T_p} \int_{T_p} D(t,t')\,dv(t)\,dv(t'). \qquad (4)$$

On est donc ramené à prouver l'inégalité

$$\int_S \int_S D(t,t')\,dv(t)\,dv(t') \ge 0 \qquad (5)$$

où S est l'un des ensembles T_p.

D'après l'inégalité $|D(t,t')|^2 \le D(t,t) \cdot D(t',t')$, la fonction D est bornée sur $S \times S$; le nombre $c = v(S)$ est positif et fini, et l'inégalité (5) est clairement vérifiée lorsque $c = 0$. Supposons donc $c > 0$ et fixons un entier $n \ge 1$. Soient $t_1, \dots, t_n$ des points de S; comme D est une covariance sur S, on a

$$\sum_{i,j=1}^n D(t_i, t_j) \ge 0. \qquad (6)$$

Posons

$$a_{ij} = \int_S \cdots \int_S D(t_i, t_j)\,dv(t_1) \cdots dv(t_n). \qquad (7)$$

Le calcul de cette intégrale est immédiat:

$$a_{ii} = c^{n-1} \int_S D(t, t)\, dv(t), \qquad (8)$$

$$a_{ij} = c^{n-2} \int_S \int_S D(t, t')\, dv(t)\, dv(t') \qquad \text{si} \quad i \neq j. \qquad (9)$$

Intégrons l'inégalité (6) par rapport à $t_1, \ldots, t_n$. Comme il y a n couples (i, j) avec $i = j$ et $n^2 - n$ couples avec $i \neq j$, on trouve aprés division par $c^{n-2}n^2$ l'inégalité

$$\frac{c}{n} \int_S D(t, t)\, dv(t) + \frac{n-1}{n} \int_S \int_S D(t, t')\, dv(t)\, dv(t') \geq 0. \qquad (10)$$

L'inégalité (5) résulte de là en faisant tendre n vers l'infini. On remarquera que la réduction au cas où D est bornée est faite pour assurer que l'intégrale $\int_S D(t, t)\, dv(t)$ est finie.

43. *Remarques.* (a) La démonstration de l'inégalité (5) est inspirée d'un artifice de Riesz [12], repris par Weil [18] dans l'étude des fonctions de type positif sur un groupe localement compact. On en donnera une "explication" au n° suivant.

(b) La démonstration précédente utilise une propriété moins forte que la définition d'une covariance, à savoir:

La matrice $(C(t_i, t_j))_{1 \leq i, j \leq n}$ est hermitienne positive pour presque tout système $(t_1, \ldots, t_n)$ de points de T, par rapport à la mesure $\mu \otimes \cdots \otimes \mu$ (n facteurs). (PC)

En effet, cette hypothèse suffit pour assurer que l'inégalité (6) a lieu presque partout sur $T \times \cdots \times T$ (n facteurs) par rapport à la mesure $v \otimes \cdots \otimes v$. De plus, l'inégalité $|D(t, t')|^2 \leq D(t, t) \cdot D(t', t')$ a lieu presque partout sur $S \times S$ par rapport à $v \otimes v$, donc D est égale presque partout sur $S \times S$ à une fonction bornée. Ceci suffit à assurer que les intégrales a_{ij} sont définies. Le reste de la démonstration est sans changement.

La condition (PC) est malheureusement assez peu utilisable. Si C est presque partout (pour $\mu \otimes \mu$) égale à une covariance, on aura encore $C[u] \geq 0$ chaque fois que cette intégrale est définie; mais C ne satisfait pas nécessairement à la condition (PC). Par exemple, avec les notations du n° 25, la fonction $-D$ est égale presque partout (pour la mesure de Lebesgue) à la covariance nulle; elle ne satisfait pas à (PC), puisque l'on a $-D(t, t) = -1$, alors que la condition (PC) exige que l'on ait $C(t, t) \geq 0$ μ-presque partout sur T.

44. Donnons maintenant une *interprétation probabiliste* des résultats précédents. Pour simplifier, on suppose que T est un espace localement compact à base dénombrable et $\mathscr{A}$ sa tribu borélienne, que C est une co-

variance continue et bornée sur T, et u une fonction continue et bornée sur T; on suppose aussi que l'on a $\mu(T) = 1$.

Soit Ω l'espace topologique produit d'une suite de copies T_n de T (pour $n \geq 1$); on munit Ω de sa tribu borélienne $\mathscr{F}$ et de la mesure de probabilité $P = \otimes_{n=1}^{\infty} \mu_n$, avec $\mu_n = \mu$ pour tout entier $n \geq 1$. Etant donné un élément $\omega = (t_n)_{n \geq 1}$ de Ω et un entier $n \geq 1$, on note $\gamma_{n,\omega}$ la mesure de probabilité sur T qui alloue la masse $1/n$ à chacun des points $t_1, \ldots, t_n$ (répétitions permises). On considère la famille $(\gamma_{n,\omega})_{\omega \in \Omega}$ comme une mesure aléatoire γ_n sur T, définie par rapport à l'espace probabilisé $(\Omega, \mathscr{F}, P)$.

Rappelons qu'une suite de mesures de probabilité μ_n sur T tend vaguement vers une mesure de probabilité μ sur T si l'on a

$$\lim_{n \to \infty} \int_T f(t)\, d\mu_n(t) = \int_T f(t)\, d\mu(t) \tag{11}$$

pour toute fonction f continue et bornée sur T.

Une des formes de la *loi forte des grands nombres* affirme que la suite des mesures aléatoires γ_n tend vaguement vers μ avec probabilité 1; autrement dit, il existe un ensemble Ω_1 dans $\mathscr{F}$ tel que $P(\Omega_1) = 1$ et que la suite des mesures $\gamma_{n,\omega}$ tende vaguement vers μ pour tout ω dans Ω_1.

Définissons une suite de variables aléatoires X_n par la formule

$$X_n = \int_T \int_T C(t, t')\overline{u(t)}u(t')\, d\gamma_n(t)\, d\gamma_n(t'). \tag{12}$$

Vu la définition d'une covariance, on a $X_n \geq 0$ [voir le cas (a) du n° 41]. La loi forte des grands nombres montre alors que la suite des variables aléatoires X_n tend avec probabilité 1 vers la constante

$$C[u] = \int_T \int_T C(t, t')\overline{u(t)}u(t')\, d\mu(t)\, d\mu(t').$$

Autrement dit, si l'on pose

$$D(t, t') = C(t, t')\overline{u(t)}u(t') \tag{13}$$

pour t, t' dans T, on a

$$X_n(\omega) = n^{-2} \sum_{i,j=1}^{n} D(t_i, t_j) \qquad \text{pour} \quad \omega = (t_1, t_2, \ldots) \tag{14}$$

et

$$C[u] = \lim_{n \to \infty} X_n(\omega) \qquad \text{pour } \omega \text{ dans } \Omega_1. \tag{15}$$

Comme on a $X_n(\omega) \geq 0$ pour tout ω dans Ω, on a bien $C[u] \geq 0$.

Remarquons que la suite des variables aléatoires X_n est majorée par une constante, et la formule (15) entraine $C[u] = \lim_{n \to \infty} E[X_n]$. Or d'après la formule (14), on a

$$E[X_n] = n^{-2} \int_T \cdots \int_T \sum_{i,j=1}^n D(t_i, t_j)\, d\mu(t_1) \cdots d\mu(t_n); \qquad (16)$$

on retrouve donc le calcul fait au n° 42.

VIII. Applications et exemples

45. *Théorie des groupes*

Montrons comment l'on retrouve les propriétés élémentaires des fonctions de type positif sur un groupe au moyen des résultats généraux sur les covariances.

Soit G un groupe topologique, d'élément unité e, opérant continuement sur un espace topologique T. On dira qu'une covariance C sur T est *invariante* par G si l'on a $C(g \cdot t, g \cdot t') = C(t, t')$ pour t, t' dans T et g dans G; supposons qu'il en soit ainsi. Si g est élément de G et f une fonction dans $\mathscr{E}_C$, le lemme du n°5 montre que la fonction $U_g f : t \mapsto f(g^{-1} \cdot t)$ appartient à $\mathscr{E}_C$ et qu'elle a même norme que f dans $\mathscr{E}_C$. Supposons de plus que C soit continue sur $T \times T$; le produit scalaire $\langle C_t | U_g C_{t'} \rangle$ est égal à $C(g^{-1} \cdot t, t') = C(t, g \cdot t')$, donc est fonction continue de g pour t, t' fixés dans T. Rappelons que les combinaisons linéaires finies des fonctions C_t sont denses dans l'espace de Hilbert $\mathscr{E}_C$. On conclut que *l'application $g \mapsto U_g$ est une représentation unitaire continue du groupe G dans l'espace de Hilbert $\mathscr{E}_C$.*

Supposons en particulier que l'on ait $T = G$, le groupe G opérant sur lui-même par translations à gauche. La formule

$$C(g, g') = c(g^{-1} g') \qquad \text{pour } g, g' \text{ dans } G \qquad (1)$$

définit une bijection de l'ensemble des covariances invariantes (à gauche) sur G sur l'ensemble des fonctions de type positif sur G. Soit c une fonction continue de type positif sur G; alors $\overline{c} = C_e$ appartient à $\mathscr{E}_C$ et l'on a

$$c(g) = \langle \overline{c} | U_g \overline{c} \rangle_C \qquad \text{pour tout } g \text{ dans } G; \qquad (2)$$

ceci précise le résultat connu selon lequel *toute fonction continue de type positif sur G est un coefficient d'une représentation unitaire continue de G.*

Un autre cas intéressant est celui où T est un espace homogène pour G. Choisissons un point t_0 de T, de stabilisateur H; l'application $gH \mapsto g \cdot t_0$ est alors un homéomorphisme de G/H sur T. La formule

$$C(g \cdot t_0, g' \cdot t_0) = c(g^{-1} g') \qquad \text{pour } g, g' \text{ dans } G \qquad (3)$$

définit alors une bijection de l'ensemble des covariances invariantes continues sur T sur l'ensemble des fonctions continues de type positif sur G, telles que $c(hgh') = c(g)$ pour $g \in G$, et h, h' dans H. L'élément $a = C_{t_0}$ de $\mathscr{E}_C$ satisfait à $U_h a = a$ pour tout $h \in H$, et l'on a $c(g) = \langle a | U_g a \rangle_C$. Ces remarques sont le point de départ de la théorie des *fonctions sphériques*, qui s'occupe du cas où G est localement compact et H compact.

Le théorème du n° 15 implique le résultat classique suivant: supposons le groupe G localement compact; une fonction continue c sur G est de type positif si et seulement si l'on a l'inégalité

$$\int_G \int_G c(g^{-1} g') \overline{u(g)} u(g') \, dg \, dg' \geq 0 \tag{4}$$

pour toute fonction u sur G, qui est continue à support compact. De même, le théorème du n° 39 est à rapprocher du théorème selon lequel toute forme linéaire continue Φ sur $L^1(G)$ telle que l'on ait $\Phi(u^* * u) \geq 0$ pour tout $u \in L^1(G)$, est de la forme

$$\Phi(u) = \int_G c(g) u(g) \, dg \tag{5}$$

où c est une fonction de type positif sur G. Mais ledit théorème du n° 39 ne suffit pas à assurer que toute fonction de type positif qui est mesurable et bornée est égale presque partout à une fonction continue de type positif.

46. *Un théorème de finitude*

Il s'agit d'un corollaire facile du théorème du n° 20, dont voici l'énoncé:

Soit $(T, \mathscr{A}, \mu)$ *un espace mesuré tel que* $\mu(T)$ *soit fini, de sorte que* L^∞ *est un sous-espace de* L^2. *Tout sous-espace vectoriel* E *de* L^∞ *qui est fermé dans* L^2 *est de dimension finie.*

La démonstration est la suivante. Si l'on pose $c = \mu(T)^{1/2}$, on a

$$\|f\|_2 \leq c \cdot \|f\|_\infty \qquad \text{pour toute fonction} \quad f \in L^\infty. \tag{6}$$

Si une suite de fonctions $f_n \in E$ converge dans L^∞ vers une fonction f, elle converge donc aussi vers f dans L^2, et l'on a $f \in E$ par hypothèse. Autrement dit, l'espace vectoriel E est fermé à la fois dans L^∞ et dans L^2, et il est complet pour les deux normes $\|.\|_2$ et $\|.\|_\infty$. Il résulte alors de l'inégalité (6) et du théorème du graphe fermé que ces deux normes sont équivalentes sur E; il existe donc une constante positive M telle que l'on ait

$$\|f\|_\infty \leq M \cdot \|f\|_2 \qquad \text{pour toute fonction} \quad f \in E. \tag{7}$$

On peut appliquer le théorème du n° 20 en prenant pour h la constante M; par suite, l'inclusion j de E dans L^2 est un opérateur de Hilbert–Schmidt et l'on a $\text{Tr}(j^* j) \leq M^2$. Soit $(f_\alpha)_{\alpha \in A}$ une base orthonormale de E; on a $\|j(f_\alpha)\|_2 = 1$ pour tout $\alpha \in A$, d'où $\sum_{\alpha \in A} 1 = \text{Tr}(j^* j) \leq M^2$. On a prouvé que E est de dimension finie au plus égale à M^2.

Le théorème de finitude précédent a été utilisé dans la théorie des fonctions automorphes. Montrons comment, dans la théorie des équations aux dérivées partielles elliptiques, il permet de *déduire le théorème de finitude du théorème de régularité*. Supposons que T soit une variété compacte de classe C^∞, que $\mathscr{A}$ soit la tribu borélienne sur T, et que la mesure μ ait localement une densité de classe C^∞ (dans un système de coordonnées locales). Soit D un opérateur différentiel elliptique défini sur T, et soit D^* son adjoint; notons E l'espace vectoriel des fonctions f de classe C^∞ telles que $Df = 0$. Le théorème de régularité affirme que E est aussi l'espace des solutions faibles, c'est-à-dire des fonctions $f \in L^2$ telles que $\int_T \overline{f(t)} D^* g(t)\, d\mu(t) = 0$ pour toute fonction g de classe C^∞ sur T. Par suite, E est un sous-espace vectoriel fermé de L^2; toute fonction $f \in E$ est continue sur l'espace compact T, donc bornée, et le théorème de finitude ci-dessus montre que E est de dimension finie.

47. *Processus aléatoires du second ordre*

Soit $(\Omega, \mathscr{F}, P)$ un espace probabilisé, et soit $X = (X_t)_{t \in T}$ une famille de variables aléatoires relative à $(\Omega, \mathscr{F}, P)$. Supposons pour simplifier que l'espérance de chaque X_t soit nulle. On dit que X est un *processus du second ordre* si $E[|X_t|^2]$ est fini pour tout $t \in T$. La *covariance* C du processus X est alors définie par

$$C(t, t') = E[\overline{X}_t X_{t'}]; \tag{8}$$

autrement dit, c'est la covariance associée à l'application $t \mapsto X_t$ de T dans l'espace de Hilbert $L^2(\Omega, \mathscr{F}, P)$.

Soit $\mathscr{H}_X$ le plus petit sous-espace vectoriel fermé de $L^2(\Omega, \mathscr{F}, P)$ contenant les X_t. Les propriétés du second ordre du processus X sont celles qui ne dépendent que de l'espace de Hilbert $\mathscr{H}_X$ et de la famille des éléments X_t de $\mathscr{H}_X$. Par exemple, la continuité du second ordre du processus X signifie la continuité de l'application $t \mapsto X_t$ de T dans $\mathscr{H}_X$; comme on l'a vu au n° 9, ceci équivaut à la continuité de C. Soit $\mathscr{E}_C$ l'espace de Hilbert de fonctions sur T associé à la covariance C. Comme au n° 5, on définit un isomorphisme H de l'espace de Hilbert $\mathscr{H}_X$ sur l'espace de Hilbert $\mathscr{E}_C$ qui associe à une variable aléatoire $Y \in \mathscr{H}_X$ la fonction $t \mapsto E[\overline{X}_t Y]$. Cet isomorphisme transforme X_t en C_t, de sorte que la famille des éléments C_t de l'espace de Hilbert $\mathscr{E}_C$ fournit un "modèle analytique" pour l'étude des propriétés du second ordre du processus X.

Par exemple, supposons que $(T, \mathscr{A}, \mu)$ soit un espace mesuré σ-fini et que la covariance C soit mesurable sur $(T, \mathscr{A})$. Soit u une fonction $\mathscr{A}$-mesurable sur T telle que l'intégrale $\int_T C(t, t)^{1/2} \cdot |u(t)|\, d\mu(t)$ soit finie. En raisonnant comme au n° 35, on montre que la formule

$$Ku(t) = \int_T C(t, t') u(t')\, d\mu(t') \tag{9}$$

définit une fonction $Ku \in \mathscr{E}_C$, et qu'on a

$$\langle Ku | f \rangle_C = \int_T \overline{u(t)} f(t) \, d\mu(t) \qquad \text{pour tout} \quad f \in \mathscr{E}_C. \qquad (10)$$

Par l'isomorphisme H de $\mathscr{H}_X$ sur $\mathscr{E}_C$, la fonction Ku correspond à une variable aléatoire $X[u] \in \mathscr{H}_X$; elle est caractérisée par la formule suivante, qui ne fait que traduire la formule (10):

$$E[X[u] \cdot Y] = \int_T u(t) E[X_t \cdot Y] \, d\mu(t) \qquad \text{pour tout} \quad Y \in \mathscr{H}_X. \qquad (11)$$

Ce résultat justifie la notation intégrale $\int_T u(t) \cdot X_t \, d\mu(t)$ pour $X[u]$. Prenant $Y = \overline{X[u]}$, on obtient le cas particulier

$$E\left[\left| \int_T u(t) \cdot X_t \, d\mu(t) \right|^2\right] = \int_T \int_T C(t, t') \overline{u(t)} u(t') \, d\mu(t) \, d\mu(t'). \qquad (12)$$

L' intégrale au second membre a été notée $C[u]$ précédemment; la formule (12) rend évident le fait qu'elle soit positive. Voici expliqués les mystères des intégrales aléatoires du second ordre!

48. *Covariances séparables et développement de Karhunen* [9]

Avec les notations précédentes, supposons que la covariance C du processus du second ordre X soit *séparable et mesurable* sur $(T, \mathscr{A})$. Interprétons d'abord le théorème du n° 36: supposons qu'il existe une constante positive M telle que l'on ait $|C(t, t')| \leq M$ presque partout sur $T \times T$ (par rapport à $\mu \otimes \mu$); il existe alors un processus du second ordre $X' = (X'_t)_{t \in T}$ tel que l'on ait $E[|X'_t|^2] \leq M$ pour tout $t \in T$, et $X'_t = X_t$ pour presque tout $t \in T$ (par rapport à μ).

Soit $(f_n)_{n \in I}$ une suite de fonctions sur T, dont les conjuguées $\overline{f_n}$ forment une base orthonormale de l'espace de Hilbert $\mathscr{E}_C$. Il lui correspond donc par H une base orthonormale $(Z_n)_{n \in I}$ de $\mathscr{H}_X$ telle que l'on ait

$$f_n(t) = E[\overline{Z_n} X_t] \qquad \text{pour} \quad n \in I \quad \text{et} \quad t \in T. \qquad (13)$$

On a donc dans l'espace de Hilbert $\mathscr{H}_X$ le développement orthogonal

$$X_t = \sum_{n \in I} f_n(t) \cdot Z_n \qquad (14)$$

généralement attribué à Karhunen [9]. La covariance C se développe comme suit:

$$C(t, t') = \sum_{n \in I} \overline{f_n(t)} \cdot f_n(t'). \qquad (15)$$

Les intégrales aléatoires du second ordre admettent le développement en série

$$\int_T u(t) \cdot X_t \, d\mu(t) = \sum_{n \in I} c_n Z_n, \tag{16}$$

dont les coefficients sont donnés par les intégrales absolument convergentes

$$c_n = \int_T u(t) f_n(t) \, d\mu(t). \tag{17}$$

Examinons le problème des *versions* du processus X. Par définition, une version de X est une fonction $\hat{X}$ sur $T \times \Omega$ telle que, pour tout $t \in T$, la fonction $\omega \mapsto X(t, \omega)$ soit un représentant de la classe $X_t \in L^2(\Omega, \mathscr{F}, P)$. Une version $\hat{X}$ de X est *mesurable* si c'est une fonction mesurable par rapport à la tribu $\mathscr{A} \otimes \mathscr{F}$. Le lemme du n° 24 admet l'interprétation suivante: *si le processus X admet une version mesurable, sa covariance C est séparable et mesurable*. Réciproquement, supposons que la covariance C soit séparable et mesurable et introduisons le développement de Karhunen. Si l'espace $\mathscr{E}_C$ est de dimension finie, on définit une version mesurable du processus X par la formule:

$$\hat{X}(t, \omega) = \sum_{n \in I} f_n(t) \cdot \hat{Z}_n(\omega); \tag{18}$$

l'ensemble I est fini, les fonctions f_n sont mesurables sur $(T, \mathscr{A})$, et $\hat{Z}_n$ est un représentant de la classe $Z_n \in L^2(\Omega, \mathscr{F}, P)$. Lorsque $\mathscr{E}_C$ est de dimension infinie, la série (18) ne converge pas telle quelle. Supposons pour simplifier que I soit l'ensemble des entiers $n \geq 1$; il existe alors une suite d'entiers $n(1) < n(2) < n(3) < \cdots$ telle que la suite des fonctions

$$X_p(t, \omega) = \sum_{n=1}^{n(p)} f_n(t) \cdot \hat{Z}_n(\omega) \tag{19}$$

converge presque partout sur $T \times \Omega$ par rapport à la mesure $\mu \otimes P$. Utilisant le théorème de Lebesgue–Fubini, on conclut qu'il existe un processus du second ordre $X' = (X'_t)_{t \in T}$ admettant une version mesurable, et tel que l'on ait $X'_t = X_t$ pour presque tout $t \in T$ (par rapport à μ).

Supposons le processus X *gaussien*. Alors $(Z_n)_{n \in I}$ est une suite indépendante de variables aléatoires gaussiennes de moyenne nulle et variance 1. Si l'ensemble I est infini, on sait que l'on a $\sum_{n \in I} |\hat{Z}_n(\omega)|^2 = +\infty$ avec probabilité 1, quelles que soient les versions $\hat{Z}_n$ des variables aléatoires Z_n. Par suite, la probabilité qu'une trajectoire appartienne à l'espace $\mathscr{E}_C$ est nulle; il faut choisir pour les trajectoires du processus X un espace fonctionnel plus grand que $\mathscr{E}_C$, dont les éléments seront en général plus "irréguliers" que ceux de $\mathscr{E}_C$. Par exemple, dans le cas du mouvement brownien, toute fonction de

$\mathscr{E}_C$ satisfait à une condition de Hölder d'ordre $\frac{1}{2}$, soit

$$|f(t) - f(t')| \leq M.|t - t'|^{1/2}. \tag{20}$$

Or Wiener [19] a prouvé que le processus du mouvement brownien satisfait à une condition de Hölder d'ordre $\frac{1}{2}$ avec probabilité 0, mais à une condition de Hölder d'ordre $c < \frac{1}{2}$ arbitraire avec probabilité 1.

49. La covariance du mouvement brownien

Le mouvement brownien est un processus $X = (X_t)_{t \in \mathbf{R}}$ satisfaisant aux conditions formulées par Einstein en 1905:

(a) si $t < t'$, la variable aléatoire réelle $X_{t'} - X_t$ est gaussienne, de moyenne 0 et de variance $t' - t$;

(b) si $t_0 < t_1 < \cdots < t_n$, les variables aléatoires $Y_i = X_{t_i} - X_{t_{i-1}}$ (pour $1 \leq i \leq n$) sont indépendantes.

Les hypothéses précédentes ne concernant que les différences $X_{t'} - X_t$, il faut ajouter une condition de normalisation; nous conviendrons qu'on a $X_0 = 0$.

D'aprés l'hypothèse (a), on a

$$E[(X_t - X_{t'})^2] = |t - t'|; \tag{21}$$

la fonction d_C associée comme au n° 9 à la covariance C de X est donc la distance sur $\mathbf{R}$ définie par $d_C(t, t') = |t - t'|^{1/2}$. La covariance C s'obtient par la relation

$$2C(t, t') = E[X_t^2] + E[X_{t'}^2] - E[(X_t - X_{t'})^2], \tag{22}$$

et l'on a donc

$$C(t, t') = \tfrac{1}{2}[|t| + |t'| - |t - t'|]. \tag{23}$$

On a en particulier $C(t, t') = 0$ si t et t' sont de signes contraires, et l'on a

$$C(t, t') = C(-t, -t') = \inf(t, t') \qquad \text{si} \quad t \geq 0, \quad t' \geq 0. \tag{24}$$

Du point de vue "analytique", pour montrer que C est effectivement une covariance, il suffit de construire un espace de Hilbert $\mathscr{H}$ et une application φ de $\mathbf{R}$ dans $\mathscr{H}$ satisfaisant aux conditions suivantes:

(a') le vecteur $\varphi(t) - \varphi(t')$ est de norme $|t - t'|^{1/2}$;

(b') si $t_1 < t_1' < t_2 < t_2'$, les vecteurs $\varphi(t_1') - \varphi(t_1)$ et $\varphi(t_2') - \varphi(t_2)$ sont orthogonaux;

(c') on a $\varphi(0) = 0$.

Voici une solution: on prend pour $\mathcal{H}$ l'espace $L^2(\mathbf{R}, \lambda)$ où λ est la mesure de Lebesgue; si $t \leq t'$, on note $I_{t,t'}$ l'indicatrice de l'intervalle $]t, t']$, et l'on pose $I_{t',t} = -I_{t,t'}$; enfin, on pose $\varphi(t) = I_{0,t}$. On a alors $I_{t,t'} = \varphi(t') - \varphi(t)$ et la vérification des conditions (a') à (c') est immédiate.

L'espace $\mathcal{E}_C$ correspondant à la covariance du mouvement brownien sera noté W_1. D'après la construction générale du n° 5, on définit un isomorphisme H de $L^2(\mathbf{R}, \lambda)$ sur W_1 par la formule $Hu(t) = \langle \varphi(t) | u \rangle$, c'est-à-dire

$$Hu(t) = \begin{cases} \int_0^t u(x)\,dx & \text{si} \quad t \geq 0, \\ -\int_t^0 u(x)\,dx & \text{si} \quad t \leq 0. \end{cases} \tag{25}$$

Par suite, W_1 se compose des fonctions absolument continues f sur $\mathbf{R}$, telles que $f(0) = 0$, et dont la dérivée f' (définie presque partout) est de carré intégrable pour λ; la norme dans W_1 est donnée par

$$\|f\|_1 = \left\{ \int_{\mathbf{R}} |f'(t)|^2\,dt \right\}^{1/2}. \tag{26}$$

On peut définir toute une suite d'espaces de Hilbert W_m analogues. On pose $W_0 = L^2(\mathbf{R}, \lambda)$, et pour tout entier $m \geq 1$, on note W_m l'espace des fonctions f sur $\mathbf{R}$ satisfaisant aux conditions suivantes:

(α)   *la fonction f est $m - 1$ fois continûment différentiable, et l'on a $f^{(j)}(0) = 0$ pour $0 \leq j \leq m - 1$;*

(β)   *la dérivée $f^{(m-1)}$ d'ordre $m - 1$ est absolument continue, et sa dérivée $f^{(m)}$* (définie presque partout) *est de carré intégrable pour λ.*

La norme dans W_m est donnée par

$$\|f\|_m = \left\{ \int_{\mathbf{R}} |f^{(m)}(t)|^2\,dt \right\}^{1/2}. \tag{27}$$

On définit un isomorphisme H_m de $L^2(\mathbf{R}, \lambda)$ sur W_m en généralisant la formule (25):

$$(m-1)! \cdot H_m u(t) = \begin{cases} \int_0^t (t-x)^{m-1} u(x)\,dx & \text{si} \quad t \geq 0, \\ -\int_t^0 (t-x)^{m-1} u(x)\,dx & \text{si} \quad t \leq 0. \end{cases} \tag{28}$$

Grâce à cette formule, on montre facilement que W_m est l'espace $\mathcal{E}_{C_m}$ associé à une covariance continue C_m caractérisée par les propriétés suivantes: on a $C_m(t, t') = 0$ si t et t' sont de signes contraires, on a $C_m(t, t') = C_m(t', t) =$

$C_m(-t, -t') = C_m(-t', -t)$, et enfin, on a

$$C_m(t, t') = \frac{1}{(2m-1)!} \sum_{i=0}^{m-1} (-1)^{m-i+1} \binom{m}{i} t^{2m-i-1} t'^i \tag{29}$$

lorsque $0 \le t \le t'$.

50. *Problème de Sturm-Liouville*

Nous considérons un intervalle compact $I = [a, b]$ de **R**, et deux fonctions mesurables positives p et q sur I; on suppose qu'il existe une constante $c > 0$ telle que l'on ait $p(x) \ge c$ pour tout x dans I. Notons $\mathscr{E}$ l'espace vectoriel des fonctions (complexes) f sur I, nulles en a et b, absolument continues de dérivée f' (définie presque partout), pour lesquelles l'intégrale $\int_I \{p(x)|f'(x)|^2 + q(x)|f(x)|^2\} \, dx$ soit finie. C'est un espace de Hilbert dans lequel le produit scalaire s'exprime ainsi:

$$\langle f | g \rangle = \int_I \{p(x)\overline{f'(x)}g'(x) + q(x)\overline{f(x)}g(x)\} \, dx. \tag{30}$$

Si une fonction $f \in \mathscr{E}$ satisfait à $\langle f | f \rangle \le 1$, on a les inégalités

$$|f(x)| \le (b-a)^{1/2}/c^{1/2}, \qquad |f(x) - f(y)| \le |x - y|^{1/2}/c^{1/2} \tag{31}$$

pour x, y dans I. Autrement dit, la boule unité fermée dans l'espace de Hilbert $\mathscr{E}$ est un ensemble équicontinu uniformément borné. D'après le critère général des nᵒˢ 6 et 11, il existe donc une covariance continue G sur l'intervalle $I = [a, b]$ telle que $\mathscr{E} = \mathscr{E}_G$.

Soit $L^2(I)$ l'espace de Hilbert formé des fonctions sur I de carré intégrable pour la mesure de Lebesgue; comme on a $\int_I G(x, x) \, dx < +\infty$, on peut appliquer les résultats généraux de la partie V. En particulier, l'espace de Hilbert $\mathscr{E}$ est contenu dans $L^2(I)$ et l'application j qui associe à chaque fonction $f \in \mathscr{E}$ sa classe $\tilde{f} \in L^2(I)$ est un opérateur de Hilbert-Schmidt de $\mathscr{E}$ dans L^2. L'opérateur $G = jj^*$ est hermitien, positif, de trace finie dans $L^2(I)$; c'est l'opérateur intégral de noyau G:

$$Gu(x) = \int_I G(x, y)u(y) \, dy. \tag{32}$$

Comme l'espace $\mathscr{E}$ contient les fonctions continument dérivables dans I, nulles en a et b, il est dense dans $L^2(I)$, et l'opérateur G est donc injectif et d'image dense dans $L^2(I)$. Notons D l'inverse de G; le domaine de D se compose des fonctions $f \in \mathscr{E}$ pour lesquelles il existe une fonction $Df \in L^2(I)$ telle que l'on ait

$$\langle f | g \rangle = \int_I \overline{Df(x)} \cdot g(x) \, dx \qquad \text{pour tout} \quad g \in \mathscr{E}. \tag{33}$$

Compte tenu de la définition (30) du produit scalaire dans $\mathscr{E}$, on voit que le domaine $\mathscr{D}$ de D se compose des fonctions f satisfaisant aux hypothèses suivantes:

(a) on a $f(a) = f(b) = 0$;

(b) la fonction f est absolument continue, de dérivée f' (définie presque partout);

(c) la fonction pf' est absolument continue, et si $(pf')'$ est sa dérivée (définie presque partout), la fonction $-(pf')' + qf$ appartient à $L^2(I)$.

L'opérateur D est donné par la formule:

$$Df = -(pf')' + qf; \tag{34}$$

ce n'est autre que l'opérateur auto-adjoint non borné associé par la méthode de Friedrichs à l' "intégrale d'énergie" (30); la fonction G n'est autre que la *fonction de Green* de l'opérateur différentiel D du second ordre.

En appliquant de manière classique le théorème de Mercer, on obtient le développement de la fonction de Green sous la forme

$$G(x, y) = \sum_{n=1}^{\infty} \lambda_n^{-1} f_n(x) \cdot \overline{f_n(y)}, \tag{35}$$

où les λ_n sont les valeurs propres de D et f_n ses fonctions propres. La suite $(\lambda_n)_{n \geq 1}$ est strictement croissante et tend vers $+\infty$, les fonctions f_n appartiennent à $\mathscr{D}$ et l'on a $Df_n = \lambda_n f_n$ pour tout entier $n \geq 1$, avec la normalisation $\int_I |f_n(x)|^2 \, dx = 1$.

Prenons par exemple $a = 0$, $b = 1$, $p(x) = 1$, $q(x) = 0$ pour tout $x \in [0, 1]$. L'espace $\mathscr{E}$ se compose des fonctions absolument continues f sur $[0, 1]$ telles que $f(0) = f(1) = 0$, et dont la dérivée f' est de carré intégrable sur $[0, 1]$; le produit scalaire dans $\mathscr{E}$ est défini par

$$\langle f | g \rangle = \int_0^1 \overline{f'(x)} g'(x) \, dx. \tag{36}$$

La covariance G se déduit facilement des résultats du n° 49, d'où

$$G(x, y) = \inf(x, y) - xy. \tag{37}$$

L'opérateur D a pour domaine l'espace $\mathscr{D}$ des fonctions f sur $[0, 1]$ telles que $f(0) = f(1) = 0$, qui admettent une dérivée absolument continue f', et dont la dérivée seconde $f'' = (f')'$ soit de carré intégrable sur $[0, 1]$; on a $Df = -f''$. Les valeurs propres et fonctions propres sont données par

$$\lambda_n = \pi^2 n^2, \qquad f_n(x) = 2^{1/2} \sin(\pi n x). \tag{38}$$

Par application de la formule (35), on retrouve le développement connu

$$\inf(x, y) = xy + \sum_{n=1}^{\infty} \frac{2}{(\pi n)^2} \sin(\pi n x) \cdot \sin(\pi n y) \tag{39}$$

pour x, y dans l'intervalle $[0, 1]$.

Compte tenu de la formule (24), on a donc un développement en série de la covariance de la restriction du processus du mouvement brownien à l'intervalle $[0, 1]$. Le développement de Karhunen correspondant s'écrit :

$$X_t = tZ_0 + 2^{1/2} \sum_{n=1}^{\infty} Z_n \cdot \sin(\pi n t)/(\pi n) \qquad \text{pour} \quad 0 \leq t \leq 1. \tag{40}$$

Cette représentation du mouvement brownien comme série de Fourier aléatoire a été introduite par Wiener [19] et souvent utilisée depuis.

51. *Autres Applications*

Nous contenterons d'indiquer quelques thèmes. Tout d'abord, la méthode indiquée au n° 50 pour construire la fonction de Green d'un opérateur différentiel ordinaire se généralise à des intervalles infinis, des conditions aux limites plus générales, et des opérateurs d'ordre plus grand que 2.

Les espaces de Hilbert associés à une covariance se prêtent bien aux problèmes d'interpolation et d'approximation. Par exemple, l'espace W_1 est adapté à l'interpolation linéaire, alors que les espaces W_m sont liés aux fonctions-splines. En traduisant en termes de processus du second ordre, on rencontre la théorie de la prédiction linéaire de Wiener–Kolmogoroff . . . .

REMERCIEMENTS

Ce travail est dédié à Laurent Schwartz, à qui je dois une part importante de ma formation comme mathématicien et comme homme. Les thèmes étudiés ici ont de nombreux points de contact avec les préoccupations mathématiques de L. Schwartz; en particulier, ses séminaires sur la thèse de Grothendieck en 1954, et sur les applications radonifiantes vers 1970, et ses travaux sur les sous-espaces hilbertiens [15, 16] m'ont été l'occasion de m'intéresser à ce sujet, ou d'y revenir.

Le théorème de Mercer généralisé a été mis au point en collaboration avec Xavier Fernique, qui l'a utilisé dans un travail [5] publié en 1970. Je le remercie des fructueuses discussions poursuivies dans le cadre du Séminaire de Probabilités de Strasbourg.

Enfin, Nicolas Bourbaki a fait sans fard de nombreuses remarques pertinentes sur une version préliminaire de ce travail, et m'a encouragé à trouver le "bon cadre" dans lequel exposer la théorie. J'espère que la présentation actuelle est conforme à ses canons mathématiques.

RÉFÉRENCES

1. N. ARONSZAJN, Theory of reproducing kernels, *Trans. Amer. Math. Soc.* **68** (1950), 337–404.
2. N. BOURBAKI, "Espaces vectoriels topologiques," nouvelle éd., Masson, Paris, 1981.
3. P. CARTIER, Classes de formes bilinéaires sur les espaces de Banach [d'après Grothendieck], Séminaire Bourbaki, 13ᵉ année 1960, 1961, exp. 211, 14 pp., Benjamin, New York, 1966.
4. N. DUNFORD ET B. J. PETTIS, Linear operations on summable functions, *Trans. Amer. Math. Soc.* **47** (1940), 323–392.
5. X. FERNIQUE, Régularité des processus gaussiens, *Invent. Math.* **12** (1971), 304–320.
6. F. GANTMACHER, "Théorie des matrices," Tome 1, Dunod, Paris, 1966.
7. A. GROTHENDIECK, "Produits tensoriels topologiques et espaces nucléaires", Memoirs Amer. Math. Soc. Vol. 16, 1955.
8. A. GROTHENDIECK, Résumé de la théorie métrique des produits tensoriels topologiques, *Bol. Soc. Mat. Sao Paulo* **8** (1956), 1–79.
9. K. KARHUNEN, Über lineare Methoden in der Wahrscheinlichkeitsrechnung, *Ann. Acad. Sci. Fenn.* no 37 (1947).
10. J.-L. KRIVINE, Constante de Grothendieck et fonctions de type positif sur les sphères, *Adv. in Math.* **31** (1979), 16–30.
11. M. LOEVE, "Probability theory," 2nd ed., Van Nostrand Reinhold, New York, 1960 (voir la section 34 plus particulièrement).
12. F. RIESZ, Über Sätze von Stone und Bochner, *Acta Sci. Math. (Szeged)* **6** (1933), 184–198.
13. F. RIESZ ET B. S. NAGY, "Leçons d'analyse fonctionnelle", Akad. Kiado, Budapest, 1953.
14. E. SCHMIDT, Auflösung der allgemeinen linearen Integralgleichung, *Math. Ann.* **64** (1907), 161–174.
15. L. SCHWARTZ, Sous-espaces hilbertiens d'espaces vectoriels topologiques et noyaux associés (noyaux reproduisants), *J. Analyse Math.* **13** (1964), 115–256.
16. L. SCHWARTZ, Sous-espaces hilbertiens et antinoyaux associés, Séminaire Bourbaki, 14ᵉ année 1961, 1962, exp. 238, 18 pages, Benjamin, New York, 1966.
17. G. SZEGÖ, "Orthogonal Polynomials," Colloquium Publications 4th ed., Amer. Math. Soc., Providence, Rhode Island, 1975.
18. A. WEIL, "L'intégration dans les groupes topologiques et ses applications", Actualités Scientifiques et Industrielles, Hermann, Paris, 1940.
19. N. WIENER, Differential space, *in* "Selected Papers," MIT Press, Cambridge, Massachusetts, 1964.

MATHEMATICAL ANALYSIS AND APPLICATIONS, PART A
ADVANCES IN MATHEMATICS SUPPLEMENTARY STUDIES, VOL. 7A

Topological Properties Inherited by Certain Subspaces of Holomorphic Functions

SEÁN DINEEN

Department of Mathematics
University College Dublin
Dublin, Ireland

DEDICATED TO LAURENT SCHWARTZ ON THE OCCASION OF HIS SIXTY-FIFTH BIRTHDAY

It is generally recognized that Laurent Schwartz has played one of the more creative roles in the shaping of modern analysis and thus it is a great honor and privilege to be invited to contribute to these proceedings. The influence of Laurent Schwartz on infinite-dimensional holomorphy through his research on topological vector spaces, distribution theory, the ε product, and his more recent work on Radon measures is extensive and shall, no doubt, continue to be so for many years to come.

In this article, the roles played by $s \approx \mathscr{S}$ (the rapidly decreasing $\mathscr{C}^{\infty}$ functions of Schwartz) and by $\mathscr{D}$ (the space of test functions of Schwartz) are crucial both because of the motivation they provided in our approach and because they are the principle examples to which our results apply.

Before discussing our results we give a brief introduction to infinite-dimensional holomorphy. A function $f: U \subset E \to F$, U an open subset of a locally convex space E and F a locally convex space, is said to be holomorphic if it is continuous and $f|_{U \cap G}$ is holomorphic as a function of several complex variables for each finite-dimensional subspace G of E. We let $H(U)$ denote the space of all holomorphic functions on U. Infinite-dimensional holomorphy is the study of $H(U)$.

One aspect of this theory is the study of various locally convex topological spaces structures on $H(U)$. This has been a fruitful object of study both in itself and as a source of interaction between infinite-dimensional holomorphy and other branches of functional analysis. For instance, this study for E, a Banach space, has led to interesting observations on the geometry of Banach spaces. When E is a fully nuclear space with a basis (defined below) we also obtain a rather satisfactory theory. For example, we obtain a monomial expansion which converges normally with respect to the compact open topology and using this and the Borel transform we obtain a duality theory

317

between analytic functionals on $H(E)$ and germs of holomorphic functions at the origin in E'_β. This in turn has led to an identification of various interesting classes of Fréchet nuclear spaces and fully nuclear spaces with a basis. The initial point of departure is the comparison of the three usual topologies on $H(E)$, τ_0, τ_ω, and τ_δ. The first of these topologies, τ_0, is just the usual compact open topology. A seminorm p on $H(E)$ is τ_ω continuous if there exists a compact subset K of E such that for every V open, $K \subset V$, there exists $C(V) > 0$ such that $p(f) \leq C(V)\|f\|_V$ for every f in $H(E)$. Naturally τ_ω is the topology generated by all τ_ω continuous seminorms.

A seminorm p on $H(E)$ is τ_δ continuous if for each increasing countable open cover of E there exists a positive integer n and $C > 0$ such that $p(f) \leq C\|f\|_{V_n}$ for every f in $H(E)$. The compact open topology is probably the most natural topology on $H(E)$ but can, however, lack useful topological properties. The τ_δ topology has good topological properties (for instance it is ultrabornological) but may be difficult to describe in a concrete fashion. The τ_ω topology, whose definition was motivated by certain properties of analytic functionals in finite dimensions, is intermediate between τ_0 and τ_ω and acts in many cases in place of one or other of τ_0 and τ_δ as the demand arises.

A locally convex space E is said to be fully nuclear if E and E'_β are both complete reflexive nuclear spaces. If E also has a Schauder basis we say it is fully nuclear with a basis. We refer to Dineen [6] for the notation we use and also for a survey of results on holomorphic functions on fully nuclear spaces. In this article we consider three different aspects of the same theory.

1.   In [5] the concept of B nuclear space was introduced and it was shown that $\tau_0 = \tau_\delta$ on $H(E)$ whenever E is a B nuclear space. In Section 1, we show that a Fréchet nuclear space with a basis is a B nuclear space if and only if it is isomorphic to a subspace of s. This places B nuclear spaces, a class of spaces arising naturally in the study of holomorphic functions, in a category which is important and which has been studied extensively in recent years by functional analysts.

2.   Using the above result on B nuclear spaces we showed in [4] that $\tau_0 = \tau_\delta$ on $H(\mathscr{D})$ and in Section 2 we extend this result to nuclear spaces with a basis which can be represented as a strict inductive limit of B nuclear spaces.

3.   Many of the results obtained in [2–5] depend on the existence of a basis in the domain space. By using entire functions, a density argument, and representation results for polynomials on fully nuclear spaces, we remove this basis requirement in a number of situations. As a bonus we also establish a connection between holomorphic extension theorems (of the Hahn–Banach type) and topological properties of quotients of $(H(E), \tau_0)$.

1. Let E denote a metrizable locally convex space with generating family of seminorms $(p_n)_{n=1}^{\infty}$, $p_n \leq p_{n+1}$ for all n. E is a dominated norm (DN) space if there exists a continuous norm p on E such that for any positive integer k there exists a positive integer n and $c > 0$ such that

$$p_k \leq rp + (c/r)p_{k+n} \qquad \text{for all} \quad r > 0.$$

The fundamental result concerning nuclear DN spaces is the following:

THEOREM 1. [7,9,10]. *A metrizable nuclear locally convex space is a DN space if and only if it is isomorphic to a subspace of s.*

Now suppose E is a Fréchet nuclear space with a basis. Then E is isomorphic to $\Lambda(P)$ where we may suppose $P = (w_m)_{m=1}^{\infty}$, $w_m = (w_{m,n})_{n=1}^{\infty}$ for all m, $w_{m+1,n} \geq w_{m,n}$ for all m and n, and

$$\frac{w_m}{w_{m+1}} = \sum_{n, w_{m,n} \neq 0} \frac{w_{m,n}}{w_{m+1,n}} < \infty \qquad \text{for all } m.$$

With this notation we have the following characterization of complete DN spaces with a basis.

THEOREM 2. *Let $E \approx \Lambda(P)$ denote a Fréchet nuclear space with a basis. The following are equivalent:*

(a) *E is a DN space;*
(b) *E is isomorphic to a subspace of s;*
(c) *$(w_{m+1,n})^2 \leq w_{m,n}w_{m+2,n}$ for all m and n;*
(d) *For each positive integer m there exists a positive integer k and $c > 0$ such that $(w_{m,n})^2 \leq cw_{1,n}w_{k,n}$ for all n;*
(e) *E is a B nuclear space, i.e., if $\beta_{m,n} = w_{m+1,n}/w_{m,n}$ then $(w_{m,n}(\beta_{m,n})^p)_{n=1}^{\infty}$ is a continuous weight of E for all positive integers m and p.*

Proof. (a) and (b) are equivalent by Theorem 1. The equivalence of (b), (c), and (d) can be found in [9] (see also [7,11]). We show that (c) $\Rightarrow$ (e) and (e) $\Rightarrow$ (d).

(c) $\Rightarrow$ (e). Since $(w_{m+1,n})^2 \leq w_{m,n}w_{m+2,n}$ for all m and n we have $w_{m+1,n}/w_{m,n} \leq w_{m+2,n}/w_{m+1,n}$ for all m and n and hence $w_{m+1,n}/w_{m,n} \leq w_{m+j+1,n}/w_{m+j,n}$ for all positive integers m, n and j. Thus

$$w_{m,n}(\beta_{m,n})^p = w_{m,n}\left(\frac{w_{m+1,n}}{w_{m,n}}\right)^p \leq w_{m,n}\prod_{j=0}^{p-1}\frac{w_{m+j+1,n}}{w_{m+j,n}} = w_{m+p,n}$$

and so (c) $\Rightarrow$ (e).

(e) $\Rightarrow$ (d). We first show by induction on m, assuming (e), that $(w_{1,n}(w_{m,n}/w_{1,n})^p)_{n=1}^{\infty}$ is a continuous weight on E for any positive integer p. The case

$m = 1$, p arbitrary is trivial. Now suppose the above is true for the positive integer m and for all p. By our induction hypothesis there exists $c_1 > 0$ and j a positive integer such that $w_{1,n}(w_{m,n}/w_{1,n})^{2p} \le c_1 w_{j,n}$ for all n. By condition (e) there exists $c_2 > 0$ and k a positive integer such that $w_{m,n}(w_{m+1,\,n}/w_{m,n})^{2p} \le c_2 w_{k,n}$ for all n. Hence

$$\left[w_{1,n}\left(\frac{w_{m+1,\,n}}{w_{1,n}}\right)^p \right]^2 = w_{1,n}\left(\frac{w_{m+1,\,n}}{w_{m,n}}\right)^p w_{1,n}\left(\frac{w_{m,n}}{w_{1,n}}\right)^p \le c_1 c_2 w_{j,n} w_{k,n} \le c^2 w_{l,n}^2,$$

where $c = \sqrt{c_1 c_2}$ and $l = j + k$. Thus $w_{1,n}(w_{m+1,\,n}/w_{1,n})^p \le c w_{l,n}$ for all n and $(w_{1,n}(w_{m,n}/w_{1,n})^p)_{n=1}^\infty$ is a continuous weight on E for all m and p. If we let $p = 2$ we obtain (d) and hence (e) $\Rightarrow$ (d). This completes the proof.

Many examples of DN spaces are to be found in [9]. In particular a nuclear power series space is a DN space if and only if it is of infinite type.

2.   We now look at holomorphic functions on fully nuclear spaces with a Schauder basis which can be represented as a strict inductive limit of Fréchet nuclear spaces. A countable direct sum of Fréchet nuclear spaces with a basis falls into this category and the following result shows that this is, in fact, the only possible type of example.

LEMMA 3.   *Let $E = \varinjlim_n E_n$ denote a strict inductive limit of Fréchet nuclear spaces and suppose E has a Schauder basis. Then $E \approx \sum_{n=1}^\infty F_n$, where each F_n is a Fréchet nuclear space with a Schauder basis. Moreover, if each E_n is a DN space then we may also suppose that each F_n is a DN space with a basis.*

Proof.   Let $(e_n)_{n=1}^\infty$ denote a Schauder basis for E. Since E is a fully nuclear space $(e_n)_{n=1}^\infty$ is an absolute basis for E. For each positive integer n let F_n denote the closed subspace of E generated by $\{e_m \,;\, e_m \in E_n, e_m \notin E_j \text{ for } j < n\}$. Since $F_n \subset E_n$ for all n and E is a strict inductive limit it follows that F_n is a Fréchet nuclear space with a Schauder basis for each n and if each E_n is a DN space, then so also is each F_n. Let $F = \sum_{n=1}^\infty F_n$. Using the natural injection from F_n into E_n and the construction of an inductive limit we obtain a continuous injection Π from F into E. If $x = \sum_{n=1}^\infty x_n e_n \in E$ then the sequence $\{x_n e_n\}_{n=1}^\infty$ is a bounded sequence in E and, since a strict inductive limit is a regular inductive limit, there exists a positive integer N such that $x_n e_n \in E_n$ for all n. Hence $\{e_n \,;\, x_n \ne 0\} \subset E_N$ and, as the basis is absolute, $(\sum_{n=1}^m x_n e_n)_{m=1}^\infty$, is a Cauchy sequence in $\sum_{n=1}^N F_n$ and $x \in F$. Thus Π is a surjective mapping. By the open mapping theorem (between countable inductive limits of Baire spaces) Π is a homeomorphism and $E \cong \sum_{n=1}^\infty F_n$. This completes the proof.

A particular example of the above type of space is $\mathscr{D}(\Omega)$, Ω an open subset of R^n, which has recently been shown [8,11] to be isomorphic to $s^{(N)}$. A modification of the proof given in [4] for $\mathscr{D}(\Omega)$ yields the following result.

THEOREM 3'.   If $E = \sum_{n=1}^{\infty} E_n$, where each E_n is a DN space with a basis, then $\tau_o = \tau_\delta$ on $H(E)$.

Proof. Since the monomials form an absolute basis for $(H(E), \tau_0)$ and $(H(E), \tau_\delta)$ it suffices to show $(H(E), \tau_\delta)' = (H(E), \tau_o)'$. Let $T \in (H(E), \tau_\delta)'$. For each n choose a neighborhood of zero in E_n, V_n, such that each monomial on E_n is bounded on V_n. This is possible since each E_n admits a continuous norm. If $z \in E$ then $z = (z_n)_{n=1}^{\infty}$ where $z_n \in E_n$ for all n and $z_n = 0$ for all n sufficiently large. For each n, $z_n = (z_{n,m})_{m=1}^{\infty}$, and so $z = \{(z_{n,m})_{m=1}^{\infty}\}_{n=1}^{\infty} = (z_{n,m})_{(n,m) \in N^2}$. Hence each monomial on E is indexed by $(N^2)^{(N)}$. If $j = (j_{n,m})_{n,m=1}^{\infty} \in (N^2)^{(N)}$ we let $h(j) = \sup\{n \in N; \exists m \text{ such that } j_{n,m} \neq 0\}$. We claim that there exists a positive integer n_0 such that if $j \in (N^2)^{(N)}$, $h(j) > n_0$, then $T(z^j) = 0$. If not, then by restricting ourselves to some $\sum_{k=1}^{\infty} E_{n_k}$ if necessary, we can find a sequence in $(N^2)^{(N)}$, $(j_n)_{n=1}^{\infty}$, such that $h(j_n) = n$ and $T(z^{j_n}) \neq 0$ for all n. Let $(\alpha_n)_{n=1}^{\infty}$ denote any sequence of complex numbers. We first show that $\{\alpha_n z^{j_n}\}_{n=1}^{\infty}$ is a locally bounded, and hence a τ_δ bounded subset of $H(E)$. Let $z \in E$ be arbitrary, say, $z \in \sum_{n=1}^{l} E_n$.

Let $M = 1 + \sup_{n=1,\ldots,l} \|\alpha_n z^{j_n}\|_V$ where $V = z + V_1 \times V_2 \times \cdots \times V_l$. Now suppose $l' > l$ and $c_{l+1}, \ldots, c_{l'}$ are positive real numbers such that $\sup_{n=1,\ldots,l'} \|\alpha_n z^{j_n}\|_{V + c_{l+1} V_{l+1} \times \cdots \times c_{l'} V_{l'}} = M$. If $c > 0$ then

$$\|\alpha_{l'+1} z^{j_{l'+1}}\|_{V \times c_{l+1} V_{l+1} \times \cdots \times c_{l'} V_{l'} \times c V_{l'+1}}$$
$$= |\alpha_{l'+1}| c^{\theta(l')} \|z^{j_{l'+1}}\|_{V \times c_{l+1} V_{l+1} \times \cdots \times c_{l'} V_{l'} \times V_{l'+1}},$$

where $\theta(l') > 0$ since $h(j_{l'+1}) = l' + 1$.

Hence, by choosing c sufficiently small and positive, we get

$$\|\alpha_{l'+1} z^{j_{l'+1}}\|_{V \times c_{l+1} V_{l+1} \times \cdots \times c_{l'} V_{l'} \times c V_{l'+1}} \leq M.$$

Since $h(j_n) = n$ the same estimate also holds for all $\alpha_n z^{j_n}$, $n \leq l'$.

Thus we can choose a sequence of positive real numbers $(c_l)_{l=1}^{\infty}$ such that

$$\sup_n \|\alpha_n z^{j_n}\|_{z + \sum_{l=1}^{\infty} c_l V_l} < \infty.$$

This shows that the sequence $(\alpha_n z^{j_n})_{n=1}^{\infty}$ is locally bounded. If we let $\alpha_n = n/|T(z^{j_n})|$ for all n then $\sup_n |T(\alpha_n z^{j_n})| = \infty$ and this contradicts the fact that T is τ_0 continuous. Hence there exists a positive integer n_0 such that $T(z^{j_n}) = 0$ if $h(j_n) > n_0$.

Let $F = \sum_{n=1}^{n_0} E_n$. F is a DN space with a basis. If $f = \sum_{j \in (N^2)^{(N)}} a_j z^j \in H(E)$ then by the above $T(f) = \sum_{j \in (N^2)^{(N)}, h(j) \leq n_0} a_j T(z^j)$ and thus, since E is a complemented subspace of E, we may define $\tilde{T}$ on $H(F)$ by the formula

$$\tilde{T}(f|_F) = T(f)$$

for every f in $H(E)$.

Since the basis in F extends to a basis in E we have

$$\sum_{m \in N^{(N)}} |a_m \tilde{T}(z^m)| < \infty \qquad \text{for every} \qquad \sum_{m \in N^{(N)}} a_m z^m \in H(F),$$

and hence $\tilde{T}$ is a τ_δ continuous seminorm on $H(F)$. Since F is a DN space with a basis, we have $\tau_o = \tau_\delta$ on $H(F)$ and $\tilde{T}$ is a τ_o continuous seminorm on $H(F)$. Hence there exist $c > 0$ and K a compact subset of F (and hence also a compact subset of E) such that for any f in $H(E)$

$$|T(f)| = |\tilde{T}(f|_F)| \le c\|f|_F\|_K \le c\|f\|_K.$$

This completes the proof.

3. We now consider holomorphic functions on closed subspaces of fully nuclear spaces. We first prove a lemma concerning linear functionals on a subspace of a fully nuclear space.

LEMMA 4. *Let F denote a closed subspace of the fully nuclear space E then $E'_\beta/F \cong F'_\beta$.*

Proof. Since F is a closed subspace of a fully nuclear space, it is a complete nuclear space and hence it is semireflexive. Thus F'_β is a barreled space and $(F'_\beta)' \cong F$. Also E'_β is a complete reflexive space and so $E'_\beta/F^\perp$ is a barreled locally convex space. Thus $E'_\beta/F^\perp$ and F'_β are both Mackey spaces, and to complete the proof it suffices to show $(E'_\beta/F^\perp)' \cong F$.

If $T \in (E'_\beta/F^\perp)'$ then $T \circ \Pi \in E'$ where Π is the canonical projection from E'_β onto $E'_\beta/F^\perp$. Since E is reflexive there exists an x in E such that $T \circ \Pi(\phi) = \phi(x)$ for every ϕ in E'_β. Suppose $x \notin F$. By the Hahn–Banach theorem there exists a ϕ in E' such that $\phi(F) = 0$ and $\phi(x) \neq 0$. Since $\phi \in F^\perp$, $\Pi(\phi) = 0$ and thus $T \circ \Pi(\phi) = 0$. This contradicts the fact that $\phi(x) \neq 0$ and so $F'_\beta \cong E'_\beta/F^\perp$. This completes the proof.

If F is a subspace of a locally convex space E we let $H(F)^\perp = \{f \in H(E); f|_F = 0\}$ and $H_E(F) = \{g \in H(F); \exists \tilde{g} \in H(E) \text{ such that } \tilde{g}|_F = g\}$. If we let r_F^E denote the natural restriction mapping from $H(E)$ into $H(F)$ then $H(F)^\perp = \text{kernel } (r_F^E)$ and $H_E(F) = \text{Range}(r_F^E)$.

THEOREM 5. *Let F denote a closed subspace of a fully nuclear space E. Then*

$$(H(E), \tau_o)/H(F)^\perp \cong (H_E(F), \tau_o) \qquad \text{by} \quad r_F^E.$$

Proof. The mapping $r_F^E : H(E) \to H_E(F)$ is a continuous surjection and has kernel $H(F)^\perp$. We complete the proof by showing that the inverse mapping from $H_E(F)$ onto $(H(E), \tau_0)/H(F)^\perp$ is continuous. Thus we must show that for each compact (convex balanced) subset K of E there exists a compact subset L_1 of F such that

$$q(f) = \inf_{g \in H(F)^\perp} \|f + g\|_K \le c\|f\|_{L_1}$$

for every f in $H(E)$.

If K is a compact subset of E then Lemma 4 implies that

$$p_K(\phi) = \inf\{\|\tilde{\phi}\|_K ; \tilde{\phi} \in E', \tilde{\phi}|_F = \phi\}, \qquad \phi \in F'$$

is a continuous seminorm on F'_β and hence there exists a compact subset L of F such that

$$p_K(\phi) \le \|\phi\|_L \qquad \text{for every } \phi \text{ in } F'$$

If $\psi \in E'$ and n is a positive integer then

$$q(\psi^n) \le \inf\{\|\phi^n\|_K : \phi \in E', \psi|_F = \phi|_F\} = (\inf\{\|\phi\|_K : \phi \in E', \psi|_F = \phi|_F\})^n$$
$$= p_K(\psi|_F)^n \le (\|\psi\|_L)^n = \|\psi^n\|_L.$$

Now suppose $P \in P(^nE)$. If $P|_F = \sum_{i=1}^\infty \psi_i^n$ where $\psi_i \in F'$, all i and $\sum_{i=1}^\infty \|\psi_i\|_V^n < \infty$ for some neighborhood V of 0 in F, then by the Hahn–Banach theorem, there exists a neighborhood W of zero in E and $\tilde{\psi}_i \in E'$, all i, such that

$$\sum_{i=1}^\infty \|\tilde{\psi}_i\|_W^n < \infty \qquad \text{and} \qquad \tilde{\psi}_i|_F = \psi_i \qquad \text{for all } i.$$

Since $\lim_{m \to \infty} \sum_{i=1}^m (\tilde{\psi}_i)^n = \tilde{p}$ in $(H(E), \tau_0)$ and $\tilde{P}|_F = P|_F$ we find that

$$q(p) = q\left(\sum_{i=1}^\infty (\tilde{\psi}_i)^n\right) \le \sum_{i=1}^\infty q((\tilde{\psi}_i)^n) \le \sum_{i=1}^\infty \|\tilde{\psi}_i\|_L^n = \sum_{i=1}^\infty \|\psi_i\|_L^n.$$

Hence

$$q(P) \le \inf\left\{\sum_{i=1}^\infty \|\psi_i\|_L^n ; p|_F = \sum_{i=1}^\infty (\psi_i)^n,\right.$$

$$\left. \sum_{i=1}^\infty \|\psi_i\|_V^n < \infty \text{ for some neighborhood } V \text{ of } 0 \text{ in } F\right\}$$

$$= \Pi_L(P) \qquad \text{for every } P \text{ in } P(^nE).$$

If

$$f = \sum_{n=0}^\infty \frac{\hat{d}^n f(0)}{n!} \in H(E)$$

then

$$q(f) \leq \sum_{n=0}^{\infty} q\left(\frac{\hat{d}^n f(0)}{n!}\right) \leq q(f(0)) + \sum_{n=1}^{\infty} \Pi_L\left(\frac{\hat{d}^n f(0)}{n!}\right).$$

Hence, by a result of [1], there exists $c > 0$ and a compact subset L_1 of F such that $q(f) \leq c\Pi_L(f) \leq c\|f\|_{L_1}$ for every f in $H(E)$.
This completes the proof.

We now apply Theorem 5 to the following problems:

(a)   When is $H_E(F) = H(F)$? That is, when can every holomorphic function on F be extended to a holomorphic function on E?
(b)   What topological properties of $H(E)$ are inherited by $H(F)$?

Since every continuous linear form on F extends, by the Hahn–Banach theorem, to a continuous linear form on E and the polynomials of finite type on F are τ_0 dense in $H(F)$ it follows that $H_E(F)$ is a dense subspace of $H(F)$. We immediately obtain the following.

COROLLARY 6.   *Let F denote a closed subspace of a fully nuclear space E.*

(a)   *If $(H(E), \tau_o)/_{H(F)^\perp}$ is complete, then every holomorphic function on F extends to a holomorphic function on E.*
(b)   *If $(H(F), \tau_o)$ is complete, then every holomorphic function on F extends to a holomorphic function on E if and only if $(H(E), \tau_o)/_{H(F)^\perp}$ is complete.*

If we let E denote a $\mathcal{DFN}$ space, we recover a result of Boland [1].

COROLLARY 7.   *If F is a closed subspace of a $\mathcal{DFN}$ space E, then every holomorphic function on F extends to a holomorphic function on E.*

Proof. $(H(E), \tau_o)$ is a Fréchet nuclear space and $H(F)^\perp$ is a closed subspace of $(H(E), \tau_o)$. Hence $(H(E), \tau_o)/_{H(F)^\perp} \cong H_E(F)$ is a Fréchet space and an application of Corollary 6a completes the proof.

We now turn to problem (b).

THEOREM 8.   *If E is a fully nuclear space and $\tau_o = \tau_\delta$ on $H(E)$ then $\tau_o = \tau_\delta$ on $H(F)$ for any closed subspace F of E.*

Proof.   Let p denote a τ_δ continuous seminorm on $H(F)$. Let $\tilde{p}: H(E) \to R$ be defined by the formula $\tilde{p}(f) = p(f|_F)$. If $(V_n)_{n=1}^{\infty}$ is an increasing countable open cover of E then $(V_n \cap F)_{n=1}^{\infty}$ is an increasing countable open cover of F. Hence there exists $c > 0$ and N a positive integer such that

$$p(f) \leq c\|f\|_{V_N} \qquad \text{for every } f \text{ in } H(E).$$

Hence $\tilde{p}$ is τ_δ and consequently τ_0 continuous on $H(E)$. Since $\tilde{p}/_{H(F)^\perp} = 0$, $\tilde{p}$ induces a continuous seminorm q on $(H(E), \tau_0)/_{H(F)^\perp}$. By Theorem 5, there exists a positive number c and a compact subset K of F such that

$$\tilde{p}(f) = p(f|_F) = q(f + H(F)^\perp) \le c\|f\|_K$$

for every f in $H(E)$. Hence for every f in $H_E(F)$, $p(f) \le c\|f\|_K$. In particular, $p(P) \le c\|P\|_K$ for every P in $P(^nF)$. If $f \in H(F)$ then

$$p\left(\sum_{n=0}^{\infty} \frac{\hat{d}^n f(0)}{n!}\right) \le \sum_{n=0}^{\infty} p\left(\frac{\hat{d}^n f(0)}{n!}\right) \le \sum_{n=0}^{\infty} \left\|\frac{\hat{d}^n f(0)}{n!}\right\|_K$$

and p is a τ_0 continuous seminorm on $H(F)$.

This completes the proof.

A similar result holds for the τ_ω topology.

THEOREM 9.   *If F is a closed subspace of a fully nuclear space E and $\tau_0 = \tau_\omega$ on $H(E)$ then $\tau_0 = \tau_\omega$ on $H(F)$.*

Proof.   Let p denote a τ_ω continuous seminorm on $H(F)$. Suppose p is ported by the compact subset K of F. We define $\tilde{p}$ on $H(E)$ by the formula

$$\tilde{p}(f) = p(f|_F) \qquad \text{for every } f \text{ in } H(E).$$

If W is a neighborhood of K in E then $W \cap F$ is a neighborhood of K in F. Hence there exists $c(W \cap F) > 0$ such that

$$\tilde{p}(f) = p(f|_F) \le c(W \cap F)\|f|_F\|_{W \cap F} \le c(W \cap F)\|f\|_W$$

for every f in $H(E)$.

Thus $\tilde{p}$ is a τ_ω continuous seminorm on $H(E)$. Since $\tilde{p}/_{H(F)^\perp} = 0$ and $\tau_\omega = \tau_0$ on $H(E)$ $\tilde{p}$ defines a continuous seminorm q on $(H(E), \tau_0)/_{H(F)^\perp}$. By Theorem 5, there exists a positive number c and a compact subset L of F such that

$$\tilde{p}(f) = p(f|_F) = q(f + H(F)^\perp) \le c\|f\|_L$$

for every f in $H(E)$. A density argument, as in the previous theorem, now completes the proof.

COROLLARY 10.   *If E is a complete DN space or a closed subspace of $\mathcal{D}$ then $\tau_0 = \tau_\delta$ on $H(E)$.*

In effect, Theorems 8 and 9 allow us to drop the basis hypothesis on a number of the results proved in [2–5].

ACKNOWLEDGMENTS

The author would like to thank D. Vogt for noticing the equivalence of B nuclear spaces and complete DN spaces with a basis and communicating the proof of Theorem 1.

REFERENCES

1. P. J. BOLAND, Holomorphic functions on nuclear spaces, *Trans. Amer. Math. Soc.* **209**, (1975), 275–281.
2. P. J. BOLAND AND S. DINEEN, Holomorphic functions on fully nuclear spaces. *Bull. Soc. Math. France* **106**, (1978), 311–335.
3. P. J. BOLAND AND S. DINEEN, Duality theory for spaces of germs and holomorphic functions on nuclear spaces, *in* "Advances in Holomorphy" (J. A. Barroso, ed.), North-Holland Mathematical Studies, Vol. 34, pp. 179–207, North-Holland Publ., Amsterdam, 1979.
4. P. J. BOLAND AND S. DINEEN, Holomorphic functions on spaces of distributions, *Pacific J. Math.* to appear.
5. S. DINEEN, Analytic functionals on fully nuclear spaces, *Studia Math.*, to appear.
6. S. DINEEN, Holomorphic functions on nuclear sequence spaces, *in* "Functional Analysis, Surveys and Recent Results, II" (K. D. Bierstedt and B. Funchssteiner, eZs.), North-Holland Mathematical Studies, vol. 38, pp. 239–256, North-Holland Publ., Amsterdam, 1979.
7. E. DUBINSKY, Basic sequences in (s), *Studia Math.* **39** (1977), 283–293.
8. M. VALDIVIA, Representaciones de los espacios $\mathscr{D}(\Omega)$ y $\mathscr{D}'(\Omega)$, *Rev. Real Acad. Ciênc. Exact. Fis. Natur. Madrid* **12** (1978), 385–414.
9. D. VOGT, Subspaces and quotients of (s), *in* "Functional Analysis Surveys and Recent Results, I" (K. D. Bierstedt and B. Fuchssteiner, eds.), North-Holland Matematical Studies, Vol. 27 pp. 167–187, North-Holland Publ., Amsterdam, 1977.
10. D. VOGT, Charakterisierung der Unterraume von s, *Math. Z.* **155** (1977), 109–117.
11. D. VOGT, Über die isomorphie lokalkonvexer Raume der Analysis mit folgenraumen, preprint, Univ. of Wuppertal.

Sur le quotient d'une variété algébrique
par un groupe algébrique

J. Dixmier

Département de Mathématiques
Université de Paris VI
Paris, France

ET

M. Raynaud

Département de Mathématiques
Université de Paris XI
Orsay, France

En hommage à Laurent Schwartz

Introduction

Dans tout l'article, k désigne un corps algébriquement clos, X une variété algébrique sur k, G un groupe algébrique sur k qui opère régulièrement dans X. Les variétés sont prises au sens de Serre [15] ou Borel [2].

Il existe différentes notions de quotient de X par G. Selon certaines de ces notions, le quotient de k^n par le groupe des homothéties est la variété réduite à un point. On préférerait se restreindre à $k^n - \{0\}$ et considérer que le quotient est l'espace projectif. Peut-on poser des définitions générales dans ce sens?

Dans tout l'article, la seule notion de quotient utilisée sera la suivante. Une variété quotient de X par G est un couple (Y, φ), où Y est une variété algébrique sur k et $\varphi : X \to Y$ un morphisme, tels que: (i) φ est ouvert, constant sur les G-orbites, et définit une bijection de l'ensemble X/G sur Y; (ii) si U est une partie ouverte de Y, le morphisme de $k[U]$ dans $k[\varphi^{-1}(U)]^G$ défini par φ est bijectif. (Ici et dans la suite, l'exposant G signifie qu'on prend les invariants par G. Rappelons d'autre part que la condition (ii) est conséquence de la condition (i) si k est de caractéristique 0 et si Y est normale [2, p. 174].

D'après [2, p. 174], (Y, φ) possède la propriété universelle évidente, de sorte que (Y, φ), s'il existe, est unique à isomorphisme près. On peut imposer que Y admette l'ensemble X/G comme ensemble sous-jacent et que φ soit

327

l'application canonique de X sur X/G; alors (Y, φ) est vraiment unique. Dans la suite, nous n'imposerons pas cette condition, et nous noterons cependant X/G la variété Y.

Dans l'exemple cité plus haut, la variété k^n/G n'existe pas; la variété $(k^n - \{0\})/G$ existe et est l'espace projectif.

Revenons au cas général. D'après [14], *il existe* une sous-variété ouverte dense X' de X, G-stable, telle que la variété quotient X'/G existe. Nous allons définir une telle sous-variété X' *canoniquement* associée à X munie de l'action de G (ce qui n'est pas le cas dans [14]). Sous certaines hypothèses, nous pourrons même obtenir que X'/G soit quasi-affine.

En fait, cet article est surtout un article d'exposition, qui reprend divers résultats, plus ou moins déjà publiés. Quelques exemples sont peut-être nouveaux.

Notations.   On a déjà signalé les notations k, X, G, $k[X]^G$. On note Γ le graphe de la relation d'équivalence définie par l'action de G dans X; c'est une partie constructible de $X \times X$. Si $x \in X$, on note Gx la sous-variété de X égale à l'orbite de x dans X.

1.   L'OUVERT $\Omega_1(X, G)$

1.1.   LEMME.   *Soit E l'ensemble des $x \in X$ tels que*

$$\dim((\{x\} \times X) \cap (\bar{\Gamma} - \Gamma)) \geq \dim Gx.$$

Alors E est une partie constructible de X, G-stable, dont l'adhérence $\bar{E}$ ne contient aucune composante irréductible de X. La relation d'équivalence définie par G dans $X - E$ a un graphe fermé dans $(X - E) \times (X - E)$.

La démonstration du lemme est analogue à celle donnée dans [4, pp. 162–163], qui traite du cas où G et X sont irréductibles.

1.2.   Ainsi, $X - \bar{E}$ est une partie ouverte dense de X, G-stable, définie canoniquement à partir de X munie de l'action de G; et la relation d'équivalence définie par l'action de G sur $X - \bar{E}$ a un graphe fermé dans $(X - \bar{E}) \times (X - \bar{E})$. *Nous noterons $\Omega(X, G)$ la sous-variété ouverte de X, d'espace $X - \bar{E}$* [le couple (X, G) est mis pour désigner X munie de l'action de G].

1.3.   REMARQUES.   L'ouvert $\Omega(X, G)$ est contenu dans le plus grand ouvert X' de X formé des points x où la dimension de Gx est localement constante. En particulier, $\Omega(X, G)$ est somme disjointe de sous-variétés ouvertes et fermées, où la dimension des orbites est constante. L'ouvert $\Omega(X', G)$ est un autre ouvert G-stable, canoniquement associé à (X, G), tel que le graphe de l'action de G soit fermé; on a $\Omega(X', G) \supset \Omega(X, G)$, et l'inclusion peut être stricte. On a $\Omega(X, G) = X$ si et seulement si Γ est fermé dans $X \times X$.

1.4. EXEMPLE. Nous allons montrer que: (a) l'ensemble E n'est pas nécessairement fermé; (b) il n'existe pas nécessairement un plus grand ouvert Z de X, G-stable, tel que le graphe de G opérant dans Z soit fermé dans $Z \times Z$; (c) il n'existe pas nécessairement un plus grand ouvert Z' de X, G-stable, tel que la variété Z'/G existe. [Les assertions (b) et (c) résultent aussi de ce qui est brièvement signalé dans [6, 1.17], p. 72.]

Soit X l'ensemble des $(\alpha, \beta, \gamma, \delta) \in k^4$ tels que $\alpha\gamma = \delta(\beta\delta - 1)$. C'est une sous-variété lisse de k^4. Soit Y_1 (resp. Y_2) l'ensemble des $(\alpha, \beta, \gamma, \delta) \in k^4$ tels que $\alpha = \delta = 0$ (resp; $\alpha = \beta\delta - 1 = 0$); les Y_i sont des sous-variétés de X.

L'application $\varphi:(\alpha, \beta, \gamma) \mapsto (\alpha, \beta, \gamma(\alpha\beta\gamma - 1), \alpha\gamma)$ de k^3 dans k^4 est une immersion ouverte de k^3 dans X; son image est $X_1 = X - Y_2$; on a $\varphi(\{0\} \times k^2) = Y_1$. Soit U l'ensemble des $(\alpha, \beta, \gamma) \in k^3$ tels que $\beta \neq 0$. L'application $\psi:(\alpha, \beta, \gamma) \mapsto (\alpha, \beta, \gamma(\alpha\beta\gamma + 1), \alpha\gamma + \beta^{-1})$ de U dans k^4 est une immersion ouverte de U dans X; soit X_2 son image; on a $\psi(\{0\} \times (k - \{0\}) \times k) = Y_2$, donc $X = X_1 \cup X_2$. Soit U' l'ensemble des $(\alpha, \beta, \gamma) \in k^3$ tels que $\alpha\beta \neq 0$. Si $(\alpha, \beta, \gamma) \in U'$, on a $\varphi(\alpha, \beta, \gamma) = \psi(\alpha, \beta, \gamma - \alpha^{-1}\beta^{-1})$. Donc, si θ désigne l'automorphisme $(\alpha, \beta, \gamma) \mapsto (\alpha, \beta, \gamma - \alpha^{-1}\beta^{-1})$ de U', X s'obtient en recollant k^3 et U le long de U' à l'aide de θ.

Faisons opérer le groupe additif $G = k$ sur k^3 et U par $\lambda.(\alpha, \beta, \gamma) = (\alpha, \beta, \gamma + \lambda)$. Cette action laisse stable U' et commute à θ, donc définit une action de G dans X. On a, pour $(\alpha, \beta, \gamma, \delta) \in X$,

$$\lambda.(\alpha, \beta, \gamma, \delta) = (\alpha, \beta, \gamma + \lambda(2\beta\delta - 1) + \lambda^2\alpha\beta, \delta + \lambda\alpha).$$

Il est clair que les variétés k^3/G et U/G existent, donc aussi les variétés X_1/G et X_2/G. On a $\lambda.(0, \beta, 0) = (0, \beta, \gamma - \lambda, 0)$ et $\lambda.(0, \beta, \gamma, \beta^{-1}) = (0, \beta, \gamma + \lambda, \beta^{-1})$, de sorte que Y_1 et Y_2 sont des réunions d'orbites. Pour $\alpha, \beta \in k$, posons

$$x_{\alpha\beta} = (\alpha, \beta, 0, 0) \in X, \qquad x_\beta = (0, \beta, 0, 0) \in Y_1$$

et, si $\beta \neq 0$,

$$x'_{\alpha\beta} = (\alpha, \beta, 0, \beta^{-1}) \in X, \qquad x'_\beta = (0, \beta, 0, \beta^{-1}) \in Y_2.$$

Si $\alpha\beta \neq 0$, on a $(\alpha\beta)^{-1}.x_{\alpha\beta} = x'_{\alpha\beta}$. Donc, pour $\beta \neq 0$, $(x_\beta, x'_\beta) \in \bar{\Gamma}$. Mais $(x_\beta, x'_\beta) \notin \Gamma$. Donc Γ est non fermé, d'où (b) et (c). On voit aussi que $x_\beta \in E$ pour $\beta \neq 0$.

Par ailleurs, $(0, 0, \gamma, \delta) \in X$ si et seulement si $\delta = 0$. Les fonctions $(\alpha, \beta, \gamma, \delta) \mapsto \alpha$ et $(\alpha, \beta, \gamma, \delta) \mapsto \beta$ sur X sont G-invariantes; donc, si $((0, 0, \gamma, 0), (\alpha', \beta', \gamma', \delta')) \in \bar{\Gamma}$, on a $\alpha' = \beta' = 0$, puis $\delta' = 0$. Or l'ensemble des $(0, 0, \gamma, 0)$ pour $\gamma \in k$ est une G-orbite. Ainsi, $x_0 \notin E$, de sorte que E est non fermé.

1.5. DÉFINITION.   On notera $\Omega_1(X, G)$ la réunion des parties ouvertes G-stables U de $\Omega(X, G)$ telles que la variété U/G existe.

1.6. PROPOSITION.   *L'ensemble $\Omega_1(X, G)$ est ouvert, dense, G-stable dans X. La variété $\Omega_1(X, G)/G$ existe.*

Il existe des parties ouvertes G-stables U denses dans X telles que U/G existe [14]. Donc $\Omega_1(X, G)$ est dense dans X. Comme la relation d'équivalence définie par G dans $\Omega(X, G)$ a un graphe fermé, la variété $\Omega_1(X, G)/G$ existe [13, p. 215].

1.7. On peut avoir $\Omega_1(X, G) \neq \Omega(X, G)$. Par exemple, prenons pour X la variété U' de [7, Exemple 4.3]; alors $\Omega(X, G) = X$ et $\Omega_1(X, G) \neq X$. En fait, comme on ne dispose pas de critère satisfaisant d'existence de quotient en géométrie algébrique, on n'a pas en général de description simple de $\Omega_1(X, G)$.

Si l'on élargit la catégorie des variétés algébriques à celle des espaces algébriques de Artin [1], on peut définir un espace algébrique quotient séparé $\Omega(X, G)/G$.

Si $X = k^n$ et si $G = k - \{0\}$ opérant par homothéties, on a $\Omega(X, G) = \Omega_1(X, G) = k^n - \{0\}$. S'il existe dans X une orbite dense X', on a $\Omega(X, G) = \Omega_1(X, G) = X'$.

1.8. On définit par récurrence les parties ouvertes croissantes G-stables $U_0, U_1, \ldots$ de X de la manière suivante: $U_0 = \varnothing$; U_i étant défini, on pose $U_{i+1} = U_i \cup \Omega_1(X - U_i, G)$. Pour des raisons de dimension, il existe un plus petit entier q tel que $U_q = X$. Alors $U_1 - U_0, \ldots, U_q - U_{q-1}$ forment *une partition finie canonique de (X, G) en sous-ensembles localement fermés G-stables, tels que les variétés quotients $(U_{i+1} - U_i)/G$ existent.*

1.9. PROPOSITION.   *Si $X/G = Y$ existe, le morphisme canonique $\varphi: X \to Y$ est universellement ouvert.*

(Rappelons que cela signifie que, pour toute Y-variété Z, la projection $X \times_Y Z \to Z$ est ouverte. La proposition 1.9 prouve que les quotients considérés ici sont les quotients géométriques au sens de [6, Définition 0.6, p. 4].)

L'assertion à démontrer est locale sur Y. On peut donc supposer que la dimension des orbites de G dans X est constante de valeur d. Alors les fibres de φ sont équidimensionnelles, de dimension d, et toute composante irréductible de X domine une composante irréductible de Y. Lorsque Y est normale, en particulier si X est normale, le fait que φ soit universellement ouvert résulte alors du théorème de Chevalley (EGA IV 14.4.4). Dans le cas général, soit Y' la normalisée de Y, et soit $X' = X \times_Y Y'$. Si φ n'est pas universellement ouvert, il existe des composantes irréductibles de X' qui ne dominent pas une composante irréductible de Y' (EGA IV 14.4.9). La réunion de ces composantes a une image fermée Z dans X. L' image de Z dans Y a une adhérence F dans Y qui ne contient aucune composante irréductible de Y, et φ est universellement ouvert au-dessus de $V = Y - F$. Soit s un point générique d'une composante irréductible de F (dans le schéma associé à Y), et soit $\tilde{T}$ le spectre d'un hensélisé strict $\tilde{\mathcal{O}}$ de l'anneau local $\mathcal{O}$ du schéma Y au point s (EGA IV 18.8.7). Notons $\tilde{s}$ le point fermé de $\tilde{T}$. Par le changement de base canonique $\tilde{T} \to Y$, on déduit $\tilde{X}$ et $\tilde{\varphi}$ de X et φ. Notons $\tilde{X}(\tilde{s})$ la fibre de $\tilde{\varphi}$ au-dessus de $\tilde{s}$. Soit $(\tilde{T}_i)_{i \in I}$ la famille (finie) des composantes irréductibles de $\tilde{T}$.

Comme $\tilde{\varphi}$ est universellement ouvert au-dessus de $\tilde{T} - \tilde{s}$ et que G opère transitivement sur les fibres de $\tilde{\varphi}$, pour chaque $i \in I$ il n'y a que deux cas possibles: (a) toute composante irréductible de $\tilde{X} \times_{\tilde{T}} \tilde{T}_i$ domine $\tilde{T}_i$; (b) $\tilde{X} \times_{\tilde{T}} \tilde{T}_i = \tilde{X}(\tilde{s}) \amalg \tilde{X}'_i$, décomposition en somme disjointe de deux ouverts G-stables, non vides.

Comme $\tilde{\varphi}$ n'est pas universellement ouvert et que les $\tilde{T}_i$ sont géométriquement unibranches en s (EGA IV 18.8.6), il existe des $i \in I$ pour lesquels on est dans le cas (b) (EGA IV 14.4.1). Alors la réunion des $\tilde{X}'_i$ pour les indices i en question est une partie non vide de $\tilde{X}$, à la fois ouverte et fermée, G-stable, ne contenant pas $\tilde{X}(\tilde{s})$, donc définie par un idempotent e de $\tilde{\mathcal{O}}[\tilde{X}]^G$, distinct de 0 et 1. Mais le morphisme $\tilde{T} \to Y$ étant plat (EGA IV 18.8.8), on a $\tilde{\mathcal{O}}[\tilde{X}]^G = \tilde{\mathcal{O}}$ et l'on obtient une contradiction puisque, $\tilde{\mathcal{O}}$ étant local, $\tilde{\mathcal{O}}$ ne contient pas d'idempotent non trivial. Par suite φ est universellement ouvert.

2. L'OUVERT $\Omega_2 (X, G)$

2.1. *Orbites séparées*

2.1.1. DÉFINITION. L'orbite Gx d'un point x de X est dite *séparée* (par les fonctions sur X invariantes par G) si pour tout $y \in X - Gx$, il existe $u \in k[X]^G$ tel que $u(x) \neq u(y)$.

Considérons le schéma affine $S = \operatorname{Spec} k[X]^G$ et le morphisme canonique de k-schémas $f : X \to S$. (Rappelons que $k[X]^G$ n'est pas nécessairement une k-algèbre de type fini et donc S n'est pas nécessairement une variété.) Les orbites séparées de X sont exactement les orbites de X qui sont des *fibres* du morphisme de schémas f.

2.1.2. Soit Δ l'ensemble des $(x, y) \in X \times X$ tels que $f(x) = f(y)$ [i.e., $u(x) = u(y)$ pour tout $u \in k[X]^G$]. C'est un fermé de $X \times X$, qui contient

Γ et donc $\bar{\Gamma}$. Soit $x \in X$; Gx est séparée si et seulement si

$$\Delta \cap (\{x\} \times X) = \{x\} \times Gx = \Gamma \cap (\{x\} \times X).$$

En particulier, la réunion des orbites séparées a pour complémentaire dans X la projection de $\Delta - \Gamma$, donc est une partie *constructible* de X.

2.1.3.   On note $\Omega_2(X, G)$ *l'intérieur* de la réunion des orbites séparées (cf. 2.2.4 et 2.4.6).

2.1.4.   Nous verrons en 2.2.3 que la variété $\Omega_2(X, G)/G$ existe et est quasi-affine. Alors que $\Omega_1(X, G)$ est toujours dense dans X, $\Omega_2(X, G)$ peut fort bien être vide. (Cela se produit par exemple si $X = k^n$ dans lequel $k - \{0\}$ opère par homothéties, ou s'il existe dans X une G-orbite dense distincte de X.) Nous verrons cependant des cas généraux où $\Omega_2(X, G)$ est dense (cf. 2.4.2, et 2.5.2).

2.2   *Un critère d'existence de quotients*

2.2.1.   Soient $A = k[X]$, $X' = \mathrm{Spec}(A)$ et $i : X \to X'$ le morphisme canonique de schémas. Rappelons que X est quasi-affine si et seulement si i est une immersion ouverte. Comme A est limite inductive de ses sous-k-algèbres de type fini, X est quasi-affine si et seulement si X est isomorphe à une sous-variété ouverte d'une variété affine.

Si X/G existe et est quasi-affine, toutes les orbites de X sont séparées. Les énoncés qui suivent fournissent une réciproque à cette assertion. Ils reposent essentiellement sur le Main theorem de Zariski. Quelques précautions sont indispensables du fait que $k[X]^G$ n'est pas nécessairement de type fini sur k, et X pas nécessairement normale.

2.2.2.   PROPOSITION.   *Soient S, f comme en 2.1.1, $x \in X$, et $y = f(x)$. On fait l'une des deux hypothèses suivantes*: (1) Γ *est fermé et l'orbite Gx est séparée*; (2) $x \in \Omega_2(X, G)$.
Alors il existe un voisinage ouvert V de y dans S tel que, si $U = f^{-1}(V)$, V soit la variété quotient U/G.

Avant de démontrer 2.2.2, donnons quelques corollaires.

2.2.3.   COROLLAIRE.   *L'ouvert $\Omega_2(X, G)$ est contenu dans $\Omega_1(X, G)$. La variété $\Omega_2(X, G)/G$ existe et est une variété quasi-affine, ouverte dans le schéma affine S. En particulier, si toutes les orbites sont séparées, la variété X/G existe et est quasi-affine.*

Posons $\Omega_2 = \Omega_2(X, G)$. Alors, avec les notations de 2.1, on a $(\Omega_2 \times X) \cap \Delta = (\Omega_2 \times X) \cap \Gamma$, donc $\Omega_2 \subset \Omega(X, G)$ (cf. 1.2). Cela étant, il résulte de 2.2.2 que $f(\Omega_2)$ est un ouvert de S qui est le quotient de Ω_2 par G, d'où les autres assertions de 2.2.3.

2.2.4.   COROLLAIRE.   *Si le graphe Γ est fermé dans $X \times X$, la réunion des orbites séparées est ouverte dans X et par suite est égale à $\Omega_2(X, G)$.*

C'est clair à partir de 2.2.2.

2.2.5.   COROLLAIRE.   *Supposons que Γ soit fermé dans $X \times X$. Pour que X/G existe, il faut et il suffit que, pour tout $x \in X$, il existe un ouvert G-stable de X contenant Gx comme orbite séparée.*

2.2.6.   COROLLAIRE.   *Soient T un k-schéma sur lequel G opère trivialement et $g: X \to T$ un k-morphisme G-équivariant. Soient $x \in X$ et $y = g(x)$. On suppose que:*

(1)   *l'orbite Gx est ouverte dans $g^{-1}(y)$;*
(2)   *Γ est fermé.*

Alors il existe un voisinage ouvert G-stable U de x tel que U/G existe.

(Ce corollaire explicite, dans le cas des actions de groupes, le résultat annoncé dans [10, Théorème 1] et jamais publié.)

Quitte à remplacer X par l'ouvert G-stable $X - (g^{-1}(y) - Gx)$, on peut supposer que $g^{-1}(y) = Gx$. Par ailleurs, on peut supposer T affine d'anneau A. Alors le morphisme de k-algèbres $A \to k[X]$ se factorise à travers $k[X]^G$. D'où si $T' = \operatorname{Spec}(k[X]^G)$, une factorisation de g:

Alors, si $y' = g'(x)$, on a $g'^{-1}(y') = Gx$, donc Gx est une orbite séparée dans X, et l'on termine grâce à 2.2.2.

2.2.7.   COROLLAIRE.   *Supposons qu'il existe un quotient Y de X par G, et soit U une sous-variété G-stable de X. Alors:*

(i)   *l'image de U dans Y est une sous-variété V de Y;*
(ii)   *il existe un quotient Z de U par G;*
(iii)   *le morphisme canonique $Z \to V$ est un homéomorphisme fini (cet homéomorphisme n'est pas toujours un isomorphisme).*

La sous-variété U de X est un ouvert du fermé G-stable $\bar{U}$. Comme $Y = X/G$, l'image de $\bar{U}$ dans Y est un fermé W de Y. Mais $\varphi: X \to Y$ est universellement ouvert (1.9), donc l'image V de U dans W est un ouvert de W, d'où (i). L'application canonique $U \to V$ sépare les orbites de G dans U; l'existence de $Z = U/G$ résulte alors de 2.2.6. Vu la propriété universelle du quotient U/G, on a un diagramme commutatif:

où h est bijectif. Comme $\varphi \,|\, U$ est universellement ouvert et ψ surjectif, h est universellement ouvert. En particulier, h est un homéomorphisme. On conclut que h est fini grâce au lemme suivant:

2.2.8. LEMME. *Soit $h: Z \to V$ un morphisme de variétés qui est bijectif et universellement ouvert. Alors h est fini.*

L'assertion à démontrer est locale sur V. Par passage à la limite, on se ramène au cas où V est un schéma, spectre d'un anneau local $\mathcal{O}$; puis par platitude au cas où $\mathcal{O}$ est hensélien. Soit s le point fermé de V. Comme au départ h est une bijection sur les points à valeurs dans k, h a des fibres géométriques non vides réduites à un point. En particulier, h est quasi-fini. Alors $Z = Z' \amalg Z''$, où Z' est fini sur V et Z'' est au-dessus de $V - \{s\}$ (EGA IV 18.12.1). Comme h est surjectif, Z' n'est pas vide. Comme h est universellement ouvert et V local, $Z' \to V$ est surjectif. Les fibres de h ayant un seul point, $Z'' = \varnothing$ et donc h est fini.

2.3. *Démonstration de 2.2.2*

2.3.1. Considérons la k-algèbre $A = k[X]^G$ comme limite inductive filtrante de sous-k-algèbres A_i, $i \in I$, de type fini sur k. Soit B_i la fermeture intégrale de A_i dans A. Montrons d'abord que B_i est de type fini sur k. L'algèbre A est réduite, contenue dans $k[X]$. Par suite $S = \mathrm{Spec}(A)$ n'a qu'un nombre fini de composantes irréductibles S_λ, $\lambda \in \Lambda$, et le corps des fractions K_λ de S_λ est une extension de type fini de k. Alors, pour $i \gg 0$, l'injection $A_i \to A$ induit une bijection sur les anneaux totaux de fractions $\prod_{\lambda \in \Lambda} K_\lambda$. Dans la suite on suppose que cette condition est réalisée pour tout i. Alors B_i est contenu dans la clôture intégrale de A_i dans $\prod_{\lambda \in \Lambda} K_\lambda$, donc est fini sur A_i et a fortiori de type fini sur k. Bien sur on a aussi $A = \varinjlim B_i$. Remplaçant A_i par B_i, on suppose désormais que A_i est *intégralement fermé dans A*.

2.3.2.   Posons $S_i = \mathrm{Spec}(A_i)$ et notons $s_i: S \to S_i$ le morphisme associé à l'injection $A_i \to A$. Soient $f_i = s_i \circ f$, et Δ_i l'ensemble des $(a, b) \in X \times X$ tels que $f_i(a) = f_i(b)$. Les Δ_i, pour $i \in I$, forment une famille filtrante décroissante de fermés dans l'espace noethérien $X \times X$, d'intersection Δ. Donc, pour $i \gg 0$, on a $\Delta = \Delta_i$. On suppose désormais $\Delta = \Delta_i$ pour tout i. Par hypothèse, Gx est une orbite séparée de X, donc une fibre de f; Gx est donc une fibre de f_i. De plus, dans le cas où $x \in \Omega_2(X, G)$, il existe un ouvert W de X, G-stable, contenant x, tel que pour tout $i \in I$ les fibres de $f_i | W: W \to S_i$ sont des G-orbites et $W = f_i^{-1} f_i(W)$.

2.3.3.   Soit $d = \dim(Gx)$. Soit U l'ensemble des $z \in X$ tels que: (a) $\dim(Gz) \geq d$; (b) la dimension de la fibre de f en z est $\leq d$. Dans (b), on peut remplacer f par f_i; d'après les théorèmes de semi-continuité, U est ouvert dans X. Bien sur U est G-stable et contient Gx. Les fibres de $f | U$ sont alors réunion disjointe d'un nombre fini d'orbites de G de dimension d. Posons $U' = U$ dans le cas où Γ est fermé, et $U' = U \cap W$ (cf. 2.3.2) lorsque $x \in \Omega_2(X, G)$. Il nous suffira de montrer que $f_i: U' \to S_i$ est universellement ouvert. En effet on a le lemme suivant:

2.3.4.   LEMME.   *Soit U'' un ouvert G-stable de U'; soit $i \in I$ tel que $f_i: U'' \to S_i$ soit universellement ouvert d'image V_i. Alors:*

(a)   *On a $U'' = f_i^{-1}(V_i)$, et les fibres de $U'' \to V_i$ sont des orbites.*
(b)   *Soit $V = s_i^{-1}(V_i) \subset S$. Supposons de plus que pour tout $j \geq i$, $f_j: U'' \to S_j$ soit universellement ouvert, d'image V_j. Alors les morphismes canoniques $V_j \to V_i (j \geq i)$ et $V \to V_i$ sont des isomorphismes, et V est le quotient par G de $U'' = f^{-1}(V)$.*

Prouvons (a).   Dans le cas où $x \in \Omega_2(X, G)$, par construction U' est réunion d'orbites séparées et l'assertion est claire. Supposons maintenant Γ fermé. Soit η un point générique de V_i. Alors $f_i^{-1}(\eta) \cap U''$ est une réunion finie, non vide, d'orbites ouvertes, donc fermées, de $f_i^{-1}(\eta)$. Comme A_i est intégralement fermé dans A, $f_i^{-1}(\eta)$ est géométriquement une seule orbite, nécessairement contenue dans U''. On en déduit qu'il existe un ouvert dense de V_i au-dessus duquel $U'' = X$ et où les fibres de f contiennent une seule orbite. Etablissons le lemme intermédiaire suivant:

2.3.5.   LEMME.   *Soit $u: M \to N$ un morphisme équivariant de k-schémas noethériens, munis d'une action de G, G opérant trivialement dans N. Soit M' un ouvert dense de M, G-stable et tel que:*

(i)   *la restriction $u': M' \to N$ est universellement ouverte, surjective;*
(ii)   *il existe un ouvert dense de N au-dessus duquel $M = M'$ et les fibres géométriques de u contiennent une seule orbite.*

Alors, si le graphe Γ de l'action de G sur M est fermé dans $M \times M$, on a $M = M'$.

En effet, considérons le diagramme cartésien:

$$
\begin{array}{ccc}
M' \times_N M & \xrightarrow{p'} & M \\
p \downarrow & & \downarrow u \\
M' & \xrightarrow{u'} & N
\end{array}
$$

Alors $M \times_N M$ est un fermé de $M \times_k M$ qui contient Γ; par suite $M' \times_N M$ contient le fermé $\Gamma \cap (M' \times_k M)$ de $M' \times_k M$ qui est le graphe Γ' de G opérant sur M'. Comme u' est universellement ouvert, p' est ouvert, donc $p'(M' \times_N M - \Gamma')$ est un ouvert de M. D'après (ii) cet ouvert est vide, donc $M' \times_N M = \Gamma'$. Comme u' est surjectif, on en déduit que $M' = M$. ∎

Appliquons 2.3.5 à l'adhérence de U'' dans $f_i^{-1}(V_i)$. On voit que U'' est une partie G-stable de $f_i^{-1}(V_i)$ à la fois ouverte et fermée. Comme A_i est intégralement fermé dans A, on en déduit que $U'' = f_i^{-1}(V_i)$. Enfin, à nouveau d'après 2.3.5, les fibres de $f_i : U'' \to V_i$ sont des orbites.

Prouvons 2.3.4(b). Soit $j \geq i$ et notons $s_i^j : S_j \to S_i$ le morphisme canonique. L'application canonique $V_j \to V_i$ est un homéomorphisme universellement ouvert et birationnel (2.3.1) donc est finie (2.2.8). Il en résulte que V_j est fermé dans $(s_i^j)^{-1}(V_i)$ [EGA II 5.4.3 (i)]. Alors V_j est à la fois ouvert et fermé dans $(s_i^j)^{-1}(V_i)$, donc est égal à $(s_i^j)^{-1}(V_i)$ puisque A_i est intégralement fermé dans A_j. On voit que, au-dessus de V_i, s_i^j est fini, donc est un isomorphisme. Par passage à la limite, on en déduit un isomorphisme $V = s_i^{-1}(V_i) \to V_i$, d'où l'assertion (b).

2.3.6. Il reste à voir que $f_i : U' \to S_i$ est universellement ouvert. Notons tout de suite que si X est normale, il en est de même de S_i et toute composante irréductible de X domine une composante irréductible de S_i. Le fait que f_i soit universellement ouvert résulte alors du théorème de Chevalley (EGA IV 14.4.4). Dans le cas général, nous allons devoir "séparer les branches de S_i," aux points de $f_i(U')$, par hensélisation.

Soit R_i le normalisé de S_i. Si $f_i : U' \to S_i$ n'est pas universellement ouvert, il existe des composantes irréductibles de $U' \times_{S_i} R_i$ qui ne dominent pas des composantes de R_i (EGA IV 14.4.9). Soit F l'adhérence de l'image de ces composantes dans S_i. Alors F est rare dans S_i et $S_i - F$ est le plus grand ouvert de S_i au-dessus duquel $f_i : U' \to S_i$ est universellement ouvert. Posons $V_i = f_i(U' \cap f_i^{-1}(S_i - F))$. Alors V_i est un ouvert de $S_i - F$, et au-dessus de V_i on peut appliquer 2.3.4(a); par suite $U' \cap f_i^{-1}(V_i) = f_i^{-1}(V_i)$ et les fibres sont des orbites sous G.

Soit s un point générique d'une composante irréductible de F. Alors $s \in f_i(U')$. Soit $\tilde{T}$ le spectre d'un hensélisé strict $\tilde{\mathcal{O}}$ de l'anneau local $\mathcal{O}$ du schéma S_i au point s et soit $\tilde{s}$ son point fermé. Par le changement de base $\tilde{T} \to S_i$, on déduit $\tilde{f}$, $\tilde{X}$, $\tilde{U}'$, $\tilde{V}$ de f_i, X, U', V_i. Alors $\tilde{f} : \tilde{U}' \to \tilde{T}$ n'est pas universellement ouvert, mais est universellement ouvert au-dessus de $\tilde{T} - \{\tilde{s}\}$, d'image $\tilde{V}$. On a $\tilde{U}' \cap \tilde{f}^{-1}(\tilde{V}) = \tilde{f}^{-1}(\tilde{V})$, et les fibres de $\tilde{f}$ au-dessus de $\tilde{V}$ sont des orbites. Notons $\tilde{X}(\tilde{s})$ et $\tilde{U}'(\tilde{s})$ les fibres de $\tilde{X}$ et $\tilde{U}'$ au-dessus de $\tilde{s}$.

Soit $\tilde{X}_1$ une réunion minimale de composantes irréductibles de $\tilde{X}$, stable par G et non contenue dans $\tilde{U}'(\tilde{s})$. Deux cas sont possibles:

(i) $\tilde{X}_1 \cap \tilde{U}'(\tilde{s}) = \varnothing$;
(ii) $\tilde{X}_1 \cap \tilde{U}'(\tilde{s}) \neq \varnothing$.

Plaçons-nous dans le cas (ii). Alors $\tilde{X}_1$ domine une composante irréductible $\tilde{T}_1$ de $\tilde{T}$. Comme $\tilde{T}_1$ est géométriquement unibranche en $\tilde{s}$ (EGA IV 18.8.6) et que le morphisme $\tilde{X}_1 \to \tilde{T}_1$ est équidimensionnel aux points de $\tilde{X} \cap \tilde{U}'(\tilde{s})$, $\tilde{X}_1 \cap \tilde{U}' \to \tilde{T}_1$ est universellement ouvert (EGA IV 14.4.4), en particulier surjectif (T_1 étant local). Montrons que $\tilde{X}_1 \subset \tilde{U}'$. C'est clair au-dessus de $\tilde{T} - \{\tilde{s}\}$, car $\tilde{T} - \{\tilde{s}\}$ est contenu dans $\tilde{V}$. Au-dessus de $\tilde{s}$, c'est clair également si $x \in \Omega_2(X, G)$, car $\tilde{X}(\tilde{s})$ est constitué d'une seule orbite. Enfin, dans le cas où Γ est fermé, il résulte de 2.3.5 que $\tilde{X}_1 \cap \tilde{U}' = \tilde{X}_1$.

Les considérations précédentes entraînent que la réunion des composantes irréductibles de $\tilde{X}$ qui rencontrent $\tilde{U}'(\tilde{s})$ est une partie à la fois ouverte et fermée de $\tilde{X}$, évidemment non vide et G-invariante, donc définie par un idempotent e de $k[X]^G$. Comme A_i est intégralement fermé dans A, $\tilde{\mathcal{O}}$ est intégralement fermé dans $\tilde{\mathcal{O}}[\tilde{X}]^G$ (EGA IV 6.14.4). Par suite $e = 1$, $\tilde{X} = \tilde{U}'$, et $\tilde{f}$ est universellement ouvert, en contradiction avec la définition de s. Ceci achève la démonstration de 2.2.2.

2.4.   *Cas d'un groupe unipotent opérant sur une variété quasi-affine*

2.4.1.   PROPOSITION.   *Soient* $A = k[X]$, $S = \mathrm{Spec}(A^G)$, *et* $f : X \to S$ *le morphisme canonique de schémas. Supposons que X soit quasi-affine et que G soit unipotent. Alors il existe un ouvert V de S, tel que $U = f^{-1}(V)$ soit dense dans X et que $V = U/G$.*

Soit X^* une réunion minimale de composantes irréductibles de X qui soit stable sous G, et soit F la réunion des autres composantes de X. Notons U un ouvert dense de $X - F$, G-stable, tel que U/G existe et soit affine. Alors le sous-schéma fermé réduit de $X' = \mathrm{Spec}(A)$, d'espace $X' - U$, est défini par un idéal I non nul de A, G-invariant. Comme G est unipotent et I réunion de k-espaces vectoriels de dimension finie, stables par G, I contient un élément a non nul, G-invariant. L'ouvert X_a de X où a est inversible est alors un ouvert non vide, G-invariant, contenu dans U, donc dense dans U. De plus X_a/G existe et est un ouvert de U/G, donc est quasi-affine. C'est dire que X_a/G est un ouvert de $\text{Spec}(k[X_a]^G)$. Mais $k[X_a]$ est le localisé de A par a (EGA I 9.3.3), donc $k[X_a]^G$ est le localisé de A^G par a. Finalement X_a/G est l'ouvert image de X_a dans S, par f. En recommencant ce raisonnement dans F, on obtient la proposition.

2.4.2.   COROLLAIRE.   *On suppose que X est quasi-affine et G unipotent. Alors $\Omega_2(X, G)$ est dense dans X; la variété $\Omega_2(X, G)/G$ existe et est quasi-affine.*

(Pour X affine et k de caractéristique 0, le fait que $\Omega_2(X, G)$ soit dense dans X résulte de [11, p. 57].)

2.4.3.   Sous les hypothèses de 2.4.1, on définit, comme en 1.8, des parties ouvertes croissantes G-stables canoniques $V_0, V_1, \ldots, V_r$ de X telles que $V_0 = \varnothing$, $V_r = X$, et que les variétés $(V_{i+1} - V_i)/G$ existent et soient quasi-affines.

2.4.4.   PROBLÈME.   Sous les hypothèses de 2.4.1, $\Omega_1(X, G)/G$ est-il quasi-affine? Ou même a-t-on $\Omega_1(X, G) = \Omega_2(X, G)$?

2.4.5.   *Application.*   Soient N un groupe de Lie réel nilpotent simplement connexe, $\mathfrak{n}$ son algèbre de Lie, $\mathfrak{n}^*$ l'espace vectoriel dual de $\mathfrak{n}$, dans lequel N opère par la représentation coadjointe. Soit $\hat{N}$ l'ensemble des classes de représentations unitaires irréductibles de N. Il existe une bijection canonique de $\hat{N}$ sur $\mathfrak{n}^*/N$. Pour définir une structure "algébrique" sur $\hat{N}$, il suffit de le faire sur $\mathfrak{n}^*/N$. Soient $N_{\mathbf{C}}$, $\mathfrak{n}_{\mathbf{C}}$, $\mathfrak{n}_{\mathbf{C}}^*$ les complexifiés de N, $\mathfrak{n}$, $\mathfrak{n}^*$. D'après [4, pp. 163–164], deux points de $\mathfrak{n}^*$ conjugués par $N_{\mathbf{C}}$ sont conjugués par N. D'après [12, Théorème 10], si une $N_{\mathbf{C}}$-orbite dans $\mathfrak{n}_{\mathbf{C}}^*$ est stable par conjugaison complexe, elle possède un point réel.

Posons $\mathfrak{n}^* = E_1$, $\mathfrak{n}_{\mathbf{C}}^* = F_1$, $\Omega_2(F_1, N_{\mathbf{C}}) = V_1$, $V_1 \cap \mathfrak{n}^* = U_1$. Alors $V_1/N_{\mathbf{C}}$ existe et est quasi-affine. La conjugaison complexe dans $\mathfrak{n}_{\mathbf{C}}^*$ laisse stable V_1 et induit une conjugaison dans $V_1/N_{\mathbf{C}}$. D'après ce qu'on a dit plus haut, U_1/N s'identifie à l'ensemble des points de $V_1/N_{\mathbf{C}}$ invariants par conjugaison.

Soit $E_2 = E_1 - U_1$, qui est une partie de $\mathfrak{n}^*$ fermée pour la topologie de Zariski. Soit F_2 l'adhérence de E_2 dans $\mathfrak{n}_{\mathbf{C}}^*$ pour la topologie de Zariski. Soient $V_2 = \Omega_2(F_2, N_{\mathbf{C}})$, $U_2 = V_2 \cap \mathfrak{n}^*$. Alors $V_2/N_{\mathbf{C}}$ existe et est quasi-affine, et U_2/N s'identifie à l'ensemble des points de $V_2/N_{\mathbf{C}}$ invariants par conjugaison.

Par récurrence, on définit les E_i, F_i, V_i, U_i, qui sont vides à partir d'un certain rang. Les U_i forment une partition finie canonique de $\mathfrak{n}^*$ en sous-ensembles localement fermés N-stables, et U_i/N est l'ensemble des points invariants par conjugaison dans la variété quasi-affine $V_i/N_{\mathbf{C}}$.

2.4.6. EXEMPLE. Même pour X affine irréductible et G unipotent, la réunion des G-orbites séparées n'est pas toujours ouverte, comme on va le voir.

On reprend l'exemple 1.4 et ses notations. Puisque $(x_\beta, x'_\beta) \in \overline{\Gamma} - \Gamma$ pour $\beta \neq 0$, l'orbite Gx_β est non séparée. Les deux premières fonctions coordonnées sont G-invariantes. Soit $x \in X$. Si la première ou la deuxième coordonnée de x est non nulle, Gx et Gx_0 sont séparées par un élément de $k[X]^G$. Si les deux premières coordonnées de x sont nulles, on a vu en 1.4 que $x \in Gx_0$. Donc l'orbite Gx_0 est séparée, ce qui prouve le résultat annoncé.

2.4.7. EXEMPLES. On suppose k de caractéristique 0, X affine, et G unipotent. Soit $X' = \Omega_2(X, G)$. Il peut arriver que $k[X]^G$ ne soit pas de type fini [8]. Toutefois, dans notre contexte, l'algèbre intéressante est $k[X'/G]$, c'est-à-dire $k[X']^G$. Or $k[X']^G$, qui contient $k[X]^G$ comme sous-algèbre, peut en être distincte. (Par exemple, prenons $G = k$ opérant dans $X = k^2$ par $\lambda.(\alpha, \beta) = (\alpha, \beta + \lambda\alpha)$; on voit tout de suite que X' est l'ensemble des $(\alpha, \beta) \in k^2$ tels que $\alpha \neq 0$; notant encore α, β les fonctions coordonnées sur k^2, on a $k[X]^G = k[\alpha]$, $k[X']^G = k[\alpha, \alpha^{-1}]$.) Il convient donc de donner un exemple où $k[X']^G$ n'est pas de type fini. Or nous allons voir que, dans l'exemple de [8], on a $k[X']^G = k[X]^G$.

Dans cet exemple, $X = k^{32}$, et G est le sous-groupe additif de k^{16} formé des $(\lambda_1, \ldots, \lambda_{16})$ tels que $\sum_{j=1}^{16} \alpha_{ij}\lambda_j = 0$ pour $i = 1, 2, 3$ (les α_{ij} étant des éléments fixés de k, algébriquement indépendants sur $\mathbf{Q}$). L'action de G sur X est définie par

$$(\lambda_1, \ldots, \lambda_{16}).(\tau_1, \ldots, \tau_{16}, \xi_1, \ldots, \xi_{16})$$
$$= (\tau_1, \ldots, \tau_{16}, \xi_1 + \lambda_1\tau_1, \ldots, \xi_{16} + \lambda_{16}\tau_{16}).$$

Pour $i, j \in \{1, \ldots, 16\}$ et $i \neq j$, soit X_{ij} l'ensemble des $(\tau_1, \ldots, \tau_{16}, \xi_1, \ldots, \xi_{16}) \in X$ tels que $\tau_i = \tau_j = 0$. Soit Y la réunion des X_{ij}. On va montrer que $X - Y \subset \Omega_2(X, G)$. Il en résultera que $X - \Omega_2(X, G)$ est de codimension ≥ 2 dans X, d'où notre assertion.

Notons encore $\tau_1, \ldots, \tau_{16}, \xi_1, \ldots, \xi_{16}$ les fonctions coordonnées sur X. On a $\tau_1, \ldots, \tau_{16} \in k[X]^G$. Pour $j = 1, \ldots, 16$, soit $f_j = \xi_j \prod_{t \neq j} \tau_t$. Soit

$$g_i = \sum_{j=1}^{16} \alpha_{ij} f_j \qquad \text{pour} \quad i = 1, 2, 3.$$

On vérifie aisément que $g_1, g_2, g_3 \in k[X]^G$. D'autre part, $\alpha_{21}g_1 - \alpha_{11}g_2$ et $\alpha_{31}g_1 - \alpha_{11}g_3$ sont divisibles par τ_1; soient h_1, h_2 leurs quotients par τ_1, qui sont des éléments de $k[X]^G$.

Soient

$$x = (\tau_1, \ldots, \tau_{16}, \xi_1, \ldots, \xi_{16}) \in X \text{ et } y = (\sigma_1, \ldots, \sigma_{16}, \eta_1, \ldots, \eta_{16}) \in X.$$

On suppose $x \notin Y$ et $y \notin Gx$. Il s'agit de montrer que x et y sont séparés par un élément de $k[X]^G$. C'est clair si $(\sigma_1, \ldots, \sigma_{16}) \neq (\tau_1, \ldots, \tau_{16})$. Supposons désormais que $\sigma_1 = \tau_1, \ldots, \sigma_{16} = \tau_{16}$.

Envisageons d'abord le cas ou $\tau_1 \tau_2 \cdots \tau_{16} \neq 0$. Définissons $\lambda_1, \ldots, \lambda_{16} \in k$ par $\lambda_j = \tau_j^{-1}(\eta_j - \xi_j)$. Pour $i = 1, 2, 3$, on a

$$\sum_{j=1}^{16} \alpha_{ij}\lambda_j = \sum_{j=1}^{16} \alpha_{ij}\tau_j^{-1}\eta_j - \sum_{j=1}^{16} \alpha_{ij}\tau_j^{-1}\xi_j = (\tau_1\tau_2 \cdots \tau_{16})^{-1}(g_i(y) - g_i(x)).$$

Si les trois scalaires $\sum_{j=1}^{16} \alpha_{ij}\lambda_j$ étaient nuls, on aurait $y \in Gx$ contrairement à l'hypothèse. Donc il existe $i \in \{1, 2, 3\}$ tel que $g_i(x) \neq g_i(y)$.

Envisageons maintenant le cas ou par exemple $\tau_1 = 0$. Alors $\tau_2\tau_3 \cdots \tau_{16} \neq 0$ puisque $x \notin Y$. On a

$$g_1(x) = \alpha_{11}\xi_1\tau_2 \cdots \tau_{16}, \qquad g_1(y) = \alpha_{11}\eta_1\tau_2 \cdots \tau_{16}.$$

Si $\xi_1 \neq \eta_1$, x et y sont séparés par g_1. Supposons donc désormais que $\xi_1 = \eta_1$. Définissons $\lambda_2, \ldots, \lambda_{16}$ par $\lambda_j = \tau_j^{-1}(\eta_j - \xi_j)$. Si l'on peut déterminer λ_1 tel que $\sum_{j=1}^{16} \alpha_{ij}\lambda_j = 0$ pour $i = 1, 2, 3$, on a $y \in Gx$ contrairement à l'hypothèse. Or ces trois équations linéaires en λ_1 sont compatibles si et seulement si

$$-\alpha_{11}\left(\sum_{j=2}^{16} \alpha_{2j}\tau_j^{-1}(\eta_j - \xi_j)\right) + \alpha_{21}\left(\sum_{j=2}^{16} \alpha_{1j}\tau_j^{-1}(\eta_j - \xi_j)\right) = 0,$$

$$-\alpha_{11}\left(\sum_{j=2}^{16} \alpha_{3j}\tau_j^{-1}(\eta_j - \xi_j)\right) + \alpha_{31}\left(\sum_{j=2}^{16} \alpha_{1j}\tau_j^{-1}(\eta_j - \xi_j)\right) = 0,$$

c'est-à-dire $h_1(y) - h_1(x) = h_2(y) - h_2(x) = 0$. Il résulte de ce qui précède que x et y sont séparés par h_1 ou par h_2.

2.4.8. EXEMPLE. Reprenons les ensembles Γ, $\bar{\Gamma}$, Δ de 2.1. Si par exemple X est irréductible et projective, on a évidemment $\Delta = X \times X$. Mais, si X est affine, on peut espérer que Δ est proche de $\bar{\Gamma}$. En fait, même pour X affine irréductible et G unipotent, on va montrer qu'on peut avoir $\bar{\Gamma} \neq \Delta$ (ce qui revient à dire, comme on le prouve aisément, que Δ peut être réductible), et même que $\bar{\Gamma}$ n'est pas toujours le graphe d'une relation d'équivalence.

On prend $G = k$, $Y = k^4$, G opérant dans Y par

$$\lambda.(\alpha, \beta, \gamma, \delta) = (\alpha, \beta + \lambda\alpha, \gamma, \delta + \lambda\gamma).$$

Soit X l'ensemble des $(\alpha, \beta, \gamma, \delta) \in k^4$ tels que $\alpha\delta - \beta\gamma = 0$. Alors X est G-stable. Soient $x = (\alpha, \beta, \gamma, \delta) \in X$ et $y = (\alpha', \beta', \gamma', \delta') \in X$. Montrons que

$$(x, y) \in \Gamma \Rightarrow \delta\beta' - \beta\delta' = 0.$$

On a $\alpha = \alpha'$, $\gamma = \gamma'$. Si $\alpha \neq 0$, on a $\delta = \alpha^{-1}\beta\gamma$, $\delta' = \alpha'^{-1}\beta'\gamma' = \alpha^{-1}\beta'\gamma$ d'où notre assertion. De même si $\gamma \neq 0$. Si $\alpha = \alpha' = \gamma = \gamma' = 0$, l'hypothèse $(x, y) \in \Gamma$ entraîne $y = x$, d'où encore le résultat.

Par suite, $(x, y) \in \bar{\Gamma} \Rightarrow \delta\beta' - \beta\delta' = 0$.

Soient $\beta, \delta \in k$. On a, pour tout $\varepsilon \in k - \{0\}$,

$$(-\varepsilon^{-1}).(\varepsilon\beta, \beta, \varepsilon\delta, \delta) = (\varepsilon\beta, 0, \varepsilon\delta, 0))$$

donc $((\varepsilon\beta, \beta, \varepsilon\delta, \delta), (\varepsilon\beta, 0, \varepsilon\delta, 0)) \in \Gamma$. On en déduit que

$$((0, \beta, 0, \delta), (0, 0, 0, 0)) \in \bar{\Gamma}.$$

Soient de même $\beta', \delta' \in k$; on a

$$((0, \beta', 0, \delta'), (0, 0, 0, 0)) \in \bar{\Gamma}.$$

Mais $((0, \beta, 0, \delta), (0, \beta', 0, \delta')) \notin \bar{\Gamma}$ dès que $\delta\beta' - \beta\delta' \neq 0$ d'après ce qu'on a vu plus haut. Donc $\bar{\Gamma}$ n'est pas le graphe d'une relation d'équivalence.

On peut aussi construire un groupe G unipotent opérant linéairement dans un espace vectoriel X de telle sorte que $\bar{\Gamma}$ ne soit pas le graphe d'une relation d'équivalence. Un tel exemple a été obtenu avec l'aide de V. Kač.

L'étude de ces exemples conduit à poser le problème suivant: Si X est affine, le plus petit graphe de relation d'équivalence contenant $\bar{\Gamma}$ est-il Δ? (On voit facilement qu'il en est bien ainsi quand G est réductif.)

2.5. *Cas d'une variété quasi-affine factorielle*

2.5.1. On a vu que $\Omega_2(X, G)$ peut être vide. Même dans ce cas, il existe parfois des parties ouvertes denses G-stables X' de X telles que $\Omega_2(X', G) \neq \varnothing$. La proposition suivante peut concerner une situation de ce genre.

2.5.2. PROPOSITION. *Supposons que X soit une variété quasi-affine factorielle et que G soit connexe et n'ait pas de caractère rationnel non trivial.*

(i) *Si U est un ouvert G-stable de X tel que la variété U/G existe, U/G est un ouvert de* $\mathrm{Spec}(k[X]^G)$.

(ii) *En particulier, si l'on pose $\Omega_1(X, G) = \Omega_1$, on a $\Omega_2(\Omega_1, G) = \Omega_1$; la variété Ω_1/G est quasi-affine.*

(Le cas où X est affine factorielle et ou $k[X]^G$ est de type fini est étudié dans [7, p. 57].)

Soient $A = k[X]$, $B = A^G$, $S = \mathrm{Spec}(B)$, $f : X \to S$ le morphisme canonique. Notons que X, étant factorielle, est irréductible et normale.

Soit U un ouvert G-stable de X tel que $U/G = V$ existe. Considérons le diagramme commutatif d'applications canoniques:

$$\begin{array}{ccc} U & \xrightarrow{\;i\;} & X \\ {\scriptstyle\varphi}\downarrow & & \downarrow{\scriptstyle f} \\ V & \xrightarrow{\;j\;} & S \end{array}$$

où i est l'immersion ouverte canonique et j est déduit de i par passage aux anneaux d'invariants sous G. Nous allons voir que j est une immersion ouverte. Soit V' un ouvert affine de V et soit $U' = \varphi^{-1}(V')$. Alors U' est un ouvert G-stable de U et $k[V'] = k[U']^G$. Notons F' le fermé complémentaire de U' dans X. Soit F'' le fermé de F' réunion des composantes irréductibles de F qui sont de codimension 1 dans X et soit $U'' = X - F''$. Alors U'' est un ouvert G-invariant de X, contenant U', et tel que $U'' - U'$ soit de co-dimension ≥ 2. Comme X est normale, l'application canonique $k[U''] \to k[U']$ est un isomorphisme (EGA IV 20.4.12). Comme X est factorielle et $X - U''$ de codimension ≤ 1 dans X, il existe $a \in k[X]$ tel que $U'' = X_a$, plus grand ouvert de X ou a est inversible. Par ailleurs U'' est G-invariant; G est connexe et n'a pas de caractères non triviaux; donc a est G-invariante [9, Théorème 1]. Finalement $k[V'] = k[U']^G = k[U'']^G$ est le localisé de A par a. C'est dire que j induit un isomorphisme de V' sur S_a. L'image de V dans S est donc un ouvert et j est localement une immersion ouverte. Comme j est évidemment birationnel sur son image et que V est séparé, j est une immersion ouverte.

2.6. *Cas d'un groupe réductif*

2.6.1. Soit X un espace vectoriel de dimension finie sur k. On suppose que G est connexe réductif, et opère dans X par une représentation ration-nelle. Pour les définitions et résultats utilisés en 2.6.1 et 2.6.2, cf. [6], notam-ment le théorème 1.10 (et le fait que, grâce au théorème de Haboush, ces résultats sont utilisables en caractéristique > 0).

Soit $x \in X$. Le point x est dit semi-stable s'il existe $f \in k[X]^G$ tel que $f(x) \neq f(0)$. Le point x est dit stable si son orbite est fermée de dimension maximale. Soit X_{ss} (resp. X_s) l'ensemble des points semi-stables (resp. stables) de X. On sait que X_s et X_{ss} sont des parties ouvertes G-stables de X, que $X_s \subset X_{ss}$, et que la variété X_s/G existe. Mais la variété X_{ss}/G n'existe pas toujours, et X_s peut être vide.

2.6.2. PROPOSITION. *Sous les hypothéses de* 2.6.1, *on a* $X_s = \Omega_2(X, G)$.

Soient Y la variété affine telle que $k[Y] = k[X]^G$, et $\varphi : X \to Y$ le morphisme canonique. On sait qu'il existe une partie ouverte Y' de Y telle que: (1) $X_s = \varphi^{-1}(Y')$; (2) Y' est la variété quotient X_s/G. Soient ω, ω' des G-orbites, avec $\omega \subset X_s$. Si $\varphi(\omega) = \varphi(\omega')$, on a $\omega' \subset \varphi^{-1}(Y') = X_s$, donc $\omega = \omega'$. Ainsi, ω est séparée dans X, et par suite $X_s \subset \Omega_2(X, G)$.

Puisque $\Omega_2(X, G)/G$ existe et que X est irréductible, toutes les orbites contenues dans $\Omega_2(X, G)$ ont même dimension, et donc ont la dimension maximale des orbites de X. Enfin il est clair qu'une orbite contenue dans $\Omega_2(X, G)$ est fermée dans X. Donc $\Omega_2(X, G) \subset X_s$.

2.6.3. Supposons désormais que G soit simple, que k soit de caractéristique 0, et que X soit l'espace d'une représentation linéaire rationnelle de G. L'ensemble $X_s = \Omega_2(X, G)$ est vide si et seulement si le stabilisateur d'un point générique de X n'est pas réductif. Lorsque la représentation de G dans X est simple, cela se produit si et seulement s'il existe dans X une orbite ouverte. Mais si la représentation de G dans X n'est pas simple, et même si l'on exclut les facteurs triviaux, X_s peut être vide dans beaucoup de cas intéressants.

Par exemple, prenons $G = \mathrm{SL}(3, k)$. Soit V l'espace de la représentation évidente de dimension 3 de G. Prenons $X = V \times V \times V^*$ muni de la représentation naturelle de G. Les faits suivants sont connus ou faciles. L'algèbre $k[X]^G$ est engendrée par les deux fonctions

$$(v, v', v^*) \mapsto \langle v^*, v' \rangle, \qquad (v, v', v^*) \mapsto \langle v^*, v' \rangle.$$

La dimension maximale des orbites est 7. La réunion X' des orbites de dimension 7 est l'ensemble des $(v, v', v^*) \in X$ vérifiant les conditions suivantes: (1) v et v' sont linéairement indépendants; (2) $\langle v^*, v \rangle$ et $\langle v^*, v' \rangle$ ne sont pas nuls tous les deux. Le stabilisateur d'un point de X' est unipotent de dimension 1, et les orbites contenues dans X' sont non fermées dans X. On a $X_s = \emptyset$. Mais deux orbites distinctes dans X' sont séparées par un élément de $k[X]^G$, de sorte que $\Omega_1(X, G) = X'$. La variété X'/G est isomorphe à $k^2 - \{0\}$.

2.6.4. Même si $X_s \neq \emptyset$, il peut arriver que $\Omega_1(X, G)$ contienne strictement X_s. Cela se produit par exemple pour $G = \mathrm{SL}(2, k)$ opérant dans les formes binaires de degré n pour $n = 2, 3, 4$ (mais pas pour $n > 4$).

2.6.5. Prenons $G = \mathrm{SL}(2, k)$ et $X = k^2 \times k^2$, G opérant dans chaque facteur k^2 par la représentation évidente. Alors $k[X]^G$ est engendrée par la fonction $(v, v') \mapsto \det(v, v')$. La dimension maximale des orbites est 3. La réunion X' des orbites de dimension 3 est l'ensemble des $(v, v') \in X$ tels que v, v' soient linéairement indépendants. On a $X' = \Omega_2(X, G) = X_s$, et X'/G est isomorphe à $k - \{0\}$. Soit $N = X - X'$, qui est un cône de dimension 3. Si $(v, v') \in N - \{0\}$, v et v' sont proportionnels, et le rapport de v à v' est un point de la droite projective $k \cup \{\infty\}$ qui caractérise l'orbite de (v, v'). On a $\Omega_2(N, G) = \varnothing$, $\Omega_1(N, G) = N - \{0\}$, et la variété $\Omega_1(N, G)/G$ s'identifie à $k \cup \{\infty\}$.

2.6.6. PROBLÈME. Supposons que G soit réductif et que X soit l'espace d'une représentation linéaire rationnelle de G. Les quotients canoniques $(U_{i+1} - U_i)/G$ de 1.8 sont-ils quasi-projectifs? (Dans l'exemple 2.6.5, ces quotients sont: la variété affine $k - \{0\}$, la variété projective $k \cup \{\infty\}$, et un point. Pour la représentation adjointe de SL(n, k), il résulte de ce qui est annoncé dans [5] que ces quotients sont des variétés affines.)

RÉFÉRENCES

1. M. ARTIN, The implicit function theorem in algebraic geometry, *Proc. Bombay Colloq. Algebraic Geom., 1968.* 13–34, *Oxford Univ. Press, London, 1968.*
2. A. BOREL, "Linear Algebraic Groups," Benjamin, New York, 1969.
3. J. DIEUDONNÉ ET A. GROTHENDIECK, Eléments de géométrie algébrique, étude locale des schemas (cité EGA IV), *Inst. Hautes Études Sci. Publ. Math.* N° 20, 24, 28, 32, 1964–1968.
4. J. DIXMIER, Sur les représentations unitaires des groupes de Lie nilpotents, I, *Amer. J. Math.* 81 (1959), 160–170.
5. H. KRAFT, Parametrisierung von Konjugationsklassen in sl_n, *Math. Ann.* 234 (1978), 209–220.
6. D. MUMFORD, "Geometric Invariant Theory," Springer-Verlag, Berlin and New York, 1965.
7. M. NAGATA, Note on orbit spaces, *Osaka Math. J.* 14 (1962), 21–31.
8. M. NAGATA, On the 14th problem of Hilbert, *Proc. Internat. Cong. of Math., 1958, Edinburgh*, pp. 459–462. Cambridge Univ. Press, London and New York, 1960.
9. V. L. POPOV, On the stability of the action of an algebraic group on an algebraic variety, *Izv. Akad. Nauk SSSR Ser. Mat.* 36 (1972), 367–379.
10. M. RAYNAUD, Sur le passage au quotient par un groupoïde plat, *C. R. Acad. Sci. Paris Sér. A–B* 265 (1967), 384–387.
11. R. RENTSCHLER, L'injectivité de l'application de Dixmier pour les algèbres de Lie résolubles, *Invent. Math.* 23 (1974), 49–71.
12. M. ROSENLICHT, Some basic theorems on algebraic groups, *Amer. J. Math.* 78 (1956), 401–443.
13. M. ROSENLICHT, On quotient varieties and the affine embedding of certain homogeneous spaces, *Trans. Amer. Math. Soc.* 101 (1961), 211–223.
14. M. ROSENLICHT, A remark on quotient spaces, *An. Acad. Brasil. Ciênc.* 35 (1963), 487–489.
15. J.-P. SERRE, Faisceaux algébriques cohérents, *Ann. of Math.* 61 (1955), 197–278.

MATHEMATICAL ANALYSIS AND APPLICATIONS, PART A
ADVANCES IN MATHEMATICS SUPPLEMENTARY STUDIES, VOL. 7A

Forced Oscillations of Nonlinear Hamiltonian Systems, II

Ivar Ekeland[†]

Centre de Recherche de Mathématiques de la Décision
Université de Paris IX Dauphine
Paris, France

Dedicated to Laurent Schwartz

We study periodic solutions of the nonlinear Hamiltonian system with n degrees of freedom:

$$\dot{x}_i = \frac{\partial H}{\partial p_i}(x, p) + g_i(t),$$

$$\dot{p}_i = -\frac{\partial H}{\partial x_i}(x, p) + h_i(t),$$

$(\mathcal{H})$

the Hamiltonian H being convex and superquadratic in both variables, and the forcing terms being T-periodic with mean value zero. We prove that, if these forcing terms lie within bounds which we explicitly compute, system $(\mathcal{H})$ has some T-periodic solution, which we also locate explicitly.

I. Introduction

Hamilton's equations, for a system with n degrees of freedom, are

$$(-\dot{p}, \dot{x}) - H'(t, x, p) = f(t) \in \mathbb{R}^{2n}. \qquad (\mathcal{H})$$

We refer to periodic solutions of $(\mathcal{H})$ as oscillations. They are free if f is identically zero, forced otherwise. Of course, the forcing term f itself will be required to be periodic, although this by itself will not imply that $(\mathcal{H})$ has a periodic solution.

This paper studies forced oscillations for a particular type of nonlinear Hamiltonian system. The origin will be an equilibrium, and the Hamiltonian itself will be convex and superquadratic in all variables (x, p) together. For instance, $H(x, p) = (\sum x_i^2 + \sum p_i^2)^\theta$, with $\theta > 1$, will do. An example of such a

[†] Sponsored by the United States Army under Contract No. DAAG29-75-C-0024.

345

system is a taut spring, which does not follow Hooke's law, $F = kl$, force proportional to length, but the law $F = kl^\theta$ with $\theta > 1$.

Free oscillations for such systems were first studied by Rabinowitz [8, 9]. The author obtained similar results [6], by using a variational method devised by Clarke and himself for convex subquadratic Hamiltonians [2, 3].

This paper, although self-contained, borrows heavily from the latter approach. It relies on a dual version of the least action principle, stated here as Proposition 2.2, but which can be found also in the papers [3, 6], and particularly [4]. The associated variational problem is shown to have a local minimum (Proposition 3.1), which gives rise to a periodic solution of the original Hamiltonian system.

Indeed, one can view this paper as a sequel to [4], which treated forced oscillations for convex subquadratic Hamiltonians. It is interesting to note that, in the latter case, periodic solutions to $(\mathscr{H})$ exist for any forcing term. This is no longer true for quadratic Hamiltonians, such as $\frac{1}{2}\sum p_i^2 + \frac{1}{2}\sum \alpha_i x_i^2$, because of the onset of resonance.

Nor is it true in the case with which we are dealing, a superquadratic Hamiltonian. We are able to show the existence of a periodic solution to system $(\mathscr{H})$ only if the forcing term f has zero mean and is smaller than some bound, not necessarily small, which we compute explicitly (Propositions 3.1 and 4.2).

We do not know what happens beyond this bound. A more detailed analysis shows the following. The periodic solutions we find give local minima of the dual action integral and converge towards the equilibrium solution when the forcing term goes to zero. On the other hand, with a few more assumptions on H, there will be another kind of periodic solution, which corresponds to saddle points of the dual action integral, and which converges to a nonconstant solution when the forcing term goes to zero. Now, when the forcing term increases, it may well be that these two kinds of periodic solutions, which are well apart when f is small, begin interfering, and finally destroy each other. That, at least, is what the behavior of the dual action integral would suggest.

II.   THE DUAL ACTION INTEGRAL

We are investigating the differential system:

$$\dot{u}(t) \in \sigma \, \partial H(t, u(t)) + f(t). \qquad (\mathscr{H})$$

The function $H: \mathbb{R} \times \mathbb{R}^{2n} \to \mathbb{R}$ is the Hamiltonian. In the Sections II and III it shall always be assumed that $H(t, u)$ is measurable in t, convex and continuous in u, with

$$H(t, u) \geq H(t, 0) = 0 \qquad \text{for all } (t, u) \tag{1}$$

and that there are constants $k > 0$ and $\beta > 2$ such that

$$H(t, u) \leq \frac{k^\beta}{\beta} |u|^\beta \qquad \text{for all } (t, u), \tag{2}$$

$$\forall t \quad r^{-1} \text{Min}\{H(t, u) \,|\, |u| = r\} \to +\infty \quad \text{as} \quad r \to \infty. \tag{3}$$

The symbol $\partial H(t, u)$ denotes the subdifferential of the function $u' \to H(t, u')$ at the point $u \in \mathbb{R}^{2n}$, in the sense of convex analysis (see [10] or [7]). It is defined by

$$v \in \partial H(t, u) \Leftrightarrow \forall u', \ H(t, u') \geq H(t, u) + (u' - u, v). \tag{4}$$

We denote by σ a linear operator which, in some appropriate base of $\mathbb{R}^{2n} = \mathbb{R}^n \times \mathbb{R}^n$, can be written as

$$\sigma = \begin{pmatrix} 0 & Id \\ -Id & 0 \end{pmatrix}. \tag{5}$$

In other words, in that particular base, $u \in \mathbb{R}^{2n}$ is written $u = (x, p)$, and $\sigma(u) = (p, -x)$. The x_i, $1 \leq i \leq n$, are position variables, and the p_i, $1 \leq i \leq n$, momentum variables. We have of course ${}^t\sigma = \sigma^{-1} = -\sigma$.

The function $f: \mathbb{R} \to \mathbb{R}^{2n}$ is the forcing term. It will be assumed to be integrable over every bounded interval.

A solution of the differential inclusion $(\mathcal{H})$ is an absolutely continuous function $u: \mathbb{R} \to \mathbb{R}^{2n}$, with derivative $\dot{u}$, such that relation $(\mathcal{H})$ holds for almost every t. Of course, if f is continuous and H is differentiable in u, the derivative H' being continuous in (t, u), then u becomes a classical C^1 solution of the differential equation:

$$\dot{u} = \sigma H'(t, u) + f(t).$$

Setting $u = (x, p)$ and $f = (g, h)$, we get the familiar form of the Hamiltonian system $(\mathcal{H})$:

$$\begin{aligned} \dot{x}_i &= \frac{\partial H}{\partial p_i}(t, x, p) + g_i(t), \\[2mm] \dot{p}_i &= -\frac{\partial H}{\partial x_i}(t, x, p) + h_i(t). \end{aligned} \tag{6}$$

Our investigation of system $(\mathcal{H})$ will rely on the dual approach initiated in [2] and developed in [3, 4, 6]. First, we have to introduce the Fenchel conjugate $G(t, \cdot)$ of the convex function $H(t, \cdot)$ (for the time being, t is just a parameter, pegged at some given value):

$$G(t, v) = \text{Sup}\{(v, u) - H(t, u) \mid u \in \mathbb{R}^{2n}\}. \tag{7}$$

In the case where $H(t, \cdot)$ is differentiable, this definition reduces to the familiar formula for the Legendre transform: $G(t, v) = (v, u) - H(t, u)$, with $v = H'(t, u)$. Formula (7), however, will hold even in nondifferentiable cases, with the following results:

LEMMA 2.1.   *The function $G: \mathbb{R} \times \mathbb{R}^{2n} \to \mathbb{R}$ is measurable in t, convex and continuous in v. It satisfies the following inequalities, where $\alpha = \beta/(\beta - 1)$ is the conjugate exponent of β:*

$$G(t, v) \geq G(t, 0) = 0 \qquad \text{for all } (t, v), \tag{8}$$

$$G(t, v) \geq (1/\alpha k^{\alpha})|v|^{\alpha} \qquad \text{for all } (t, v). \tag{9}$$

Moreover, the three following relations are equivalent:

$$u \in \partial G(t, v), \tag{10}$$

$$v \in \partial H(t, u), \tag{11}$$

$$G(t, v) + H(t, u) = (u, v). \tag{12}$$

Proof.   Convexity and lower semicontinuity of $G(t, \cdot)$, as well as the Fenchel reciprocity formulas $(10) \Leftrightarrow (11) \Leftrightarrow (12)$, are classical properties, which can be found in [10] or [7]. Measurability with respect to t is proved in [7, Chapter 7] or in the paper [11]. We check the remaining properties by using the definition (7) of G.

Because of condition (2), the function $G(t, \cdot)$ is finite at every point $u \in \mathbb{R}^{2n}$. Since it also is convex, it must be continuous.

We have by condition (1):

$$G(t, v) \geq (v, 0) - H(t, 0) = 0,$$

$$G(t, 0) = \sup_{u} - H(t, u) = 0.$$

Hence formula (8). Now for (9), using inequality (2)

$$G(t, v) \geq \text{Sup}\{(v, u) - (k^{\beta}/\beta)|u|^{\beta} \mid u \in \mathbb{R}^{2n}\} = \text{Sup} \underset{s \geq 0}{\text{Sup}} \{(v, u) - (k^{\beta}/\beta)s^{\beta}\}$$

$$= \underset{s \geq 0}{\text{Sup}} \{s|v| - (k^{\beta}/\beta)s^{\beta}\} = \frac{1}{k^{\beta/(\beta - 1)}} (1 - (1/\beta))|v|^{\beta/(\beta - 1)}. \quad \blacksquare$$

The dual action integral can now be written:

$$I(v) = \int_{0}^{T} \left\{ \frac{1}{2}(\sigma \dot{v}(t), v(t)) + G(t, -\sigma \dot{v}(t) + \sigma f(t)) \right\} dt. \tag{13}$$

Equation (13) was introduced in [4], where its critical points where related to solutions of system ($\mathscr{H}$). We shall investigate it anew, both to make this chapter self-contained and to account for some differences in the analytical setting.

We shall be working in the function spaces $L^{\alpha}(0, T; \mathbb{R}^{2n})$, and $L^{\beta}(0, T; \mathbb{R}^{2n})$, and denoting by $\|\cdot\|_{\alpha}$ and $\|\cdot\|_{\beta}$ the corresponding norms. We are particularly interested in the Sobolev space $W^{1,\alpha}(0, T; \mathbb{R}^{2n})$ and its closed subspace E defined by

$$v \in E \Leftrightarrow \dot{v} \in L^{\alpha}(0, T; \mathbb{R}^{2n}) \qquad \text{and} \qquad \int_0^T \dot{v}(t)\, dt = 0 = \int_0^T v(t)\, dt. \quad (14)$$

The norm of v in E shall be defined to be $\|\dot{v}\|_{\alpha}$. The dual action integral (13) now defines a functional I on E. Our next results relate local minima of I to solutions of ($\mathscr{H}$) satisfying $u(0) = u(T)$.

Note that the integrand in formula (13), i.e., the function

$$L(t, v, w) = \tfrac{1}{2}(\sigma w, v) + G(t, -\sigma w + \sigma f(t)) \quad (15)$$

on $\mathbb{R} \times \mathbb{R}^{2n} \times \mathbb{R}^{2n}$ is neither convex nor differentiable in $(v, \dot{v})$. In the search for local minima of the integral I over E, the classical Euler–Lagrange and transversality conditions will not be applicable. However, the necessary conditions of Clarke [1] will be available. We state:

PROPOSITION 2.2.   *Assume v is a local minimum of the dual action integral I over E. Then there is some vector $\xi \in \mathbb{R}^{2n}$ such that the translate u defined by $u(t) = v(t) + \xi$ is a solution of the Hamiltonian system ($\mathscr{H}$) on the time interval $[0, T]$ satisfying the boundary condition $u(0) = u(T)$.*

Proof.   Let v be a local minimum for I over E. Using the terminology of [1], this implies that v is a weak local minimum of the variational problem:

$$\text{Inf} \int_0^T L(t, v(t), \dot{v}(t))\, dt; \qquad v(0) = v(T). \quad (16)$$

The integrand L is given in formula (15); it is locally Lipschitz in $(v, \dot{v})$, so that the necessary conditions of [1] hold. They tell us that there exists an absolutely continuous function $\lambda: [0, T] \to \mathbb{R}^{2n}$ such that

$$\dot{\lambda}(t) = \tfrac{1}{2}\sigma\dot{v}(t) \qquad\qquad\qquad \text{a.e.,} \quad (17)$$

$$\lambda(t) = -\tfrac{1}{2}\sigma v(t) + \sigma\, \partial G(t, -\sigma\dot{v}(t) + \sigma f(t)) \qquad \text{a.e.,} \quad (18)$$

$$\lambda(0) = \lambda(T). \quad (19)$$

In the particular case when G is C^1, differentiating (18) with respect to time and comparing with (17) yields the usual Euler–Lagrange equations, while (19) is the usual transversality condition.

We then define a function $u:[0, T] \to \mathbb{R}^n$ by

$$u(t) = \tfrac{1}{2}v(t) - \sigma\lambda(t). \tag{20}$$

Equation (17) tells us that $\dot{u} = \dot{v}$, so that $u = v + \xi$ for some constant $\xi \in \mathbb{R}^{2n}$. We rewrite Eq. (18) as follows:

$$u(t) \in \partial G(t, -\sigma\dot{v}(t) + \sigma f(t)) \qquad \text{a.e.} \tag{21}$$

We invert this equation by Fenchel's reciprocity formula $(10) \Leftrightarrow (11)$:

$$-\sigma\dot{v}(t) + \sigma f(t) \in \partial H(t, u(t)) \qquad \text{a.e.} \tag{22}$$

Since $\dot{u} = \dot{v}$, this gives equation $(\mathscr{H})$ for u:

$$\dot{u}(t) \in \sigma\,\partial H(t, u(t)) + f(t) \qquad \text{a.e.} \tag{23}$$

Finally, conditions (19) and (20) yield $u(0) = u(T)$.

III.   Periodic Solutions

We recall the differential system we are dealing with:

$$\dot{u}(t) \in \sigma\,\partial H(t, u(t)) + f(t) \qquad \text{a.e.} \tag{$\mathscr{H}$}$$

under the assumptions that $H(t, u)$ is measurable in t, convex continuous in u, and

$$H(t, u) \geq H(t, 0) = 0 \qquad \text{for all } (t, u), \tag{1}$$

$$H(t, u) \leq (k^\beta/\beta)|u|^\beta \qquad \text{with } k > 0 \quad \text{and} \quad \beta > 2, \tag{2}$$

$$\lim_{r \to \infty} \operatorname*{Min}_{|u|=r} H(t, u) = +\infty. \tag{3}$$

From now on, the forcing term f and the Hamiltonian $H(\cdot, u)$ (for each fixed $u \in \mathbb{R}^{2n}$) are assumed to be T-periodic functions of t, for some given $T > 0$. We shall show that, if the forcing term f is not too large and has mean zero, there is a T-periodic solution to system $(\mathscr{H})$.

Recall that α, is the conjugate exponent of β, i.e., $\alpha^{-1} + \beta^{-1} = 1$. We then introduce some constants, the actual values of which can be computed from β as follows.

$$b(\beta) = \tfrac{1}{2}\pi^{-2/\beta}, \tag{4}$$

$$c(\beta) = (2 - \alpha)(2\alpha - 2)^{(\alpha-1)/(2-\alpha)}(\alpha b(\beta))^{-1/(2-\alpha)}, \tag{5}$$

$$d(\beta) = (2/\beta)^{1/(2-\alpha)}b(\beta)^{-1/(2-\alpha)}. \tag{6}$$

PROPOSITION 3.1.   *Assume* $\|f\|_\alpha \leq c(\beta)k^{-\beta/(\beta-2)}T^{-(2/\beta)(\beta-1)/(\beta-2)}$,   *and* $\int_0^T f(t)\,dt = 0$. *Consider the ball B in E defined by*

$$v \in B \Leftrightarrow \|\dot{v} - f\|_\alpha \leq d(\beta)k^{-\beta/(\beta-2)}T^{-(2/\beta)(\beta-1)/(\beta-2)}. \tag{7}$$

The dual action integral I then attains its minimum relative to B. Any point $v \in B$ *where this minimum is attained satisfies the sharper estimate*:

$$\|\dot{v} - f\|_\alpha \leq [d(\beta)/c(\beta)]\|f\|_\alpha = [2/(\beta - 2)]\|f\|_\alpha. \tag{8}$$

Before starting the proof, we bring the origin to f in the space L^α, so as to get more convenient expressions for the dual action integral I and the ball B. We set $\dot{w} = \dot{v} - f$, and hence $w = v - F$, where $F(t)$ is the primitive of $f(t)$ with mean zero:

$$\dot{F}(t) = f(t) \qquad \text{and} \qquad \int_0^T F(t)\,dt = 0. \tag{9}$$

The ball B now is defined by $\|\dot{w}\| \leq d(\beta)k^{-\beta/(\beta-2)}T^{-(2/\beta)(\beta-1)/(\beta-2)}$. The functional I becomes, in the new coordinates for E:

$$I(w) = \int_0^T \left\{ \frac{1}{2}\,(\sigma\dot{w}(t) + \sigma f(t),\, w(t) + F(t)) + G(t,\, -\sigma\dot{w}(t)) \right\} dt$$

$$= \left\{ \frac{1}{2}\,\langle \sigma f, F \rangle + \langle \sigma\dot{w}, F \rangle + \frac{1}{2}\,\langle \sigma\dot{w}, w \rangle + \int_0^T G(t,\, -\sigma\dot{w}(t)) \right\} dt.$$

The angle brackets denote the duality pairing between $L^\alpha(0, T; \mathbb{R}^{2n})$ and $L^\beta(0, T; \mathbb{R}^{2n})$. The first term, a constant, can be disregarded for minimization purposes. The functional to investigate thus becomes

$$\bar{I}(w) = \langle \sigma\dot{w}, F \rangle + \frac{1}{2}\,\langle \sigma\dot{w}, w \rangle + \int_0^T G(t,\, -\sigma\dot{w}(t))\,dt. \tag{10}$$

We now proceed in two steps:
Step 1. $\bar{I}(w) > 0$ for all $w \in \partial B$.
We estimate I from below. Formula (10) yields:

$$\bar{I}(w) \geq -\|\sigma\dot{w}\|_\alpha\|F\|_\beta - \frac{1}{2}\|\sigma\dot{w}\|_\alpha\|w\|_\beta + \int_0^T G(t,\, -\sigma\dot{w}(t))\,dt. \tag{11}$$

Using Lemma 2.1, we go further:

$$\bar{I}(w) \geq -\|\sigma\dot{w}\|_\alpha\|F\|_\beta - \tfrac{1}{2}\|\sigma\dot{w}\|_\alpha\|w\|_\beta + (1/\alpha k^\alpha)\|\sigma\dot{w}\|_\alpha^\alpha. \tag{12}$$

Because of formula (2.5), we see that σ is an isometry, so that $\|\sigma\dot{w}\|_\alpha = \|\dot{w}\|_\alpha$. The evaluation of $\|w\|_\beta$ in terms of $\|\dot{w}\|_\alpha$ and of $\|F\|_\beta$ in terms of $\|f\|_\alpha$ presents us with a special problem.

The map $\dot{w} \rightarrow w$, with $\int_0^T w(t)\,dt = 0$, from L^α to L^β, is a linear continuous operator. When $\alpha = 1$ and $\beta = \infty$, it is readily seen to have norm $\frac{1}{2}$. When $\alpha = 2$ and $\beta = 2$, it is known to have norm $T/2\pi$ (see [3] or [4]). Using the convexity theorem of M. Riesz (see [5], for instance), we conclude that in the case where $1 \le \alpha \le 2$, its norm is at most $(T/2\pi)^{2/\beta}(\frac{1}{2})^{1-2/\beta}$ which is $b(\beta)T^{2/\beta}$.

Inequality (12) now becomes

$$\bar{I}(w) \ge -b(\beta)T^{2/\beta}\|f\|_\alpha\|\dot{w}\|_\alpha - \tfrac{1}{2}b(\beta)T^{2/\beta}\|\dot{w}\|_\alpha^2 + (1/\alpha k^\alpha)\|\dot{w}\|_\alpha^\alpha. \tag{13}$$

Now consider the function $\varphi(s)$ of the real variable s:

$$\varphi(s) = -b(\beta)T^{2/\beta}\|f\|_\alpha s - \tfrac{1}{2}b(\beta)T^{2/\beta}s^2 + (1/\alpha k^\alpha)s^\alpha. \tag{14}$$

Clearly $\varphi(0) = 0$, and $\varphi'(0) < 0$. We want to solve the equation $\varphi(s) = 0$ with $s > 0$. After simplifying by s, this becomes:

$$\|f\|_\alpha + \frac{1}{2}s = \frac{1}{\alpha k^\alpha}\frac{T^{-2/\beta}}{b(\beta)}s^{\alpha-1}. \tag{15}$$

In other words, we seek the intersection of the curve

$$s \rightarrow \frac{1}{\alpha k^\alpha}\frac{T^{-2/\beta}}{b(\beta)}s^{\alpha-1}$$

with the straight line $s \rightarrow \|f\|_\alpha + \frac{1}{2}s$. This is easily done. We first seek out the point s_0 on the curve where the slope of the tangent is $\frac{1}{2}$; there will be two points of intersection s_1 and s_2 if the given line lies under the tangent, none if it lies above (see Fig. 1).

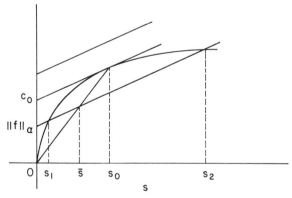

FIGURE 1

The computation gives

$$\frac{T^{-2/\beta}}{b(\beta)}\frac{\alpha-1}{\alpha k^\alpha}s_0^{\alpha-2}=\frac{1}{2}, \qquad \text{hence} \quad s_0=\left|\frac{\alpha k^\alpha b(\beta)}{2(\alpha-1)}\right|^{1/(\alpha-2)}T^{2/\beta(\alpha-2)}, \qquad (16)$$

$$c_0=\frac{1}{\alpha k^\alpha}\frac{T^{-2/\beta}}{b(\beta)}s_0^{\alpha-1}-\frac{\alpha-1}{\alpha k^\alpha}\frac{T^{-2/\beta}}{b(\beta)}s_0^{\alpha-1}=\frac{2-\alpha}{\alpha k^\alpha}\frac{T^{-2/\beta}}{b(\beta)}s_0^{\alpha-1}. \qquad (17)$$

The equation $\varphi(s)=0$ will have two different roots $s_2>s_1>0$ provided that $\|f\|_\alpha\le c_0$, that is,

$$\|f\|_\alpha\le(2-\alpha)(\alpha b(\beta))^{1/(\alpha-2)}(2\alpha-2)^{(\alpha-1)/(\alpha-2)}k^{\alpha/(\alpha-2)}T^{2/\beta(\alpha-2)}. \qquad (18)$$

The graph of φ is shown in Fig. 2.

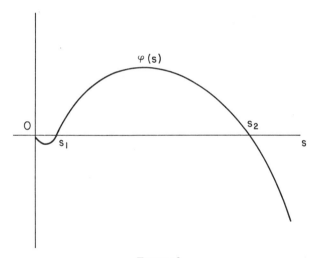

FIGURE 2

The result now follows immediately from the estimate $\bar{I}(w)\ge\varphi(\|w\|_\alpha)$, and the fact that s_0 lies between s_1 and s_2.

Step 2. $\bar{I}$ attains its minimum on B.

Let w_n be a minimizing sequence in B:

$$\bar{I}(w_n)\to\text{Inf}\{\bar{I}(w)|w\in B\} \qquad (19)$$

The sequence $\dot{w}_n$ belongs to the ball with radius

$$d(\beta)k^{-\beta/(\beta-2)}T^{-(2/\beta)(\beta-1)/(\beta-2)}$$

and center 0 in L^α. Since this space is reflexive, there is a subsequence $w_{n'}$, which converges weakly to some w belonging to the same ball:

$$\dot{w}_{n'} \to \dot{w} \quad \text{weakly in} \quad L^{\alpha}, \tag{20}$$

$$w_{n'} \to w \quad \text{strongly in} \quad L^{\beta}, \tag{21}$$

$$w \in B. \tag{22}$$

The strong convergence of the $w_{n'}$ follows from the compactness of the map $\dot{w} \to w$ from L^{α} to L^{β}. The weak convergence (20) yields immediately:

$$\langle \sigma \dot{w}_{n'}, F \rangle \to \langle \sigma \dot{w}, F \rangle. \tag{23}$$

Moreover, we have

$$\langle \sigma \dot{w}_{n'}, w_{n'} \rangle - \langle \sigma \dot{w}, w \rangle = \langle \sigma \dot{w}_{n'} - \sigma \dot{w}, w \rangle + \langle \sigma \dot{w}_{n'}, w_{n'} - w \rangle. \tag{24}$$

The first term goes to zero because the $\dot{w}_{n'}$ converge weakly, and the second one goes to zero because the $\dot{w}_{n'}$ are uniformly bounded and the $w_{n'}$ converge strongly:

$$\tfrac{1}{2}\langle \sigma \dot{w}_{n'}, w_{n'} \rangle \to \tfrac{1}{2}\langle \sigma \dot{w}, w \rangle. \tag{25}$$

Finally, by known properties of nonnegative convex integrands (see [7] or [11]), we have

$$\liminf \int_0^T G(t, -\sigma \dot{w}_n(t))\, dt \geq \int_0^T G(t, -\sigma \dot{w}(t))\, dt. \tag{26}$$

Adding up (23), (25), (26), and comparing them with formulas (10) and (19), we get

$$\bar{I}(w) \leq \text{Inf}\{\bar{I}(w') | w' \in B\}. \tag{27}$$

Since $w \in B$, equality must hold in (27), proving that w is a minimizer in B. *Conclusion*: The minimizers satisfy estimate (8).

From formula (10), we see that $\bar{I}(0) = 0$. Since $0 \in B$, we see that $I(w) \leq 0$ for any minimizer w of $\bar{I}$ on B. This implies that $\|w\|_{\alpha}$ must lie between 0 and the first positive root s_1 of φ. Figure (1) gives us by inspection the desired estimate:

$$s_1 < \bar{s} = s_0 \|f\|_{\alpha}/c_0. \tag{28}$$

COROLLARY 3.2.   *Assume f and $H(\cdot, u)$ are T-periodic, with*

$$\int_0^T f(t)\, dt = 0 \qquad and \qquad \|f\|_{\alpha} \leq c(\beta) k^{-\beta/(\beta-2)} T^{-(2/\beta)(\beta-1)/(\beta-2)}. \tag{29}$$

Then the Hamiltonian system $(\mathcal{H})$ has at least one T-periodic solution u such that:

$$\|\dot{u} - f\|_{\alpha} \leq [2/(\beta - 2)]\|f\|_{\alpha}. \tag{30}$$

Proof. In Proposition 3.1, we have found some $v \in E$ which minimize I on B, and which are interior to B because of estimate (8). Clearly v is a local minimum for $\bar{I}$ on E, so that we can apply Proposition 2.2. The result follows immediately; estimate (30) follows from (8), and the relation $\dot{u} = \dot{v}$. ∎

We will refer to the T-periodic solutions found in this way as *solutions of type* (E). When there is no forcing term, $f = 0$, this type (E) solution is simply $u = 0$, rest at equilibrium. When the forcing term f is small, estimate (29) tells us that the solution u is almost constant. With a few more assumptions on H, and the equation $u \in \partial G(-\sigma\dot{u} + \sigma f)$, it can be proved in fact that u is small. For instance:

COROLLARY 3.3.   *Assume moreover that there is some constant $c > 0$ such that $|v| \geq c|u|^{\beta-1}$ for all $v \in \partial H(t, u)$. Then, in addition to (30), the T-periodic solutions of type (E) satisfy the following estimate:*

$$\|u\|_\beta \leq \left| \frac{1}{c} \frac{2}{\beta-2} \|f\|_\alpha \right|^{1/(\beta-1)}. \tag{31}$$

Proof. We have $-\sigma(\dot{u}(t) - f(t)) \in \partial H(t, u(t))$, so that

$$|u(t)|^\beta \leq c^{-\beta/(\beta-1)}|\dot{u}(t) - f(t)|^{\beta/(\beta-1)}. \tag{32}$$

Integrating over $[0, T]$ yields the desired result. ∎

We conclude this argument with two remarks. First, note that the estimate (8) is very rough, and more elaborate calculations will yield better ones. For instance, it is clear from Fig. 1 that $s_1 = O(\|f\|_\alpha^{1/(\alpha-1)})$ (using Landau's symbol O), so that, when $\|f\|_\alpha \to 0$, we have the estimate $\|\dot{u} - f\|_\alpha = O(\|f\|_\alpha^{\beta-1})$, which is certainly better than (30).

Note also that the preceding argument will carry over, with suitable modifications, to the case $\beta = 2$. However, we then fall within the scope of the paper [4], to which we refer for results.

IV.   OTHER HAMILTONIANS

We now wish to extend the preceding results to other Hamiltonians, which do not satisfy the inequality $H(t, u) \leq k^\beta |u|^\beta \beta^{-1}$ over all of $\mathbb{R}^{2n}$.

We begin with Hamiltonians which satisfy this condition in a neighborhood of the origin only. For the sake of simplicity, we shall assume that the Hamiltonian H does not depend on t. Throughout, we assume that $H'(0, 0) = 0$.

PROPOSITION 4.1.   *Assume that there is a neighborhood $\mathcal{U}$ of the origin in $\mathbb{R}^{2n}$ such that H is C^2 on $\mathcal{U}$, the second derivative $H''(u)$ being positive* *definite for* $u \neq 0$, *and satisfying, for some constants* $b > a > 0$ *and* $\beta > 2$:

$$a|u|^{\beta - 2}|v|^2 \leq (H''(u)v, v) \leq b|u|^{\beta - 2}|v|^2, \qquad \text{for all } u \in \mathcal{U}, \quad v \in \mathbb{R}^{2n}. \quad (1)$$

Then, for any $T > 0$, *there is some* $\varepsilon > 0$ *such that, whenever* $\|f\|_\alpha \leq \varepsilon$, *with* $\int_0^T f \, dt = 0$, *the Hamiltonian system* $(\mathcal{H})$ *has some periodic solution lying inside* $\mathcal{U}$.

Proof. We can always assume that $\mathcal{U}$ is a ball with radius $\eta > 0$. It follows from the assumptions that H is convex on $\mathcal{U}$. Now consider a ray $t \to tu$ from the origin, with $|u| = 1$. As long as $0 < t < \eta$, we have $tu \in \mathcal{U}$, and, setting $\gamma = \beta - 2$:

$$(H'(tu), v) = \int_0^t (H''(su)u, v) \, ds \geq \int_0^t as^\gamma |v| \, ds = \frac{a}{\gamma + 1} t^{\gamma + 1}|v|.$$

$$\leq \int_0^t bs^\gamma |v| \, ds = \frac{b}{\gamma + 1} t^{\gamma + 1}|v|.$$

Hence, for all $u \in \mathcal{U}$:

$$a(\gamma + 1)^{-1}|u|^{\gamma + 1} \leq |H'(u)| \leq b(\gamma + 1)^{-1}|u|^{\gamma + 1}. \quad (2)$$

Integrating once more yields, for all $u \in \mathcal{U}$:

$$a \frac{|u|^{\gamma + 2}}{(\gamma + 1)(\gamma + 2)} \leq H(u) \leq b \frac{|u|^{\gamma + 2}}{(\gamma + 1)(\gamma + 2)}. \quad (3)$$

It is now simply a matter of finding a convex function $\bar{H} : \mathbb{R}^n \to \mathbb{R}$, satisfying (2) and (3) over all of $\mathbb{R}^n$, and coinciding with H on $\mathcal{U}$. This being done, we apply to $\bar{H}$ Corollaries 3.2 and 3.3, with $\beta = \gamma + 2$. It follows from estimates (3.30) and (3.31) that if $\|f\|_\alpha$ is small enough, the periodic solution u we have found for

$$\dot{u}(t) = \sigma \bar{H}'(u(t)) + f(t) \quad (4)$$

will lie entirely inside $\mathcal{U}$, so that $\bar{H}(u(t)) = H'(u(t))$ for all t. It follows that it is actually a solution of

$$\dot{u}(t) = \sigma H'(u(t)) + f(t). \quad \blacksquare \quad (5)$$

We now turn to another class of Hamiltonians, of particular importance for applications. These are the Hamiltonians which split as:

$$H(t, x, p) = \tfrac{1}{2}p^2 + V(t, x). \quad (6)$$

Such Hamiltonians are common in classical mechanics. The first term is kinetic energy, the second one potential energy. Because the first term is quadratic, they cannot satisfy growth conditions such as (2.2)–(3.2), even locally.

However, results similar to Proposition 3.1 and its corollaries still hold, with slight modifications. The growth assumptions now will be made directly on the potential $V: \mathbb{R} \times \mathbb{R}^n \to \mathbb{R}$. We shall assume it to be

$$\text{measurable in } t, \quad \text{convex continuous in } x, \tag{7}$$

$$V(t, x) \geq V(t, 0) = 0, \quad \text{for all } (t, x), \tag{8}$$

$$V(t, x) \leq (k^\beta/\beta)|x|^\beta, \quad \text{for all } (t, x),$$
$$\text{for some } k > 0 \text{ and } \beta > 2, \tag{9}$$

$$\forall t \quad \underset{|x|=r}{\text{Min}} \, V(t, x) \to +\infty \quad \text{when } r \to \infty. \tag{10}$$

We introduce new constants:

$$c'(\beta) = (2 - \alpha)(\alpha b(\beta)^2)^{-1/(2-\alpha)} 2^{-1/\alpha}(2\alpha - 2)^{(\alpha-1)/(2-\alpha)}, \tag{11}$$

$$d'(\beta) = |2/\beta b(\beta)^2|^{1/(2-\alpha)} 2^{-1/\alpha}. \tag{12}$$

PROPOSITION 4.2.   *Assume V satisfies conditions (7)–(10), and*

$$\int_0^T f(t) \, dt = 0 \quad \text{and} \quad \|f\|_\alpha \leq c'(\beta) k^{-\beta/(\beta-2)} T^{-(\beta+2)(\beta-1)/\beta(\beta-2)}. \tag{13}$$

Then the Hamiltonian system:

$$(\dot{x}, \dot{p}) \in (p, -\partial V(t, x)) + (0, f(t)) \tag{14}$$

has at least one solution (x, p) such that

$$x(0) = x(T) \quad \text{and} \quad p(0) = p(T), \tag{15}$$

$$\|\dot{p} - f\|_\alpha \leq \frac{d'(\beta)}{c'(\beta)} \|f\|_\alpha = \frac{2}{\beta - 2} \|f\|_\alpha. \tag{16}$$

Proof.   Denote by $v = (y, q)$ the dual variable of $u = (x, p)$. We can easily compute $G(y, q)$ for this case:

$$G(y, q) = \underset{x,p}{\text{Sup}} \left\{ xy + pq - \frac{1}{2} p^2 - V(t, x) \right\}$$

$$= \underset{x}{\text{Sup}} \{ xy - V(t, x) \} + \underset{p}{\text{Sup}} \left\{ pq - \frac{1}{2} p^2 \right\} = V^*(t, y) + \frac{1}{2} q^2. \tag{17}$$

Here $V^*(t, \cdot)$ is the Legendre transform of the function $V(t, \cdot)$. It is convex, continuous, minimum at the origin, and satisfies the estimate:

$$V^*(t, u) \geq (1/\alpha k^\alpha)|y|^\alpha. \tag{18}$$

The dual action integral now is

$$I(y, q) = \int_0^T \left\{ -\dot{y}q + \frac{1}{2}\dot{y}^2 + V^*(t, -\sigma\dot{q} + \sigma f) \right\} dt. \tag{19}$$

on the space E defined by

$$\dot{y} \in L^2(0, T; \mathbb{R}^N), \qquad \int_0^T \dot{y}\, dt = 0 = \int_0^T y\, dt, \tag{20}$$

$$\dot{q} \in L^\alpha(0, T; \mathbb{R}^N), \qquad \int_0^T \dot{q}\, dt = 0 = \int_0^T q\, dt. \tag{21}$$

We wish to prove that the functional I has a local minimum on E. For this, we have to estimate it from below. We first write it slightly differently:

$$I(y, q) = \frac{1}{2}\int_0^T (\dot{y} - q)^2\, dt + \int_0^T \left\{ -\frac{1}{2}q^2 + V^*(t, -\dot{q} + f) \right\} dt$$

$$= \tfrac{1}{2}\|\dot{y} - q\|_2^2 + J(q). \tag{22}$$

Here

$$J(q) = \int_0^T \left\{ -\frac{1}{2}q^2 + V^*(t, -\dot{q} + f) \right\} dt. \tag{23}$$

Using inequality (18) we get an estimate for J:

$$J(q) \geq -\tfrac{1}{2}\|q\|_2^2 + (1/\alpha k^\alpha)\|\dot{q} - f\|_\alpha^\alpha. \tag{24}$$

Call F the primitive of f which has mean value zero, and set $q' = q - F$. We get, denoting by angle brackets the duality pairing between $L^\alpha(0, T; \mathbb{R}^n)$ and $L^\beta(0, t; \mathbb{R}^n)$:

$$J(q) \geq -\tfrac{1}{2}\langle F, F \rangle + \langle F, q' \rangle - \tfrac{1}{2}\langle q', q' \rangle + (1/\alpha k^\alpha)\|\dot{q}'\|_\alpha^\alpha. \tag{25}$$

To go further, we need to estimate $\|q'\|_\alpha$ and $\|F\|_\alpha$ in terms of $\|\dot{q}'\|_\alpha$ and $\|f\|_\alpha$. Taking into account the normalization conditions, $\int_0^T q'\, dt = 0 = \int_0^T \dot{q}'\, dt$, we get $\|q'\|_2 \leq (T/2\pi)\|\dot{q}'\|_2$ and $\|q'\|_\infty \leq (T/4)\|\dot{q}'\|_\infty$. Using the convexity theorem of M. Riesz again, we get $\|q'\|_\beta \leq T(2\pi)^{-2/\beta}(\tfrac{1}{4})^{1-2/\beta}\|\dot{q}'\|_\beta$. Transposing, we get, for $1 < \alpha < 2$:

$$\|q'\|_\alpha \leq (T/2)\pi^{-2/\beta}(\tfrac{1}{2})^{1-2/\beta}\|\dot{q}'\|_\alpha. \tag{26}$$

Similarly

$$\|F\|_\alpha \leq (T/2)\pi^{-2/\beta}(\tfrac{1}{2})^{1-2/\beta}\|f\|_\alpha \tag{27}$$

Writing this back into inequality (25) yields:

$$J(q) \geq \psi(\|\dot{q}'\|_\alpha) - \tfrac{1}{2}\langle F, F \rangle, \tag{28}$$

where the function ψ of the real variable s is given by

$$
\begin{aligned}
\psi(s) &= -(T/2)\pi^{-2/\beta}\|f\|_\alpha T^{2/\beta}b(\beta)s - \tfrac{1}{2}(T/2)\pi^{-2/\beta}T^{2/\beta}b(\beta)s^2 + (1/\alpha k^\alpha)s^\alpha \\
&= -\pi^{-2/\beta}(b(\beta)/2)T^{(\beta+2)/\beta}\|f\|_\alpha s \\
&\quad - \tfrac{1}{2}\pi^{-2/\beta}(b(\beta)/2)T^{(\beta+2)/\beta}s^2 + (1/\alpha k^\alpha)s^\alpha.
\end{aligned}
\tag{29}
$$

This is the same function as $\varphi(s)$ of formula (3.14), with the coefficient $b(\beta)$ changed to $b(\beta)^2(\tfrac{1}{2})^{1-2/\beta}$ and the exponent $2/\beta$ changed to $(\beta+2)\beta^{-1}$. It follows that, provided

$$
\|f\|_\alpha \le (2-a)(\alpha b(\beta)^2(\tfrac{1}{2})^{1-2/\beta})^{1/(\alpha-2)}(2\alpha-2)^{-(\alpha-1)/(\alpha-2)}k^{\alpha/(\alpha-2)}T^{(\beta+2)/\beta(\alpha-2)},
\tag{30}
$$

we will have

$$
\psi(s_0') > 0 \quad \text{with} \quad s_0' = \left|\frac{\alpha k^\alpha b(\beta)^2 2^{2/\beta}}{4(\alpha-1)}\right|^{1/(\alpha-2)}T^{(\beta+2)/\beta(\alpha-2)}
\tag{31}
$$

From then on, we proceed as in the proof of Proposition 3.1. Going back to Eq. (22), we have

$$
I(y,q) \ge \tfrac{1}{2}\|\dot{y}-q\|_2^2 + \psi(\|\dot{q}-f\|_\alpha).
\tag{32}
$$

Provided condition (30) is satisfied, the functional I will attain its minimum relative to the cylinder C in E defined by

$$
C = \{(y,q) \in E\,|\,\|\dot{q}-f\|_\alpha \le s_0'\}.
\tag{33}
$$

We show as in Proposition 3.1 that this minimum is attained at an interior point, indeed that condition (16) holds. The proof of Proposition 2.2 then carries over to this case, showing that some translate (x,p) of (y,q) is a T-periodic solution of the Hamiltonian system (14). ∎

Of course, in the case where the potential $V(t,x)$ is differentiable with respect to x, a more compact way to write system (14) is Newton's equation:

$$
\ddot{x} + V'(t,x) = f(t).
\tag{34}
$$

Let us give a simple example to illustrate Proposition 4.2. The n-dimensional system of differential equations:

$$
\ddot{x}_i + ax_i \sum_{j=1}^n x_j^2 = f_i(t), \quad 1 \le i \le n, \quad a > 0
\tag{35}
$$

will have a T-periodic solution, provided the forcing terms f_i all are T-periodic, have mean zero, and satisfy

$$
\|f\|_{4/3} = \left|\int_0^T \left(\sum_{i=1}^n f_i(t)^2\right)^{2/3}dt\right|^{3/4} \le c'(4)a^{-1/2}T^{-9/4}.
\tag{36}
$$

This solution will satisfy the estimate:

$$\|\ddot{x} - f\|_{4/3} \leq \|f\|_{4/3}. \tag{37}$$

Another estimate follows immediately by substituting Eq. (35):

$$\|x\|_4 = \left| \int_0^T \left(\sum_{i=1}^n x_i(t)^2 \right)^2 dt \right|^{1/4} \leq \left(\frac{1}{a} \|f\|_{4/3} \right)^{1/3} \tag{38}$$

All this follows from Proposition 4.2, with the exponent $\beta = 4$ and the potential $V(x) = a4^{-1}(\sum_{i=1}^n x_i^2)^2$. Here $c'(4) = d'(4) = (\pi\sqrt{2})^{3/2} \simeq 6.06$.

Finally, we can adapt the proof of Proposition 4.1 to the particular case of classical Hamiltonians, to get the following result:

PROPOSITION 4.3. *Assume $V'(0) = 0$ and there is a neighborhood $\mathcal{V}$ of the origin in $\mathbb{R}^n$ such that v is C^2 on $\mathcal{V}$, the second derivative $V''(x)$ being positive definite for $x \neq 0$, and satisfying, for some constants $b' > a' > 0$ and $\beta > 2$:*

$$a'|x|^{\beta - 2}|y|^2 \leq (V''(x)y, y) \leq b'|x|^{\beta - 2}|y|^2, \qquad \text{for all} \quad x \in \mathcal{V}, \quad y \in \mathbb{R}^n. \tag{39}$$

Then for any $T > 0$, there is some $\varepsilon > 0$ such that, whenever $\|f\|_\alpha \leq \varepsilon$, with $\int_0^T f \, dt = 0$, the Hamiltonian system:

$$\ddot{x} \in \partial V(x) + f(t) \qquad \text{a.e.} \tag{40}$$

has some periodic solution lying inside $\mathcal{U}$.

REFERENCES

1. F. CLARKE, The Euler-Lagrange differential inclusion, *J. Differential Equations* **19** (1975), 80–90.
2. F. CLARKE, Periodic solutions to Hamiltonian inclusions, *J. Differential Equations*, to appear.
3. F. CLARKE AND I. EKELAND, Hamiltonian trajectories having prescribed minimal period, *Comm. Pure Appl. Math*, **33** (1980), 103–116.
4. F. CLARKE AND I. EKELAND, Nonlinear oscillations and boundary-value problems for Hamiltonian systems, *Arch. Rational Mech. Anal.*, to appear.
5. N. DUNFORD AND J. SCHWARTZ, "Linear Operators," Wiley, New York, 1958, 1971.
6. I. EKELAND, Periodic Hamiltonian trajectories and a theorem of P. Rabinowitz, *J. Differential Equations*, **34** (1979), 523–534.
7. I. EKELAND AND R. TEMAM, "Convex Analysis and Variational Problems," Elsevier/North-Holland, Amsterdam, 1976.
8. P. RABINOWITZ, Periodic solutions of Hamiltonian systems, *Comm. Pure Appl. Math.* **31** (1978), 157–184.
9. P. RABINOWITZ, A variational method for finding periodic solutions of differential equations, *in* "Nonlinear Evolution Equations" (M. Crandall, ed.). Academic Press, New York, 1979.
10. R. T. ROCKAFELLAR, "Convex Analysis," Princeton Univ. Press, Princeton, New Jersey, 1970.
11. R. T. ROCKAFELLAR, Integrals which are convex functionals, *Pacific J. Math.* **24** (1968), 525–539.

AMS (MOS) 1970 Subject Classifications: 34C15, 34C25, 49A10, 70H30, 70K99.

MATHEMATICAL ANALYSIS AND APPLICATIONS, PART A
ADVANCES IN MATHEMATICS SUPPLEMENTARY STUDIES, VOL. 7A

Von Neumann's Uniqueness Theorem Revisited

Gérard G. Emch

Departments of Mathematics and Physics
The University of Rochester,
Rochester, New York

To Laurent Schwartz, a Tribute at the Occasion of His Sixty-Fifth Birthday

For finitely many degrees of freedom, we construct a representation of the Weyl CCR acting on a nonseparable Hilbert space in which plane waves are vector states. Though irreducible, this representation is not equivalent to the Schroedinger representation.

1.

The *Schroedinger representation* of the Weyl canonical commutation relations (CCR) for n degrees of freedom is defined on the Hilbert space $\mathfrak{H} = \mathfrak{L}^2(R^n, d^n x)$ as the pair $\{U(R^n), V(R^n)\}$ where for every a,b in R^n, and Ψ in $\mathfrak{H}$:

$$(U(a)\Psi)(x) = \Psi(x - a), \qquad (V(b)\Psi)(x) = \exp(-i<x,b>)\Psi(x).$$

These unitary operators satisfy the following characteristic relation, known as the *Weyl form of the CCR*:

$$U(a)V(b) = \exp(i<a,b>)V(b)U(a).$$

To simplify both the notation and the exposition, we will restrict our attention to the case $n = 1$, which already exhibits the phenomenon we are interested in.

The weakly continuous one-parameter group $U(R)$ of unitary operators is interpreted physically (see also Section 9) as representing space translations. Its generator $P(= -id/dx)$ is a self-adjoint operator, interpreted as the *momentum*, which has continuous simple spectrum $\{k \mid k \in R\}$.

2.

The eigenfunctions (*not* eigenvectors) of P are the locally square-integrable functions $\{\Psi_k | k \in R\}$ defined on R by:

$$\Psi_k(x) = (2\pi)^{-1/2} \exp(ikx).$$

These functions are referred to as *plane waves* in the physical literature. Appearing in the theory as the complete set of all eigenfunctions of the momentum operator, the Ψ_ks satisfy the following two physically important conditions which, to a mathematician, simply amount to the statement of the spectral theorem for P: (i) for every k in R there exists exactly one Ψ_k satisfying the eigenfunction equation

$$P\Psi_k = k\Psi_k;$$

and (ii) every *wave packet* Ψ in $\mathfrak{H}$ can be expanded in a unique way as the coherent superposition of plane waves:

$$\Psi(x) = \int dk \, \tilde{\Psi}(k)\Psi_k(x).$$

3.

We next introduce the abelian *C*-algebra* $\mathfrak{B}$ generated by $\{V(b) | b \in R\}$, which is the closure, in the norm topology of $\mathfrak{B}(\mathfrak{H})$, of the *-algebra

$$\mathfrak{W} = \left\{ \sum_{i \in I} \lambda_i V(b_i) \Big| \lambda_i \in C, \, b_i \in R, \, \text{card}(I) < \infty \right\}.$$

The group $\alpha(R)$ of automorphisms of $\mathfrak{B}$ defined by

$$\alpha(a)[B] = U(a)BU(-a) \quad \forall a, B \in R \times \mathfrak{B}$$

is completely specified by its action on $V(R)$, namely,

$$\alpha(a)[V(b)] = \exp(iab)V(b) \quad \forall a, b \in R \times R.$$

We note that for every B in $\mathfrak{B}$ and every $\varepsilon > 0$, there exist V in $\mathfrak{W}$ and a_0 in R^+ such that

$$\|B - V\| < \varepsilon/3, \qquad \|\alpha(a)[V] - V\| < \varepsilon/3, \qquad \forall a \text{ with } |a| < a_0,$$

and thus

$$\|\alpha(a)[B] - B\| < \varepsilon,$$

which is to say that the maps, defined for every B in $\mathfrak{B}$ by

$$a \in R \mapsto \alpha(a)[B]$$

are continuous in the norm topology of $\mathfrak{B}$.

4.

A state ψ on $\mathfrak{B}$ is a positive linear functional on $\mathfrak{B}$, normalized to $\langle \psi; I \rangle = 1$. ψ is said to be invariant under the action of the dual $\alpha(R)^*$ of $\alpha(R)$ if $\psi \circ \alpha(a) = \psi$ for all a in R; it is further said to be extremal invariant if it cannot be decomposed into a convex sum of such states. The purpose of this section is to show the *existence of at least one state ϕ_0 on $\mathfrak{B}$ which is extremal under the action of the dual $\alpha(R)^*$ of $\alpha(R)$.* Let ψ be an arbitrary state on $\mathfrak{B}$. Since $\alpha(\cdot)[B]$ is continuous for every B in $\mathfrak{B}$ (see Section 3), the maps

$$a \in R \mapsto \langle \psi; \alpha(a)[B] \rangle \in C$$

are continuous, and bounded (e.g., by $\|B\|$). Since on the other hand R is abelian, it is amenable, i.e., there exists at least one invariant mean η on R, that is to say, a state η on the C^*-algebra $\mathscr{C}(R)$ of all complex-valued continuous bounded functions on R, such that

$$\eta(f_a) = \eta(f) \qquad \forall a, f \in R \times \mathscr{C}(R),$$

where f_a is the element of $\mathscr{C}(R)$ defined by

$$f_a : x \in R \mapsto f(x - a) \in C.$$

With these tools in hand, we define the map

$$\eta \psi : B \in \mathfrak{B} \mapsto \eta(\langle \psi; \alpha(\cdot)[B] \rangle) \in C.$$

$\eta \psi$ is a state on $\mathfrak{B}$; this indeed follows from the facts that both η and ψ are positive normalized linear functionals, and that $\mathfrak{B}$ has a unit. From the invariance of the mean η, it furthermore follows that

$$\eta \psi \circ \alpha(a) = \eta \psi \quad \forall a \in R.$$

Hence the set $\mathfrak{S}_R$ of all states on $\mathfrak{B}$ which are invariant under the action of the dual $\alpha(R)^*$ of $\alpha(R)$ is *not* empty. This was the only delicate part of the proof. Indeed $\mathfrak{S}_R$ is bounded (namely, by 1) as a subset of $\mathfrak{B}^*$, the latter being equipped with its natural norm; since moreover $\mathfrak{S}_R$ is closed in the weak-$*$ topology of $\mathfrak{B}^*$, it is compact in this topology; since $\mathfrak{S}_R$ is also convex, we conclude therefore from the Krein–Milman theorem that $\mathfrak{S}_R$ is the closed convex hull of its extreme points. This establishes the existence of at least one state ϕ_0 satisfying the claim made in the beginning of this section.

5.

Let ϕ_0 be a state on $\mathfrak{B}$, which is extremal under the action of the dual $\alpha(R)^*$ of the continuous group $\alpha(R)$ of automorphisms of $\mathfrak{B}$. By the standard Gelfand–Naimark–Segal [3, 7] construction, there exists a unique (up to unitary equivalence) triple $\{\mathfrak{H}_0, \Phi_0, \pi_0\}$ where $\mathfrak{H}_0$ is a Hilbert space, Φ_0 is a vector, of $\mathfrak{H}_0$, normalized to 1, and π_0 is a $*$-representation of $\mathfrak{B}$ as a C^*-subalgebra of $\mathfrak{B}(\mathfrak{H}_0)$, such that $\pi_0(\mathfrak{B})\Phi_0$ is dense in $\mathfrak{H}_0$ and

$$(\pi_0(B)\Phi_0, \Phi_0) = \langle \phi_0; B \rangle \quad \forall B \in \mathfrak{B}.$$

The invariance

$$\phi_0 \circ \alpha(a) = \phi_0 \quad \forall a \in R$$

moreover implies that for every a in R the map

$$\pi_0(B)\Phi_0 \in \mathfrak{H}_0 \mapsto \pi_0(\alpha(a)[B])\Phi_0 \in \mathfrak{H}_0$$

is well defined and extends to a unitary operator $U_0(a)$ on $\mathfrak{H}_0$ such that

$$U_0(a)\pi_0(B)U_0(a)^* = \pi_0(\alpha(a)[B]) \quad \forall a, B \in R \times \mathfrak{B}.$$

In particular upon writing $V_0(b)$ for $\pi_0(V(b))$, we have for all a, b, a_1, a_2, b_1, b_2 in R

$$U_0(a)V_0(b) = \exp(iab)V_0(b)U_0(a)$$

and

$$U_0(a)^* = U_0(-a), \qquad V_0(b)^* = V_0(-b),$$
$$U_0(a_1)U_0(a_2) = U_0(a_1 + a_2), \qquad V_0(b_1)V_0(b_2) = V_0(b_1 + b_2).$$

Hence $\{U_0(a), V_0(b) \,|\, a, b \in R\}$ is a $*$-algebraic representation of the Weyl CCR. We now show that this representation is irreducible. Let

$$\mathfrak{N} = \{N \in \mathfrak{B}(\mathfrak{H}_0) \,|\, [N, U_0(a)] = 0 = [N, V_0(b)] \;\forall a, b \in R\}.$$

Since $\mathfrak{N}$ is a von Neumann algebra, it is generated by its positive elements. Let thus $X \in \mathfrak{N}$ with $X \geq 0$ and $X \neq 0$. Since $\mathfrak{B}$ is the norm closure of $\mathfrak{W}$, and since $\pi_0(\mathfrak{B})\Phi_0$ is dense in $\mathfrak{H}_0$, $X\Phi_0 \neq 0$. We can therefore assume, without loss of generality, that $\|X\Phi_0\| = 1$. Clearly the map

$$\phi_1 : B \in \mathfrak{B} \mapsto (\pi_0(B)X\Phi_0, X\Phi_0) \in C$$

is a state on $\mathfrak{B}$, and satisfies

$$\langle \phi_1; \alpha(a)[B] \rangle = (U_0(a)\pi_0(B)U_0(-a)X\Phi_0, X\Phi_0)$$
$$= (\pi_0(B)X\Phi_0, X\Phi_0) = \langle \phi_1; B \rangle \quad \forall a, B \in R \times \mathfrak{B},$$

i.e.,

$$\phi_1 \circ \alpha(a) = \phi_1 \quad \forall a \in R.$$

Furthermore for every B in $\mathfrak{B}$, we have

$$\langle \phi_1; B^*B \rangle = \|\pi_0(B)X\Phi_0\|^2 = \|X\pi_0(B)\Phi_0\|^2$$
$$\leq \|X\|^2 \|\pi_0(B)\Phi_0\|^2 = \|X\|^2 \langle \phi_0; B^*B \rangle,$$

i.e.,

$$\phi_1 \leq \lambda \phi_0 \quad \text{with} \quad \lambda \geq 1,$$

since $\|X\Phi_0\| = 1$. Suppose $\lambda > 1$, and define ϕ_2 by

$$\phi_0 = \mu \phi_1 + (1 - \mu)\phi_2 \quad \text{with} \quad \mu = 1/\lambda.$$

Clearly ϕ_2 is a positive linear functional on $\mathfrak{B}$, normalized to $\langle \phi_2; I \rangle = 1$, i.e., ϕ_2 is a state on $\mathfrak{B}$. Moreover

$$\phi_i \circ \alpha(a) = \phi_i, \quad \forall a \in R, \quad i = 0, 1, 2,$$

which contradicts the extremality of ϕ_0 under the action of the dual $\alpha(R)^*$ of $\alpha(R)$. Hence $\lambda = 1$, i.e., $\phi_1 = \phi_0$ and thus

$$(\pi_0(B)\Phi_0, (X^2 - I)\pi_0(C)\Phi_0) = 0 \quad \forall B, C \in \mathfrak{B}.$$

Since $\pi_0(\mathfrak{B})\Phi_0$ is dense in $\mathfrak{H}_0$, this implies $X^2 = I$, and thus $X = I$ since X was chosen to be positive. Consequently $\mathfrak{R} = CI$, and thus $\{U_0(a), V_0(b) | a, b \in R\}$ is indeed irreducible.

Note that the above argument, trod in the opposite direction, can be also used to show that if π is an irreducible representation of the Weyl CCR on some Hilbert space $\mathfrak{H}$, and if there exists Ψ in $\mathfrak{H}$ such that $U(a)\Psi = \Psi$ for all a in R, then $\psi : B \in \mathfrak{B} \mapsto (\mathfrak{B}\Psi, \Psi)/(\Psi, \Psi)$ is an extremal $\alpha(R)^*$-invariant state on $\mathfrak{B}$; this remark thus emphasizes how crucial the existence proof of Section 4 is to the argument developed in this chapter.

6.

With $\{U_0(a), V_0(b) | a, b \in R\}$ and Φ_0 defined in the preceding section we now form, for every k in R,

$$\Phi_k = V_0(-k)\Phi_0,$$

and we notice that

$$U_0(a)\Phi_k = U_0(a)V_0(-k)U_0(-a)\Phi_0 = \exp(-ika)\Phi_k \quad \forall a \in R$$

Hence

$$P\Phi_k = k\Phi_k \quad \forall k \in R$$

where P is the generator of the weakly continuous group of unitary operators $\{U(a) = \exp(-iPa) | a \in R\}$. Note further that $\|\Phi_k\| = 1$, so that the states on

$$\mathfrak{B}(\mathfrak{H}_0) = \{U_0(a), V_0(b) | a, b \in R\}'',$$

defined for every k in R by

$$\phi_k : A \in \mathfrak{B}(\mathfrak{H}_0) \mapsto (A\Phi_k, \Phi_k) \in C,$$

are dispersion-free on the abelian von Neumman algebra

$$\mathfrak{A} = \{U(a) | a \in R\}''$$

to which P is affiliated. Consequently, each ϕ_k (or Φ_k if one prefers) is to be interpreted as a *plane wave* of momentum k.

7.

$\mathfrak{H}_0$ is *not* separable. To see this we notice that for $k \neq k'$ we have

$$(\Phi_k, \Phi_{k'}) = (V_0(-k)\Phi_0, V_0(-k')\Phi_0) = \langle \phi_0; V(k'-k) \rangle$$
$$= \langle \phi_0; \alpha(a)[V(k'-k)] \rangle = \exp[i(k'-k)a]\langle \phi_0; V(k'-k) \rangle \quad \forall a \in R.$$

Hence $(\Phi_k, \Phi_{k'}) = 0$ whenever $k \neq k'$. Since we already know that $\|\Phi_k\| = 1$ for all k in R, we obtain that $\{\Phi_k | k \in R\}$ is an orthonormal basis in $\mathfrak{H}_0$, indexed by a noncountable set, namely R, so that $\mathfrak{H}_0$ is indeed not separable.

8.

Since $\mathfrak{H}_0$ is not separable $\{U_0(a), V_0(b) | a, b \in R\}$, though irreducible, is *not* equivalent to the Schroedinger representation of the CCR described in Section 1. Consequently, at least one of the assumptions of von Neumann's uniqueness theorem [6] must be violated by the representation just constructed; the culprit indeed is that, although

$$a \in R \mapsto U_0(a) \in \mathscr{U}(\mathfrak{H}_0)$$

is weakly continuous, this is not the case for

$$b \in R \mapsto V_0(b) \in \mathscr{U}(\mathfrak{H}_0).$$

From a computation similar to that carried out in Section 7, one obtains indeed that for any k, k' in R

$$(V_0(b)\Phi_k, \Phi_{k'}) = \begin{cases} 1 & \text{if} \quad b = k - k', \\ 0 & \text{otherwise.} \end{cases}$$

As a consequence, no position operator can be defined as the putative generator of the unitary group $\{V_0(b)|b \in R\}$, in contrast with the situation encountered in the Schroedinger representation where we had $V(b) = \exp(-iQb)$ for all b in R. From a physical point of view, this result should have been expected to occur in the present situation where the representation space is spanned by (normalized!) plane waves.

9.

Although this is not emphasized enough in the physical literature, a close look at the proof of von Neumann's uniqueness theorem reveals that it does indeed crucially depend on the assumption that both $U(R)$ and $V(R)$ be weakly continuous. The following statement of von Neumann's theorem, in terms of Mackey's systems of imprimitivity [4,5], will satisfy both the physicist and the mathematician.

Let Q be a self-adjoint operator acting on some Hilbert space $\mathfrak{H}$ and $\{E(\lambda)|\lambda \in R\}$ be the corresponding resolution of the identity, i.e.,

$$Q = \int \lambda E(d\lambda).$$

Let further $U(R)$ be a weakly continuous one-parameter group of unitary operators acting on $\mathfrak{H}$ in such a manner that for every a in R and every Borel subset Δ of R

$$U(a)E(\Delta)U(-a) = E(\Delta - a).$$

Then, the weakly continuous one-parameter unitary group $\{V(b)|b \in R\}$ generated by Q satisfies the Weyl canonical commutation relation

$$U(a)V(b) = \exp(iab)V(b)U(a) \quad \forall a, b \in R.$$

Furthermore, if $\{U(a), V(b)|a, b \in R\}$ is irreducible, then $\mathfrak{H}$ is separable. Conversely, if $\mathfrak{H}$ is separable, then $\{U(a), V(b)|a, b \in R\}$ is unitarily equivalent to a direct sum of countably many copies of the Schroedinger representation. In particular if $\{U(a), V(b)|a, b \in R\}$ is irreducible, then it is unitarily equivalent to the Schroedinger representation.

This formulation leads directly to the physical interpretation of Q and P [= the generator of $U(R)$] as the canonically conjugate observables *position* and *momentum*; $E(\Delta)$ for instance is the observable corresponding to the proposition: the particle is in Δ.

The usual form of the canonical commutation relations found in most physics texts, namely

$$[P, Q] = -iI$$

appears as a consequence of the assumptions of the above theorem; to make it equivalent to these assumptions, this relation must be supplemented with precise conditions on the domain $\mathcal{D}$ where it is assumed to hold (see, e.g. [1, 2]).

10.

All the constructions and results presented in this paper extend readily (at the cost of only minor changes in the notation) to the case of n ($< \infty$) degrees of freedom.

REFERENCES

1. J. DIXMIER, Sur la relation $i(PQ - QP)$, *Compositio Math.* **13** (1958), 263–270.
2. C. FOIAS, L. GEHER, and B. SZ.-NAGY, On the permutability condition of quantum mechanics, *Acta Sci. Math.* (*Szeged*) **21** (1960), 78–89.
3. I. M. GELFAND AND M. A. NAIMARK, On the imbedding of normed rings into the ring of operators in Hilbert space, *Mat. Sb.* **12**(*54*) (1943), 197–217.
4. G. W. MACKEY, "Mathematical Foundations of Quantum Mechanics," Benjamin, New York, 1963.
5. G. W. MACKEY, "Induced Representations and Quantum Mechanics," Benjamin, New York, 1968.
6. J. VON NEUMANN, Die Eindeutigkeit der Schroedingerschen Operatoren, *Math. Ann.* **104** (1931), 570–578.
7. I. E. SEGAL, Postulates for general quantum mechanics, *Ann. of Math.* **48** (1947), 930–948.

MATHEMATICAL ANALYSIS AND APPLICATIONS, PART A
ADVANCES IN MATHEMATICS SUPPLEMENTARY STUDIES, VOL. 7A

Symmetry of Positive Solutions of Nonlinear Elliptic Equations in R^n

B. Gidas

Institute for Advanced Study
Princeton, New Jersey

Wei-Ming Ni[†‡]

Institute for Advanced Study
Princeton, New Jersey

AND

L. Nirenberg[§]

Courant Institute of Mathematical Sciences
New York University
New York, New York

DEDICATED TO LAURENT SCHWARTZ ON THE OCCASION OF HIS 65TH BIRTHDAY

We prove symmetry of positive solutions of two general types of nonlinear elliptic equations on R^n:

(a) $-\Delta u = g(u)$, $g(u) = O(u^\alpha)$ near $u = 0$ for suitable $\alpha > 1$, and
(b) $-\Delta u + u = g(u)$, $g(u) = O(u^\alpha)$ near $u = 0$, $\alpha > 1$.

Under certain additional assumptions on the nonlinearities, we show that positive solutions tending to zero at infinity are spherically symmetric about some point. In the case of equations of type (a), we also establish symmetry of positive solutions with isolated singularities.

† Present address: Department of Mathematics, University of Pennsylvania, Philadelphia, Pennsylvania.
‡ Supported in part by the U.S. Army Research Office,
Grant No. DAA29-78-G-0127 and NSF Grant No. MCS 77-18723(02).
§ Supported in part by U.S. Army Research Office
Grant No. DAA29-78-G-0127 and NSF Grant No. NSF-Int 77-20878.

369

1. INTRODUCTION

In Gidas *et al.* [4], some special forms of the maximum principle and a device of Alexandroff were employed to study symmetry and related properties of positive solutions of nonlinear elliptic (and parabolic) equations in bounded domains and in the entire space. In bounded domains, the class of nonlinearities we could treat was quite general.[†] In contrast, in the entire space our results were rather special. The following is typical: Let $u(x)$ be positive, $C^{2+\mu}$, $0 < \mu < 1$, solution of

$$-\Delta u = g(u) \qquad \text{in } R^n, \quad n > 2,$$

such that $u(x) = O(|x|^{2-n})$ at infinity. Assume that for some $k \geq (n + 2)/(n - 2)$, $u^{-k}g(u)$ is Hölder continuous on $0 \leq u \leq u_0$, where u_0 is the maximum of $u(x)$. Then $u(x)$ is spherically symmetric about some point, and if that point is taken as the origin, then $u_r < 0$ for $r > 0$.

In this chapter we study spherical symmetry of positive solutions in the entire space and treat several classes of nonlinearities as well as some singular solutions. Our main results are

THEOREM 1.   *Let* $u(x)$ *be a positive* C^2 *solution of*

$$-\Delta u = g(u) \qquad \text{in } R^n, \quad n \geq 3 \tag{1.1}$$

with

$$u(x) = O(1/|x|^m) \qquad \text{at infinity,} \quad m > 0. \tag{1.2}$$

Assume (i) *on the interval* $0 \leq u \leq u_0$, $u_0 = \max u$, $g = g_1 + g_2$ *with* $g_1 \in C^1$, g_2 *continuous and nondecreasing,* (ii) *for some*

$$\alpha > \max((n + 1)/m, (2/m) + 1),$$

$g(u) = O(u^\alpha)$ *near* $u = 0$. *Then* $u(x)$ *is spherically symmetric about some point in* R^n, *and* $u_r < 0$ *for* $r > 0$ *where* r *is the radial coordinate about that point. Furthermore*

$$\lim_{x \to \infty} |x|^{n-2}u(x) = k > 0.$$

THEOREM 2.   *Let* $u(x)$ *be a positive,* C^2 *solution of*

$$-\Delta u + m^2u = g(u) \qquad \text{in } R^n, \quad n \geq 2, \quad m > 0 \tag{1.3}$$

[†] Since it may prove useful, we wish to point out that in [4, Theorem 2.1], for general nonlinear equations, except in the case that $g(x) < 0$ in condition (b), it suffices that $u \in C^2(\Omega) \cap C^1(\bar{\Omega})$.

with $u(x) \to 0$ as $|x| \to \infty$ and g continuous, $g(u) = O(u^\alpha)$, $\alpha > 1$ near $u = 0$. On the interval $0 \le s \le u_0 = \max u(x)$, assume

$$g(s) = g_1(s) + g_2(s)$$

with g_2 nondecreasing and $g_1 \in C^1$ satisfying, for some $C > 0$, $p > 1$,

$$|g_1(u) - g_1(v)| \le C|u - v|/|\log \min(u, v)|^p, \qquad 0 \le u, v \le u_0.$$

Then $u(x)$ is spherically symmetric about some point in R^n and $u_r < 0$ for $r > 0$, where r is the radial coordinate about that point. Furthermore,

$$\lim_{r \to \infty} r^{(n-1)/2} e^r u(r) = \mu > 0.$$

Remark. In particular, if $u > 0$ tends to zero at infinity and satisfies

$$\Delta u + f(u) = 0,$$

$f \in C^{1+\mu}$, $\mu > 0$ and $f(0) = 0$, $f'(0) < 0$ then Theorem 2 applies.

With no loss in generality we shall always take $m = 1$ in (1.3). We use the same techniques to study certain isolated singularities. More precisely, we prove

THEOREM 3. *Let $u(x)$ be a positive C^2 solution of*

$$-\Delta u = g(u) \qquad in \quad R^n - \{0\} \tag{1.4}$$

satisfying

$$u(x) = O(1/|x|^m) \qquad at \ \infty \ for \ some \quad m > 0, \tag{1.5}$$

$$u(x) \to +\infty \qquad as \quad x \to 0. \tag{1.6}$$

Furthermore, assume:

(i) $g(u)$ *is continuous, nondecreasing in u for $u \ge 0$, and for some $\alpha > (n+1)/m$, $g(u) = O(u^\alpha)$ near $u = 0$,*

(ii) $\underline{\lim}_{u \to +\infty} [g(u)/u^p] > 0$, *for some $p \ge n/(n-2)$.*

Then $u(x)$ is spherically symmetric about the origin, and $u_r < 0$ for $r > 0$.

A generalization, Theorem 3', is given in Section 3, and variants of Theorem 1 are presented at the end of Section 2.

THEOREM 4. *Let $u(x)$ be a positive solution of*

$$\Delta u + u^{(n+2)/(n-2)} = 0 \qquad in \quad R^n - \{0, \infty\} \tag{1.7}$$

*with two singularities located at the origin and at infinity. More precisely,
assume*

$$u(x) \to +\infty \qquad as \quad x \to 0,$$
$$|x|^{n-2}u(x) \to +\infty \qquad as \quad x \to \infty.$$

Then $u(x)$ is spherically symmetric about the origin.

A particular equation covered by Theorem 2 is

$$-\Delta u + m^2 u = \lambda u^3 \qquad on \ R^3, \quad \lambda > 0, \quad m > 0. \tag{1.8}$$

This equation gives rise to solitary wave solutions of the nonlinear Klein–Gordon equation. Existence of positive spherically symmetric solutions of (1.8) has been known for some time [8].

An equation covered by Theorem 3 is

$$\Delta u + u^\alpha = 0 \tag{1.9}$$

in R^n, with $(n+1)/(n-2) < \alpha$. For $(n+1)/(n-2) < \alpha < (n+2)/(n-2)$ Eq. (1.9) is known [5] to have spherically symmetric solutions satisfying (1.5) and (1.6). Certain equations in astrophysics [2] are related to (1.9) with $n = 3$ and $\alpha < 5$. See also [5] for further analysis of spherically symmetric solutions of such equations. It follows from the analysis in [5] that for $\alpha > (n+2)/(n-2)$, (1.9) does not have spherically symmetric solutions satisfying (1.5), (1.6) for $m = n - 2$; therefore Theorem 3 implies there are no positive solutions satisfying these conditions. For $\alpha = (n+2)/(n-2)$ [Eq. (1.7)], all spherically symmetric solutions are explicitly known [1, 3]. Among them there exist a two-parameter family of solutions which are singular both at the origin and at infinity. Theorem 4 asserts that these are the only solutions with two isolated singularities. Since Eq. (1.7) is conformally invariant, the two singular points could be brought to any two finite points of R^n. An example of an equation covered by Theorem 1 with $m = n - 2$ and $\alpha > (n+2)/(n-2)$ was given in [4, p. 212]. Equation (1.7) has the regular solution

$$u(x) = \left(\frac{\lambda\sqrt{n(n-2)}}{\lambda^2 + |x|^2}\right)^{(n-2)/2}, \qquad \lambda > 0, \tag{1.10}$$

and it provides an example for Theorem 1 with $m = n - 2$, and $\alpha = (n+2)/(n-2)$. It is natural to ask whether there is an equation which satisfies the conditions of Theorem 1 with $\alpha < (n+2)/(n-2)$. Here is an example, for $n > 3$; we also have an example for $n = 3$. With r as polar coordinate, and $\alpha \geq (n+1)/(n-2)$, consider the function $u(r)$ defined as follows:

$$u(r) = (1/r^{n-2}) - (a/r^s) \qquad for \quad r \geq 1,$$

where

$$s = \alpha(n-2) - 2, \qquad as = (s+2-n)^{-1} = [\alpha(n-2) - n]^{-1}.$$

For $0 \le r \le 1$ we define

$$u(r) = A - (b/2)r^2 + (c/4)r^4,$$

where

$$2c = n(n-2) - \frac{\alpha(n-2)}{\alpha(n-2) - n}, \qquad 2b = n^2 - 4 - \frac{\alpha(n-2) + 2}{[\alpha(n-2) - n]},$$

$$A = (b/2) - (c/4) + 1 - a.$$

With these choices one verifies that $u \in C^2$ on $0 \le r < \infty$, u is positive, decreasing, and

$$u_{rr} + [(n-1)/r]u_r + g(u) = 0,$$

where g is Lipschitz continuous and

$$g(u) = O(u^{\alpha}) \qquad \text{near} \quad u = 0.$$

Our proofs are modeled on the proofs in [4] of similar results in balls. An important tool there was the Hopf boundary lemma, [4, Lemma (H)] and in this chapter we derive suitable forms of it for positive solutions (tending to zero at infinity) of equations in R^n. Here are two forms (in which $\partial_j = \partial/\partial x_j$). We consider only $n > 2$.

LEMMA (H_1').   Let $u > 0$ be of class C^2 in $|x| \ge R > 0$, tend to zero at infinity, and satisfy

$$Lu \equiv (\Delta + \textstyle\sum b_j(x)\partial_j + c(x))u \le 0 \qquad \text{in} \quad |x| \ge R, \tag{1.11}$$

with

$$b_j = O(|x|^{1-p}), \qquad c(x) = O(|x|^{-p}), \qquad p > 2. \tag{1.12}$$

Then, for some $\mu > 0$,

$$u(x) \ge \mu/|x|^{n-2}.$$

LEMMA (H_2'):   Let $u > 0$ be of class C^2 in $|x| \ge R > 0$, tend to zero at infinity, and satisfy

$$Lu \equiv (\Delta - 1 + \textstyle\sum b_j(x)\partial_j + c(x))u \le 0 \tag{1.13}$$

with

$$b_j, c = O(|x|^{-p}), \qquad p > 1.$$

Then, for some $\mu > 0$,

$$u(x) \ge \mu e^{-|x|}/|x|^{(n-1)/2}.$$

These and some related results are proved in Section 5.

Remark.   In$|x| \geq 1$ the function

$$u = 1/|x|^m, \qquad m > n - 2,$$

satisfies

$$(\Delta - m(m + 2 - n)/|x|^2)u = 0$$

showing that the condition $p > 2$ cannot be omitted in Lemma (H'$_1$).

A similar example concerning Lemma (H'$_2$) is given in Section 5.

It would be of interest in Theorems 1 and 2 to permit the nonlinear term to depend on derivatives of u. Here is an extension of Theorem 2, having a very similar proof—which we leave to the reader.

THEOREM 2'.   *Let $u > 0$ be a C^2 solution of*

$$\Delta u - u + g_1(u, u_1, \ldots, u_n) + g_2(u) = 0$$

with $u, \nabla u = O(e^{-|x|})$ at infinity. Assume (i) *$g_2(s)$ is continuous and nondecreasing on the interval $0 \leq s \leq u_0 = \max u(x)$, and $g(u) = O(u^\alpha)$, $\alpha > 1$ near the origin and* (ii) *$g_1 \in C^{1+\varepsilon}$, $\varepsilon > 0$, and g_1 and its first derivatives vanish at the origin in R^{n+1}. Furthermore, for $u > 0$, g_1 is symmetric in the second argument u_1. Then u is symmetric about some hyperplane $x_1 = \lambda$ and $u_{x_1} < 0$ for $x_1 > \lambda$.*

Clearly if, in addition, $g_1(u, u_1, \ldots, u_n)$ depends only on u and $|\text{grad } u|^2$ one infers that u is spherically symmetric about some point and that $u_r < 0$ if that point is chosen as the origin.

2.   THE PROOF OF THEOREM 1

We begin with some preliminary observations.

Equation (1.1) may be written in the form

$$\Delta u + c(x)u = 0, \tag{2.1}$$

where $c(x) = g(u(x))/u(x) = O(|x|^{-p})$ near infinity. Here $p = m(\alpha - 1) > 2$ by assumption. We may apply Lemma (H'$_1$) and infer that

$$u(x) \geq \mu |x|^{2-n}, \qquad \mu > 0. \tag{2.2}$$

It is easily seen, on the other hand, that for some constant $c > 0$,

$$u = c \int \frac{1}{|x - y|^{n-2}} f(y) \, dy, \tag{2.3}$$

where $f(y) = g(u(y)) = O(|y|^{-\alpha m})$ near infinity; recall $\alpha m > n + 1$.

For each $x \in R^n$ we use x^λ to denote the reflection of x in the hyperplane $x_1 = \lambda$. Throughout the paper, C, C_1, c, c_1, etc., denote positive constants independent of u, λ, x, and y. We write $u_j = \partial u/\partial x_j$, etc. Unless indicated otherwise, all integrals extend over R^n.

We will need some simple technical facts about integrals of the form (2.3) for f continuous and satisfying

$$f(y) = O(|y|^{-q}) \quad \text{near infinity}, \qquad q > n + 1. \tag{2.4}$$

LEMMA 2.1. Let $u \in C^2$ be given by (2.3) with f satisfying (2.4). Then

(a) $$\lim_{|x| \to \infty} |x|^{n-2} u(x) \to c \int f(y) \, dy, \tag{2.5}$$

and

(b) $$\frac{|x|^n}{x_1} u_1(x) \to -(n-2)c \int f(y) \, dy \quad \text{as } |x| \to \infty \text{ with } x_1 \to \infty. \tag{2.6}$$

(c) If $\lambda^i \in R \to \lambda$ and $\{x^i\}$ is a sequence of points going to infinity, with $x_1^i < \lambda^i$, then

$$\frac{|x^i|^n}{\lambda^i - x_1^i} (u(x^i) - u(x^{i^{\lambda^i}})) \to 2(n-2)c \int f(y)(\lambda - y_1) \, dy. \tag{2.7}$$

Postponing the proof to Appendix A, we proceed with the arguments to prove Theorem 1. In the following, u satisfies the conditions of Theorem 1 and $f = g \circ u$. Combining (2.2) and (2.5), we infer that

$$\int f(y) \, dy > 0. \tag{2.8}$$

Furthermore, using (2.6) and (2.8), we may infer that

$$u_1(x) < 0 \qquad \text{for} \quad x_1 \geq \text{some constant } \bar{\lambda}. \tag{2.9}$$

We now fix the origin so that

$$\int f(y) y_j \, dy = 0, \qquad j = 1, \dots, n;$$

we shall prove spherical symmetry about the origin.

LEMMA 2.2. There exists $\lambda_0 > 0$ such that $\forall \lambda \geq \lambda_0$

$$u(x) > u(x^\lambda) \qquad \text{if} \quad x_1 < \lambda. \tag{2.10}$$

Proof. Assume not. Then $\exists \lambda^i \to +\infty$ and a sequence $\{x^i\}$, with $x_1^i < \lambda^i$ and

$$u(x^i) \leq u(z^i), \tag{2.11}$$

where $z^i = x^{i^{\lambda^i}}$. Since $z^i \to \infty$ we have $u(z^i) \to 0$ and so, necessarily, $|x^i| \to \infty$. Furthermore, in view of (2.9), we see that $x_1^i < \bar{\lambda}$. Fix some $\lambda > 0$, $\lambda > \bar{\lambda}$. Using (2.9) once more we see that for $\lambda^i > \lambda$, i.e., for i sufficiently large, we have

$$u(x^i) \leq u(z^i) \leq u(x^{i^\lambda}),$$

since $\bar{\lambda} < x_1^{i^\lambda} < z_1^i$.

We now apply Lemma 2.1c with λ in place of λ^i and find, recalling $\int f y_1 \, dy = 0$,

$$0 \geq \frac{|x^i|^n}{\lambda - x_1^i} (u(x^i) - u(x^{i^\lambda})) \to 2(n-2)c\lambda \int f(y) > 0 \tag{2.12}$$

This is a contradiction; the lemma is proved. ∎

LEMMA 2.3. *Suppose condition (2.10) holds for some $\lambda > 0$. Then it holds for all $\tilde{\lambda}$ in a neighborhood.*

Proof. We observe first that according to [4, Lemma 2.2] and Remark 1 of Section 2.3 there (see also [4, Lemma 4.3]):

$$u_1 < 0 \quad \text{on the hyperplane} \quad x_1 = \lambda. \tag{2.13}$$

If (2.10) does not hold for all neighboring $\tilde{\lambda}$, there exists a sequence $\lambda^j \to \lambda$ and sequence of points $\{x^j\}$ with $x_1^j < \lambda^j$ such that

$$u(x^j) \leq u(x^{j^{\lambda^j}}). \tag{2.14}$$

Either a subsequence, which we again call x^j, converges to some x with $x_1 \leq \lambda$ and $u(x) \leq u(x^\lambda)$ or else $x^j \to \infty$. In the first case, in view of (2.10), we must have $x_1 = \lambda$—but then (2.14) implies $u_1(x) \geq 0$, contradicting (2.13). So $x^j \to \infty$. But if we now apply Lemma 2.1c exactly as in the preceding proof, using the sequence λ^i, we reach a contradiction. ∎

Completion of the proof of Theorem 1. Lemmas 2.2 and 2.3 imply that the set of positive λ for which property (2.10) holds is open and includes all sufficiently large λ. Thus for all λ in some *maximal* open interval $0 \leq \lambda_1 < \lambda < +\infty$, property (2.10) holds. Applying [4, Lemmas 2.2 and 4.3] and Remark 1 of Section 2.3 there, we find

$$u_{x_1} < 0 \quad \text{on} \quad x_1 = \lambda \quad \text{for} \quad \lambda > \lambda_1. \tag{2.15}$$

In particular

$$u_{x_1} < 0 \quad \text{for} \quad x_1 > \lambda_1$$

and, by continuity,

$$u(x) \geq u(x^{\lambda_1}) \qquad \text{for} \quad x_1 < \lambda_1.$$

We will now show that $\lambda_1 = 0$. If $\lambda_1 > 0$, then by [4, Lemmas 2.2 and 4.3] and Remark 1 of Section 2.3 there, we either have

$$u(x) \equiv u(x^{\lambda_1}) \qquad \text{for} \quad x_1 < \lambda_1, \qquad (2.16)$$

or else properties (2.10) and (2.15) hold for λ_1. The latter cannot occur, by Lemma 2.3. The former also cannot occur, because if it did, choose a sequence of $\{x^i\}$, $|x^i| \to \infty$, $x_1^i < \lambda_1$ and apply Lemma 2.1c as before to obtain a contradiction. Thus $\lambda_1 = 0$, and we have proved $u(x) \geq u(x^0)$ for $x_1 < 0$. Reversing the x_1 axis, we conclude that u is symmetric about the plane $x_1 = 0$, and $u_{x_1} < 0$ for $x_1 > 0$. Since the x_1 direction was chosen arbitrarily, we conclude that $u(x)$ is spherically symmetric about the origin and $u_r < 0$ for $r > 0$. The last assertion in Theorem 1 follows from (2.15). This completes the proof of Theorem 1. ∎

Remark 2.1   Our choice of origin so that $\int f(y) y_j \, dy = 0, j = 1, \ldots, n$, was used in going from (2.7) to (2.12) in the proof of Lemma 2.3. In case $g(u)$ is nondecreasing, i.e., $g = g_2$, then Lemma 2.3 holds independently of choice of origin, i.e.,

LEMMA 2.3′.   *If $g(s)$ in Theorem 1 is nondecreasing in s on $0 \leq s \leq \max u$, then the set of λ for which property (2.10) holds is open.*

Proof.   We proceed as in the proof of Lemma 2.3. Assuming (2.10) holds for some λ, we want to show that it holds for neighboring $\tilde{\lambda}$. Assuming the contrary we find as in the proof of Lemma 2.10 that

$$0 \geq \int dy \, f(y)(\lambda - y_1) = -\int dy \, f(y^\lambda)(\lambda - y_1)$$

thus

$$0 \geq \int dy (f(y) - f(y^\lambda))(\lambda - y_1). \qquad (2.17)$$

Note that $f(y) = g(u(y))$, and g is nondecreasing in u. Since u satisfies (2.10), it follows that $f(y) \geq f(y^\lambda)$ for $y_1 < \lambda$. Furthermore it is not possible that $f(y) \equiv f(y^\lambda)$, for in that case, as is easily seen, we would necessarily have $g(s) = \text{constant} = 0$ on $0 \leq s \leq \max u$, which would contradict the fact that $u > 0$. Thus the integrand in (2.17) is nonnegative and not identically zero—a contradiction. Lemma 2.3′ is proved. ∎

Remark 2.2   Lemma 2.3′ will be used in the proofs of Theorems 3 and 4.

In Theorem 1 the solutions decay exactly like $c|x|^{2-n}$ near infinity. It may happen that solutions decay differently. For example, in R^3, the function

$$u = (1 + |x|^2)^{-k}, \qquad k > 0,$$

satisfies $\Delta u + g(u) = 0$ with

$$g(u) = -2k(2k-1)u^{1+1/k} + 4k(k+1)u^{1+2/k}.$$

So in this case $m = 2k$, $\alpha = 1 + 1/k$ and $m(\alpha - 1) = 2$, and of course Theorem 1 does not apply. The following is a variant of Theorem 1—which however also does not cover this example.

THEOREM 1′.   *Let $u(x)$ be a positive C^2 solution of*

$$-\Delta u = g(u) \qquad \text{in } R^n, \quad n \geq 3 \tag{2.18}$$

with

$$u(x) = O(1/|x|^m) \quad \text{at infinity,} \qquad m > 0. \tag{2.19}$$

Assume (i) $g(u) \geq 0$ *on* $0 \leq u \leq u_0$, $u_0 = \max u$, $g = g_1 + g_2$ *with* $g_1 \in C^1$, g_2 *continuous and nondecreasing,* (ii) *for some* $\alpha > (n+1)/m$, $g(u) = O(u^\alpha)$ *near* $u = 0$. *Then $u(x)$ is spherically symmetric about some point in R^n, and $u_r < 0$ for $r > 0$ where r is the radial coordinate about that point.*

Proof.   As before, Lemma 2.1 applies, as does Lemma 2.3. But we have to give a new proof of Lemma 2.2. The rest of the proof of Theorem 1 then applies.

Proof of Lemma 2.2 for Theorem 1′.   From (2.3) we have

$$c^{-1}(u(x) - u(x^\lambda)) = \int dy\, f(y) \left\{ \frac{1}{|x-y|^{n-2}} - \frac{1}{|x^\lambda - y|^{n-2}} \right\}$$

$$= I_1 - I_2,$$

where

$$I_1 = \int_{y_1 < \lambda} dy\, f(y) \left\{ \frac{1}{|x-y|^{n-2}} - \frac{1}{|x^\lambda - y|^{n-2}} \right\},$$

$$I_2 = \int_{y_1 > \lambda} dy\, f(y) \left\{ \frac{1}{|x^\lambda - y|^{n-2}} - \frac{1}{|x-y|^{n-2}} \right\}.$$

For $r \leq s$ we use

$$(n-2)\frac{1}{s^{n-1}}(s-r) \leq \frac{1}{r^{n-2}} - \frac{1}{s^{n-2}} \leq (n-2)\frac{1}{r^{n-1}}(s-r). \tag{2.20}$$

Note that

$$|x^\lambda - y| - |x - y| = \frac{4(\lambda - x_1)(\lambda - y_1)}{|x^\lambda - y| + |x - y|}, \tag{2.21}$$

Thus

$$I_1 \geq (n - 2) \int_{y_1 < \lambda} dy\, f(y) \frac{1}{|x^\lambda - y|^{n-1}} \frac{4(\lambda - x_1)(\lambda - y_1)}{|x^\lambda - y| + |x - y|}$$

$$\geq 4(n - 2)(\lambda - x_1) \int_{y_1 < \lambda,\, |y| < R} dy\, f(y) \frac{1}{|x^\lambda - y|^{n-1}} \frac{\lambda - y_1}{|x^\lambda - y| + |x - y|},$$

where R is a fixed number so that $\int_{|y| < R} dy\, f(y) \geq c_0 > 0$. Using $|x^\lambda - y| \leq |x^\lambda| + |R|$, and $|x^\lambda - y| \geq |x - y|$, valid for $y_1 < \lambda$, we obtain for large λ,

$$I_1 \geq c_1 \frac{1}{|x^\lambda|^{n-1}} \frac{\lambda - x_1}{|x^\lambda|} \int_{y_1 < \lambda,\, |y| \leq R} dy\, f(y)(\lambda - y_1)$$

$$\geq c \frac{1}{|x^\lambda|^{n-1}} \frac{(\lambda - x_1)(\lambda - R)}{|x^\lambda|}. \tag{2.22}$$

Next we derive an upper bound for I_2. Using (2.20) and (2.21), we obtain

$$I_2 \leq 2(n - 2)(\lambda - x_1) \int_{y_1 > \lambda} dy\, f(y) \frac{(y_1 - \lambda)}{|x^\lambda - y|^{n-1}} \frac{1}{|x^\lambda - y| + |x - y|}$$

$$\leq I_2' + I_2'',$$

where

$$I_2' = 2(n - 2)(\lambda - x_1) \int_{y_1 > \lambda,\, |y| > 2|x^\lambda|} dy\, f(y) \frac{1}{|x^\lambda - y|^{n-1}}$$

$$I_2'' = 2(n - 2)(\lambda - x_1) \int_{y_1 > \lambda,\, |y| < 2|x^\lambda|} dy\, f(y) \frac{(y_1 - \lambda)}{|x^\lambda - y|^{n-1}} \frac{1}{|x^\lambda - y| + |x - y|}.$$

To bound I_2', we use (i) $|x^\lambda - y| > |x^\lambda|$ implied by $|y| > 2|x^\lambda|$, and (ii) $f(y) = O(|y|^{-\alpha m})$ valid for large $|y|$. Thus, for large λ

$$I_2' \leq c \frac{1}{|x^\lambda|^{n-1}} (\lambda - x_1) \int_{y_1 > \lambda,\, |y| > 2|x^\lambda|} dy\, f(y)$$

$$\leq c \frac{1}{|x^\lambda|^{n-1}} (\lambda - x_1) \frac{1}{|x^\lambda|^{\alpha m - n}}, \tag{2.23}$$

since $\alpha m > n$. To bound I_2'', we write

$$I_2'' \leq J_1 + J_2,$$

$$J_1 = 2(n-2)(\lambda - x_1) \int_{y_1 > \lambda, \, |y| < 2|x^\lambda|, \, |x^\lambda - y| < |x^\lambda|/2} dy \, f(y) \frac{1}{|x^\lambda - y|^{n-1}},$$

$$J_2 = 2(n-2)(\lambda - x_1) \int_{y_1 > \lambda, \, |y| < 2|x^\lambda|, \, |x^\lambda - y| > |x^\lambda|/2} dy \, f(y) \frac{|y_1 - \lambda|}{|x^\lambda - y|^n}.$$

Using $|y| > |x^\lambda|/2$, and $f(y) = O(|y|^{-\alpha m})$, for large $|y|$, we bound J_1 (for large λ) by

$$J_1 \leq c \frac{1}{|x^\lambda|^{n-1}} (\lambda - x_1) \frac{1}{|x^\lambda|^{\alpha m - n}}. \tag{2.24}$$

In the integral for J_2 we have $|x^\lambda|/2 < |x^\lambda - y| < 3|x^\lambda|$, so that

$$J_2 \leq c \frac{1}{|x^\lambda|^n} (\lambda - x_1) \int_{y_1 > \lambda} dy \, f(y) |y|$$

$$\leq c \frac{1}{|x^\lambda|^n} (\lambda - x_1) \frac{1}{\lambda^{\alpha m - n - 1}} \qquad \text{since} \quad \alpha m > n + 1.$$

Combining this with (2.23) and (2.24), we obtain

$$I_2 \leq c \frac{1}{|x^\lambda|^n} (\lambda - x_1) \frac{1}{\lambda^{\alpha m - n - 1}}. \tag{2.25}$$

Comparing (2.22) and (2.25), we see that $I_1 - I_2 > 0$ for sufficiently large λ, since $\alpha m > n + 1$. This completes the new proof of Lemma 2.2 and of Theorem 1'. ∎

With the aid of a similar argument and the argument used to complete the proof of Theorem 1, one establishes

THEOREM 1''. *Let $u(x)$ be a positive C^2 solution of*

$$-\Delta u = g(|x|, u(x)) \text{ in } R^n, \qquad n \geq 3$$

with $u = O(|x|^{-m})$, $m > 0$, near infinity. Assume:

(i)' *for $r \geq 0$ and $0 < s \leq u_0 = \max u$, $g(r, s)$ is continuous, positive, non-decreasing in s and strictly decreasing in r;*

(ii)' *for some $\alpha > (n+1)/m$, and some constant $C > 0$,*

$$g(r, u) \leq C u^\alpha \qquad \text{for} \quad u \leq u_0.$$

Then u is spherically symmetric about the origin, and $u_r < 0$ for $r > 0$.

3. The Proofs of Theorems 3 and 4

The following proposition asserts that the singular solutions in Theorems 3 and 4 are distribution solutions.

PROPOSITION 3.1. *Let* $D = \{x \in R^n : |x| \leq R\}$, *and* $u(x)$ *be a positive* C^2 *solution of*

$$-\Delta u = g(u) \qquad in \quad \Omega = D - \{0\} \tag{3.1}$$

with $u(x) \to +\infty$ *as* $x \to 0$.
  Assume $g(u) \geq 0$ *for* $0 \leq u$ *and*

$$\lim_{u \to +\infty} \frac{g(u)}{u^p} > 0, \qquad for\ some \quad p \geq \frac{n}{n-2}. \tag{3.2}$$

Then $g(u(\cdot)) \in L_1(D)$, $u \in L_p(D)$, *and* u *is a distribution solution of* (3.1).

Proof.   First we show that $g(u(\cdot)) \in L_1(D)$. Let $\zeta(u)$ be a C^∞ nonnegative function defined on $u \geq 0$ which equals 1 for $u \leq c$, vanishes for $u \geq c$ and is nonincreasing. Here $c > \max_{|x|=R} u$. Multiplying (3.1) by $\zeta(u)$ we find, using Green's theorem,

$$\int \zeta(u) g(u(x)) - \int \dot\zeta(u) |\nabla u|^2 = -\int_{|x|=R} \frac{\partial u}{\partial r}\, dS.$$

Since $g(u) \geq 0$ and $\dot\zeta \leq 0$ we conclude that

$$\int_{u<c} g(u(x)) \leq -\int_{|x|=R} \frac{\partial u}{\partial r}\, dS$$

Letting $c \to \infty$ we conclude that

$$\int_{|x|<R} g(u(x)) \leq \int_{|x|=R} \frac{\partial u}{\partial r}\, dS. \tag{3.3}$$

By (3.2) we see that $u \in L_p(D)$. To show that $u(x)$ is a distribution solution, we prove that for $\zeta \in C_0^\infty(D)$,

$$\int_D dx\, (u\,\Delta\zeta + \zeta g(u)) = 0. \tag{3.4}$$

Let $\tau(r)$ be a C^∞ function on $r \geq 0$ vanishing on a neighborhood of the origin and equal to 1 for $r \geq R$. Let $\tau_\varepsilon(r) = \tau(r/\varepsilon)$. It suffices to prove

$$\int_D \tau_\varepsilon(r)(u\,\Delta\zeta + \zeta g(u)) \to 0 \qquad as \quad \varepsilon \downarrow 0. \tag{3.5}$$

By Green's theorem and Eq. (3.1) we have

$$\int_D \tau_\varepsilon(r)(u\,\Delta\zeta + \zeta g(u)) = -\int_{|x|<\varepsilon} \left\{ u\zeta\,\Delta\tau_\varepsilon + 2u\frac{\partial\zeta}{\partial x_i}\frac{\partial\tau_\varepsilon}{\partial x_i} \right\}. \qquad (3.6)$$

Note that $|\Delta\tau_\varepsilon| \le c/\varepsilon^2$ and $|\partial\tau/\partial x_i| \le c/\varepsilon$. Hence

$$\int_{|x|<\varepsilon} |u\zeta\,\Delta\tau_\varepsilon| \le c\left(\int_{|x|<\varepsilon} u^p \right)^{1/p} \varepsilon^{-2}\varepsilon^{n(1-1/p)}$$

$$= o(\varepsilon^{(n-2)-n/p}) \qquad \text{as} \quad \varepsilon \downarrow 0, \quad \text{since} \quad u \in L_p,$$

$$\to 0 \qquad\qquad \text{as} \quad \varepsilon \to 0;$$

similarly for the other term in (3.6). Thus (3.5), and hence (3.4) are proved. ∎

Proof of Theorem 3. Under the assumptions of Theorem 3, we shall prove that the set of positive λ for which property (2.10) holds contains all sufficiently large λ and is open. Using Proposition 3.1, it is easy to see that the integral equation (2.3) holds. Using condition (i) of the theorem, Lemma 2.3' and the proof of Lemma 2.2 for Theorem 1' hold. We now proceed as in the proof of Theorem 1 to show that $\lambda_1 = 0$. If $\lambda_1 > 0$, then either we have (2.16), or else property (2.10) holds. The former cannot occur, because $u(x) \to +\infty$ as $x \to 0$, and the latter cannot occur by Lemma 2.3'. This completes the proof as in Theorem 1. ∎

A suitable adaptation of this proof yields the following generalization.

THEOREM 3'. *Let $u(x)$ be a positive C^2 solution of*

$$\Delta u + g(|x|, u) = 0 \qquad \text{in} \quad R^n \backslash \{0\}$$

satisfying

$$u(x) = O(|x|^{-m}) \quad \text{at } \infty \quad \text{for some} \quad m > 0, \qquad u(x) \to +\infty \quad \text{as} \quad x \to 0.$$

Assume

(i) *$g(r, u)$ is continuous, nonincreasing in r, nondecreasing in u for $r, u \ge 0$, and for some $\alpha > (n+1)/m$, $|g(r, u)| \le C|u|^\alpha$ for u small, and*

(ii) *$\underline{\lim}_{u \to +\infty} g(r, u)u^{-p} \ge c_0 > 0$ for $r < 1$ and some $p \ge n/(n-2)$. Then u is spherically symmetric about the origin and $u_r < 0$ for $r > 0$.*

The argument used in proving Theorem 3 also yields the following result:

THEOREM 5. *Let $a_1, \ldots, a_k \in R^n$ lie on a line, e.g., the x_n axis. Let $u(x)$ be a positive solution of*

$$-\Delta u = g(u) \qquad \text{in} \quad R^n - \{a_1, a_2, \ldots, a_k\}. \qquad (3.7)$$

Assume

$$u \in C^2(R^n - \{a_1, \ldots, a_k\}),$$

$$u(x) = O(1/|x|^m) \quad at \ \infty, \qquad m > 0, \tag{3.8}$$

$$u(x) \to +\infty \qquad as \quad x \to a_j, \quad j = 1, \ldots, k. \tag{3.9}$$

Assume further that conditions (i) *and* (ii) *of Theorem 3 hold. Then* $u(x)$ *is cylindrically symmetric about the* x_n *axis and if* r *denotes distance from the axis,* $u_r < 0$ *for* $r > 0$.

Remark. The result of Theorem 5 can be strengthened for the particular equation (1.7) and for two or three singularities. This equation is invariant under conformal transformations, i.e., if u is a solution and $x \to y$ is a conformal transformation, then the function

$$u(y) = u(x)J^{-(n-2)/2n}(x), \tag{3.10}$$

where $J(x)$ is the Jacobian of the transformation, is also a solution. If $u(x)$ has three isolated singularities, then by a conformal transformation, the three singular points may be brought to lie along a straight line. Then Theorem 5 applies and yields cylindrical symmetry about that line. For two isolated singularities, we have Theorem 4 which we prove next.

Proof of Theorem 4:   Let $e_n = (0, 0, \ldots, 0, 1)$, and

$$y = \frac{x - e_n}{|x - e_n|^2} + e_n, \tag{3.11}$$

$$v(y) = |x - e_n|^{n-2} u|x|. \tag{3.12}$$

Then $v(y)$ satisfies (1.7), outside of the two singularities located at $y = 0$ and $y = e_n$, where v becomes infinite, and v is regular at infinity, i.e., $O(|y|^{2-n})$ there. Since $g(u) = u^{(n+2)/(n-2)}$ is nondecreasing. Theorem 5 applies and shows that $v(y)$ is rotationally symmetric about the y_n axis. Consequently, $u(x)$ is rotationally symmetric about the x_n axis. Since the choice of the axis was arbitrary, we conclude that $u(x)$ is spherically symmetric about the origin. ∎

4.   PRELIMINARY RESULTS FOR EQ. (1.3)

We may set $m = 1$.

We will need the Green's function of $-\Delta + 1$ on R^n. It is given by

$$0 < G(|x - y|) = (|x - y|)^{-(n-2)/2} K_{(n-2)/2}(|x - y|), \tag{4.1}$$

where $K_\nu(z)$ denotes the modified Bessel function of order ν. In Appendix C we summarize the properties of K_ν used in this section. In particular, $G(r)$

satisfies

$$G(r) \le C \frac{e^{-r}}{r^{n-2}} (1 + r)^{(n-3)/2}, \tag{4.2}$$

$$G_r'/G \to -1 \quad \text{at infinity.} \tag{4.3}$$

For $n = 3$,

$$G(r) = \sqrt{\pi/2}(e^{-r}/r),$$

and the reader may carry out the computations below without using any particular properties of the Bessel functions.

First we prove that the solutions in Theorem 2 decay exponentially at ∞.

PROPOSITION 4.1. *Let* $u(x) > 0$ *be a* C^2 *solution of* (1.3), *tending to zero at infinity and assume*

$$g(u) = O(u^\alpha) \quad \text{for some} \quad \alpha > 1, \quad \text{near} \quad u = 0.$$

Then

$$u(x), |\nabla u(x)| = O(e^{-|x|}/|x|^{(n-1)/2}) \quad \text{at} \quad \infty. \tag{4.4}$$

This result follows from the results of Kato [6, Sections 5 and 6] but we include a simple proof here.

Proof. For $r > 0$ set

$$h(r) = \int_{S^{n-1}} u(r\theta) \, d\omega(\theta).$$

Here r represents a polar coordinate and θ a point on the unit sphere; $d\omega$ is element of volume on the sphere. Since $g(u) = O(u^\alpha)$ for $\alpha > 0$, and $u \to 0$ at infinity, we see that for any $\varepsilon > 0$, $|g(u)| < \varepsilon u$ for $|x|$ sufficiently large. Hence integrating (1.3) with respect to θ we find for $r > r_0$

$$h_{rr} + [(n-1)/r]h_r > (1-\varepsilon)h. \tag{4.5}$$

Hence $r^{n-1}h_r$ is increasing. Since $h \to 0$ as $r \to \infty$ we see that $h_r \le 0$ for $r > r_0$.
Multiplying (4.5) by $2h_r$ we find, taking $r > r_0$ from now on,

$$(h_r^2 - (1-\varepsilon)h^2)_r + 2[(n-1)/r]h_r^2 \le 0,$$

and hence $w = h_r^2 - (1-\varepsilon)h^2$ is decreasing. Since $h \to 0$ at ∞ we must have $w \to 0$, for otherwise $w \to c^2 > 0$ which implies $h_r \to -c$—contradicting the fact that $h \to 0$. Thus

$$h_r^2 - (1-\varepsilon)h^2 \ge 0,$$

or

$$h_r + \sqrt{1-\varepsilon}\, h \le 0.$$

Setting $a = \sqrt{1-\varepsilon}$ we have

$$e^{ar}h \text{ is decreasing}$$

and hence

$$h \leq Ce^{-ar}$$

for some constant C. Since ε was arbitrary we conclude that

$$h(r) = O(e^{-ar}), \qquad r \to \infty$$

for any $a < 1$.

It follows that for $f(y) = -u(y) + g(u(y))$,

$$\int_{|x-y|<1} |f(y)|\,dy = O(e^{-a|x|}) \quad \forall a < 1$$

so that

$$\int_{|x-y|<1} (u + |\Delta u|)\,dy = O(e^{-a|x|}). \tag{4.6}$$

By standard interior estimates we find

$$\left(\int_{|x-y|<3/4} u^p\,dy\right)^{1/p} = O(e^{-a|x|}), \qquad p = \frac{n}{n-2},$$

and hence

$$\left(\int_{|x-y|<3/4} |\Delta u|^p\,dy\right)^{1/p} = O(e^{-a|x|}).$$

Using interior estimates again we find in case $q = np/(n-2p) > 0$,

$$\left(\int_{|x-y|<1/2} u^q\,dy\right)^{1/q} \subset O(e^{-a|x|})$$

while if $n - 2p = 0$, a similar estimate holds for any L^s norm of u, $s < \infty$. If $n < 2p$, the estimate holds for the L^∞ norm.

Continuing in this way we finally infer that

$$u(x) = O(e^{-a|x|}), \quad \forall a < 1,$$

and hence

$$g(u) = O(e^{-b|x|}) \qquad \text{for some} \quad b > 1. \tag{4.7}$$

Next we make use of the formula

$$u(x) = \int G(|x-y|)g(u(y))\,dy$$

$$\leq C \int \frac{e^{-|x-y|}}{|x-y|^{n-2}} (1 + |x-y|)^{(n-3)/2} e^{-b|y|}\,dy,$$

by (4.2) and (4.7). Consequently

$$(1 + |x|)^{(n-1)/2} e^{|x|} u(x)$$

$$\leq C \int e^{(1-b)|y|} \frac{(1 + |x|)^{(n-1)/2}}{|x - y|^{n-2}} (1 + |x - y|)^{(n-3)/2} \, dy \qquad (4.8)$$

$$\leq C \int_{|y-x|<|x|/2} + C \int_{|y-x|>|x|/2} \qquad (4.9)$$

$$= I_1 + I_2.$$

Now for $|x|$ large,

$$I_1 = C \int_{|y-x|<|x|/2} \leq C_1 e^{(1-b)|x|/2} |x|^n \leq C_2 \qquad (4.10)$$

for some constants C_1, C_2 independent of x. For $|x|$ large, using $1 + |x| \leq (1 + |x - y|)(1 + |y|)$ we find next

$$I_2 \leq C_1 \int_{R^n} e^{(1-b)|y|} (1 + |x|)^{(n-1)/2} \left(\frac{1 + |y|}{1 + |x|} \right)^{(n-1)/2} \, dy \leq C_2. \qquad (4.11)$$

Combining (4.8)–(4.11) we obtain the desired bound (4.4) for $u(x)$. Using interior estimates as before we obtain the same result for $|\nabla u(x)|$. The proposition is proved. ■

It follows from Proposition 4.1 that $u > 0$ is an exponentially decaying solution of

$$\Delta u - u + f(x) = 0, \qquad f = O(e^{-\alpha|x|}), \qquad \alpha > 1. \qquad (4.12)$$

We will have need of some further asymptotic results near infinity for such a solution. The solution u is given by

$$u(x) = \int G(|x - y|) f(y) \, dy \qquad (4.12')$$

with

$$f(y) = O(e^{-\alpha|y|}), \qquad \alpha > 1. \qquad (4.12'')$$

PROPOSITION 4.2. *Let $u > 0$ be given by (4.12′) with f satisfying (4.12″). Suppose $\{x^i\} \in R^n$ is a sequence going off to infinity in R^n and*

$$\xi = \lim_i \frac{x^i}{|x^i|}, \qquad \xi \equiv (\xi_1, \xi') = (\xi_1, \xi_2, \dots, \xi_n).$$

Then

(a) $$\lim_{i \to \infty} |x^i|^{(n-1)/2} e^{|x^i|} u(x^i) = \sqrt{\frac{\pi}{2}} \int dy \, f(y) e^{\xi \cdot y}. \qquad (4.13)$$

(b)   *Suppose $\xi_1 = 0$ and that λ^i is a sequence of real numbers converging to a number λ and $\lambda^i > x_1^i$. Let z^i be the reflection of x^i in the plane $x_1 = \lambda^i$, i.e., $z^i = (2\lambda^i - x_1^i, x_2^i, \ldots, x_n^i)$. Then*

$$\lim_{i \to \infty} \frac{|x^i|}{\lambda^i - x_1^i} \, |x^i|^{(n-1)/2} e^{|x^i|}(u(x^i) - u(z^i)) = \sqrt{2\pi} \int dy \, e^{\xi' \cdot y'} f(y)(\lambda - y_1). \quad (4.13')$$

(c)   *If $x_1^i \to +\infty$ then*

$$\frac{|x^i|^{(n+1)/2}}{x_1^i} \, e^{|x^i|} u_1(x^i) \to -\sqrt{\frac{\pi}{2}} \int dy \, e^{\xi \cdot y} f(y). \quad (4.14)$$

Proposition 4.2 will be proved in Appendix B.

5.   SOME FORMS OF THE HOPF LEMMA

In this section we prove Lemmas (H_1') and (H_2'). In addition we shall have need of a form of the second in a half-space:

LEMMA (H_2''):   *Let $u > 0$ be of class C^2 in $\{|x| \geq R\} \cap \{x_1 < 0\}$ and continuous in the closure. Assume u vanishes on $x_1 = 0$, tends to zero at infinity, and satisfies a differential inequality:*

$$Lu \equiv \left(\Delta - 1 + x_1 \beta_1 \partial_1 + \sum_2^n b_\alpha \partial_\alpha + c(x) \right) u \leq 0 \quad (5.1)$$

with

$$x_1 \beta_1, b_\alpha, c = O(|x|^{-p}) \quad \text{for some} \quad p > 1, \quad \alpha = 2, \ldots, n.$$

Then for some $\mu > 0$

$$u(x) \geq -\mu x_1 e^{-|x|} / |x|^{(n+1)/2}. \quad (5.2)$$

In our application we will have $\beta_1 = b_2 = \cdots = b_n = 0$. There is a similar half-space version of Lemma (H_1') which may prove useful:

LEMMA (H_1'').   *Let $u > 0$ be of class C^2 in $\{|x| \geq R\} \cap \{x_1 < 0\}$ and continuous in the closure. Assume u vanishes on $x_1 = 0$ and tends to zero at infinity. Assume u satisfies*

$$Lu = \left(\Delta u + x_1 \beta_1 \partial_1 + \sum_2^n b_\alpha \partial_\alpha + c(x) \right) u \leq 0 \quad (5.3)$$

with

$$\beta_1, c = O(|x|^{-p}), \qquad b_\alpha = O(|x|^{1-p}), \qquad \alpha = 2, \ldots, n, \quad p > 2.$$

Then for some $\mu > 0$

$$u(x) \geq -\mu x_1/|x|^n.$$

Proof of Lemma (H$'_1$).   We know that

$$c \leq K|x|^{-p}.$$

Then, since $u \geq 0$,

$$\tilde{L}u \equiv (L - K|x|^{-p})u \leq 0,$$

and the maximum principle holds for $\tilde{L}$—for the coefficient of the zero-order term is nonpositive. We may therefore suppose that our original coefficient c is nonpositive, and we may suppose the same in the other lemmas.

Without loss of generality we may take R as large as we like.

We will use the following comparison function

$$z = (1/r^{n-2}) + (1/r^s), \tag{5.4}$$

where $r = |x|$, with

$$n - 2 < s < n - 4 + p. \tag{5.5}$$

Such a value of s exists since $p > 2$. Using (1.12), a computation yields

$$Lz \geq -Cr^{2-n-p} + s(s - 2 + n)r^{-s-2}.$$

Because of (5.5) we see that the second term dominates, for r large, and thus

$$Lz > 0 \qquad \text{for } R \text{ large.}$$

Let $t > 0$ be so small that

$$u \geq tz \qquad \text{on } |x| = R.$$

Then $u_t = u - tz \geq 0$ on $|x| = R$, vanishes at infinity, and satisfies

$$Lu_t < 0 \qquad \text{in } |x| > R.$$

We may therefore apply the minimum principle and conclude that $u \geq tz$ in $|x| \geq R$. The conclusion of Theorem (H$'_1$) follows—with $\mu = t$.   ∎

Proof of Lemma (H$'_2$).   The proof is the same as the preceding, with a new comparison function:

$$z = G(r)(1 + (1/r^s)), \qquad 0 < s < p - 1; \tag{5.6}$$

here G is the Green's function for $(\Delta - 1)$ given in (4.1). According to (4.1), (4.3), and (C.3) in Appendix C, for r large,

$$G \sim \sqrt{\pi/2}(e^{-r}/r^{(n-1)/2}), \qquad -G_r = G(1 + O(1/r)),$$
$$G_{rr} = G(1 + O(1/r)). \tag{5.7}$$

A direct calculation then shows that

$$Lz \geq -C\frac{G(r)}{r^p} - \frac{2sG_r}{r^{s+1}} + \frac{G}{r^{s+2}} s(s+2-n) = G\left(\frac{2s}{r^{s+1}} - \frac{C}{r^p} - \frac{C}{r^{s+2}}\right)$$

$$> 0 \qquad \text{for } R \text{ sufficiently large,}$$

since $s + 1 < p$. The preceding argument then yields $u \geq tz$ for some $t > 0$—proving the lemma. ∎

Remark.   The function $u = G/r$ satisfies

$$(\Delta - 1 + c(r))u = 0$$

with $c = O(1/r)$, showing that the condition $p > 1$ may not be omitted.

Proof of Lemma (H_2'').   The proof is the same, only now we work in $\Omega_R = \{|x| \geq R\} \cap \{x_1 < 0\}$, R large, and must find a corresponding comparison function. In fact the following will do:

$$w = z_1 = z_{x_1} = -\sqrt{\pi/2}(e^{-r}/r^{(n+1)/2}) x_1[1 + O(1/r)], \qquad (5.8)$$

where z is the function in (5.6). A simple calculation shows that

$$(\Delta - 1)w = \partial_1(\Delta - 1)z = -2sx_1 \frac{e^{-r}}{r^{s+1+(n+1)/2}}\left(1 + O\left(\frac{1}{r}\right)\right).$$

On the other hand

$$|x_1\beta_1\partial_1 w|, |b_\alpha\partial_\alpha w|, |cw| \leq -Cx_1 e^{-r}/r^{(n+1)/2+p}.$$

Since $s + 1 < p$ it follows that

$$Lw > 0 \qquad \text{in } \Omega_R, \quad \text{for } R \text{ large.}$$

Now on $|x| = R$ we have $u > cx_1$ for $c < 0$, by the Hopf lemma. Hence, for $t > 0$ sufficiently small

$$u_t = u - tw > 0 \qquad \text{on } |x| = R, \quad x_1 < 0,$$

and u_t vanishes on $x_1 = 0$ and at infinity. By the maximum principle, recall that we may suppose $c(x) \leq 0$, we find

$$u \geq tw \qquad \text{in } \Omega_R, \quad R \text{ large;}$$

the desired conclusion follows from (5.8). ∎

The proof of Lemma (H_1'') is the same, using the comparison function $w = z_1$ with z given by (5.4).

6. Proof of Theorem 2

Our proof is similar to that of Theorem 1 but somewhat more complicated. By Propositions 4.1 and 4.2 we see first that $u(x), |\nabla u(x)| = O(e^{-|x|})$ and $f(x) = g(u(x)) = O(e^{-\alpha|x|})$. Furthermore if $x \to \infty$, $x/|x| \to \xi$ then

$$|x|^{(n-1)/2}e^{|x|}u(x) \to \sqrt{\frac{\pi}{2}} \int dy\, f(y)e^{\xi \cdot y} \tag{6.1}$$

We have $g(u)/u = c(x) = O(e^{-\varepsilon|x|})$, $\varepsilon = \alpha - 1$. Thus we may apply Lemma (H_2') and conclude that

$$|x|^{(n-1)/2}e^{|x|}u(x) \geq \mu_0 > 0 \quad \text{near infinity.}$$

Combining with (6.1) we see that

$$\sqrt{\frac{\pi}{2}} \int dy\, f(y)e^{\xi \cdot y} \geq \mu_0 \qquad \text{for} \quad |\xi| = 1. \tag{6.2}$$

LEMMA 6.1. *Under the assumptions of Theorem 2 there exists $\bar{\lambda} > 0$ such that*

$$u_1(x) < 0 \qquad \text{for} \quad x_1 \geq \bar{\lambda}.$$

Proof. Suppose not. Then there is a sequence x^i with $x_1^i \to +\infty$ such that

$$u_1(x^i) \geq 0.$$

Applying Proposition 4.2c we find

$$-\sqrt{\frac{\pi}{2}} \int dy\, e^{\xi \cdot y} f(y) \geq 0$$

contradicting (6.2). ■

LEMMA 6.2. *Under the assumptions of Theorem 2, there exists $\lambda_0 \geq \bar{\lambda}$ such that for all $\lambda \geq \lambda_0$,*

$$u(x) > u(x^\lambda) \qquad \text{for} \quad x_1 < \lambda. \tag{6.3}$$

Proof. Suppose not. Then there exists a sequence $\lambda^i \to +\infty$ and a sequence $\{x^i\}$ with $x_1^i < \lambda^i$ such that for $z^i = x^{i\lambda^i}$

$$u(x^i) \leq u(z^i). \tag{6.4}$$

Since $|x^{i\lambda^i}| \to \infty$, so that $u(x^{i\lambda^i}) \to 0$, we see that $|x^i| \to \infty$. Furthermore, we have necessarily $x_1^i < \bar{\lambda}$, for if $x_1^i \geq \bar{\lambda}$, then

$$u(z^i) - u(x^i) = \int_{x^i}^{z^i} u_1 < 0$$

by Lemma 6.1—contradicting (6.4).

By restricting ourselves to a suitable subsequence we may suppose

$$x^i/|x^i| \to \xi, \quad \text{and then} \quad \xi_1 \leq 0.$$

Case 1. $\xi_1 < 0$. In this case if $d_i = |z^i| - |x^i|$ we see, omitting i,

$$d + \sqrt{|x'|^2 + x_1^2} = [(2\lambda - x_1)^2 + |x'|^2]^{1/2},$$

so that, on squaring,

$$d^2 + 2d|x_1| = 4\lambda^2 - 4\lambda x_1 \geq -4\lambda x_1.$$

Since $x_1 \sim \xi_1|x|$, and $\lambda \to +\infty$ we see that $d \to \infty$, i.e., $d_i \to \infty$.

We now apply Proposition 4.2a and infer that

$$\lim|x^i|^{(n-1)/2}e^{|x^i|}u(x^i) = \sqrt{\frac{\pi}{2}} \int f(y)e^{\xi \cdot y}$$

$$> 0 \qquad \text{by (6.2).}$$

On the other hand,

$$\lim|x^i|^{(n-1)/2}e^{|x^i|}u(z^i) = \lim(|x^i|/|z^i|)^{(n-1)/2}e^{|x^i|-|z^i|} \cdot |z^i|^{(n-1)/2}e^{|z^i|}u(z^i).$$

Since $|x^i| - |z^i| = -d_i \to -\infty$ we see again from Proposition 4.2a that this tends to zero. We infer that for Case 1, (6.4) is impossible.

Case 2. $\xi_1 = 0$. By (6.4) and Lemma 6.1 we may assert that for any fixed $\lambda \geq \bar{\lambda}$,

$$u(x^i) \leq u(z^i) \leq u(x^{i\lambda}) \qquad \text{if} \quad \lambda^i \geq \lambda. \tag{6.5}$$

We now apply Proposition 4.2b, with λ in place of the λ^i in the proposition, to obtain (here $x^{i\lambda} = \zeta^i$)

$$0 \geq \lim \frac{|x^i|}{\lambda - x_1^i} |x^i|^{(n-1)/2}e^{|x^i|}(u(x^i) - u(\zeta^i)) = \sqrt{2\pi} \int dy\, e^{\xi \cdot y} f(y)(\lambda - y_1).$$

But according to (6.2) this is impossible for large λ. ∎

LEMMA 6.3. *Under the conditions of Theorem 2, the set of λ for which property (6.3) holds is open.*

Proof. As in the proof of Lemma 2.3, we note that if property (6.3) holds then

$$u_{x_1} < 0 \qquad \text{on} \quad x_1 = \lambda. \tag{6.6}$$

Here we use the fact that $g(u) = g_1(u) + g_2(u)$ with $g_1 \in C^1$ and g_2 nondecreasing. We will show that if property (6.3) holds for some λ it holds for any $\tilde{\lambda}$ near λ. Suppose the contrary. Then there exist sequences $\{x^i\}$, $\lambda^i \to \lambda$, $x_1^i < \lambda^i$ such that (here $z^i = x^{i\lambda^i}$)

$$u(x^i) \leq u(z^i).$$

Clearly $|x^i| \to \infty$ and we may choose a suitable subsequence so that

$$\frac{x^i}{|x^i|} \to \xi = (\xi_1, \xi'), \qquad \xi_1 \leq 0.$$

Then

$$|x^i|^{(n-1)/2} e^{|x^i|} u(x^i) \leq \left||x^i|/|z^i|\right|^{(n-1)/2} e^{|x^i| - |z^i|} |z^i|^{(n-1)/2} e^{|z^i|} u(z^i).$$

Using the identity (2.21) we see that $|x^i| - |z^i| \to 2\lambda\xi_1$, and using Proposition 4.2a we find

$$\int f(y) e^{\xi \cdot y} \leq e^{2\lambda\xi_1} \int f(y) e^{\xi^0 \cdot y}. \tag{6.7}$$

Now we are going to use Lemma (H_2''). In the half-space $x_1 < \lambda$ set $v(x) = u(x^\lambda)$ and

$$w(x) = u(x) - v(x) > 0 \qquad \text{by (6.3)}.$$

Then

$$(\Delta - 1)w + g(u) - g(v) = 0,$$

or

$$(\Delta - 1)w + g_1(u) - g_1(v) = g_2(v) - g_2(u) \leq 0,$$

since g_2 is increasing. Using the hypotheses on g_1 we see that

$$(\Delta - 1 + c(x))w \leq 0$$

with $c(x) = O(|x|^{-p})$ near infinity, $p > 1$. In the half-space $x_1 < \lambda$, w satisfies all the conditions of Lemma (H_2'') and we may assert that for some $\mu_0 > 0$

$$\frac{|x|^{(n+1)/2} e^{|x|}}{\lambda - x_1} w(x) \geq \mu_0 > 0. \tag{6.8}$$

Suppose now $\xi_1 < 0$. If $z^i \to \infty$ is a sequence in the half-space $x_1 < \lambda$ with $z^i/|z^i| \to \xi$ then

$$\frac{|z^i|^{(n+1)/2} e^{|z^i|}}{\lambda - z_1^i} (u(z^i) - u(z^{i\lambda})) \geq \mu_0.$$

By Proposition 4.2a

$$-\frac{1}{\xi_1} \int f(y)(e^{\xi \cdot y} - e^{2\lambda\xi_1 + \xi^0 \cdot y}) \geq \mu_0$$

which contradicts (6.7).

Suppose, then, that $\xi_1 = 0$. According to Proposition 4.2b,

$$\int e^{\xi \cdot y} f(y)(\lambda - y_1) \leq 0. \tag{6.7'}$$

On the other hand if in the half-space $x_1 < \lambda$ we let $x \to \infty$ with $x/|x| \to \xi$, then, according to (6.8),

$$\lim \frac{|x|^{(n+1)/2} e^{|x|}}{\lambda - x_1} (u(x) - u(x^\lambda)) \geq \mu_0.$$

Applying Proposition 4.2b once more, with all $\lambda^i = \lambda$ we find

$$\sqrt{2\pi} \int e^{\xi \cdot y} f(y)(\lambda - y_1) \geq \mu_0$$

contradicting (6.7'). Lemma 6.3 is proved. ■

   Completion of the proof of Theorem 2.   It follows from Lemmas 6.2 and 6.3 that there is a maximal interval

$$\lambda_1 < \lambda < \infty$$

for which (6.3) holds; clearly λ_1 is finite. By [4, Lemma 4.3], and Remark 1 of Section 2.3 there, we also have

$$u_{x_1} < 0 \qquad \text{for} \quad x_1 > \lambda_1.$$

By continuity we have

$$u(x) \geq u(x^{\lambda_1}) \qquad \text{for} \quad x_1 < \lambda_1$$

and by the same reference in [4] we have either $\equiv$ or $>$. In view of Lemma 6.3 it cannot be the latter for then the interval would not be maximal. Thus we conclude that

u is symmetric about the plane $x_1 = \lambda_1$ and $u_{x_1} < 0$ for $x_1 > \lambda_1$.

   The same conclusion holds for the other coordinate directions and we conclude that u is symmetric about each of n planes, $x_j = \lambda_j$ and grad $u = 0$ only at their intersection. We may now take their intersection as the origin.

   The same argument may be applied to any unit direction γ and we infer that u is symmetric about some plane $x \cdot \gamma = c(\gamma) = $ const. At the point on this plane where u achieves the maximum we have grad $u = 0$ (since the normal derivative to the plane is zero at every point of the plane). It follows that $c(\gamma) = 0$. Thus u is symmetric about every plane through the origin. In addition we also conclude that $u_r < 0$. The last inequalities in Theorem 2 follow from (6.1) and (6.2). Theorem 2 is proved. ∎

APPENDIX A

We shall often make use of the identity (2.21):

$$|x^\lambda - y| - |x - y| = \frac{4(\lambda - x_1)(\lambda - y_1)}{|x^\lambda - y| + |x - y|}. \tag{A.1}$$

Proof of Lemma 2.1. Observe first that Lemma 2.1 holds in case f has compact support. Indeed in this case (2.5) and (2.6) are immediate, while (2.7) follows from

$$\frac{|x^i|^n}{\lambda^i - x_1^i} \left(\frac{1}{|x^i - y|^{n-2}} - \frac{1}{|x^i - y^{\lambda^i}|^{n-2}} \right) \to 2(n-2)(\lambda - y_1).$$

This in turn is readily verified with the aid of (A.1).

To prove the lemma for any f satisfying (2.4) we observe that f is in the Banach space $B_{q'}$ for $q' < q$, defined by the completion of C_0^∞ under the norm $\| \ \|_{q'}$:

$$\|h\|_{q'} = \sup_y \{(1 + |y|)^{q'} |h(y)|\}.$$

Let us now fix q' in $n + 1 < q' < q$.

Note first that the right-hand sides of (2.5), (2.6), and (2.7) are bounded linear functionals on $B_{q'}$. To prove the lemma we have only to show that the following functionals

$$j_x(f) = |x|^{n-2} u(x),$$

$$k_x(f) = \frac{|x|^n}{x_1} u_1(x), \qquad x_1 > 1,$$

$$l_i(f) = \frac{|x^i|^n}{\lambda^i - x_1^i} (u(x^i) - u(x^{i\lambda^i}))$$

are all uniformly bounded linear functionals on $B_{q'}$. A standard argument then yields the result.

For example let us prove (2.7) using the uniform boundedness of the l_i. Let $l(f)$ denote the linear functional defined on the right-hand side of (2.7). Suppose $f_j \in C_0^\infty$ converges in $B_{q'}$ to f. Then

$$l_i(f) - l(f) = l_i(f - f_j) + (l_i - l)f_j + l(f_j - f),$$

so that

$$\left|l_i(f) - l(f)\right| \le C\|f - f_j\|_{q'} + \left|(l_i - l)f_j\right| + C\|f_j - f\|_{q'}.$$

Given $\varepsilon > 0$, choose j so large that $C\|f - f_j\|_{q'} < \varepsilon/3$ and then choose i so large that $|(l_i - l)f_j| < \varepsilon/3$. Hence $|(l_i - l)f| < \varepsilon$ and the desired result follows. C will denote various constants.

Consider first

$$\left|j_x(f)\right| = c\left|\int \frac{|x|^{n-2}}{|x - y|^{n-2}} f(y)\right|$$

$$\le c\|f\|_{q'} \int \frac{|x|^{n-2}}{|x - y|^{n-2}} \frac{1}{(1 + |y|)^{q'}}.$$

Clearly

$$\int_{|y-x| < |x|/2} \le C$$

while

$$\int_{|y-x| > |x|/2} \le C \int \frac{1}{(1 + |y|)^{q'}} \le C.$$

It follows that

$$\left|j_x(f)\right| \le C\|f\|_{q'}.$$

Next, for $x_1 > 1$,

$$\left|k_x(f)\right| \le C\|f\|_{q'} \int \frac{|x_1 - y_1|}{x_1} \frac{|x|^n}{|x - y|^n} \frac{1}{(1 + |y|)^{q'}}. \tag{A.2}$$

Now

$$\int_{|x-y| < |x|/2} \le C$$

and

$$\int_{|x-y| > |x|/2} \le C \int \frac{1}{(1 + |y|)^{q'-1}} \le C,$$

since $|x_1 - y_1|/x_1 \le 1 + |y_1|$. The desired uniform boundedness of k_x follows.

To bound the functionals l_i is more tedious. We shall use the inequalities: For $0 < r \le s$

$$(n - 2)\frac{1}{s^{n-1}}(s - r) \le \frac{1}{r^{n-2}} - \frac{1}{s^{n-2}} \le (n - 2)\frac{1}{r^{n-1}}(s - r). \tag{A.3}$$

Let

$$K = K(x; y; \lambda) = \frac{|x|^n}{\lambda - x_1} \left| \frac{1}{|x - y|^{n-2}} - \frac{1}{|x^\lambda - y|^{n-2}} \right|,$$

then

$$K \le C_1 |x|^n \frac{|\lambda - y_1|}{|x - y| + |x^\lambda - y|} \max\{|x - y|^{1-n}, |x^\lambda - y|^{1-n}\}. \quad \text{(A.4)}$$

Indeed, for $y_1 < \lambda$ we have [using (A.3)]

$$K \le (n - 2) \frac{|x|^n}{\lambda - x_1} \frac{1}{|x - y|^{n-1}} \frac{4(\lambda - x_1)(\lambda - y_1)}{|x^\lambda - y| + |x - y|},$$

similarly, for $y_1 > \lambda$, and (A.4) is proved. Therefore

$$|l_i(f)| \le C_1 |x^i|^n \int dy \, f(y) \frac{|\lambda^i - y_1|}{|x^i - y| + |x^{i\lambda^i} - y|} \max\{|x^i - y|^{1-n}, |x^{i\lambda^i} - y|^{1-n}\}.$$

We wish to prove

$$|l_i(f)| \le C\|f\|_{q'} \quad \text{with} \quad C \text{ independent of } i. \quad \text{(A.5)}$$

Let

$$A_1 = c|x|^n \int_{|y - x| < |x|/2, \, y_1 < \lambda} dy \, f(y) \frac{\lambda - y_1}{|x - y| + |x^\lambda - y|} |x - y|^{1-n},$$

$$A_2 = c|x|^n \int_{|y - x^\lambda| < |x^\lambda|/2, \, y_1 > \lambda} dy \, f(y) \frac{y_1 - \lambda}{|x - y| + |x^\lambda - y|} |x^\lambda - y|^{1-n},$$

$$A_3 = c|x|^n \int_{|y - x| > |x|/2, \, y_1 < \lambda} dy \, f(y) \frac{\lambda - y_1}{|x - y| + |x^\lambda - y|} |x - y|^{1-n},$$

$$A_4 = c|x|^n \int_{|y - x^\lambda| > |x^\lambda|/2, \, y_1 > \lambda} dy \, f(y) \frac{y_1 - \lambda}{|x - y| + |x^\lambda - y|} |x^\lambda - y|^{1-n}.$$

To prove (A.5) it suffices to prove with some constant C independent of x but which may depend on λ:

$$\sum A_j \le C\|f\|_{q'}. \quad \text{(A.6)}$$

With various constants C (which may depend on λ) we have

$$\begin{aligned}
A_1 &\le c|x|^n \|f\|_{q'} \int_{|y - x| < |x|/2, \, y_1 < \lambda} dy \, |x - y|^{1-n} (1 + |y|)^{-q'} \\
&\le C\|f\|_{q'} |x|^{n+1} (1 + |x|)^{-q'} \\
&\le C\|f\|_{q'} \quad \text{since} \quad q' > n + 1.
\end{aligned}$$

A similar estimate holds for A_2. Next

$$A_3 \leq C\|f\|_{q'}|x|^n \int_{|y-x|>|x|/2,\, y_1<\lambda} dy\, |x-y|^{-n}(1+|y|)^{1-q'}$$

$$\leq C\|f\|_{q'} \int_{R^n} dy\, (1+|y|)^{1-q'}$$

$$\leq C\|f\|_{q'} \quad \text{since} \quad q' > n+1.$$

The same argument applies to A_4. Combining these estimates we obtain (A.6) and hence (A.5).

Lemma 2.1 is proved. ∎

APPENDIX B.   THE PROOF OF PROPOSITION 4.2

We will make use of the following consequence of (C.5) in Appendix C:

$$G(r) - G(s) = -\int_r^s d\xi\, G'(\xi) = \int_r^s d\xi\, \frac{K_{n/2}(\xi)}{\xi^{(n-2)/2}}.$$

By (C.1), we have for $r \leq s$

$$\frac{K_{n/2}(s)}{s^{(n-2)/2}} (s-r) \leq G(r) - G(s) \leq \frac{K_{n/2}(r)}{r^{(n-2)/2}} (s-r). \tag{B.1}$$

First we prove (4.13), (4.13′), and (4.14) for functions f with compact support. For y in a compact set K, if $x \to \infty$ with $x/|x| \to \xi$, we see from (C.3) that

$$\lim |x|^{(n-1)/2} e^{|x|} G(|x-y|) = \sqrt{\frac{\pi}{2}} \lim \left(\frac{|x|}{|x-y|}\right)^{(n-1)/2} e^{|x|-|x-y|}$$

$$= \sqrt{\frac{\pi}{2}} e^{\xi \cdot y}, \tag{B.2}$$

since

$$|x| - |x-y| = y \cdot \frac{x}{|x|} + O(|x|^{-1}) \qquad \text{uniformly for } y \text{ in } K.$$

Using (B.2) we see that (4.13) holds for continuous functions f with compact support.

To verify (4.13′) for such functions we will use the formula $[\text{see } (C.2)]$

$$G(z) = \sqrt{\pi/2}\, e^{-|z|} H_{(n-2)/2}(|z|),$$

where

$$H_{(n-2)/2}(|z|) = |z|^{(1-n)/2} \frac{1}{\Gamma[(n-1)/2]} \int_0^\infty e^{-t} t^{(n-3)/2} \left(1+\frac{t}{2|z|}\right)^{(n-3)/2} dt. \tag{B.3}$$

We wish to prove that for y in a compact set K if $x_i \to \infty$ with $x_i/|x_i| \to \xi$ and $\xi_1 = 0$, and $x_1^i < \lambda^i$, $\lambda^i \to \lambda$, then

$$G_i = \frac{|x^i|^{(n+1)/2}}{\lambda^i - x_1^i} e^{|x^i|}(G(|x^i - y|) - G(|x^i - y^{\lambda^i}|)) \to 2\sqrt{\frac{\pi}{2}}(\lambda - y_1)e^{\xi' \cdot y'} \quad \text{(B.4)}$$

uniformly for y in K. We have

$$\sqrt{\frac{2}{\pi}} G_i = \frac{|x^i|^{(n+1)/2}}{\lambda^i - x_1^i} e^{|x^i|}e^{-|x^i - y|}(1 - e^{|x^i - y| - |x^i - y^{\lambda^i}|})H_{(n-2)/2}(|x^i - y^{\lambda^i}|)$$

$$\quad\quad\quad (\text{B.5})$$

$$+ \frac{|x^i|^{(n+1)/2}}{\lambda^i - x_1^i} e^{|x^i|}e^{-|x^i - y|}|H_{(n-2)/2}(|x^i - y|) - H_{(n-2)/2}(|x^i - y^{\lambda^i}|)|.$$

Using (A.1) we see that

$$\frac{|x^i|}{\lambda^i - x_1^i}(1 - e^{|x^i - y| - |x^i - y^{\lambda^i}|}) = \frac{|x^i|}{\lambda^i - x_1^i}\left(1 - \exp\frac{4(\lambda^i - x_1^i)(y_1 - \lambda^i)}{|x^i - y| + |x^i - y^{\lambda^i}|}\right)$$

$$\to 2(\lambda - y_1),$$

because $x_1^i/|x^i| \to 0$. Since $|x^i|/|x^{i^{\lambda^i}}| \to 1$ and $|z|^{(n-1)/2}H_{(n-2)/2}(|z - y|) \to 1$ as $z \to \infty$, the first product in the right-hand side of (B.5) tends to $2(\lambda - y_1)e^{\xi \cdot y}$.

To conclude the proof of (B.4) we must show that the other term in (B.5) tends to zero. This follows rather easily from the implicit formula (B.3) with the aid of (A.1). Namely, using (A.1) one shows first that

$$\frac{|x^i|^{(n+1)/2}}{\lambda^i - x_1^i}\left|\frac{1}{|x^i - y|^{(n-1)/2}} - \frac{1}{|x^i - y^{\lambda^i}|^{(n-1)/2}}\right| \le C\frac{|\lambda^i - y_1|}{|x^i - y| + |x^i - y^{\lambda^i}|} \to 0.$$

Next one establishes, with the aid of (A.1) that

$$\int_0^\infty e^{-t}t^{(n-3)/2} \frac{|x^i|}{\lambda^i - x_1^i}\left[\left(1 + \frac{t}{2|x^i - y|}\right)^{(n-3)/2} - \left(1 + \frac{t}{2|x^i - y^{\lambda^i}|}\right)^{(n-3)/2}\right]dt$$

$$\to 0 \quad \text{as} \quad i \to \infty.$$

These yield the desired result and (B.4) is proved.

To prove (4.14) for functions with compact support we must prove that for y in a compact set K if $x \to \infty$ with $x_1 \to +\infty$, $x/|x| \to \xi$, then uniformly in K,

$$J(x, y) = \frac{|x^i|^{(n+1)/2}}{x_1} e^{|x|}G'(|x - y|)\frac{x_1 - y_1}{|x - y|} \to -\sqrt{\frac{\pi}{2}} e^{\xi \cdot y}.$$

Now

$$G'(r) = \sqrt{\frac{\pi}{2}}\frac{e^{-r}}{r^{(n-1)/2}}\left(-1 + O\left(\frac{1}{r}\right)\right),$$

and so

$$J(x, y) = -\sqrt{\frac{\pi}{2}} \frac{|x|^{(n+1)/2}}{|x-y|^{(n+1)/2}} e^{|x|-|x-y|} \frac{x_1 - y_1}{x_1} \left(1 + O\left(\frac{1}{|x-y|}\right)\right),$$

and one obtains the desired limit immediately since, as we noted earlier, $|x| - |x - y| \to \xi \cdot y$.

We have thus proved (4.13), (4.13′), and (4.14) for continuous functions f with compact support. To establish these for our function $f(y) = O(e^{-\alpha|y|})$ we define for each γ in $1 < \gamma < \alpha$ the Banach space B_γ with norm

$$\|h\|_\gamma = \sup\{e^{\gamma|y|}|h(y)|\}.$$

Now choose a sequence of functions f_j with compact support so that $\|f - f_j\|_{\alpha'} \to 0$ as $j \to \infty$, where $1 < \alpha' < \alpha$. As in Lemma 2.1, the proofs will be completed if we prove that the functionals involved in these equations are bounded on $B_{\alpha'}$. We consider first (4.13′) [Eq. (4.13) is simpler and can be treated in a similar way]. Let

$$l_i(f) = \int dy\, f(y) J(x^i; y; \lambda^i), \tag{B.6}$$

where

$$J = J(x; y; \lambda) = [|x|/(\lambda - x_1)](|x|)^{(n-1)/2} e^{|x|} \{G(|x - y|) - G(|x - y^\lambda|)\}.$$

For $y_1 < \lambda$, we use (B.1) to bound

$$|J| \le \frac{|x|}{\lambda - x_1} (|x|)^{(n-1)/2} e^{|x|} \frac{K_{n/2}(|x - y|)}{|x - y|^{(n-2)/2}} \frac{4(\lambda - x_1)(\lambda - y_1)}{|x - y| + |x - y^\lambda|}$$

$$\le C|x|^{(n+1)/2} e^{|y|} \frac{\lambda - y_1}{|x - y| + |x - y^\lambda|} \frac{(1 + |x - y|)^{(n-1)/2}}{|x - y|^{n-1}} = J_1$$

by (C.4). Similarly, for $y_1 > \lambda$,

$$|J| \le c|x|^{(n+1)/2} e^{|y|} \frac{y_1 - \lambda}{|x - y| + |x - y^\lambda|} \frac{(1 + |x - y^\lambda|)^{(n-1)/2}}{|x - y|^{n-1}} = J_2.$$

Setting

$$A_1 = \int_{y_1 < \lambda,\, |y-x| < |x|/2} J_1 f, \qquad A_2 = \int_{y_1 > \lambda,\, |y-x^\lambda| < |x^\lambda|/2} J_2 f$$

$$A_3 = \int_{y_1 < \lambda,\, |y-x| > |x|/2} J_1 f, \qquad A_4 = \int_{y_1 > \lambda,\, |y-x^\lambda| > |x^\lambda|/2} J_2 f,$$

we wish to show

$$\sum_1^4 A_i \le C\|f\|_{\alpha'}.$$

with C independent of x, but it may depend on λ. This will yield the desired result

$$|l_i(f)| \leq C\|f\|_{\alpha'}$$

and will then complete the proof of Proposition 4.2b.

We have

$$A_1 \leq C\|f\|_{\alpha'}|x|^{(n+1)/2} \int_{|y-x|<|x|/2} e^{(1-\alpha')|y|} \frac{(1+|x-y|)^{(n-1)/2}}{|x-y|^{n-1}} dy$$

$$\leq C\|f\|_{\alpha'}|x|^{(n+1)/2} e^{(1-\alpha')|x|/2}|x|^{(n+1)/2} \leq C\|f\|_{\alpha'}.$$

A similar estimate holds for A_2. Next

$$A_3 \leq C\|f\|_{\alpha'}|x|^{(n+1)/2} \int_{|y-x|>|x|/2} \frac{(|\lambda|+|y|)e^{(1-\alpha')|y|}}{|x-y|^{(n+1)/2}} dy$$

$$\leq C\|f\|_{\alpha'} \int (|\lambda|+|y|)e^{(1-\alpha')|y|} dy \leq C\|f\|_{\alpha'}$$

with a similar result for A_4. Proposition 4.2b is proved. As we remarked, the completion of the proof of Proposition 4.2a is simpler.

To complete the proof of Proposition 4.2c we show that for $x_1 > 1$ the linear functionals

$$k_x(f) = \frac{|x|^{(n+1)/2}}{x_1} e^{|x|} u_1(x)$$

are uniformly bounded in $B_{\alpha'}$. This follows from the inequality

$$\int \frac{|x|^{(n+1)/2}}{x_1} e^{|x|}|G_r(|x-y|)| \frac{|x_1-y_1|}{|x-y|} e^{-\alpha'|y|} dy \leq C$$

for $x_1 > 1$. From (C.5) we have

$$G_r(x) = -K_{n/2}/r^{(n-2)/2},$$

so we want to prove the uniform boundedness in $x_1 > 1$ of

$$\int \frac{|x|^{(n+1)/2}}{|x-y|^{n/2}} |K_{n/2}(|x-y|)| \left|1 - \frac{y_1}{x_1}\right| e^{|x|-\alpha'|y|} dy.$$

By the estimate in (C.4), it suffices to prove

$$J = \int \frac{|x|^{(n+1)/2}}{|x-y|^n} (1+|x-y|)^{(n-1)/2} \left|1 - \frac{y_1}{x_1}\right| e^{|x|-|x-y|-\alpha'|y|} dy \leq C.$$

Since $|x| - |x-y| \leq |y|$ and $1 + |x-y| \leq (1+|x|)(1+|y|)$ we find

$$J \leq C \int \frac{|x|^n}{|x-y|^n} \frac{|x_1-y_1|}{x_1} (1+|y|)^{(n-1)/2} e^{(1-\alpha')|y|} dy.$$

The last integral is bounded by some constant times the integral occurring in (A.2), which we proved was bounded.

The proof is complete. ∎

APPENDIX C

In this appendix we list the properties of the modified Bessel functions K_ν [7] we used in Sections 4 and 6. Let $\nu \geq 0$ be half an integer and $z > 0$. Then

$$K_\nu(z) \text{ is positive and decreasing in } z. \tag{C.1}$$

The most convenient integral representation for our purpose is

$$K_\nu(z) = \frac{1}{\Gamma(\nu + \frac{1}{2})} \sqrt{\frac{\pi}{2}} \frac{e^{-z}}{\sqrt{z}} \int_0^{+\infty} e^{-t} t^{\nu - 1/2} \left(1 + \frac{1}{2} \frac{t}{z}\right)^{\nu - \frac{1}{2}} dt \tag{C.2}$$

where Γ is the gamma function. In case $\nu - \frac{1}{2} = k$ is an integer this reduces to

$$K_\nu = \sqrt{\frac{\pi}{2}} \frac{e^{-z}}{z^{1/2}} \sum_{j=0}^{k} \frac{\Gamma(k + 1 + j)}{\Gamma(k + 1 - j)} (2z)^{-j}.$$

Asymptotic expansion for large z if $\nu = $ integer:

$$K_\nu(z) = \sqrt{\frac{\pi}{2}} \frac{e^{-z}}{\sqrt{z}} \left\{ \sum_{j=0}^{M-1} \frac{\Gamma(\nu + \frac{1}{2} + j)}{j! \Gamma(\nu + \frac{1}{2} - j)} \frac{1}{(2z)^j} + O\left(\frac{1}{|z|^M}\right) \right\}. \tag{C.3}$$

The remainder after M terms does not exceed the $(M + 1)$-term in absolute value, and is of the same sign provided that $M \geq \nu - \frac{1}{2}$.

As $z \to 0$

$$K_\nu(z) \sim \begin{cases} \frac{1}{2} \Gamma(\nu)(\frac{1}{2}z)^{-\nu}, & \nu > 0, \\ -\log z, & \nu = 0. \end{cases} \tag{C.4}$$

Thus for $\nu > 0$, $z \geq 0$, $K_\nu(z) \leq C(e^{-z}/z^\nu)(1 + z)^{\nu - \frac{1}{2}}$.

$$K_\nu'(z) = -K_{\nu-1}(z) - \frac{\nu}{z} K_\nu(z) = \frac{\nu}{z} K_\nu(z) - K_{\nu+1}(z) \tag{C.5}$$

REFERENCES

1. J. CERVERNO, L. JACOBS, AND C. NOHL, Elliptic solutions of classical Yang–Mills theory, *Phys. Lett. B* **69** (1977), 351–354.
2. S. CHANDRASEKHAR, "An Introduction to the Study of Stellar Structure," Dover, New York, 1957.

3. B. GIDAS, Symmetry properties and isolated singularities of positive solutions of nonlinear elliptic equation, *in* "Nonlinear Differential Equations in Engineering and Applied Sciences" (R. L. Sternberg, ed.), pp. 255–273 Marcel Dekker, New York, 1979.

4. B. GIDAS, W-M. NI, AND L. NIRENBERG, Symmetry and related properties via the maximum principle, *Comm. Math. Phys.* **68** (1979), 209–243.

5. D. D. JOSEPH, AND T. S. LUNDGREN, Quasilinear Dirichlet problems driven by positive sources, *Arch. Rational Mech. Anal.* **49** (1973), 241–269.

6. T. KATO, Growth properties of solutions of the reduced wave equation with a variable coefficient, *Comm. Pure Appl. Math.* **12** (1959), 403–425.

7. W. MAGNUS, AND F. OBERHETTINGER, "Formulas and Theorems for the Special Functions of Mathematical Physics," Springer-Verlag, Berlin and New York, 1966.

8. Z. NEHARI, On a nonlinear differential equation arising in nuclear physics, *Proc. Roy. Irish Acad. Sect. A* **62** (1963), 117–135.

Added in proof: We have just learned that Lemma (H_1') is almost the same as the corollary on page 528 of N. Meyers and J. Serrin, The exterior Dirichlet problem for second-order elliptic partial differential equations, *J. Math. and Mech.* **9** (1960), 513–538.

MATHEMATICAL ANALYSIS AND APPLICATIONS, PART A
ADVANCES IN MATHEMATICS SUPPLEMENTARY STUDIES, VOL. 7A

A Remark on the Representation Theorems of
Frederick Riesz and Laurent Schwartz

J. Horváth

Department of Mathematics
University of Maryland
College Park, Maryland

DEDICATED TO LAURENT SCHWARTZ ON THE OCCASION OF HIS 65TH BIRTHDAY

INTRODUCTION

In 1980 we celebrate not only the 65th birthday of Laurent Schwartz but also the hundredth anniversary of the birth of Frederick Riesz. It seems therefore appropriate to make an elementary remark, which claims that at a certain point the works of these two mathematicians meet.

A classical theorem of Frederick Riesz gives an analytic representation of continuous linear forms on the space of continuous functions. It is often said that by defining measures as continuous linear forms, the theorem of Frederick Riesz has become a definition. This is not so, at least if we consider the original form of the theorem, which establishes a correspondence not between measures and continuous linear forms but between functions of bounded variation and continuous linear forms. If we change the setting of the theorem slightly, replacing a closed interval by an open one, and if we apply the integration by parts formula, then Riesz's original theorem actually states that every Radon measure is the derivative in the sense of distributions of a function, which is of bounded variation on every compact subinterval, and is in particular locally integrable: this is a version of the representation theorem of Laurent Schwartz. By imitating the proof of the one-dimensional case, one obtains a proof for this result also in the case of several dimensions.

In the first section of the paper I present the theorem of Frederick Riesz for an open interval of the real line in the form in which it is equivalent to the representation theorem of Laurent Schwartz. While most of this is, of course, well known, it may be useful to give a coherent account of the facts which will motivate in the second section the higher-dimensional case.

1. The One-Dimensional Case

Let I be a nonempty open interval $]a, b[$ of the real line $\mathbf{R}$, which may be finite or infinite, i.e., $-\infty \le a < b \le \infty$. For the sake of simplicity, we shall only consider real-valued functions and real distributions in this paper. We denote by $\mathscr{V}(I)$ the vector space of all functions $f : I \to \mathbf{R}$ whose total variation

$$q_{\alpha\beta}(f) = \sup_{\Delta} \sum_{j=1}^{l} |f(x_j) - f(x_{j-1})|$$

is finite on every compact subinterval $[\alpha, \beta]$ of I; here the least upper bound is taken with respect to all subdivisions

$$\Delta : \alpha = x_0 < x_1 < \cdots < x_l = \beta \tag{1}$$

of $[\alpha, \beta]$. Each $q_{\alpha\beta}$ is a seminorm; we consider $\mathscr{V}(I)$ equipped with the family $(q_{\alpha\beta})$ of seminorms, as $[\alpha, \beta]$ varies in I, and with the locally convex topology defined by it. There is also a natural linear preorder on $\mathscr{V}(I)$ for which the cone P of positive elements consists of all functions increasing on I.

The closure $\{\bar{0}\}$ of $\{0\}$ in the non-Hausdorff space $\mathscr{V}(I)$ consists of the constants, and the subspace $\{\bar{0}\}$ is precisely also $P \cap (-P)$. The associated Hausdorff space $\mathscr{V}_0(I) = \mathscr{V}(I)/\{\bar{0}\}$ is therefore an ordered vector space. Actually $\mathscr{V}_0(I)$ is a directed space since any function in $\mathscr{V}(I)$ is the difference of two increasing functions [5, Proposition 5.1.4], and it is easy to check that $\mathscr{V}_0(I)$ is even an order-complete Riesz-space (vector lattice) [5, Example 5.2.3].

Each element $f \in \mathscr{V}(I)$ is a locally integrable function, hence we can associate with f the Radon measure T_f which has density f with respect to Lebesgue measure, i.e., which is given by

$$\langle T_f, \varphi \rangle = \int_I \varphi(x) f(x)\, dx$$

for all functions φ belonging to the space $\mathscr{K}(I)$ of continuous functions with compact support in I. It will be sometimes convenient to choose the test functions φ from the space $\mathscr{D}(I)$ of infinitely differentiable functions with compact support, which is dense in $\mathscr{K}(I)$ [4, Proposition 4.4.1].

With $f \in \mathscr{V}(I)$ we can associate a second Radon measure, the Stieltjes measure S_f, defined by the Stieltjes integral,

$$\langle S_f, \varphi \rangle = \int_I \varphi(x)\, df(x)$$

for all $\varphi \in \mathscr{K}(I)$. The inequality

$$|\langle S_f, \varphi \rangle| \le q_{\alpha\beta}(f) \max_x |\varphi(x)|,$$

which holds for all $\varphi \in \mathcal{K}(I)$ with Supp $\varphi \subset [\alpha, \beta]$, shows that S_f is indeed a continuous linear form on $\mathcal{K}(I)$, i.e., a Radon measure on I. Clearly the map $f \mapsto S_f$ is linear from $\mathcal{V}(I)$ into the space $\mathcal{M}(I) = \mathcal{K}'(I)$ of all Radon measures on I. The integration by parts formula

$$\int_I \varphi(x)\,df(x) = -\int_I \varphi'(x)f(x)\,dx, \qquad \varphi \in \mathcal{D}(I),$$

can be rewritten as

$$\langle S_f, \varphi \rangle = -\langle T_f, \varphi' \rangle = \langle \partial T_f, \varphi \rangle,$$

and shows that the measure S_f is the derivative in the sense of distributions of the measure T_f with density f (or, as it is customarily expressed, S_f is the derivative in the distribution sense of the function f). It is very easy to show [5, Proposition 5.15.2] and not at all surprising that the measures of finite sub-intervals of I with respect to S_f are given by

$$\begin{aligned} S_f(]\alpha, \beta[) &= f(\beta - 0) - f(\alpha + 0), & S_f([\alpha, \beta]) &= f(\beta + 0) - f(\alpha - 0), \\ S_f(]\alpha, \beta]) &= f(\beta + 0) - f(\alpha + 0), & S_f([\alpha, \beta[) &= f(\beta - 0) - f(\alpha - 0). \end{aligned} \tag{2}$$

After these preparations, we can state the representation theorem of Frederick Riesz in the following form:

THEOREM 1.   *The linear map $f \mapsto S_f$ from $\mathcal{V}(I)$ into $\mathcal{M}(I)$ is surjective, i.e., for every Radon measure μ on I there exists a function f, of bounded variation on every compact subinterval of I, such that $\mu = S_f = \partial T_f$. A function $f \in \mathcal{V}(I)$ belongs to the kernel of the linear map $f \mapsto S_f$ if and only if there exists a constant C such that $f(x - 0) = f(x + 0) = C$ for every $x \in I$.*

Let us reproduce with slight modifications the proof given by Frederick Riesz for his theorem in 1914 [9]. The measure μ is, by definition, given originally as a continuous linear form on $\mathcal{K}(I)$. We need to extend μ so that it acts on linear combinations of characteristic functions of intervals. This can be achieved in three fashions:

(a)   Using the theory of integration, one obtains that μ acts on $\mathcal{L}^1(I, \mu)$; this will be the approach taken in this note.

(b)   For each compact subset K of I denote by $\mathcal{L}_K^\infty(I)$ the vector space of all essentially bounded functions with respect to Lebesgue measure whose support is contained in K. We equip $\mathcal{L}_K^\infty(I)$ with the seminorm N_∞, where $N_\infty(f) = M_\infty(|f|)$ is the essential maximum of $|f|$ [1, Chapter IV, 2nd ed., Section 6, No. 3, Definition 2]. Then $\mathcal{L}_c^\infty(I) = \bigcup_K \mathcal{L}_K^\infty(I)$ is the space of all essentially bounded functions with compact support. We equip $\mathcal{L}_c^\infty(I)$ with the finest locally convex topology for which the injections $\mathcal{L}_K^\infty(I) \hookrightarrow \mathcal{L}_c^\infty(I)$ are continuous. Then the canonical injection $\mathcal{K}(I) \hookrightarrow \mathcal{L}_c^\infty(I)$ is a strict morphism [5, Proposition 5.14.3], i.e., $\mathscr{K}(I)$ can be considered as a subspace of $\mathscr{L}_c^\infty(I)$ and we can apply the Hahn–Banach extension theorem.

(c)   Use an ad hoc method based on considering increasing and decreasing sequences of functions in $\mathscr{K}(I)$, used by Riesz in [9], and which is not substantially different from (a).

Once the extension has been accomplished, we pick c in I and set

$$f(x) = \begin{cases} \mu(]c, x]) & \text{if } c < x < b, \\ -\mu(]x, c]) & \text{if } a < x < c, \end{cases} \tag{3}$$

and $f(c) = 0$. Then for $a < x < y < b$ we have

$$\mu(]x, y]) = f(y) - f(x). \tag{4}$$

If Δ is a subdivision (1) of $[\alpha, \beta]$ and

$$\varepsilon_j = \text{sign}\{f(x_j) - f(x_{j-1})\}, \qquad 1 \le j \le l,$$

then

$$\sum_{j=1}^{l} |f(x_j) - f(x_{j-1})| = \sum_{j=1}^{l} \varepsilon_j \mu(]x_{j-1}, x_j])$$

$$= \int_I \left(\sum_{j=1}^{l} \varepsilon_j \chi_{]x_{j-1}, x_j]}(x) \right) d\mu(x) \le |\mu|([\alpha, \beta]),$$

where, in general, χ_M denotes the characteristic function of the set M. It follows that $q_{\alpha\beta}(f) \le |\mu|([\alpha, \beta])$, and in particular $f \in \mathscr{V}(I)$.

Let us now prove that $\mu = S_f$. Given $\varphi \in \mathscr{K}(I)$ with support in $[\alpha, \beta]$ and an arbitrary $\varepsilon > 0$, choose $\delta > 0$ so that on the one hand for any subdivision (1) of $[\alpha, \beta]$ with $x_j - x_{j-1} \le \delta, 1 \le j \le l$, and for any $\xi_j \in [x_{j-1}, x_j]$ we have

$$\left| \int_I \varphi(x) \, df(x) - \sum_{j=1}^{l} \varphi(\xi_j)\{f(x_j) - f(x_{j-1})\} \right| \le \frac{1}{2}\varepsilon, \tag{5}$$

and that on the other hand

$$|\varphi(y) - \varphi(x)| \le \frac{1}{2} \frac{\varepsilon}{1 + |\mu|([\alpha, \beta])}$$

whenever $|x - y| \le \delta$. Fixing such a Δ and points ξ_j, define the step function ψ by

$$\psi(x) = \begin{cases} \varphi(\xi_j) & \text{if } x_{j-1} < x \le x_j, \ 1 \le j \le l, \\ 0 & \text{if } x \le \alpha \ \text{or} \ x > \beta. \end{cases}$$

Then by (4)

$$\left| \sum_{j=1}^{l} \varphi(\xi_j)\{f(x_j) - f(x_{j-1})\} - \langle \mu, \varphi \rangle \right| = \left| \int_I \{\psi(x) - \varphi(x)\} \, d\mu(x) \right|$$

$$\leq \frac{1}{2} \frac{\varepsilon}{1 + |\mu|([\alpha, \beta])} |\mu|([\alpha, \beta]) \leq \frac{1}{2} \varepsilon. \quad (6)$$

Since $\varepsilon > 0$ is arbitrary, (5) and (6) together imply that indeed

$$\langle \mu, \varphi \rangle = \int_I \varphi(x) \, d\mu(x) = \int_I \varphi(x) \, df(x) = \langle S_f, \varphi \rangle.$$

Finally we determine the kernel of the map $f \mapsto S_f$. If $f(x - 0) = f(x + 0) = C$ for all $x \in I$, then—since f can only have countably many discontinuities—one has $f(x) = C$ except on a countable set $D \subset I$. One can therefore find subdivisions (1) of $[\alpha, \beta]$ with arbitrarily small $\max(x_j - x_{j-1})$ so that no x_j lies in D. By the definition of the Stieltjes integral this implies that $S_f = 0$. Conversely, if $S_f = 0$, then formulas (2) show that

$$f(x - 0) = f(x + 0) = f(y - 0) = f(y + 0)$$

for any two points $x, y \in I$. ∎

If f is an increasing function, then clearly S_f is a positive measure. Conversely, if μ is a positive Radon measure, then from formulas (2) we see that we can find an increasing function $f \in \mathscr{V}(I)$ such that $\mu = S_f$. This observation and Theorem 1 yield the following known result, which is the basis of one of the definitions of a function of bounded variation of several variables [3, 6].

COROLLARY Let T be a distribution on the open interval $I \subset \mathbf{R}$ such that ∂T is a Radon measure. Then there exists a function $f \in \mathscr{V}(I)$ such that $T = T_f$. If ∂T is a positive Radon measure, then f can be chosen to be increasing.

Proof. If $\partial T = \mu \in \mathscr{M}(I)$, then by Theorem 1 there exists an $f \in \mathscr{V}(I)$ such that $\mu = S_f = \partial T_f$, and if $\mu \geq 0$, then f can be chosen to be increasing. We have $\partial(T - T_f) = \mu - \mu = 0$, hence [4, Proposition 4.4.3] $T = T_{f+c}$ for some $c \in \mathbf{R}$. ∎

The map $f \mapsto S_f$ has good properties not only with respect to order but also with respect to topology:

THEOREM 2. The surjective linear map $f \mapsto S_f$ is a strict morphism from $\mathscr{V}(I)$ onto $\mathscr{M}(I)$ if we equip $\mathscr{M}(I)$ with the strong topology $\beta(\mathscr{M}(I), \mathscr{K}(I))$.

This is an immediate consequence of the formula

$$q_{\alpha\beta}(f) = \sup\left\{\int \varphi \, df : \varphi \in \mathcal{K}(I), \text{Supp } \varphi \subset [\alpha, \beta], |\varphi(x)| \leq 1 \text{ for all } x \in I\right\}$$

which can be established easily. ∎

Let us denote by $\mathscr{V}^1(I)$ the subspace of $\mathscr{V}(I)$ consisting of those $f \in \mathscr{V}(I)$ for which $q_{\alpha\beta}(f)$ is bounded as α and β vary in $a < \alpha < \beta < b$, and define

$$q(f) = \sup_{\alpha,\beta} q_{\alpha\beta}(f) = \lim_{\alpha \downarrow a, \beta \uparrow b} q_{\alpha\beta}(f).$$

q is a seminorm on $\mathscr{V}^1(I)$ and $q(f) = 0$ if and only if f is a constant. The associated normed space $\mathscr{V}_0^1(I) = \mathscr{V}^1(I)/\{\bar{0}\}$ can therefore be considered as a subspace of $\mathscr{V}_0(I)$. The canonical injections $\mathscr{V}^1(I) \hookrightarrow \mathscr{V}(I)$ and $\mathscr{V}_0^1(I) \hookrightarrow \mathscr{V}_0(I)$ are continuous.

Following Bourbaki, we denote by $\mathscr{M}^1(I)$ the space of all integrable ("bounded," cf. [1, Chapter III, 2nd ed., Section 1, No. 8] and [4, Chapter 4, Section 5]) Radon measures on I. If $f \in \mathscr{V}^1(I)$, then clearly

$$|\langle S_f, \varphi \rangle| = \left|\int \varphi(x) \, df(x)\right| \leq q(f) \max_{x \in I} |\varphi(x)|$$

for all $\varphi \in \mathscr{K}(I)$, hence $S_f \in \mathscr{M}^1(I)$ and $\|S_f\| \leq q(f)$. For integrable measures we have the following version of the Riesz representation theorem:

THEOREM 3.   *The restriction to $\mathscr{V}^1(I)$ of the linear map $f \mapsto S_f$ is a surjective map u from $\mathscr{V}^1(I)$ onto $\mathscr{M}^1(I)$. For every $\mu \in \mathscr{M}^1(I)$ there exists an $f \in \mathscr{V}^1(I)$ such that $\mu = S_f$ and $\|\mu\| = q(f)$. The associated injection $\bar{u}: \mathscr{V}^1(I)/\mathrm{Ker}\ u \to \mathscr{M}^1(I)$ is an isometric isomorphism of Banach spaces.*

Proof.   If μ is a positive measure belonging to $\mathscr{M}^1(I)$, then by the remark made after Theorem 1 there exists an increasing function f such that $\mu = S_f$. From [1, Chapter IV, 2nd ed., Section 4, No. 7, Proposition 12 and No. 5, Proposition 8] it follows that

$$\|\mu\| = \mu(I) = \sup_{[\alpha,\beta] \subset I} \mu(]\alpha, \beta[) = \sup_{[\alpha,\beta] \subset I} \{f(\beta - 0) - f(\alpha + 0)\},$$

and since f is increasing, this is equal to

$$\sup_{[\alpha,\beta] \subset I} \{f(\beta) - f(\alpha)\} = q(f).$$

If now μ is an arbitrary measure in $\mathscr{M}^1(I)$, then we decompose it into the difference of its positive and negative parts: $\mu = \mu^+ - \mu^-$. We can find increasing functions g, h such that $S_g = \mu^+$, $S_h = \mu^-$, $\|\mu^+\| = q(g)$ and $\|\mu^-\| = q(h)$. The function $f = g - h \in \mathscr{V}^1(I)$ satisfies $S_f = \mu$, and using [1,

Chapter III, 2nd ed., Section 1, No. 8, Corollary 2 of Proposition 11] we have

$$\|\mu\| \le q(f) \le q(g) + q(h) = \|\mu^+\| + \|\mu^-\| = \|\mu\|,$$

hence in particular

$$\|\mu\| = \inf_{S_f = \mu} q(f) = \min_{S_f = \mu} q(f). \quad \blacksquare$$

To conclude this section, let us show quickly how the Riesz representation theorem in its original form for a compact interval $[a, b]$ follows from Theorem 3.

THEOREM 4. *For every continuous linear form L on the Banach space $\mathscr{C}([a, b])$ there exists a function f of bounded variation on $[a, b]$ such that*

$$L(\varphi) = \int_a^b \varphi(x)\,df(x)$$

for all $\varphi \in \mathscr{C}([a, b])$ and $\|L\|$ is equal to the total variation of f.

Proof. The Lebesgue–Tietze extension theorem $[2\ (4.5.1)]$ shows that $\Phi \mapsto \Phi|[a, b]$ is a surjective strict morphism from the Banach space $\mathscr{C}_0(\mathbf{R})$ onto the Banach space $\mathscr{C}([a, b])$, whose kernel $\mathscr{N}$ consists of those functions $\Phi \in \mathscr{C}_0(\mathbf{R})$ whose restriction to $[a, b]$ is zero. The dual $\mathscr{C}([a, b])'$ is isometrically isomorphic to $\mathscr{N}^\perp$, i.e., to the space of those measures in $\mathscr{M}^1(\mathbf{R})$ whose support is contained in $[a, b]$.

Thus, given $L \in \mathscr{C}([a, b])'$, there exists $\mu \in \mathscr{M}^1(\mathbf{R})$ with Supp $\mu \subset [a, b]$ such that $\|L\| = \|\mu\|$ and $L(\varphi) = \int \Phi\,d\mu$ for all $\Phi \in \mathscr{C}_0(\mathbf{R})$ whose restriction to $[a, b]$ is $\varphi \in \mathscr{C}([a, b])$. According to Theorem 3, there exists $f \in \mathscr{V}^1(\mathbf{R})$ such that $q(f) = \|\mu\|$ and that

$$\int_{\mathbf{R}} \Phi(x)\,d\mu(x) = \int_{\mathbf{R}} \Phi(x)\,df(x)$$

for all $\Phi \in \mathscr{C}_0(\mathbf{R})$. Formulas (2) and the condition Supp $\mu \subset [a, b]$ show that

$$f(x - 0) = f(x + 0) = f(a - 0) \qquad \text{for} \quad x < a$$

and

$$f(y - 0) = f(y + 0) = f(b + 0) \qquad \text{for} \quad y > b.$$

Hence f must be a constant for $x < a$ and for $y > b$, since otherwise we could find a $g \in \mathscr{V}^1(\mathbf{R})$ such that $S_g = \mu$ and $q(g) < q(f) = \|\mu\|$, which is impossible. Thus $q(f) = q_{ab}(f) = \|\mu\| = \|L\|$, and

$$L(\varphi) = \int_{\mathbf{R}} \Phi(x)\,d\mu(x) = \int_{\mathbf{R}} \Phi(x)\,df(x) = \int_a^b \varphi(x)\,df(x),$$

for all $\varphi \in \mathscr{C}([a, b])$, where Φ is any function in $\mathscr{C}_0(\mathbf{R})$ such that $\Phi|[a, b] = \varphi$. $\quad \blacksquare$

2. THE HIGHER-DIMENSIONAL CASE

The generalization of Theorem 1 to several dimensions can be stated as follows:

THEOREM 5. *Let*

$$Q = \{x \in \mathbf{R}^n : a_j < x_j < b_j,\ 1 \le j \le n\}$$

be an open n-dimensional interval $(-\infty \le a_j < b_j \le \infty)$ *in* $\mathbf{R}^n$. *For every Radon measure* μ *on* Q *there exists a locally bounded, Lebesgue-measurable function* f *such that*

$$\mu = \partial_1 \partial_2 \cdots \partial_n f$$

in the sense of distributions (i.e., $\mu = \partial_1 \partial_2 \cdots \partial_n T_f$, *where* $T_f = f \cdot \lambda$ *is the measure with density* f *with respect to Lebesgue measure* λ).

In order to present a "civilized proof" [7, p. 33], we shall first consider the case $n = 2$, and make some remarks concerning the general case at the end. For the same reason, we shall consider first only a finite open square

$$Q = \{(x, y) \in \mathbf{R}^2 : -a < x, y < a\}$$

with center at the origin. It is no restriction of the generality to assume that μ is a positive measure, since every real measure is the difference of two positive ones.

Given $(x, y) \in Q$, we denote by $Q_{x,y}$ the set obtained when we remove from the closed rectangle with opposite vertices $(0,0)$ and (x, y) the closed left and lower sides. More precisely:

$$\begin{aligned}
Q_{x,y} &= \{(\xi, \eta) \in \mathbf{R}^2 : 0 < \xi \le x,\ 0 < \eta \le y\} && \text{if } x \ge 0,\quad y \ge 0, \\
Q_{x,y} &= \{(\xi, \eta) \in \mathbf{R}^2 : x < \xi \le 0,\ 0 < \eta \le y\} && \text{if } x < 0,\quad y \ge 0, \\
Q_{x,y} &= \{(\xi, \eta) \in \mathbf{R}^2 : x < \xi \le 0,\ y < \eta \le 0\} && \text{if } x < 0,\quad y < 0, \\
Q_{x,y} &= \{(\xi, \eta) \in \mathbf{R}^2 : 0 < \xi \le x,\ y < \eta \le 0\} && \text{if } x \ge 0,\quad y < 0.
\end{aligned}$$

In particular $Q_{x,y} = \varnothing$ whenever either $x = 0$ or $y = 0$.

Analogously to (3), and similarly to the definition of a distribution function in probability theory, but taking into account that μ is not necessarily finite on Q, we define

$$f(x, y) = \operatorname{sgn}(xy)\mu(Q_{x,y}).$$

In particular f is positive in the first and third quadrants and negative in the second and fourth quadrants. Furthermore $f(x, y) = 0$ whenever $x = 0$ or $y = 0$.

Next, let (x, y) and (u, v) be two points of Q such that $u \le x, v \le y$, and set

$$Q_{u,v}^{x,y} = \{(\xi, \eta) \in \mathbf{R}^2 : u < \xi \le x, v < \eta \le y\}.$$

Analogously to (4) we have here the formula

$$\mu(Q_{u,v}^{x,y}) = f(x, y) - f(u, y) - f(x, v) + f(u, v). \tag{7}$$

There is a computational proof for this formula, valid in n dimensions, which we shall present at the end, but in our case it is simpler to examine the different possible cases separately. We consider only a few typical ones among them and leave it to the reader to establish a complete list.

(a) $0 < u < x, 0 = v < y$.   Then $Q_{x,y} = Q_{u,y} \cup Q_{u,0}^{x,y}$, where the sets in the union are disjoint. From the definition of f we get

$$f(x, y) = f(u, y) + \mu(Q_{u,0}^{x,y}),$$

which yields (7) if we take into account that $f(x, 0) = f(u, 0) = 0$.
(b) $0 = u < x, 0 < v < y$.   Then $Q_{x,y} = Q_{x,v} \cup Q_{0,v}^{x,y}$, where the sets in the union are disjoint. It follows from the definition of f that

$$f(x, y) = f(x, v) + \mu(Q_{0,v}^{x,y}),$$

which is again (7) if we observe that $f(0, y) = f(0, v) = 0$.
(c) $0 < u < x, 0 < v < y$.   This time

$$Q_{x,y} = Q_{u,v} \cup Q_{u,0}^{x,v} \cup Q_{0,v}^{u,y} \cup Q_{u,v}^{x,y},$$

where the union is again disjoint. Using (a) and (b) we obtain

$$f(x, y) = f(u, v) + \{f(x, v) - f(u, v)\} + \{f(u, y) - f(u, v)\} + \mu(Q_{u,v}^{x,y}),$$

from which (7) follows.
(d) $u < 0 < x, v < 0 < y$.   In this case we have the disjoint union

$$Q_{u,v}^{x,y} = Q_{x,y} \cup Q_{u,y} \cup Q_{x,v} \cup Q_{u,v},$$

which yields again (7).
For $0 < r < a$ let us now write

$$Q_r = \{(x, y) \in \mathbf{R}^2 : -a + r < x, y < a - r\}.$$

The function f is clearly bounded on each closed set $\bar{Q}_r$; in particular it is locally bounded.
Since $\mu(\bar{Q}_r) < \infty$, by countable additivity of the measure μ, only a countable number of vertical segments or of horizontal segments in $\bar{Q}_r$, going from boundary to boundary, can have a strictly positive μ-measure. Let H be the collection of all horizontal segments σ, going from the left boundary to the

right boundary of Q, such that $\mu(\sigma) > 0$. If for each $n \in \mathbf{N}$, $n > 1/a$, we set

$$H_n = \{\sigma \in H : \mu(\sigma \cap Q_{1/n}) > 0\},$$

it follows from [1, Chapter IV, 2nd ed., Section 4, No. 5, Proposition 8] that $H = \bigcup_n H_n$. Since each H_n is countable, so is H. A similar assertion holds for the collection V of vertical segments going from top to bottom of Q and such that $\mu(\sigma) > 0$. Denote by S the set of all points of Q which lie either on a segment belonging to H or to V, or on one of the coordinate axes. Then S has Lebesgue measure zero.

We claim now that at every point (x, y) not belonging to S the function f is continuous. Suppose, e.g., that $x > 0$, $y > 0$. Then

$$\bigcap_{h>0} Q_{x+h, \, y+h} = Q_{x,y},$$

and $\bigcup_{h>0} Q_{x-h, \, y-h}$ is the set which is left after the segments

$$\sigma_x = \{(\xi, \eta) \in \mathbf{R}^2 : \xi = x, \, 0 < \eta \leq y\}$$

and

$$\sigma_y = \{(\xi, \eta) \in \mathbf{R}^2 : 0 < \xi \leq x, \, \eta = y\}$$

have been removed from $Q_{x,y}$. By the choice of the point (x, y) we have $\mu(\sigma_x) = \mu(\sigma_y) = 0$, and so [1, Chapter IV, 2nd ed., Section 4, No. 5, Corollary of Proposition 7 and Proposition 8] imply that

$$\lim_{h \to 0} f(x + h, \, y + h) = \lim_{h \to 0} \mu(Q_{x+h, \, y+h}) = \mu(Q_{x,y}) = f(x, y).$$

For $x - h \leq \xi \leq x + h$, $y - h \leq \eta \leq y + h$ the value $f(\xi, \eta)$ lies between $f(x - h, y - h)$ and $f(x + h, y + h)$, from which the claim follows.

Given a subdivision

$$\Delta : \begin{cases} -a + r = x_0 < x_1 < \cdots < x_l = a - r, \\ -a + r = y_0 < y_1 < \cdots < y_m = a - r \end{cases} \tag{8}$$

of $\bar{Q}_r$, denote by R_{jk} the rectangle

$$Q_{x_{j-1}, y_{k-1}}^{x_j, y_k} = \{(x, y) \in \mathbf{R}^2 : x_{j-1} < x \leq x_j, \, y_{k-1} < y \leq y_k\},$$

and by Ξ the selection, for each pair (j, k), of a point $(\xi_j, \eta_k) \in \bar{R}_{jk}$ $(1 \leq j \leq l$, $1 \leq k \leq m)$. Define the step functions $t = t_{f, \Delta, \Xi}$ by

$$t = \sum_{j=1}^{l} \sum_{k=1}^{m} f(\xi_j, \eta_k) \chi_{R_{jk}}. \tag{9}$$

We show that they converge to f almost everywhere on $\bar{Q}_r$ as $\max_{j,k}\{x_j - x_{j-1}, y_k - y_{k-1}\} \to 0$, i.e., f is "Riemann integrable". In other words, for λ-almost every $(x, y) \in \bar{Q}_r$ and for every $\varepsilon > 0$ there exists a $\delta > 0$ such that $|f(x, y) - t(x, y)| \leq \varepsilon$ whenever $x_j - x_{j-1} \leq \delta$ and $y_k - y_{k-1} \leq \delta$. Indeed, if $(x, y) \in R_{jk}$, then

$$f(x, y) - t(x, y) = f(x, y) - f(\xi_j, \eta_k),$$

and if (x, y) is a point at which f is continuous, then this difference is arbitrarily small provided that (ξ_j, η_k) is sufficiently close to (x, y). It follows in particular that f is Lebesgue measurable [1, Chapter IV, 2nd ed., Section 5, No. 4, Corollary 1 of Theorem 2 and No. 2, Proposition 6].

Let now $\varphi \in \mathscr{D}(Q)$ be a test function and pick an $\varepsilon > 0$. Assume that Supp φ is contained in the square Q_r. There exists $\delta > 0$ such that if Δ is a subdivision (8) of $\bar{Q}_r$ with $x_j - x_{j-1} \leq \delta, y_k - y_{k-1} \leq \delta (1 \leq j \leq l, 1 \leq k \leq m)$, then for every Ξ we have

$$\left| \iint_Q \varphi(x, y)\, d\mu(x, y) - \sum_{j=1}^{l} \sum_{k=1}^{m} \varphi(\xi_j, \eta_k) \mu(R_{jk}) \right| \leq \varepsilon \tag{10}$$

[1, Chapter IV, 2nd ed., Section 4, No. 3, Proposition 3]. Let us choose $\xi_j = x_j, \eta_k = y_k$. Using (7), a two dimensional Abel transformation yields

$$\sum_{j=1}^{l} \sum_{k=1}^{m} \varphi(x_j, y_k) \mu(R_{jk})$$

$$= \sum_{j=1}^{l-1} \sum_{k=1}^{m-1} \varphi(x_j, y_k)\{f(x_j, y_k) - f(x_{j-1}, y_k) - f(x_j, y_{k-1}) + f(x_{j-1}, y_{k-1})\}$$

$$= \sum_{j=0}^{l-1} \sum_{k=0}^{m-1} f(x_j, y_k)\{\varphi(x_{j+1}, y_{k+1}) - \varphi(x_j, y_{k+1}) - \varphi(x_{j+1}, y_k) + \varphi(x_j, y_k)\},$$

$$\tag{11}$$

where we took into account that φ vanishes outside Q_r, and therefore $\varphi(x_0, y_k) = \varphi(x_j, y_0) = \varphi(x_l, y_k) = \varphi(x_j, y_m) = 0$ $(0 \leq j \leq l, 0 \leq k \leq m)$. We now apply the following known elementary result (cf. [2, Section (8.12), Problem 5]):

LEMMA   *Let φ be a twice continuously differentiable function in Q. For $0 < r < a$ and for $\varepsilon > 0$ there exists $0 < \delta < r$ such that*

$$|\varphi(x + h, y + k) - \varphi(x, y + k) - \varphi(x + h, y) + \varphi(x, y) - \partial_x\partial_y\varphi(x, y)hk| \leq \varepsilon|h||k|$$

for all $(x, y) \in \bar{Q}_r$ provided that $|h| \leq \delta, |k| \leq \delta$.

With the help of this lemma, (11) yields

$$\left| \sum_{j=1}^{l} \sum_{k=1}^{m} \varphi(x_j, y_k)\mu(R_{jk}) \right.$$

$$\left. - \sum_{j=1}^{l} \sum_{k=1}^{m} f(x_{j-1}, y_{k-1})\partial_x\partial_y\varphi(x_{j-1}, y_{k-1})(x_j - x_{j-1})(y_k - y_{k-1}) \right|$$

$$\leq \varepsilon \sum_{j=1}^{l} \sum_{k=1}^{m} |f(x_{j-1}, y_{k-1})|(x_j - x_{j-1})(y_k - y_{k-1}) \leq \varepsilon \sup_{\bar{Q}_r} |f(x, y)|\lambda(Q_r),$$

$$(12)$$

provided that δ is sufficiently small. Similarly as for the step function (9), we see that

$$\sum_{j=1}^{l} \sum_{k=1}^{m} f(x_{j-1}, y_{k-1})\partial_x\partial_y\varphi(x_{j-1}, y_{k-1})\chi_{R_{jk}}$$

converges λ-almost everywhere to $f \cdot \partial_x\partial_y\varphi$ on $\bar{Q}_r$ or on Q (since $\text{Supp}\,\varphi \subset Q_r$), hence by the Lebesgue dominated convergence theorem the second sum on the left-hand side of (12) is arbitrarily close to

$$\iint_{Q} f(x, y)\partial_x\partial_y\varphi(x, y)\, dx\, dy$$

if only δ is sufficiently small. Since $\varepsilon > 0$ is arbitrary, the last remark combined with (10) and (12) implies that

$$\int_Q \varphi\, d\mu = \int_Q \partial_x\partial_y\varphi f\, d\lambda = \langle T_f, \partial_x\partial_y\varphi \rangle = \langle \partial_x\partial_y T_f, \varphi \rangle.$$

Since this is true for every $\varphi \in \mathscr{D}(Q)$, we have indeed $\mu = \partial_x\partial_y T_f$, q.e.d.

Let us now make the promised remarks concerning the general case. Obviously we may still assume that μ is positive. We choose a point $c = (c_1, \ldots, c_n) \in Q$, and for $x = (x_1, \ldots, x_n) \in Q$ we denote by $v(x)$ the number of those differences $x_j - c_j$ ($1 \leq j \leq n$) which are strictly negative. Also for $x \in Q$ we denote by Q_x the parallelepiped whose opposite vertices are c and x; more precisely:

$$Q_x = \{\xi \in \mathbf{R}^n : c_j < \xi_j \leq x_j \text{ if } x_j \geq c_j, x_j < \xi_j \leq c_j \text{ if } x_j < c_j\}.$$

We define

$$f(x) = (-1)^{v(x)}\mu(Q_x).$$

Next, let x and y be two points in Q such that $x_j \leq y_j$ for $1 \leq j \leq n$, and set

$$Q_x^y = \{\xi \in \mathbf{R}^n : x_j < \xi_j \leq y_j, 1 \leq j \leq n\}.$$

Denote by $\mathbf{E}$ the set of the 2^n vectors $e = (e_1, \ldots, e_n)$ where each e_j is equal either to 0 or to 1. For $e \in \mathbf{E}$ let $z = (x|e|y) = (z_1, \ldots, z_n)$ be the vector whose components are

$$z_j = \begin{cases} x_j & \text{if } e_j = 0, \\ y_j & \text{if } e_j = 1. \end{cases}$$

Analogously to (4) and (7), we prove here the formula

$$\mu(Q_x^y) = \sum_{e \in \mathbf{E}} (-1)^{n+|e|} f((x|e|y)), \tag{13}$$

where $|e| = e_1 + \cdots + e_n$. Indeed, denote by χ_x^y the characteristic function of Q_x^y. If for $1 \le j \le n$ we set

$$\psi_t^{(j)} = \begin{cases} \chi_{(c_j, t]} & \text{if } t > c_j, \\ -\chi_{(t, c_j]} & \text{if } t < c_j, \end{cases}$$

$\psi_t^{(j)}(c_j) = 0$, then $\chi_{(s,t]} = \psi_t^{(j)} - \psi_s^{(j)}$ is the characteristic function of the interval $(s, t] \subset \mathbf{R}$, and

$$\chi_x^y = \bigotimes_{j=1}^n (\psi_{y_j}^{(j)} - \psi_{x_j}^{(j)}).$$

Multiplying out, we obtain for $\xi = (\xi_1, \ldots, \xi_n) \in Q$ that

$$\chi_x^y(\xi_1, \ldots, \xi_n) = \psi_{y_1}^{(1)}(\xi_1) \cdots \psi_{y_n}^{(n)}(\xi_n) - \psi_{x_1}^{(1)}(\xi_1) \psi_{y_2}^{(2)}(\xi_2) \cdots \psi_{y_n}^{(n)}(\xi_n) - \cdots$$
$$+ (-1)^n \psi_{x_1}^{(1)}(\xi_1) \cdots \psi_{x_n}^{(n)}(\xi_n). \tag{14}$$

If, in general, we denote by χ_z the characteristic function of Q_z, then

$$\chi_z = (-1)^{v(z)} \bigotimes_{j=1}^n \psi_{z_j}^{(j)},$$

and therefore we can rewrite (14) in the form

$$\chi_x^y = \sum_{\mu=0}^n (-1)^{n+\mu} \sum_{|e|=\mu} (-1)^{v((x|e|y))} \chi_{(x|e|y)}.$$

It follows that

$$\mu(Q_x^y) = \sum_{\mu=0}^n (-1)^{n+\mu} \sum_{|e|=\mu} (-1)^{v((x|e|y))} \mu(Q_{(x|e|y)})$$

$$= \sum_{\mu=0}^n (-1)^{n+\mu} \sum_{|e|=\mu} f((x|e|y)) = \sum_{e \in \mathbf{E}} (-1)^{n+|e|} f((x|e|y)),$$

which proves (13).

From here on the notational modifications one has to bring to the case $n = 2$ considered above are fairly straightforward. ∎

COROLLARY.[8].   *Under the hypotheses of Theorem 5 there exists a continuous function g on Q such that*

$$\mu = \partial_1^2 \cdots \partial_n^2 g$$

in the sense of distributions.

Indeed, set

$$g(x_1, \ldots, x_n) = \int_{c_1}^{x_1} \cdots \int_{c_n}^{x_n} f(t_1, \ldots, t_n)\, dt_1 \cdots dt_n,$$

where $c = (c_1, \ldots, c_n) \in Q$. ∎

ACKNOWLEDGMENT

Discussions with Dr. P. Kolmus were very helpful in the preparation of this paper.

REFERENCES

1. N. BOURBAKI, "Eléments de Mathématiques, Livre VI, Intégration," Hermann, Paris, 1952–1969.
2. J. DIEUDONNÉ, "Eléments d'Analyse, 1. Fondements de l'Analyse Moderne," Gauthier-Villars, Paris, 1969; "Foundations of Modern Analysis," Academic Press, New York, 1960.
3. W. FLEMING, Functions whose partial derivatives are measures, *Illinois J. Math.* **4** (1960), 452–478.
4. J. HORVÁTH, "Topological Vector Spaces and Distributions, I," Addison-Wesley, Reading, Massachusetts, 1966.
5. J. HORVÁTH, "Topological Vector Spaces and Distributions, II," to appear.
6. K. KRICKEBERG, Distributionen, Funktionen beschränkter Variation und Lebesguescher Inhalt nichtparametrischer Flächen, *Ann. Mat. Pura Appl.* **44** (1957), 105–133.
7. J. E. LITTLEWOOD, "A Mathematician's Miscellany," Methuen, London, 1953.
8. A. PIETSCH, Ein elementarer Beweis des Darstellungssatzes für Distributionen, *Math. Nachr.* **22** (1960), 47–50.
9. F. RIESZ, Démonstration nouvelle d'un théorème concernant les opérations fonctionnelles linéaires, *Ann. Sci. École Norm. Sup.* **31** (1914), 9–14; "Oeuvres complètes," Vol. I, pp. 490–495, Gauthier-Villars, Acad. Sci. Hongrie, Paris–Budapest, 1960.